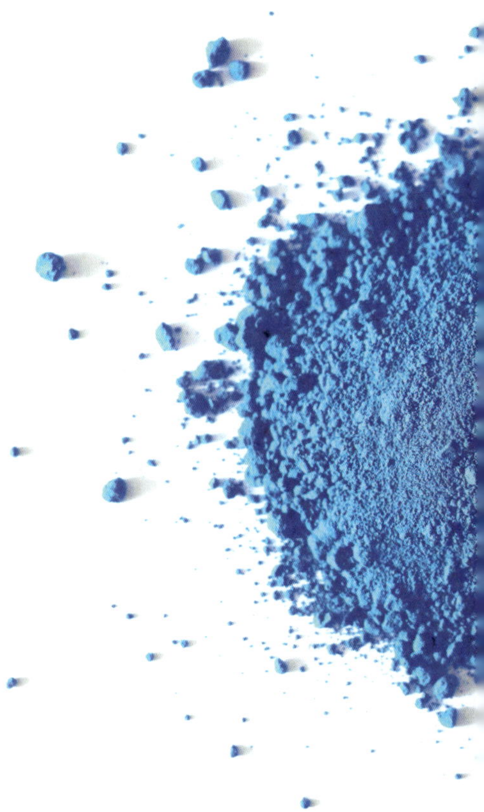

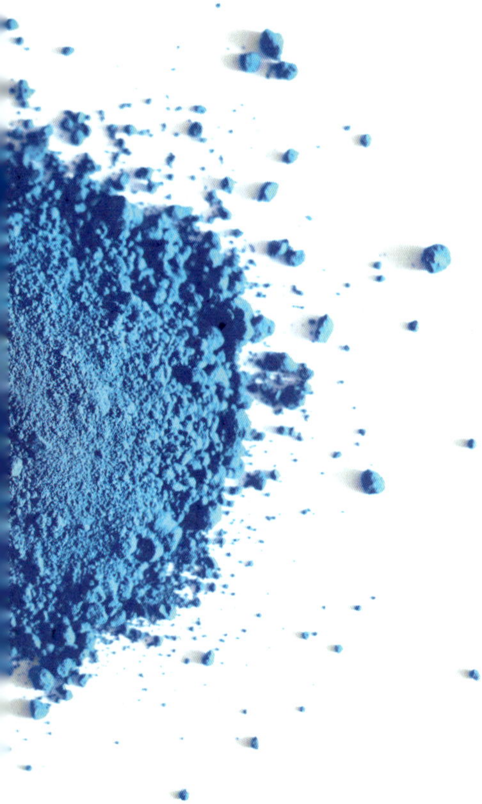

ERROR

the Art of Imperfection

Edited by Gerfried Stocker / Christine Schöpf / Hannes Leopoldseder

Ars Electronica 2018
Festival for Art, Technology, and Society

September 6–10, 2018
Ars Electronica, Linz

Editors: Hannes Leopoldseder, Christine Schöpf, Gerfried Stocker

Editing: Katia Kreuzhuber, Alexander Wöran

Translations:
German-English, all texts (unless otherwise indicated): Mel Greenwald
English-German: Ingrid Fischer-Schreiber

Copyediting: Laura Freeburn

Graphic design and production: Cornelia Prokop, Lunart Werbeagentur

Typeface: Klavika

Printed by: Gutenberg-Werbering Gesellschaft m.b.H, Linz

Paper: Magno Volume 1,1 Vol., 115 g/m², 300 g/m²

Published by
Hatje Cantz Verlag GmbH
Mommsenstrasse 27
10629 Berlin
Germany
Tel. +49 30 3464678-00
Fax +49 30 3464678-29
www.hatjecantz.com
A Ganske Publishing Group company

Hatje Cantz books are available internationally at selected bookstores. For more information about our distribution partners please visit our homepage at www.hatjecantz.com

ISBN 978-3-7757-4441-6

Printed in Austria

Climate neutral
Print product
ClimatePartner.com/53401-1808-1003

PEFC
PEFC/06-39-27

PEFC Certified
This product is from sustainably managed forests and controlled sources
www.pefc.org

ARS ELECTRONICA

HATJE
CANTZ

Ars Electronica 2018

Festival für Kunst, Technologie und Gesellschaft
Festival for Art, Technology and Society

Organization
Ars Electronica Linz GmbH & Co KG

Managing Directors
Diethard Schwarzmair, Gerfried Stocker
Ars-Electronica-Straße 1, 4040 Linz, Austria
Tel: +4373272720
Fax: +4373272722
info@ars.electronica.art

Co-organizer Prix Ars Electronica Gala and Big Concert Night
LIVA – Linzer Veranstaltungsgesellschaft mbH,
Brucknerhaus Linz
Directors: Dietmar Kerschbaum, Thomas Ziegler

Co-organizer CyberArts Exhibition
OK Offenes Kulturhaus im OÖ Kulturquartier
Directors: Martin Sturm, Gabriele Daghofer

Co-organizer Campus
Universität für künstlerische und industrielle Gestaltung
Rector: Reinhard Kannonier

Directors Ars Electronica: Gerfried Stocker, Christine Schöpf
Head of Festival: Martin Honzik
Technical Head: Karl Julian Schmidinger
Head of Finance & Organization: Veronika Liebl

Production Team: Alex Anikina, Christl Baur, Klaus Birklbauer, Amelie Brandstetter, Bernd Breitenauer, Florina Costamoling, Florian Cossee, Melanie de Jong, Anna Eder, Chris Eichenauer, Stephan Feichter, Hannes F. Franks, Marion Friedl, Benedikt Foreith, Jessica Galirow, Violeta Gil-Martinez, Eva Maria Grabmair, Dominik Greuter, Jürgen Hagler, Randolf Helmstetter, Bernhard Hinterreiter, Ferenc Hirt, David Holzweber, Nikolaus Jungwirth, Isabella Kartusch, Andrea Kohut, Katharina Kolar, Katia Kreuzhuber, Bernhard Küllinger, Tobias Leibetseder, Juliane Leitner, Michael Lahner, Magdalena Lettner, Sami Mandee, Lisa Martl, Kristina Maurer, Felix Merten, Hans Christian Merten, Claudia Moser, Fabian Mühlberger, Manuela Naveau, Michaela Obermayer, Emiko Ogawa, Hideaki Ogawa, Matthias Paul, Franz Peterseil, Veronika Platz, Moana Ponesch, Christina Radner, Isabella Reder, Katharina Richtsfeld, Elena Robles-Mateo, Alexandra Röck, Michael Samhaber, Michael Sarsteiner, Sonja Schachinger, Miro Schawalder, Thomas Schlager, Julian Schmiederer, Daniel Schöngruber, Laurin Siehs, Markus Sigl, Killian Sochor, Karla Spiluttini, Nora Spiluttini, Sebastian Sprenger, Lukas Traxler, Jochen Tuch, Tatiana Villacob, Gohar Vardanyan, Joschi Viteka, Andreas Weilguny, Laura Welzenbach, Sophie à Wengen, Nina Wenhart, Michaela Wimplinger, Alexander Wöran, Viktoria Wöss, Carla Zamora Campos

Co-curators
Architects: Jürgen Haller, Christoph Weidinger
Ars Electronica Animation Festival: Christine Schöpf, Jürgen Hagler
Campus Exhibition, Interface Cultures: Christa Sommerer, Laurent Mignonneau, Tamiko Thiel, Michaela Ortner, Fabrizio Lamoncha
Campus Exhibition, Hexagram: Chris Salter, Anna Kerekes
Expanded Animation: Jürgen Hagler, Alexander Wilhelm
Himatsubushi Trail: Himatsubushi Research Team
Music Monday: Werner Jauk
Nightline: Salon 2000
Sonic Saturday: Volkmar Klien, Se-Lien Chuang, Andreas Weixler

Prix Ars Electronica 2018

Idea: Hannes Leopoldseder
Conception: Christine Schöpf, Gerfried Stocker
Coordination: Martin Honzik, Emiko Ogawa
Technical Management: Karl Julian Schmidinger
Finance & Organization: Veronika Liebl

Production Team: Christl Baur, Florina Costamoling, Ingrid Fischer-Schreiber, Marion Friedl, Jessica Galirow, Eva Maria Grabmair, Elmar Glaubauf, Jürgen Hagler, Kristina Maurer, Hans Christian Merten, Manuela Naveau, Tatiana Villacob, Christina Radner, Jutta Schmiederer, Nana Thurner, Joschi Viteka

Press
Christopher Sonnleitner, Joan Bairam, Robert Bauernhansl, Martin Hieslmair, Vanessa Graf

Marketing
Wolfgang Königsmaier, Michael Sick-Leitner, Eva Tsackmaktsian, Christine Utz

Ars Electronica Linz GmbH & Co KG

Managing Directors
Diethard Schwarzmair, Gerfried Stocker

Ars Electronica Center
Katharina Aichinger, Viktoria Aistleitner, Joan Bairam, Andreas Bauer, Reinhard Bengesser, Carla Danzer, Ernesto Danzer, Shirin Darwish, Blanka Denkmaier, Erika Eiter, Julia Felberbauer, Melinda File, Julian Fischer, Gerhard Josef Friedl, Andrea Fröhlich, Philipp Gartlehner, Mitra Gazvini-Zateh, Christian Gerber, Elisabeth Gerhard, Vanessa Graf, Anna-Teresa Greger, Nicole Grüneis, Ulrike Gschwandtner, Thu Trang Eva Ha, Harald Haas, Birgit Hartinger, Barbara Heinzl, Katherine Heller, Thomas Hillinger, Florian Hofer, Sri Rahayu Hofstadler, Gerold Hofstadler, Eva Hofstädter, Thomas Jannke, David Jentgens, Andela Jovic, Smail Jusic, Elisabeth Kattner, Herwig Kerschner, Sarka Klafböck, Viktoria Klepp, Karin Knoll, Michael Koller, Thomas Kollmann, Christoph Kremer, Nadine Krenner, Andreas Leeb, Sabine Leidlmair, Juliane Leitner, Anna Katharina Link, Ulrike Mair, Clemens Mock, Erika Mondria, Horst Morocutti, Claudia Moser, Silvia Mukherjee, Daniel Murina, Heinrich Niederhuber, Andrea Oberfichtner, Michaela Obermayer, Benjamin Perndl, Dietmar Peter, Svetlana Petrovic, Katharina Pilar, Armin Pils, Anna Renata Polewiak, Thomas Prenn, Helmut Reinthaler, Ulrike Rieseneder, Ezan Riza, Raphaela Salhofer, Petra Saubolle-Hofmann, Alina Sauter, Birgitt Schäffer,

Mario Schmidhumer, Nina Schönberger, Fabian Schwarz, Thomas Schwarz, Elio Seidl, Manfred Seifriedsberger, Magdalena Sick-Leitner, Martin Spanka, Elisabeth Spöck, Minoosh Steininger, Margarethe Stöttner-Breuer, Thomas Straßhofer, Andreas Stürmer, Johannes Stürzlinger, Ilona Stütz, Istvan Szabo, Bhoomesh Tak, Michael Thaler, Schirin Turko, Thomas Viehböck, Florian Voggeneder, Raffaela Vornicu, Manuel Walch, Florian Wanninger, Lisa Wiesmayr, Judith Wittinghofer, Ralph Maximilian Zilian, Martin Zillner, Lilith Zschetzsche

Ars Electronica Futurelab
Roland Aigner, Julia Angerer, Florian Berger, Patrick Berger, Gloria Bope, Samuel Jakob Eckl, Marianne Eisl (Ternek), Maria Eschlböck, Peter Freudling, Matthew Gardiner, Rachel Hanlon, Roland Haring, Peter Holzkorn, Horst Hörtner, Kyoko Kunoh, Anna Kuthan, Christopher Lindinger, Elisabeth Luger, Michael Mayr, Stefan Mittlböck-Jungwirth-Fohringer, Otto Naderer, Nicolas Naveau, Ali Nikrang, Hideaki Ogawa, Benjamin Olsen, Maria Anna Pfeifer, Johannes Pöll, Daniel Rammer, Erwin Reitböck, Clemens Francis Scharfen, Raphael Elias Schaumburg-Lippe, Simon Schmid

Ars Electronica Festival / Prix / Exhibitions
Sophie á Wengen, Christl Baur, Amelie Brandstetter, Florina Costamoling, Melanie de Jong, Hannes Franks, Marion Friedl, Jessica Galirow, Violeta Gil Martinez, Eva Maria Grabmair, Martin Honzik, Andrea Kohut, Katharina Kolar, Katia Kreuzhuber, Magdalena Lettner, Veronika Christina Liebl, Lisa Martl, Kristina Maurer, Hans Christian Merten, Fabian Mühlberger, Manuela Naveau, Emiko Ogawa, Christina Radner, Alexandra Röck, Michael Samhaber, Karl Schmidinger, Karla Spiluttini, Nora Spiluttini, Sebastian Sprenger, Tatjana Villacob Melendez, Laura Welzenbach, Alexander Wöran, Viktoria Wöss, Carla Milena Zamora Campos

Ars Electronica Management Services
Robert Bauernhansl, Barbara Diesenreither, Johannes Egler, Michaela Frech, Bettina Gahleitner, Gerhard Grafinger, Anton Grünwald Belada, Haitham Harwash, Martin Hieslmair, Christoph Hofbauer, Julia Hofstätter, Andrea Huemer, Markus Jandl, Klaus Kaiser, Michael Kaiser, Elisabeth Kapeller, Isabella Kartusch, Stephan Kobler, Wolfgang Königsmaier, Fadil Kujundzic, Florian Lehner, Dominic Lengauer, Peter J. Nitzschmann, Sabine Plank, Michael Sick-Leitner, Christopher Sonnleitner, David Starzengruber, Daniela Thaller, Eva Tsackmaktsian, Christine Utz, Michaela Wimplinger

Ars Electronica Solutions
Ina Badics, Chris Bruckmayr, Stefan Dorn, Stefanie Farkashazy, Michaela Fragner, Yvonne Hauser, Barbara Hinterleitner, Martin Kaiser, Maryrose Mercado, Michael Mondria, Harald Moser, Patrick Müller, My Trinh Müller-Gardiner, Stephan Pointner, Andreas Pramböck, Gerald Priewasser-Höller, Gabriele Purdue, Dominik Trichlin, Markus Wipplinger, Claus Zweythurm

Ars Electronica SPAXELS
Harald Dirisamer

OK Offenes Kulturhaus im OÖ Kulturquartier

OK CyberArts Team
Directors: Martin Sturm, Gabriele Daghofer
Public Relations & Project Management: Maria Falkinger
Exhibition Team: Genoveva Rückert*, Andreas Kurz, Jarno Bachheimer, Maria Venzl*
Art Mediation & Texts: Rosi Grillmair, Marlies Stöger and Curators*
OK Night: Markus Reindl*
Production Manager: Aron Rynda
Production Team: Katharina Baldinger-Hackl, Attila Ferenczi, Alfred Fürholzer, Martin Haselsteiner, David Kraxberger, Bernhard Kitzmüller, Franz Quirchtmayr, Felix Pöchhacker, Andreas Steindl, André Tschinder, Hans-Jörg Weidinger, Gerhard Wörnhörer
Team: Verena Atteneder, Jocelyn Antensteiner, Renate Berger, Michael Dalpiaz, Walter Eckerstorfer, Max Fabian, Petra Fohringer, Werner Friesenecker, Michaela Fröhlich, Michael Gritzer, Josef Gruber, Stephan Hadwiger, Dominik Harrer, Sabine Haunschmid, Evi Heininger, Robert Herbst, Susi Heuschober, Maximilian Höglinger, Fritz Holzinger, Karin Huemer, Florian Hujber, Rainer Jessl, Christoph Kaltenböck, Sabrina Kamenar, Sigrid Krenner, Daniela Lacher, Simon Lachner, Katharina Lackner, Barbara Mair, Judith Maule, Franz Minichberger, Walter Mühlböck, Wolfgang Nagl, Christian Öhlinger, Kadir Özdemir, Maria Pachinger, Josef Pfarrhofer, Franz Pfifferling, Wilhelm Pichler, Angelika Pöschl, Gerda Sailer, Markus Schiller, Ulrike Schimpl, Helene Schoißengeyr

LIVA – Linzer Veranstaltungsgesellschaft mbH, Brucknerhaus Linz

Directors: Dietmar Kerschbaum, Thomas Ziegler
Team: Wolfgang Scheibner, Wolfgang Schützeneder, Ursula Kislinger

Universität für künstlerische und industrielle Gestaltung

Rector: Reinhard Kannonier
Vice-Rectors: Frank Louis, Sabine Pollak, Christine Windsteiger
Team: Nina Wenhart, Andre Zogholy

ORGANIZER

ARS ELECTRONICA

OK im OÖ kulturquartier

BRUCKNER HAUS LINZ

LINZ UNESCO City of Media Arts

Ars Electronica Linz GmbH & Co KG
is a company of the city of Linz.

COOPERATION PARTNERS

Aalto University
Academy of Fine Arts Prague
afo architekturforum oberösterreich
Animafest Zagreb
Anita Beckers Galerie
Anton Bruckner Privatuniversität
Architecture Lab
arebyte Gallery
ArtScience Museum
Art & Science Gallery Lab AXIOM
ArtTechLab Amsterdam
ASIFA Austria
Atelierhaus Salzamt
AWS – Jugend Innovativ
Badisches Staatstheater Karlsruhe
BANDITS-MAGES
Berliner Festspiele
Bernstein Innovation GmbH
Berufsförderungsinstitut OÖ
Bikuben Fonden
bitforms gallery
Bjørn Svin
Bora Bora
BORG Linz
BOZAR Center for Fine Arts Brussels
Bruckner Orchester Linz
bugnplay.ch
c3
Capital of Children Office A/S
Carl Emil Carlsen
Center for the Promotion of Science
Central Academy of Fine Arts Beijing
Central Linz
CLICK Festival
CoderDojo
Copenhagen Institute of Interaction Design
CREATE-IoT
Crossing Europe
CUBUS
Dansk Komponist Forening
Delavski dom Trbovlje
Diözese Linz
DNA Galerie
Donau-Universität Krems
Ecsite (The European network of science centres and museums)
Education Group GmbH
EP7 Paris
ETH Zürich
EU National Institutes for Culture
Europe for Festivals, Festivals for Europe EFFE
European Southern Observatory (ESO)
European Space Agency (ESA)
European Theatre Lab: Drama Goes Digital
European Theatre Convention
FAB Verein zur Förderung von Arbeit und Beschäftigung
Fachhochschule Oberösterreich – Campus Hagenberg
Fachhochschule St. Pölten
FACT (FOUNDATION FOR ART AND CREATIVE TECHNOLOGY)
Filmakademie Baden-Württemberg GmbH, Animationsinstitut
FMX – Conference on Animation, Effects, Games and Transmedia
Fraunhofer MEVIS: Institute for Medical Image Computing
FutureEverything
Galleria Artericambi
Galerie W / Galerie Albert Benamou et Véronique Maxé
Gebärdenwelt.tv
Gluon
Helsingør Kommune
HTL Leonding
IKM Institut für Kulturmanagement und Gender Studies
IMA Institut für Medienarchäologie
IMPAKT
Interface Cultures
Interio AG
International Students Creative Award (ISCA)
Johannes Kepler Universität Linz
Jugend hackt
K11 Art Foundation
Kapelica Gallery / Kersnikova Institute
Kepler Salon – Verein zur Förderung von Wissensvermittlung
Kickstarter
Kinderkulturzentrum Kuddelmuddel
King's College London
KNOWLEDGE CAPITAL
Knud Højgaards Fond
Koneensäätiö
KONTEJNER – bureau of contemporary art praxis
Kote Marjanishvili State Drama Theatre
Krmpf Krmpf Studios
Kunstuniversität Linz – Lehramt Bildnerische Erziehung
Kunstuniversität Linz – Universität für künstlerische und industrielle Gestaltung
L.A.T.R.A EE
LABoral – Centro de Arte y Creación Industrial
Landesfeuerwehrverband OÖ
Landesmusikschule Steyr
Landestheater Linz
LENTOS Kunstmuseum Linz
Linz Tourismus
maecenia Frankfurter Stiftung
Magic Shifter
MAM Mario Mauroner Contemporary Art Vienna + Salzburg
Mariendom Linz
mb21
m-cult
Memphis
mica austria
Moviemento
Mozilla
Museo nazionale della scienza e della tecnologia Leonardo da Vinci
National Culture & Arts Foundation
New Design University Privatuniversität GesmbH
New Museum
Nokia Bell Labs
Nome Gallery
Nora Hannula & The Nordic Beasts
Nordic Culture Fund
NOVA Prize
OMAi
Onassis Cultural Centre
Österreichische Akademie der Wissenschaften
Österreichischer Musikfonds
Otelo
Picaroon
Plovdiv 2018
Queen Mary University of London
QUT Queensland University of Technology
RIXC Center for New Media Culture
Robo Technologies, Inc.
Rock im Dorf Festival
Roy Ascott Studio, Shanghai
Salon 2000
School of the Art Institute of Chicago
Science Gallery Dublin
Shantou University
SILK Fluegge
Stadtpfarrkirche Urfahr
Stadtwerkstatt
STAIR Lab (Chiba Institute of Technology)
STATE Studio Berlin
STARTS Residencies
Taipeh performing arts center
Tangible Media Group / MIT Media Lab
The Tech Museum of Innovation
Théâtre de la Manufacture CDN Nancy Lorraine
Trotec Laser GmbH
UCL – The Bartlett School of Architecture
ULF – Unabhängiges Landesfreiwilligenzentrum
Universal Music Austria
Universität Innsbruck
University College London
University of Auckland
University of the Arts London
Valletta 2018
Very Mainstream Studio
Very Theatre
Waag
WEAR Sustain
WERKLEITZ
WERKPUNK
WRO Art Center
YAIR
ZERO1NE
zkm Karlsruhe
ZSI (Zenrum für Soziale Innovation)
ZusammenHelfen in Oberösterreich

ARS ELECTRONICA RECEIVES SUPPORT FROM

Land Oberösterreich

Bundeskanzleramt

Europäische Kommission
Horizon 2020

Europäische Kommission

Europäische Union

Creative Europe

Australia Council
for the Arts

KulturKontakt
Austria

Amt der NÖ Landesregierung
Abteilung Kunst und Kultur

Fonds zur Förderung der
wissenschaftlichen Forschung (FWF)

Danish Arts Council

Danish Arts Foundation

Ministry of Culture Taiwan

Chile

Creative Industries Fund NL

Bundesministerium für
Bildung, Wissenschaft
und Forschung

Bundesministerium für
Europa, Integration
und Äußeres

SPONSORS

Hakuhodo

Hyundai Motor Company

VH Award

Österreichische Post AG

WKO Oberösterreich

WKOÖ Fachgruppe für Unternehmens-
beratung, Buchhaltung und IT

Greiner Holding

netidee

RISC Software GmbH

NTT

MAXON Computer GmbH

Hutchison Drei Austria GmbH

g.tec medical engineering GmbH

Liwest Kabelmedien GmbH

NTS New Technology Systems GmbH

Ton & Bild Medientechnik GmbH

Industriellenvereinigung
Oberösterreich

Oberösterreich Tourismus GmbH

Conrad
Electronic GmbH & Co KG

Weyland GmbH

Rosenbauer International AG

NHK Japan Broadcasting Corporation

Pädagogische Hochschule OÖ

Arbeiterkammer OÖ

Ableton AG

.ART
(DOT ART DOMAINS)

Niio

Beep Electronic Art Collection

Österreichisches Rotes Kreuz

Cordial GmbH

Microsoft

Linz AG

LIVA

The Culture Yard

Hexagram

Vertretung der Regierung
von Quebec

Fonds de recherche Societe
et culture Quebec

Social Sciences and Humanities
Research Council of Canada

Innovation.ca

Concordia University

UQAM – Université du
Québec à Montréal

Lenovo Technology B.V.

Agency for Cultural Affairs,
Government of Japan

Japan Media Arts Festival

Empowerment Informatics

University of Tsukuba

Eternit Österreich GmbH

ACC • ACI

IKEA Möbelvertrieb OHG

BIO AUSTRIA

Lenox Trading Lehner GmbH

Aruba Networks, Inc.

Triple A Aqua Service GmbH

vrei media gmbh

Klangfarbe – Musikinstrumente
und tontechnische Geräte
Handelsges.m.b.H.

NEC Display Solutions
Europe GmbH

Mayr - Schulmöbel
Gesellschaft m.b.H.

S. Spitz GmbH

clever on smart GmbH

Alfred Kärcher GmbH

Vöslauer Mineralwasser GmbH

Mondi Grünburg GmbH

Autodesk, Inc.

Baumschule Ernst Junger

Norbert Pilz GmbH

VELLO GmbH

Stangl Reinigungstechnik GmbH

KAESER KOMPRESSOREN GmbH

Musik & Co Schimpelsberger GmbH

Plaspack Netze GmbH

Smurfit Kappa Wellkart GmbH

Bildrecht GmbH

4YOUgend – Verein OÖ Jugendarbeit

Papertronic GmbH

L. Bösendorfer Klavierfabrik GmbH

KUKA CEE GmbH

Wacker Neuson SE

RINGER KG

Association for Robots in Architecture

Organoid® Technologies GmbH

AURO GmbH

Christian Bauernfeind

TECHquadrat Werbetechnik GmbH

Weinhaus Wakolbinger GmbH

SSI Schäfer Shop GmbH

Seybold Handelsagentur

Grover

INVR.SPACE

PROJECT PARTNERS

STARTS Prize'18

STARTS EU

European Media Art Platform

esero Austria

IMMERSIFY

BR41N.IO

MEDIA PARTNERS

ORF OÖ

OÖ Nachrichten

Der Standard

Tips

Radio FM4

Die Presse

Ö1

Kunstradio

Artjaws

Contents

EVENTS, CONCERTS & PERFORMANCES

ARS ELECTRONICA ANIMATION

U19 – CREATE YOUR WORLD

GUEST PROJECTS

ARS ELECTRONICA

Gerfried Stocker (AT)

ERROR
the Art of Imperfection

The 2018 Ars Electronica Festival Theme: ERROR – The Art of Imperfection

At what point does an error become a mistake, a fail, and what makes it the celebrated source of unprecedented ideas and inventions? When is an error an oversight and when is it intentional deception, a fake? An error is a discrepancy from what we expect, a deviation from the norm ... but what is the norm and who establishes it? An error doesn't have to be a mistake; it can be an opportunity!

But how much tolerance can we summon up for such deviations, and is it enough for the leeway and latitude that are necessary to unleash their inherent productive power which can be harnessed for social and economic innovation? Or will we allow ourselves to be misled by the populist rhetoric of fear and social scoring? Observing the current situation, one very quickly gets the impression that something has gone terribly wrong with the Digital Revolution and the 21st century. Millions of people feel that they have been defrauded of their sovereignty over their data and their privacy. Deception and fakery have become realities of everyday life, and influence public sentiment and the public opinion formation process. And hovering above it all is a diffuse anxiety of being left behind by the swift dynamics of development. Was the dream of a beautiful digital world an error, and how can we rescue this dream?

This day and age is characterized by a compulsion to achieve perfection and a seemingly unwavering faith in technology. And amidst this drive to optimize, to increase efficiency and raise productivity, and, in even more instances, merely to enjoy the possibilities that digital technologies and social media place at our fingertips, we put ourselves at the mercy of machinery that does its utmost to make lemmings of digital consumption out of us.

Big Data surveillance takes preventative action upon detecting any deviation from our habitual ways. And it is said that in the future, social scoring will do an even better job of optimizing our behavior and attuning it to social norms and standards. The more the technologies deployed for this purpose are perfected and made more efficient, the tighter our situation becomes. Whoever doesn't fit in sticks out and gets cut.

But it is precisely this imperfection that offers the greatest potential for new solutions. Our objective should not be optimization, since this is merely a best-possible approach and adaptation to what we can now think and deem correct. Optimization leaves no leeway for the unanticipated, and thus no latitude to recognize and rectify what actually are undesirable developments or to come up with better ideas with which to set forth on alternate courses.

Effectively dealing with errors, risk tolerance and creativity are perhaps the skills that are most important for our future.

How many errors in the genetic sequences of living creatures did evolution have to make until LUCA (last universal common ancestor) became Homo sapiens 3.5 million years ago? And how many errors did Homo sapiens need to learn from in order to achieve our current state of development? And how much poorer in terms of experiences and insights would humankind now be if there had always only been "normal" people and the statistical mean ... no other kinds, deviant thinkers, people of different colors, or those with alternative beliefs?

To err is human, it's said. Could that be why we're incessantly striving for perfection and steadfastly believe we can attain it with technology and science, and in spite of the fact that there is nothing that we fear more than being eliminated by a world of machinery that functions perfectly well without us?

How can we rethink our very ambivalent relationship to technology as the driving force for configuring our future, and what errors should we perhaps not repeat in the process?

The call for social intelligence is now being juxtaposed to our enthusiasm for the digital world and artificial intelligence. We are propagating the courage to welcome imperfection, since isn't that quite possibly what will always set us apart from the machines!

Formats and Programs at the 2018 Festival

Opening up the dimension of media-art and science to the general public has always been a main concern of Ars Electronica; nevertheless, intensifying the efforts in the avant-garde media-art genre with specialized partners, particularly in selected areas of emphasis, has been intrinsic to the spirit of Ars Electronica since its very inception. In order to live up to both ends of the spectrum the festival offers a variety of well-established and new formats. The Ars Electronica Animation Festival is already an integral part and even earned the title festival within the festival. Another well-known format by now is the focus on digital music and sound art with Music Monday and Sonic Saturday in cooperation with the Anton Bruckner Private University.

Theater and Digital Media was successfully launched at the last festival. This year includes a session on digital strategies for theater and a fascinating performance. Last years' introduction of the Ars Electronica Gallery Spaces, as a setting for media artists, collectors and galleries to compare experiences and discuss core issues like the conservation of media-art project, also sparked great interest. Markus Poschner will conduct his second concert with the Bruckner Orchestra in conjunction with the Big Concert Night,

in which the program concentrates on connecting as well as juxtaposing tradition, the state-of-the-art and modernism.
A special highlight this year is the Himatsubushi Trail, which delves into the art of killing time as a means of finding new perspectives for the future in an ever-accelerating society. Fashion & Technology is a new section revolving around the newest developments in wearables and smart devices on the one hand. On the other hand, a lecture panel will shed light on different aspects of fashion with regard to interdisciplinary approaches in the Art & Science field as well as sociopolitical developments. As a festival of art, technology and society, a crucial part of Ars Electronica's mission is monitoring and mediating encounters with technological and social developments. Accordingly we are hosting the Internet of Things (IoT)-workshop Trust in Invisible Agents, an interactive conversation about ideas of transparency, clarity and governance within the realm of the IoT. The Ars Electronica Innovation Forum 2018 discusses interdependencies of culture and economy in relation to this year's festival topic *ERROR – The Art of Imperfection*.
Swarm Arena is the latest outcome of joint research efforts by the Ars Electronica Futurelab and Japanese telecommunications giant NTT with the aim of working on using unmanned aerial or ground vehicles as a means of communication. At the festival not only the current state of research will be presented, moreover experts will openly discuss the facets and future possibilities of this project.

Symposia, Workshops, Tutorials

From artworks and projects in the exhibition to symposia the recurrent theme *ERROR – The Art of Imperfection* will be explored from a multitude of angles throughout the whole festival. The theme conference on Friday takes a look at the phenomenology of erring. Various scholars reflect on conscious respectively unconscious acts of deviation from the norm. As a counterpoint, the Academy of Error on Sunday intends to celebrate the error by revealing its productive potential and examining it on a more personal level. At the Space Art – Trial and Error in Art & Science panel different speakers will expand on the mechanics and dynamics of the error when facing the venture of exploring our universe with an interdisciplinary approach.
The STARTS Day offers lectures, panels and workshops on the potential of evolving future innovators and encompasses tours through the STARTS Prize Exhibition.

Theme Exhibitions

Error stems from the latin word errō and does not only mean to be wrong but also wandering about. Wandering about – and wondering – seems to be a fitting mode when visiting a festival for art, technology and society that approaches such a multifaceted term as error. Of course, from a curatorial and certainly an organizational viewpoint, there have to be subdivisions, especially when it comes to setting up installations in premises measuring almost 100,000 square meters. The theme exhibitions have been structured according to two fundamental approaches, though there is a great deal of overlapping among them. Art and Technology projects, the origination of which was motivated primarily by research and exploration, make up the exhibits in the large halls on the 1st Upper Level; art and technology projects that are primarily means of artistic expression predominate in the large exhibition parcours arrayed in the spectacular spaces of the lower Levels.

This cursory description of some of the items on this year's lineup can provide only a rough idea of the character of this festival, and constitutes only a small part of a very extensive festival program that is jam-packed with offerings of consummate quality.

This is made possible by collaboration with the festival's countless partners and supporters throughout the domains of art, technology and society. They all contribute to this festival with their ideas, visions and commitment, and have earned our sincere thanks.

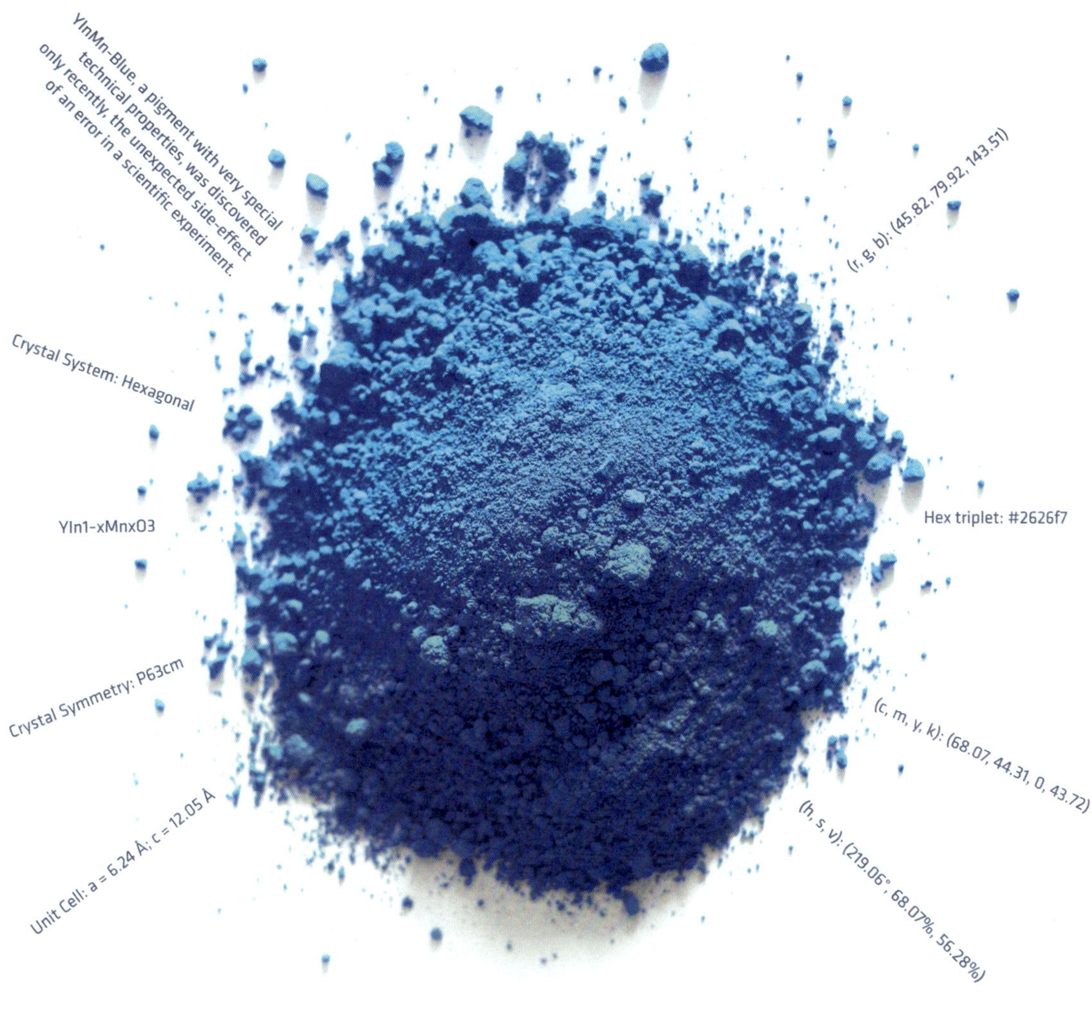
YInMn-Blue, a pigment with very special technical properties, was discovered only recently, the unexpected side-effect of an error in a scientific experiment.
Crystal System: Hexagonal
YIn1-xMnxO3
Crystal Symmetry: P63cm
Unit Cell: a = 6.24 Å; c = 12.05 Å
(r, g, b): (45.82, 79.92, 143.51)
Hex triplet: #2626f7
(c, m, y, k): (68.07, 44.31, 0, 43.72)
(h, s, v): (219.06°, 68.07%, 56.28%)

ERROR

the Art of Imperfection

CONFERENCES, LECTURES, WORKSHOPS

Conferences

Opening Symposium

It is all about ERROR – The Art of Imperfection. The Opening Symposium offers an insight into the discourse contributions of Ars Electronica's core formats and is intended to serve as a teaser that arouses curiosity for more. The speakers – all of them fundamental to the festival's conferences, lectures and workshops – will present previews of their lectures on art, innovation, science and technology, which will be held in the following symposia. With the wide range of options given by the festival program, this is a perfect opportunity to get an overview and decide on the personal key interests.

Theme Symposium: ERROR – The Art of Imperfection

Imperfection Implied

The Fragile
Errors in Culture, Body, Mind and Technology

Errors have always played an important role for us. Without errors in genetic coding, human evolution would not have happened. Mutations made our brains bigger, turned our spines upright and gave us the physical capability of advanced communication. Once we became aware of our own fragility, the trial and error mechanism enabled and possibly even forced us to learn and develop behaviors, social structures, science and sophisticated technology.
Errors are imperfections, deviations from the norm, not conscious failures or mistakes. Imperfection is an important creative and poetic force that also fosters social and technical innovation. In our longing to manage the imperfections of ourselves and our surroundings we invented all kinds of mind and body extensions. To expand our reach our ancestors thought of complicated steering mechanisms for seafaring, optic devices helped us study our solar system, and soon privatized space travel aims to colonize other planets to ensure humanity's survival. We have learned to include errors in our thinking to make substantial progress in social innovation as well as technical and industrial developments. Still, humanity remains a fragile construct. Nassim Nicholas Taleb, a Lebanese-American scholar, has coined the concept of *Antifragility* to which he refers as being fundamentally different from resilience: "Antifragility is beyond resilience or robustness. The resilient resists shocks and stays the same; the antifragile gets better." In this sense, our world has proven itself quite antifragile; mutating, varying its behavior and reactions after errors have occurred. At the same time other evolutionary errors have been carried from generation to generation and appear to be making no efforts to vanish. Every replication of a cell creates errors in its structure. It remains yet to be seen if humanity can prove to become truly antifragile at some point in the future.
Being human implies imperfection, still striving towards perfection and optimization is deeply rooted in our behavior. We are trying to optimize Artificial Intelligence by having it learn from errors it has made. With this obsession with pseudo-perfection, we are preprogramming catastrophes. It is hubris to assume that we as imperfect beings can produce perfect machines that will not commit any errors.
Luckily for us, there are those that defy our fragility and dive head-on into taking risks. Pioneers of human innovation embrace the dangers of committing grave errors to imagine and eventually realize complex projects, like having robots plot out steel bridges on the go or engineering an over-the-counter kit for homemade custom remedies for vaginal health.

Fakes, Responsibilities and Strategies

Conscious erring and the accountability question

Apart from randomly occurring errors, we have been experiencing intentional man-made deviations that are using current technology to produce lifelike deep fake videos and spread misinformation in digital media. In a mediascape where the average user couldn't care less about the veracity of sources, who is responsible for regulating the distinction between true and false? Observing the international political landscape and attitudes, the trend appears to be leading away from affirmations of altruistic humanist values towards self-centered, short-sighted decision-making processes. Thanks to rhetorics of fear being spread without rational evaluation or any hint of ethical conscience via social media, absolutist rulers are being voted into power and employ authoritarian methods of public and private monitoring.

As huge amounts of data are being generated via our monitored and quantified selves and surroundings, then collected and processed by automated systems and learning machines, we demand more regulations of who is doing what with our personal data. But we have to understand that by using and interacting with digital technologies, we are becoming digital citizens ourselves and therefore need to take over responsibilities as well. It should be an important goal for us to actively develop powerful strategies of data autonomy. It seems that the next big leap in dealing with our errors needs to proceed from delegating problem solving to artificial intelligence or official entities towards developing and promoting a sophisticated social intelligence and ethical responsibility. Part of this responsibility is to encourage the emergence of polymaths and cross-disciplinary approaches. As a counterpart to hyper-specialized experts staying well within the bounds of their respective fields, polymaths could be viewed as symbolic for the essence of aforementioned imperfection. Yet conscious imperfection plays a crucial role of defying our mania for perfection in the fields of technological, economical and industrial innovation. Especially in the design of emerging technologies like intelligent machines, these allegedly abnormal but consequently creative thinking processes and alternative viewpoints are sorely needed. There is a strong expectation that artistic participation provides an advanced strategy for enhancing and strengthening ethical conscience and responsible innovation in the European path towards artificial intelligence.

Text: Karla Spiluttini

Academy of Error

The Academy of Error sets the counterweight to Friday's Error conference. It is a celebration of errors, mistakes, and failures. It seeks to reveal the potential that lies in committing errors, turning the unexpected experience of failing into a productive and constructive part of our processes. It shifts the reflection on the act of erring from an outward perspective to a more personal level. Professionals from different fields of science will give insight into the handling of errors in their respective disciplines.

Space Art – Trial and Error in Art & Science

Art might not be the first category that comes to mind when thinking of actual space travel and exploration; the technical aspect – rockets, spacesuits, space platforms – would probably be more dominant. At first glance, it almost seems impossible to combine these two. Yet, exploring is a vital part of both domains: searching for the formerly unknown or unconscious or, at least, looking at the dark corners of our universe. Space Art is a category which vibrantly participates in the field of Art and Science. However, the conference will delve into the mechanics of trial and error in Art and Science in general, with a particular focus on Space Art.

Lecture Program

Expanded Animation-Hybrid Technologies in Animation

In collaboration with the Upper Austria University of Applied Sciences' Hagenberg Campus, the 6th Expanded Animation symposium carries on a process launched in 2013 – mapping the wide-ranging domain of animated worlds of imagery beyond the well-trodden paths. The symposium stays the course originally set at its inception, and presents theoretical positions and perspectives from the art world, the R&D field and the industrial sector. The mission: To function as a driving force advancing an interdisciplinary discourse. This year's symposium is an inquiry into future interfaces in animation. *Interfaces in Motion* will focus on animation technology at the manifold interfaces where humans, computers and interaction meet.

Digital Theater Network Meeting

The European Theatre Lab is an international group of artists, scholars and scientists from theaters and research institutions. Together with 7 theaters, their work is guided by a distinguished Advisory Board of leading experts from the arts and technology.
After two years of applied research, the results in the fields of app-based virtual & augmented reality, 3D sound, auralization, psycho-acoustic effects and voice recognition surtitling will be introduced during Ars Electronica 2018.

PRIX FORUM & ART TALKS

One of the absolute highlights of every Ars Electronica is the opportunity to meet Prix Ars Electronica prizewinners and to attend Prix forums to hear the artists elaborate on their oeuvre and current work. In the Prix Forum IV – Visionary Pioneers of Media Art session, the talk from Derrick de Kerckhove, and a lecture from Roger Malina is planned and afterwards we celebrate the Leonardo's 50th Anniversary together! The Media Art Market Symposium on Saturday as part of the Prix Forum & Art Talks addresses questions around the Media Art Market. It still suffers from less visibility on the common contemporary art market, even if the latest changes in art, science and technology are the essence of media artworks.
The Media Art Market Symposium on Saturday is situated in the Prix Forum + Art Talks, which will be held in the OK. The discussion shows the current and future situation of the Media Art Market. It still suffers from less visibility on the common contemporary art market, even if the latest changes in art, science and technology are the essence of media artworks.

Gallery Space Panels

The Gallery Space Panels are situated in the Säulenhalle. Deep in the core of PostCity gallerists, artists, theorists and collectors dig deep into the complexity of the new media market out of their own experience. From business models for young artists, through the contemporary art aspects of new media artworks, to the conservation of past and present: the Gallery Space Panels create a comprehensive picture of the media art market 2018. Two of the panels are co-curated, one by kickstarter and the other by .art.

Music Monday

The Sound Art Parcours "Music Monday" is a longstanding Ars Electronica tradition. The tour passes through the diversity and plurality of the relationship spaces of music / sound art and the media arts. Theoretical reflection and discussions with the artists, personal as well as individual physical experience, immersion in the mediality of sound: this is what Music Monday stands for, because it enables intensive guided encounters with the sound worlds that we find at the festival.

Music Education Day

A coalition of Austrian musical institutions and initiatives is producing an encounter with the future prospects of young Austrian musicians. This event is an outgrowth of Music Summit, a working group formed at the 2017 Ars Electronica Festival; now, one year later, they've put together a program for musicians, teachers and musicologists.

The emphasis is on the future of the music industry and musical training. Considering these two systems jointly is meant to provide a better overview and more profound insights for musicians. Music Education Day will lay down the beat and set the tone

Open Futurelab

As in past years, the "Open Futurelab" means a stage within the festival, where projects and project partners of the Ars Electronica Futurelab – like NTT, the Japanese telecommunications giant – are presented. Selected industrial, educative, artistic and research collaborations will be positioned next to each other in one spot, displaying the variety of projects and wide range of activities going on at the Futurelab in recent months.

Campus Forum

For the very first time this year, a selection of universities will not only present their work in a special Campus exhibition, but also host their own discursive format - the Campus Forum Conferences. Hexagram will present their research network, dedicated to research-creation in the fields of media arts, design, technology and digital culture. Other panels will discuss teaching a digital native generation for a sustainable future technology, the relation of error and education, or the developments of social activism, digital communities and responsible innovation in today's media art.

The Politics of Fashion: Fashion as a Social Bot

In the 21st century, fashion interwoven with technology has been affirmed as an innovative field of development. What has been emerging here—tools, material & textile technologies such as 3-D printing and digitization, software developments, bio-materials and interactive wearables—promises to open up new areas of deployment in medicine, sportswear and everyday attire. At the same time, this evokes the promise of more sustainable fashion production in the future, the prospect of more efficiently outfitting and accessorizing constructions of a person's body, labor and identity, and enabling them to function more efficiently and live better thanks to new materials. Fashion operates as a social bot, an ideological tool programmed algorithmically to optimize the human body and retail consumption.
What's the connection between sociopolitical developments and these new, optimized materials, human sources and natural resources? Who or what produces substantive texture? And doesn't this synergy between research and fashion, fashion and technology, run the risk of reproducing this structure?
The concept of expanded fashion research permits us to shift our perspective, to look beyond the fast fashion system to an approach that's not yet marketable—rethinking the positioning of fashion as social, processual sculpture at the interface of technology and art, and enabling it to manifest political, social and aesthetic forms. Evolving fashion design and production processes call for alternative formal and substantive design concepts, methodological strategies, and artistic, social and aesthetic work and design practices in order to meet new social challenges.

Austrian Center for Fashion Research (ACfFR)
The host of the roundtable discussion & keynotes is the Austrian Center for Fashion Research (ACfFR), an interdisciplinary research facility financed by the Republic of Austria's Federal Ministry of Digital and Economic Affairs. ACfFR is operated jointly by the Academy of Fine Arts Vienna and Linz Art University's Department of Fashion & Technology. Its mission is to internationalize and strategically develop fashion research in Austria. To accomplish this, ACfFR dovetails artistic research and practice-led research in the field of fashion with innovative research in cultural studies and art history. The Center gathers and strengthens existing expertise and research at universities and museums and in networks, and nurtures the emergence of research projects at the nexus of fashion, design, culture, art and business.

Text: Christiane Luible-Bär

Guest Conferences

Prospects for Political Education

Is the world coming unhinged? You would think so now that atomic threat scenarios are once again discussed on a daily basis and we observe the metamorphosis of conventional political systems. But do young people see things that way too? This symposium – produced jointly by the Upper Austria Teacher-Training College – will attempt to come up with answers to such questions by staging speeches as catalysts, workshops as settings for discussions of ideas and methods, and a Dialog of Disobedience as a wake-up call.

Deep Fake or Rendering the Truth

The ability of computers to fake reality convincingly is going to become more and more of a critical problem as hackers, extremist news organizations and politicians seek to control the media narrative through increasingly convincing visuals. This panel, co-organized with Tobias Revell, Natalie Kane and the Impakt Festival Utrecht, will consider and speculate on possible futures for rendered realities and suggest strategies for regulating or countering artificial realities created by computation.

Co-funded by Creative Industries Fund NL

Sonic Saturday

This year the Anton Bruckner Private University is organizing for the third time the Sonic Saturday Symposium, which this year is dedicated to the topic "... under control of music, music under control of... composing (in) digital worlds." Visitors can look forward, together with the Music Monday, to two days dedicated to the most advanced forms of digital music and sound art, which are not only of interest to specialists in this genre.

ZusammenHelfen Conference

This is the fourth consecutive year that ZusammenHelfen–Working Together in Upper Austria for Refugees is hosting a conference for all those who are committed to and interested in helping people forced to flee, or are affected by refugees and integration. This year's conclave entitled "Day of Encouragement" will scrutinize new prospects, discuss the latest developments and challenges, and elaborate on successful projects.

New Business Models for Artists / Kickstarter Panel

The Kickstarter Panel will discuss the basic question of how artists get paid. Digital art has always run counter to traditional institutional modes, and sought direct engagement with audiences. Digital artworks have explored have explored some of the most urgent concerns of our time, while the artists who create them have often struggled to find consitent funding support. How are artists experimenting with payment and distribution models in ways that are both creative and practical?

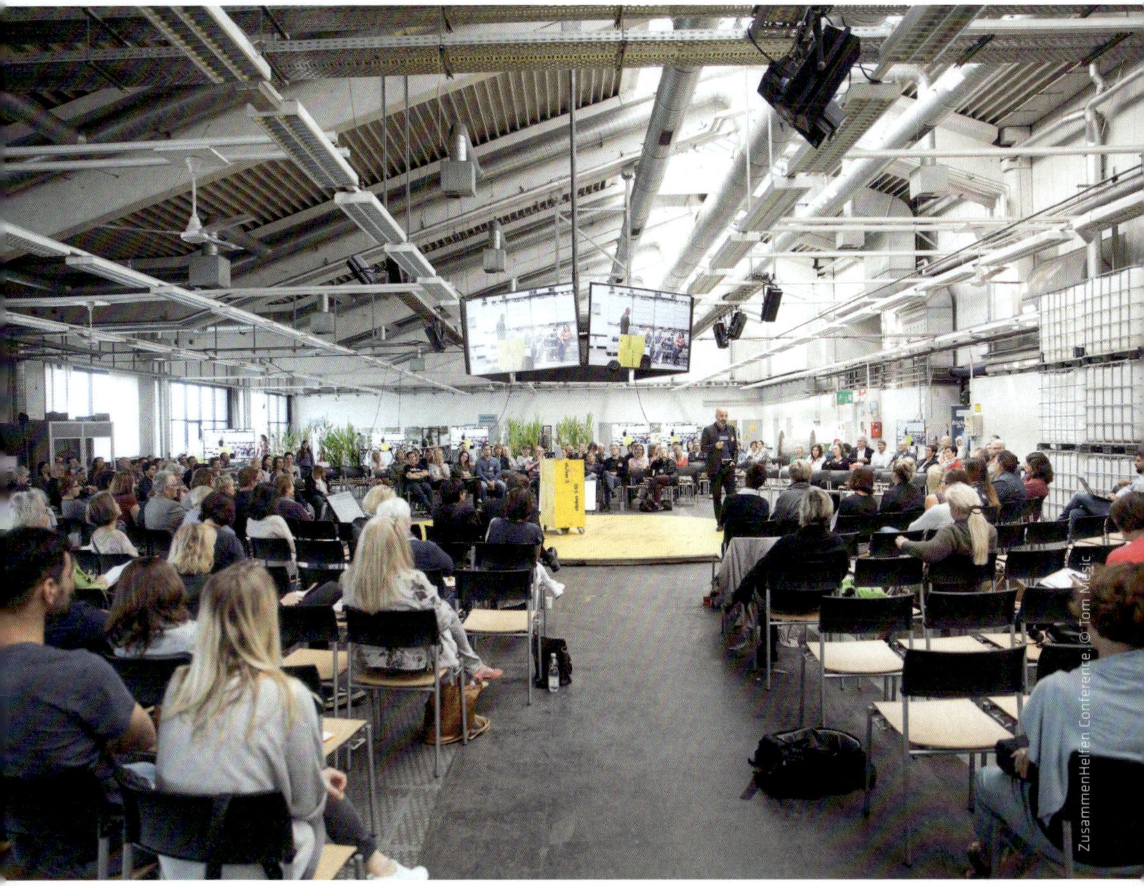

ZusammenHelfen Conference. © Tom Mesic

Innovation Forum

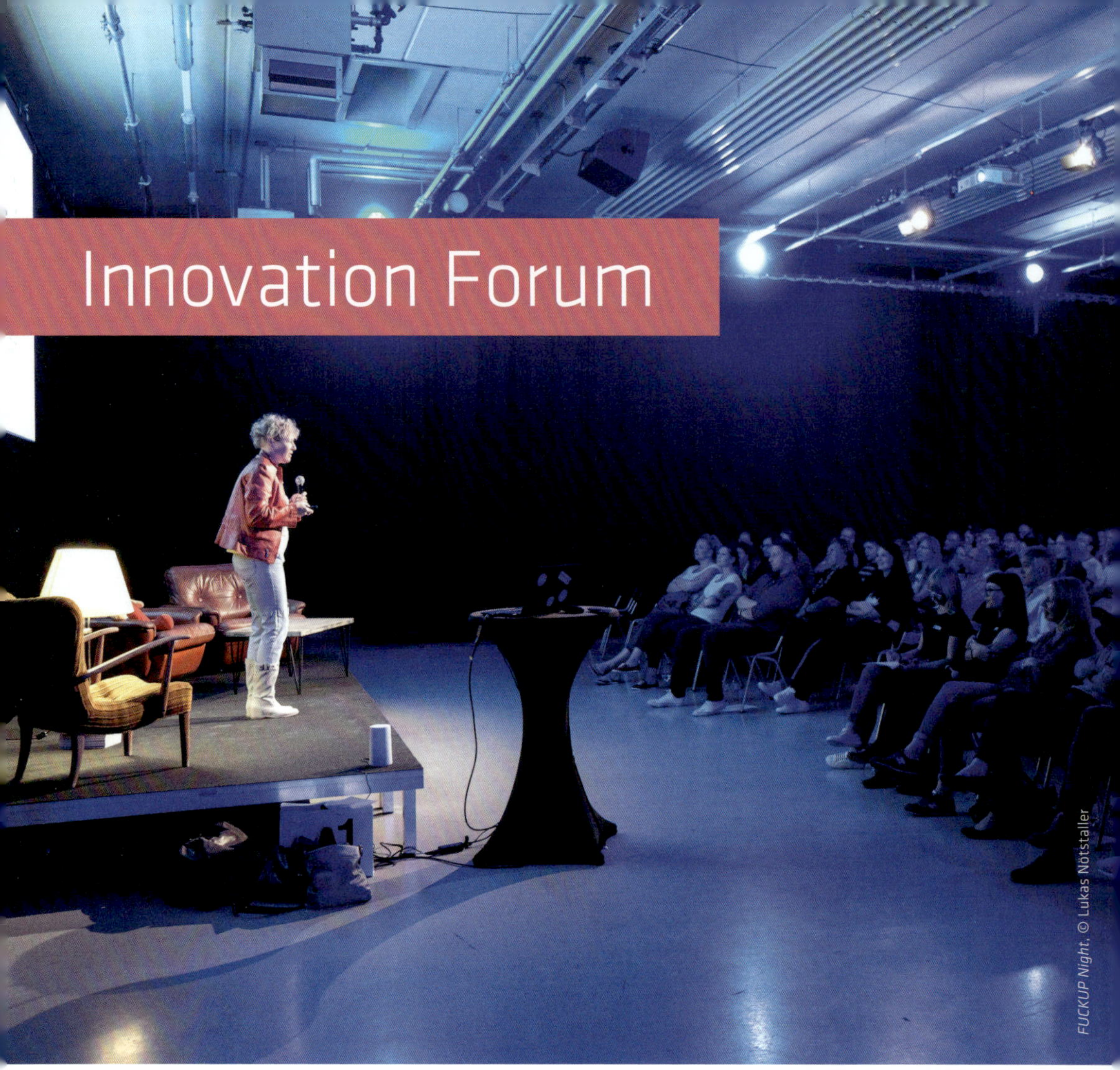

FUCKUP Night, © Lukas Nötstaller

The Ars Electronica Innovation Forum 2018 discusses the interdependencies of culture and economy on the basis of this year's festival topic ERROR – The Art of Imperfection.

The two-day event gives some insight into the role of error management and the links of innovation in art, science and industry from various perspectives.

Innovation Forum Get Inspired

(in collaboration with Wirtschaftskammer Oberösterreich)

The Thursday program with the title Innovation Forum Get Inspired starts with a focus on the local perspective of economy and innovation in comparison to the famed Silicon Valley economic scene.

The Get Inspired Symposium will examine the constructive potential of failures and fallacies and possibilities of error management. In the afternoon of Get Inspired participants may choose between specially designed guided tours through the festival venue followed by Get Inspired Presentations of established, evolving and start-up businesses or a workshop program on the topic of Internet of Things, discussing the value of artistic intervention as part of innovation practice.

FUCKUP NIGHT Linz Special

At the FUCKUP NIGHT Linz Special, selected speakers provide their personal insights into the world of occupational failure. With the help of 10 images, they'll give an account of a project that flopped big-time and the lesson they learned from this. We – an alliance of the University of Linz, Linz Art University and Tabakfabrik - are pursuing a vision: destigmatizing failure, socially and personally. We're open to all sorts of tales of botch jobs, blunders and screw-ups. The field of endeavor doesn't Matter – art, culture, business, social, technological, scientific, whatever. There are no restrictions: the more diverse, the better! The FUCKUP NIGHT Special is being staged as a lead-in to the Future Innovators Summit at the Ars Electronica Festival. Visitors will receive valuable inputs to take away from this event. Learning from errors makes something totally new possible; it enables innovation. Whoever talks about it in front of others might also help them avoid making the same mistake.

Fuckup Nights Linz: Birgit Wimmer-Wurm, Kathrin Anzinger, Nina Fuchs (organisation team)
Fuckup Nights Vienna: Dejan Stojanovic (facilitator)

STARTS Day

In collaboration with the European Commission

STARTS day will emphasize the potential of evolving future innovators and show extraordinary samples of innovation at the nexus of art, science and technology. The day concentrates on the STARTS program and will lead visitors through several different formats of discussion, presentation of outstanding projects as well as theoretical reviews of the role of STARTS collaborations.
The STARTS Day will bring artists, creative professionals, scientists and industry representatives together to present STARTS collaborations and art-science residency programs from around the world. In "Embracing the Risk – STARTS Talks" prominent practitioners and winners of the STARTS Prize 2018 will present their works, visions and experiences. The STARTS Day will close with a conference on Artificial Intelligence in Art & Science.

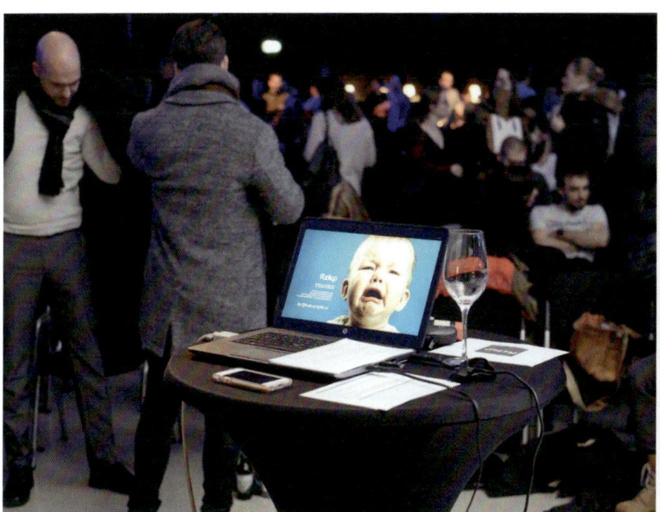

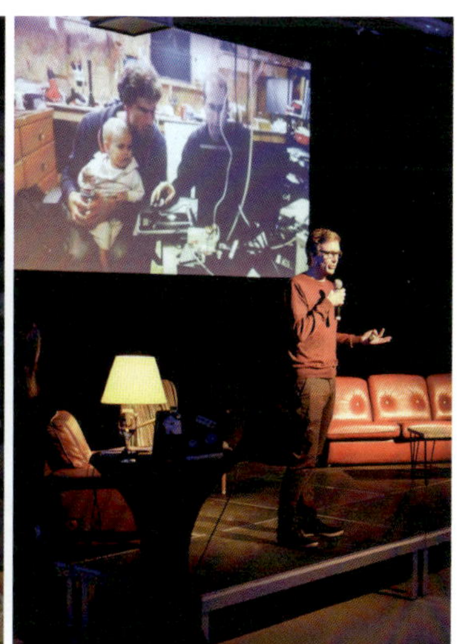

FUCKUP Night, © Lukas Nötstaller

Workshops

EMAP Jam-Session

EMAP (European Media Art Platform), supported by the Creative Europe programme of the European Union, annually awards production grants to outstanding European media artists and supports research, production, presentation and distribution of media art in Europe and beyond. In this session the partner institutions will present themselves and artists will have the chance to obtain information about the program.

BR41N.IO Hackathon

Brain Hackathons are brainstorming and collaborative marathons designed to rapidly produce working prototypes. At Ars Electronica 2018, BR41N.IO, organized by g.tec brings developers, technologists, engineers, students, artists, and scientists together in teams of 5 participants each over 2 days to cram and build solutions that they can present. Hence, the Hackathon provides an environment for innovation and entrepreneurship. By putting creative minds from multiple disciplines together for a short period of time, we have the opportunity to discover and uncover new possibilities for using BCI-related hardware and software.

Future Sessions: Trust in Invisible Agents

Expert Workshop Series on Internet of Things

The panel organized by FutureEverything, as part of Create-IoT, will stimulate an interactive conversation around ideas of transparency, clarity, ethics and governance within the realm of the Internet of Things. This roundtable session positions artistic intervention as a practical and conceptual toolset to challenge current thinking and explore possible futures.

Future Innovators Summit 2018

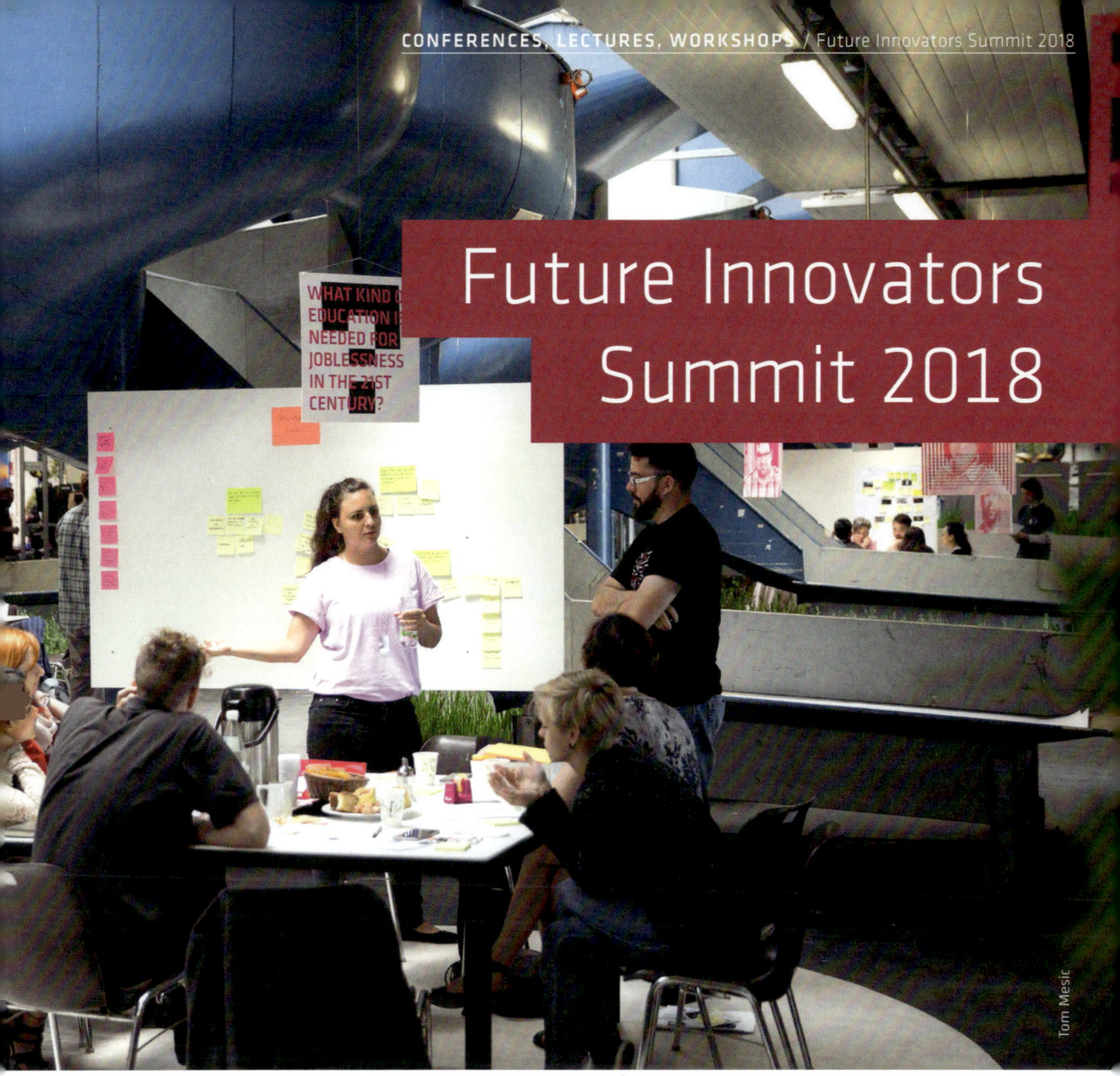

Tom Mesic

The Future Innovators Summit (FIS) is a creative workshop format created by Ars Electronica Futurelab and Hakuhodo. Now after five years FIS has gathered over 100 experienced professionals as well as young entrepreneurs and social activists, technicians and scientists, and of course artists and designers at the Ars Electronica Festival in September in Linz for mutual inspiration and the exchange of ideas and know-how. Future Innovators are invited to explore new ways of collective brainstorming and creative prototyping on the crucial questions of the future. The FIS program during the festival includes a broad range of lectures, presentations and exhibitions as well as ample opportunity for participants to engage in dialogue with each other and the public.

The goal of FIS is not to create solutions, but to create so-called Creative Questions, which means missions for tomorrow corresponding to the theme of each Ars Electronica Festival – this year the main focus is "ERROR – The Art of Imperfection."

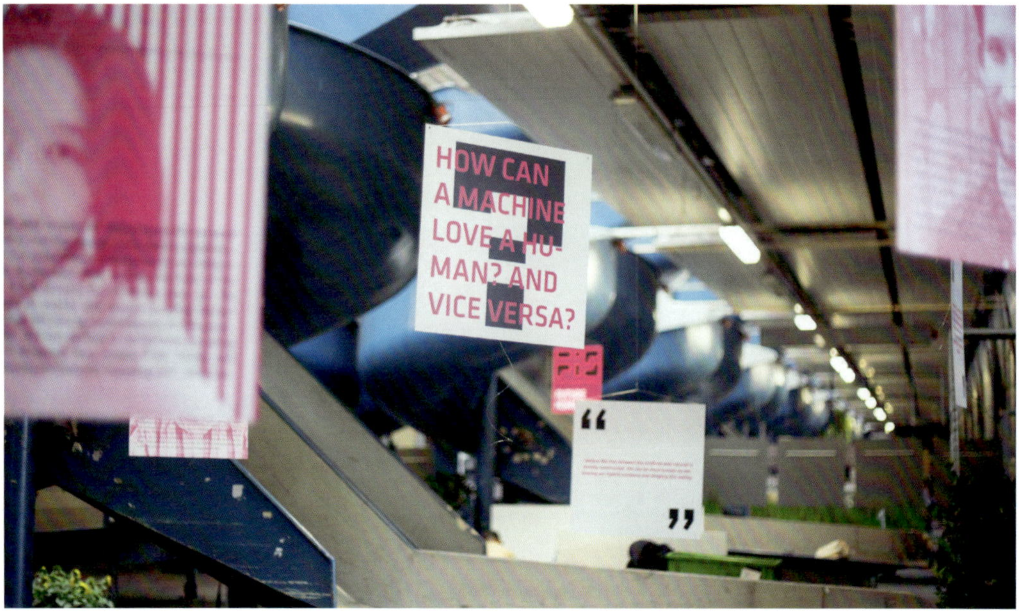

Vanessa Graf

FROM THINK-TANK TO DO-TANK

Since 2018, the Future Innovators Summit has not just been held during the festival in Linz, but also for the first time in Tokyo, in the framework of Ars Electronica Tokyo Initiative (AETI). FIS Tokyo (for more information see p. 367) was realized in Tokyo Midtown in May 2018 together with Hakuhodo and focused on themes that originated in Tokyo, such as:

DEATH-LIFE
TECH-SKIN
PUBLIC-PRIVATE

Thus, the Future Innovators Summit has now become an annual activity of Ars Electronica to raise questions as missions for tomorrow—globally and locally. We aim to create not just a think-tank but also a "do-tank" for social innovation inspired by the cutting-edge festival topics and site-specific topics.

The Themes of FIS2018

FUTURE HUMANITY

When we are facing a world where machines may evolve to be better thinkers and doers than humans, how does the meaning of error change? How do errors make us more human? *Future Humanity* seeks out our visions and missions in correlation with technology, other beings and ourselves.

1. What errors are we attracted to? And why?
2. How can we co-exist with imperfections?
3. How can we become more human?

FUTURE DIGNITY

Everyone is aware of errors such as failing to treat others with dignity or the rise of political and economic uncertainty that attend the digital progress in society. These errors are caused by a lack of respect and tolerance for the co-existence of tradition and the future. Would the knowledge and experience that we – the human tribe – have accumulated through our thousands of years of history, be of any use when the future is so unpredictable and highly digitized? *Future Dignity* seeks future beliefs and confidence for the digital society, especially when we are behind the wheels to make big decisions for tomorrow now.

1. How would dignity be created in the digital society?
2. What is tradition in an unpredictable future?
3. How might arts of imperfection enhance our future dignity?

FUTURE SHARING

Digitization in our daily lives, between humans (and also between humans and non-humans) is evolving very quickly. In this new stage of community building and sharing, what will we need or want to share? What would be the errors entailed? *Future Sharing* explores our actions for tomorrow as beings who exist in a community of some kind, be it physical or digital.

1. What errors could occur in our future shared society?
2. What would humans need to share? And how would it be realized?
3. How could we arrange for tolerance of errors in society?

The Future Innovators Summit is an initiative from Ars Electronica and Hakuhodo, and supported by netidee / Internet Foundation Austria.

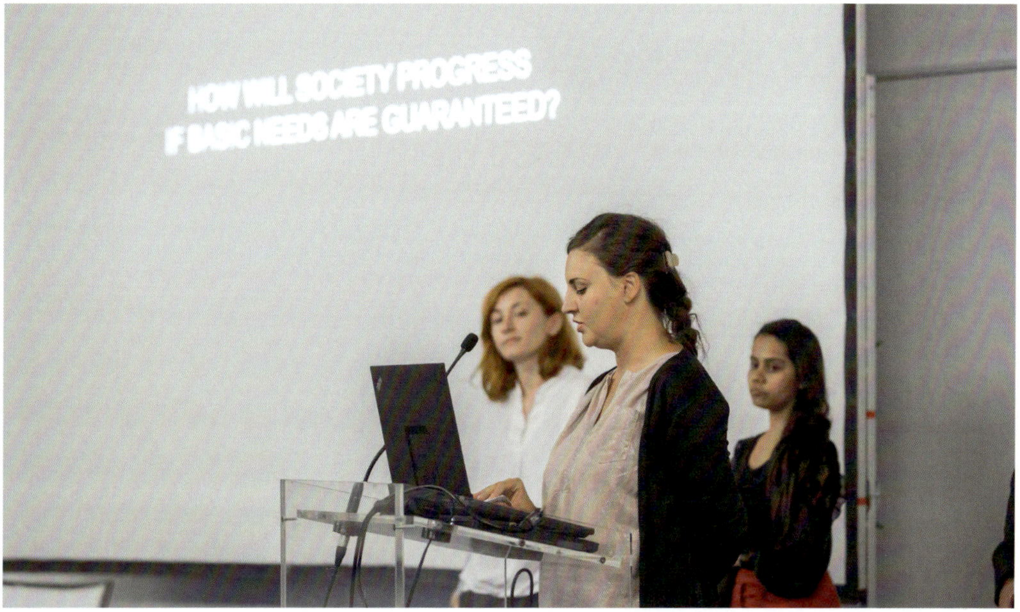

Florian Voggeneder

ERROR

the Art of Imperfection

EXHIBITIONS

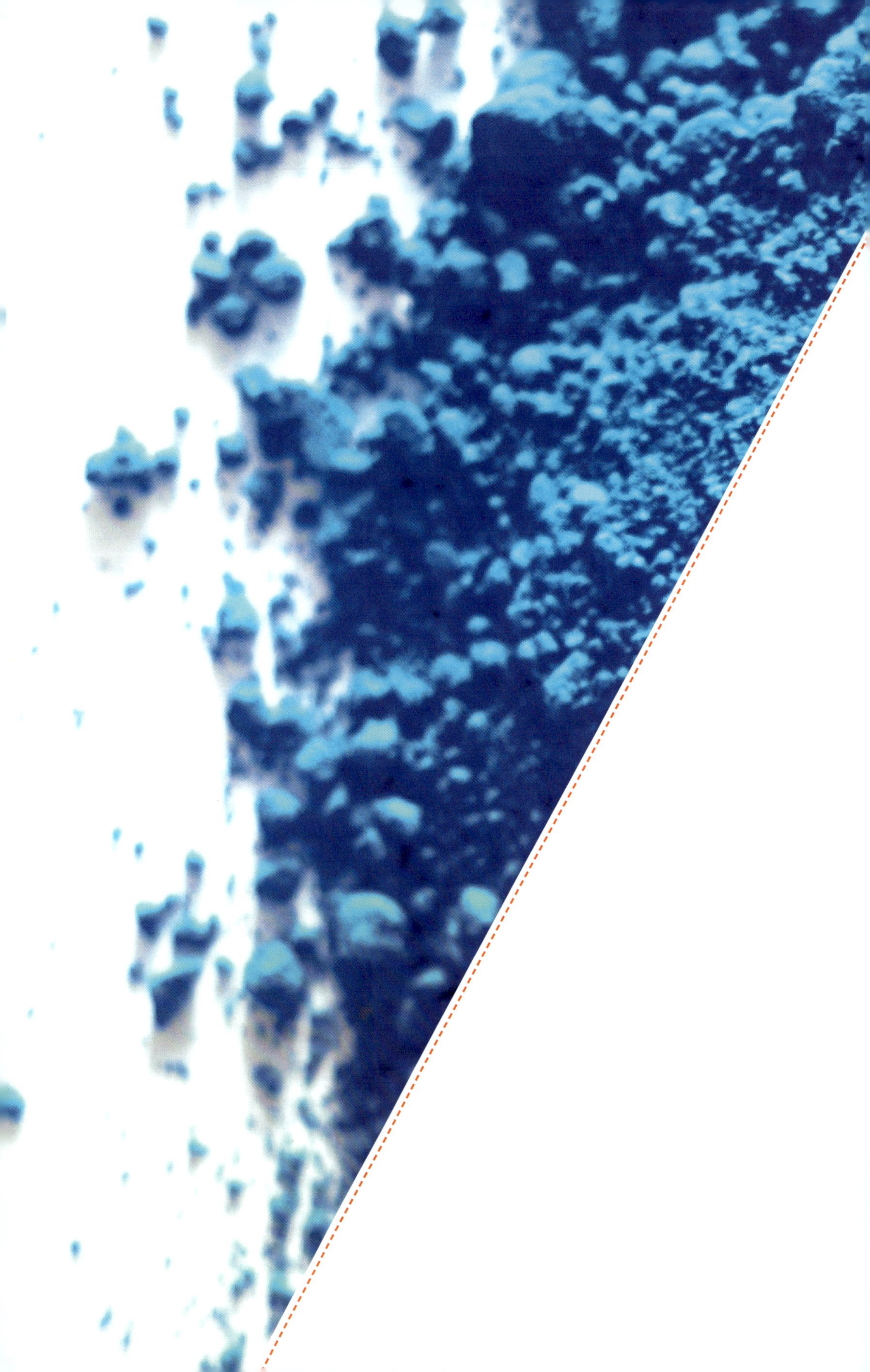

ERROR, FAKE & FAILURE

ERROR IN PROGRESS

The theme exhibitions "Error, Fake & Failure" on the upper floor, "Error in Progress" in the basement and the "Himatsubushi Trail" on or under the roof landscape of Postcity occupy a central position in Ars Electronica's extensive exhibition program. Error, Fake & Failure traces the phenomenon of error, spanning the interpretive frame from error as deviation from the norm and as failure or disappointment on the one hand and intentional forgery or deception on the other. Error in Progress expresses the ambiguity of its title, which not only alludes to the processual nature of trial and error and the open dynamics of the imperfect, but also the question of the errors, mistakes and aberrations of our progress - politically, socially and technologically. The exhibition reflects the broad, increasing unease with current developments and presents artistic work as an effective way of reformulating problems, changing perspectives and developing alternative concepts.

Forensic Oceanography (CH/IT) and Forensic Architecture (UK)

Mare Clausum

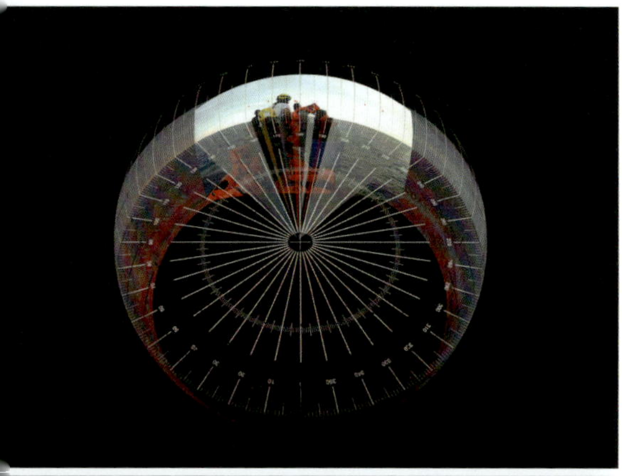

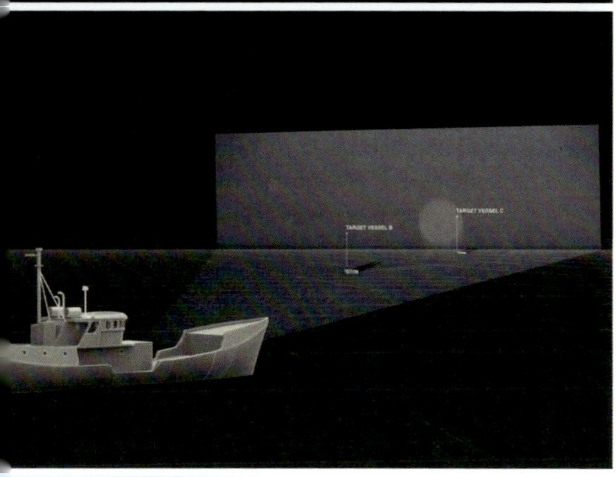

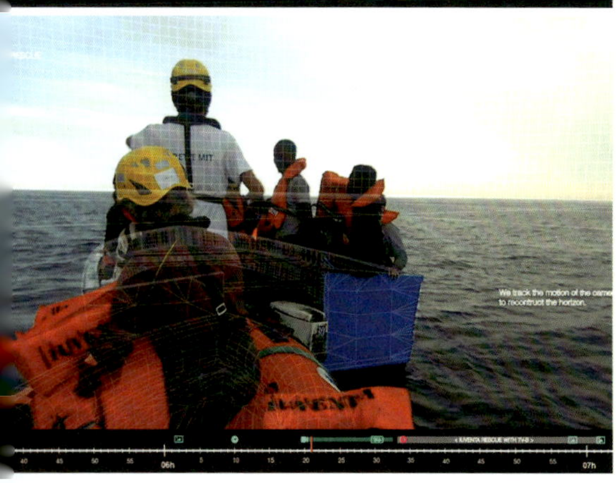

As the EU's policies of deterrence deployed since the Arab uprisings failed to stem migrants' crossings of the Mediterranean, the Italian government – in collaboration with other European governments and agencies – deployed a two-pronged strategy to close off the central Mediterranean: on the one hand, by criminalizing and limiting the rescue activities of the NGOs that have stepped in to make up for the lack of state rescue operations; and on the other, by reinforcing the collaboration with Libyan authorities and militias to prevent and intercept departures, thus physically containing migrants on the Mediterranean's southern shore without requiring the direct involvement of Italian or EU authorities. This undeclared operation, which dramatically escalated throughout 2017, is what Forensic Oceanography has called *Mare Clausum* ("closed sea" in Latin). Forensic Oceanography and Forensic Architecture have investigated two cases central to this ongoing *Mare Clausum* research, *The Iuventa* and *Sea Watch vs. Libyan Coast Guard*, each concerning one of the dimensions of this policy which entails migrants being brought back to a country where their lives are endangered, and their human rights are systematically violated. These investigations have been made possible by an exponential increase in video documentation by the different actors involved, allowing for a unique form of 3-D-modelling of incidents developed together with Forensic Architecture.

The Iuventa, June 18, 2017

Since the end of 2016, culminating in summer 2017, a growing campaign of delegitimization and criminalization has systematically targeted NGOs engaged in search and rescue. On August 2nd, the ship Iuventa, of the German NGO Jugend Rettet ("Youth Rescue"), was seized by the Italian judiciary under suspicion of "assistance to illegal migration" and collusion with smugglers during three different rescue operations: the first on September 10, 2016, the second and third on June 18, 2017. The seizure came only days after the NGO, along with several others, had refused to sign

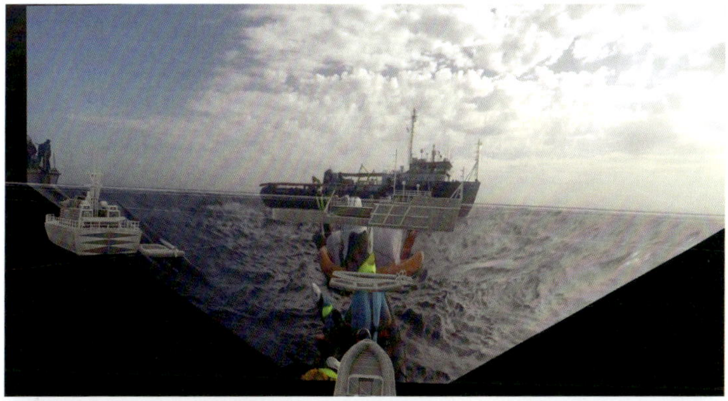

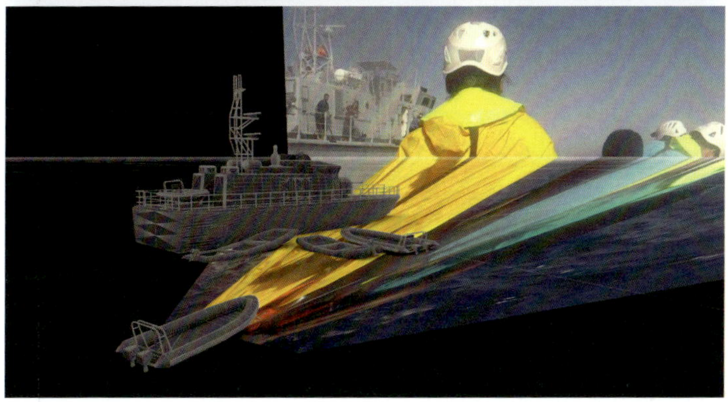

a "code of conduct" that would have dangerously limited their activities. The video presented here offers a counter-investigation of the authorities' version of these three episodes, and a refutation of their accusations.

With the support of Borderline Europe, the WatchTheMed platform and Transmediale

The SeaWatch vs Libyan Coast Guard Case, November 6, 2017

On November 6, 2017, the rescue NGO Sea Watch (SW) and a patrol vessel of the Libyan Coast Guard (LYCG) simultaneously directed themselves towards a migrants' boat in distress in international waters. The boat, which had departed from Tripoli a few hours earlier, carried between 130 and 150 passengers. A confrontational rescue operation ensued, and while SW was eventually able to rescue and bring to safety in Italy 59 passengers, at least 20 people died before or during these events, while 47 passengers were ultimately pulled back to Libya, where several faced grave human rights violations—including being detained, beaten, and sold to another captor who tortured them to extract ransom from their families. The unfolding of this incident has been reconstructed in a video by Forensic Oceanography in collaboration with Forensic Architecture. To reconstruct the circumstances of this particular incident, however, Forensic Oceanography has produced a detailed written report which argues it is also necessary to understand the policies that shaped the behavior of the actors involved, and the patterns of practices of which this event was only a particular instantiation.

Investigation by Forensic Oceanography and Forensic Architecture
Video reconstruction by Forensic Oceanography and Forensic Architecture
Report by Forensic Oceanography

Forensic Architecture (UK)

Ecocide in Indonesia

In 2015, fires in the Indonesian territories Kalimantan and Sumatra consumed over 21,000 square kilometers of forest and peat lands. Fumes from about 130,000 local sources combined into a massive cloud, a few hundred kilometers long and a few kilometers thick. It contained more carbon, methane, ammonium and cyanide than those produced by the entire annual emissions of the German, British or Japanese industries. As the acrid cloud drifted north and westwards, it engulfed a zone that extended from Indonesia across Malaysia, Singapore, southern Thailand and Vietnam. Scientists estimate that this resulted in more than a hundred thousand premature deaths, and that the fires might push the world beyond 2°C of global warming – and into the realm of potential and unpredictable calamities – faster than expected. The cloud can be understood as the harbinger of a new international crime of ecocide, one likely to become more relevant in the years to come.

Forensic Architecture
Undertaken in collaboration with FIBGAR (Baltasar Garzón and Manuel Vergara)
Project team:
Eyal Weizman (Principal Investigator), Samaneh Moafi (Project Coordinator), Jason Men, Christina Varvia, Nichola Czyz, Nabil Ahmed and Paulo Tavares

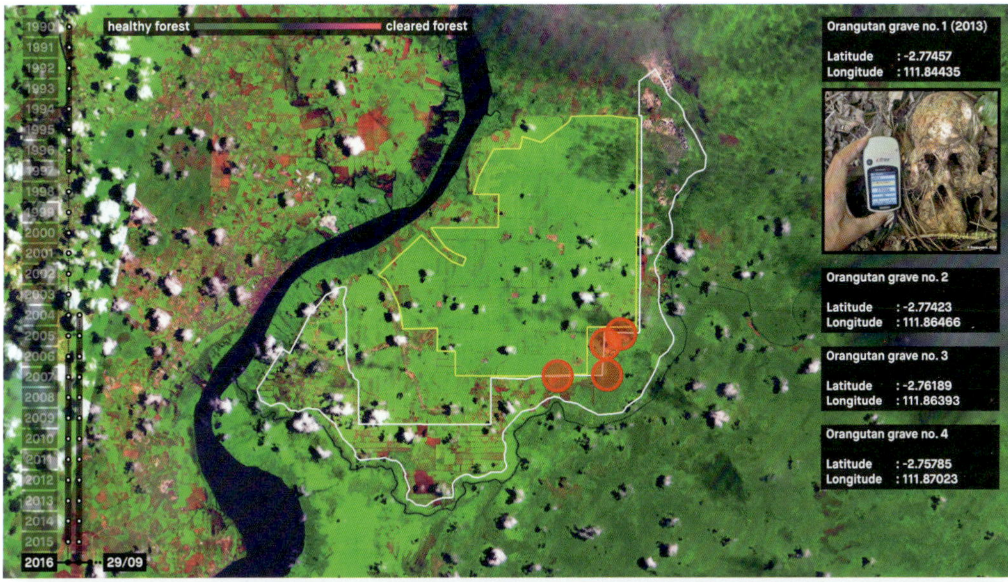
healthy forest
cleared forest
1990
1891
1992
1993
1994
1995
1996
1997
1998
1999
2000
2001
2002
2003
2004
2005
2006
2007
2008
2009
2010
2011
2012
2013
2014
2015
2016
29/09
Orangutan grave no. 1 (2013)
Latitude : -2.77457
Longitude : 111.84435
Orangutan grave no. 2
Latitude : -2.77423
Longitude : 111.86466
Orangutan grave no. 3
Latitude : -2.76189
Longitude : 111.86393
Orangutan grave no. 4
Latitude : -2.75785
Longitude : 111.87023

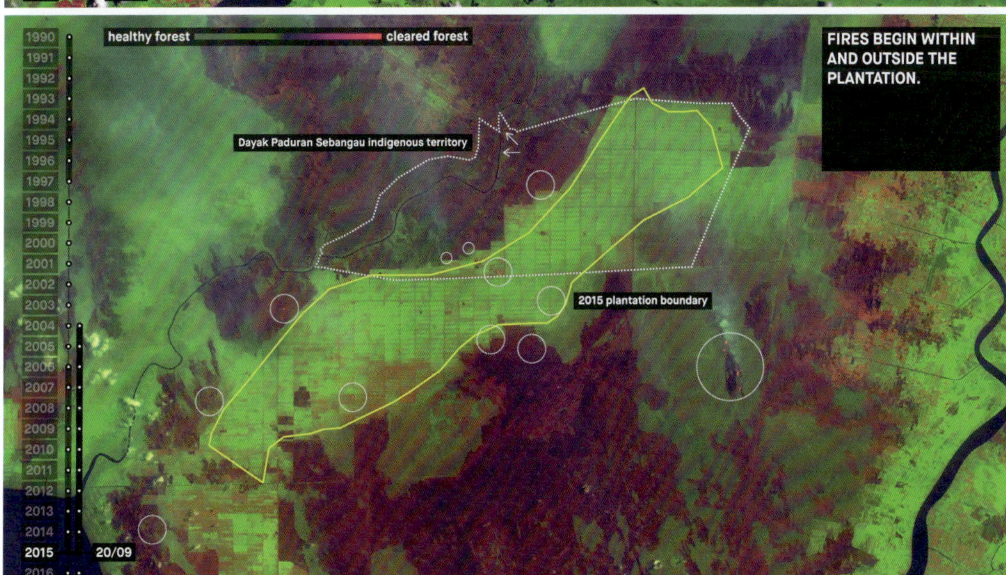
healthy forest
cleared forest
1990
1991
1992
1993
1994
1995
1996
1997
1998
1999
2000
2001
2002
2003
2004
2005
2006
2007
2008
2009
2010
2011
2012
2013
2014
2015
20/09
2016
FIRES BEGIN WITHIN AND OUTSIDE THE PLANTATION.
Dayak Paduran Sebangau indigenous territory
2015 plantation boundary

Anatol Bogendorfer (AT), Hörstadt (AT)

Die Tonfälle

Can the tones of voice of an online linguistic culture be examined and portrayed artistically? In this project, Hörstadt scrutinizes the inflections and the culture of conversation and argumentation manifested in internet forums. This is done primarily pursuant to an acoustic-phenomenological analysis. In concrete terms, what we are investigating here is hate speech posted to the sites of various Austrian journalists – e.g. Armin Wolf (ORF – Austrian Broadcasting Company), Hanna Herbst (Vice Magazine), Rainer Nowak (Die Presse newspaper).

Postings containing hate speech are usually formulated in writing. Linguistically, however, they are closely related to an oral–and thus to an acoustic–culture. This is so not only because they can be associated with the so-called *Affektsprache* (language of the emotions). Things like anonymity, invisibility and spatial distance also have corresponding expressions in acoustic space. What can likewise be expressed is the differentiation between what is said and how it's said – for instance, in terms of the parameters of the tone of voice such as pronunciation, sound, timbre, volume and intonation.

On the basis of the postings containing hate speech that we have been given access to, Anatol Bogendorfer of Hörstadt and actress/voice-over artist Maria Fliri jointly investigate the phenomenon of language changing its medium from writing to sound. The result takes an audiovisual form.

Concept: Anatol Bogendorfer (AT) / Hörstadt
Producer: Hörstadt Team (Androsch, Bogendorfer, Knipp, Saftic)
Director: Anatol Bogendorfer
Cast: Maria Fliri
Cameraman: Christian Dietl
Sound Engineer: Alex Jöchtl
Set Designer: Tabea Cray
Location: Quartier
Editor: Anatol Bogendorfer
Support: Las Gafas Film, Retro Goldmine Film, Vilo, Quartier

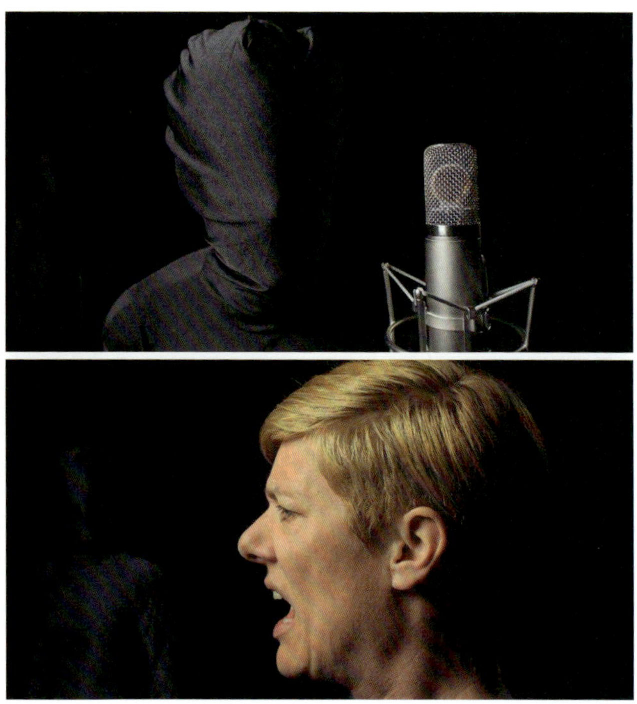

Esther Hovers (NL)

False Positives

The project *False Positives* is about intelligent surveillance systems. These are cameras that are able to detect deviant behavior within public space. *False Positives* revolves around the question of normal behavior. It aims to raise this question by basing the project on eight different 'anomalies'. These so-called anomalies are signs of body-language and movement that could indicate criminal intent. It is through these anomalies the algorithms are built and cameras are able to detect deviant behavior. The eight different anomalies were pointed out to me by several intelligent surveillance experts with whom I collaborated for this project. The work consists of several approaches, photographs and pattern drawings. All together, these form an analysis of different settings in and around the business district of the de facto European capital: Brussels.

Supported by Mondriaan Fund

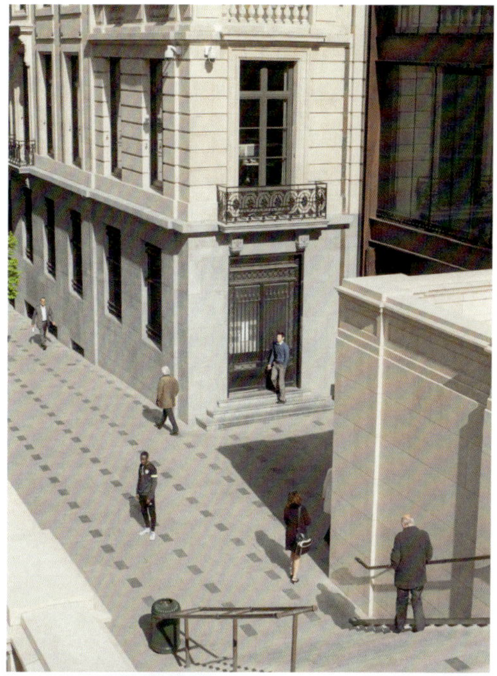

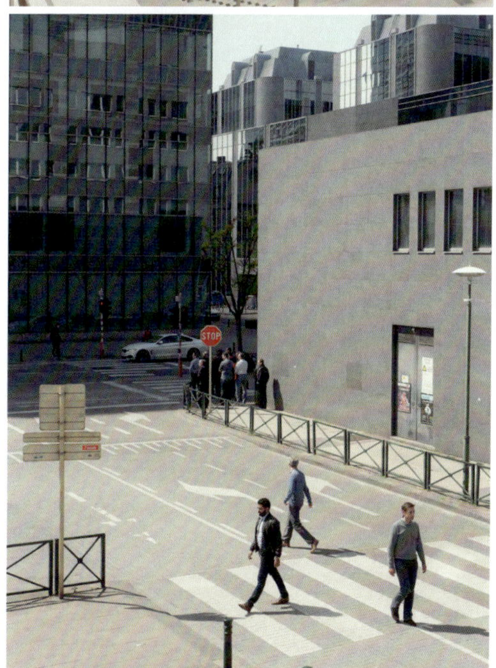

Michael Saup (DE), Andrea Winter (DE), Andreas Erhart (DE)

+b (ORBIT)

+b (ORBIT) shows Buckminster Fuller's Dymaxion Map of the earth built from multiple layers of white industrial sugar cubes and illuminated by the complete sequence of all nuclear explosions from 1945 until now. Using the cubes as three-dimensional pixels, *+b* emphasizes the intimate relationship between information, energy, resources and their impact on society and nature. *+b* stages the most extreme power released by humankind, irreversibly transforming the atmosphere and igniting the epoch of the nuclear Anthropocene with its application and supposed mastery of atomic power. The work illustrates how this mastery is really the reiteration of a profound error and the subsequent compounding of that error. We keep on making mistakes. Some of these errors are extraordinarily beautiful and useful, some are terrifyingly destructive with long-term planetary impact, and many are both.

Co-produced by Trans-Media-Akademie Hellerau e.V.
with the help of The Buckminster Fuller Institute, die wellenmaschine, Silke and Uwe Buhrdorf, Shuichi Fukazawa, Ulf Langheinrich, David McConville, Robert W. Gray, Knut Bressgott, Karolina Funk, Acci Baba, Endre Ketzel, Li Alin, Nicole Pesant Méalin, Rosa-Lee Sendlinger, Nadine Bors and openframeworks

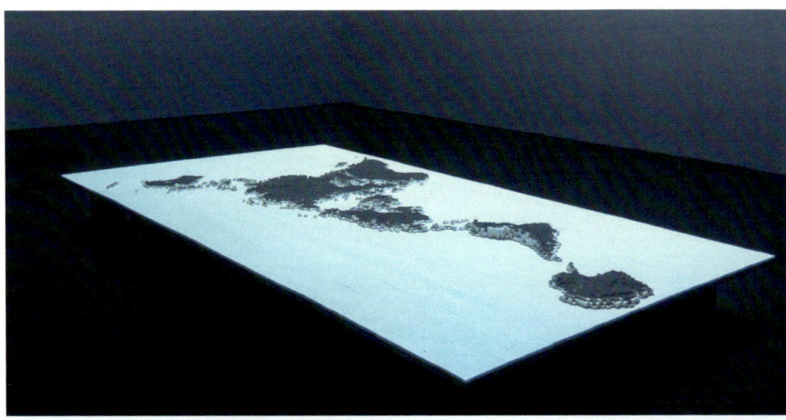

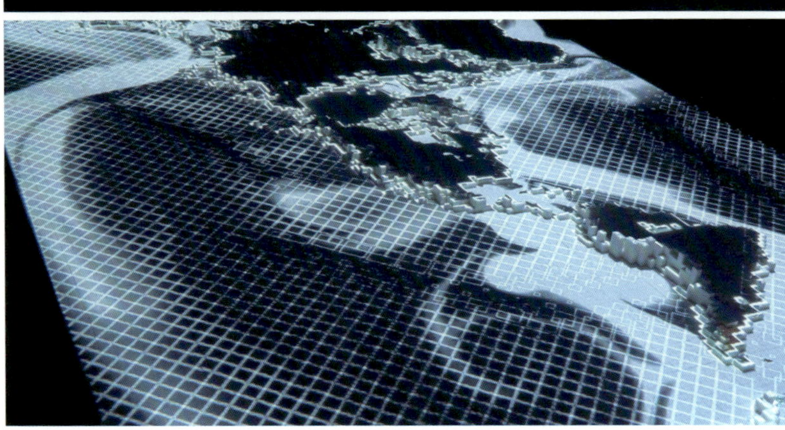

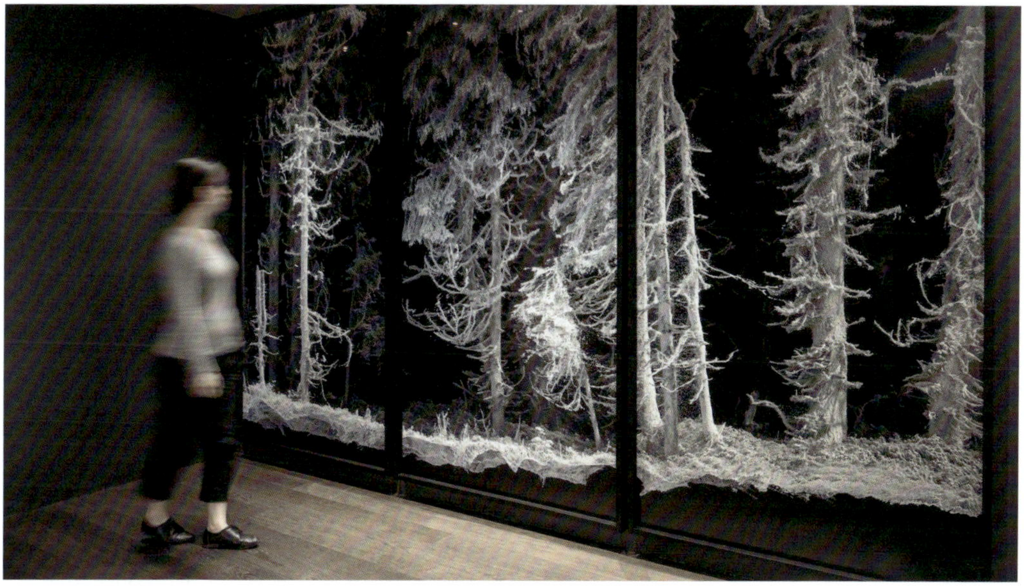

Quayola (UK/IT)

Remains

Remains is an ongoing project focusing on nature and the tradition of landscape paintings. High-precision laser scanners are used to capture natural landscapes at vast resolutions, resulting in complex digital renderings printed on large-format archival paper. The combination of highly detailed geometric reconstructions and the imperfections of the 3-D-scanning process create hybrid formations, somewhere in between the real and the artificial. While recreating similar conditions to 'en plein air' painters of the late 19th century, the natural landscapes are actually observed and analyzed through extensive technological apparatuses, and re-purposed through new modes of visual synthesis.

A project by Quayola
Commissioned by Audemars Piguet
Supported by TECHquadrat Werbetechnik GmbH
Courtesy of the artist and Audemars Piguet

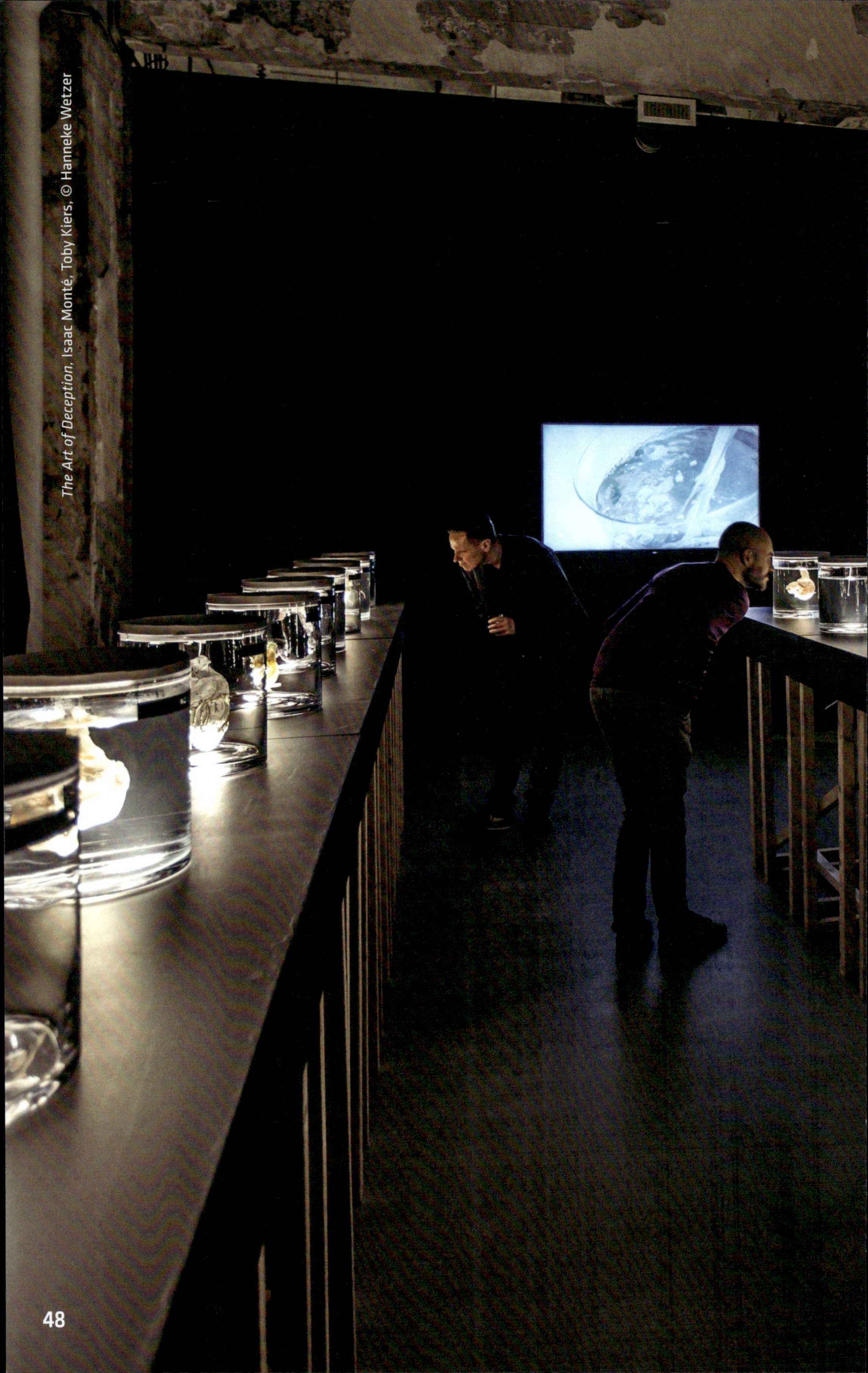

The Art of Deception, Isaac Monté, Toby Kiers, © Hanneke Wetzer

Isaac Monté (BE), Toby Kiers (US)

The Art of Deception

Humans use deception to achieve perfection in society, art and science. Reacting to this through art, we have taken discarded pig hearts and transformed them into elegant vessels for new life by decellularizing them and re-populating them with various techniques, into aesthetically improved hearts for humans.

Decellularization marks a new era of synthetic biology—organs are stripped of their cellular contents, leaving behind a sterile scaffold that can be repopulated with stem cells. While the medical utilization of this resource is being realized, the artistic and creative value of ghost organs represents unexplored territory. With this collection of 21 transformed hearts we explore how biological interventions and aesthetic manipulation can be used as tools for the ultimate deception: the transformation of inner beauty, from grotesque to perfect. Can the ghost organ be a blank canvas for designers? Can organs be objects of design? Will humans be able to manipulate organs for aesthetic purposes? The discarded dead hearts will not function as canonical organs, but rather as a representation of how far science can manipulate the human body.

In collaboration with Professor Toby Kiers (Free University Amsterdam), Commissioned by Bio Art & Design Awards, with the support of ZonMw (The Netherlands Organization for Health Research and Development).

Special thanks to: ZonMw, NWO, BioArt Laboratories, MU, Waag Society, Dr. Renée van Amerongen - Swammerdam Institute for Life Sciences, Dr. Monique Verstegen - Erasmus Medical Center Rotterdam, Yvonne Steinvoort - Erasmus Medical Center Rotterdam, Dr. ir. Jos Malda & Kim van Dorenmalen - University Medical Center Utrecht, Professor Paul van der Valk - Vrije Universiteit Amsterdam, Simon Dupin, Riet Vooijs, Galerie PUUR, Lise Lefebvre, Suzanne van Beest, Bram Geers, Rodrigo Leite de Oliveira, Gonçalo Sousa Pinto, Myriah Lesko, Dario Tortorelli, Rosita van Audenhove, Monica Monté, Job Taks, Gera Bikker, Iza Stepska, Elise Marcus, Levi Baetens, Pushpi Bagchi, Gabrielle Kennedy, Remon van Droffelaar, Willem Velthoven and everyone who has supported me in one way or another.

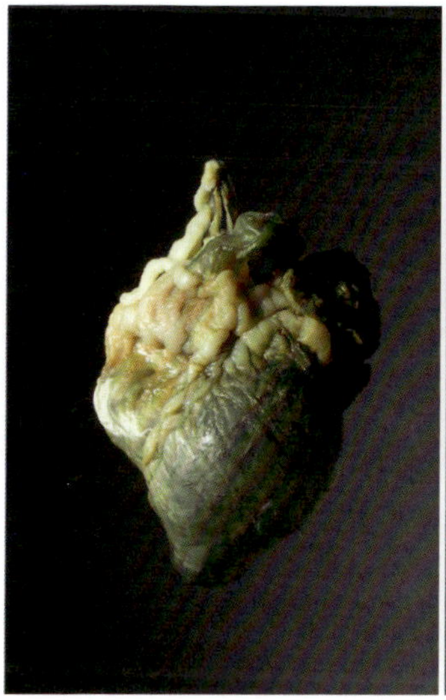

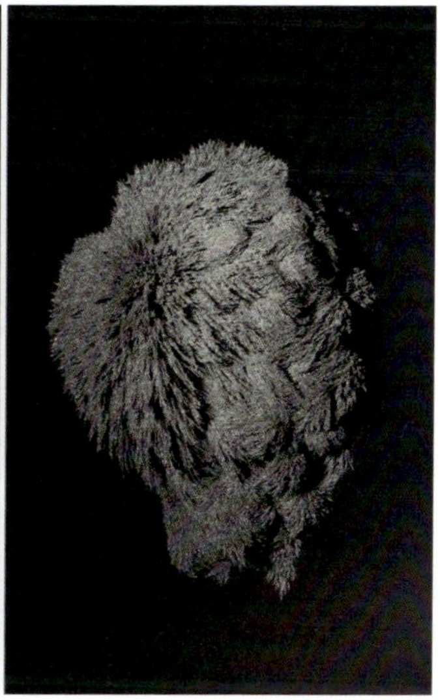

Monica Monté

Marco Donnarumma (DE/IT), in collaboration with Neurobiotics Research Laboratory (DE) and Ana Rajcevic (DE/UK)

Amygdala

Amygdala is an installation exploring the essence of humans' expectations and anxieties over artificial intelligence (AI) and robotics. It reanimates a key symbol of human collective history – an ancient ritual of purification – through the glare of today's technocratic society.

An AI robot in the form of an uncannily human-like limb is nested inside an industrial-grade server cabinet. The same cabinet used in computer server farms. Named *Amygdala*, the robot uses a sharp steel knife to sculpt a large piece of skin. Its labor is careful and never-ending. The robot's only aim is, in fact, to learn an animistic ritual of purification known as "skin-cutting."

The robot's movements are not pre-programmed, but emerge interactively from particular neural networks called "biomimetic adaptive algorithms".

These algorithms, used in humanoid robotics development and programmed by Donnarumma, mimic the sensorimotor system of mammals. Thus, the robot learns by doing; it teaches itself the cognitive and physical discipline required to perform the ritual.

An artwork by Marco Donnarumma in collaboration with Neurobotics Research Laboratory (DE) and Ana Rajcevic Studio (DE/UK).
Concept, research, artistic direction and programming: Marco Donnarumma
Additional programming and research: Prof. Alberto de Campo
Scientific advisor: Prof. Manfred Hild
3D modeling and printing: Christian Schmidts
Exhibit design research: Rosalie Laurin
Photography: Margherita Pevere, William Veder, Marco Donnarumma
Co-production: Retune Festival
Funding: Berlin University of the Arts, Graduiertenschule
Funding: Einstein Stiftung
In kind support: Berlin Center for Advanced Studies in Arts and Sciences
In kind support / Dissemination: Baltan Laboratories

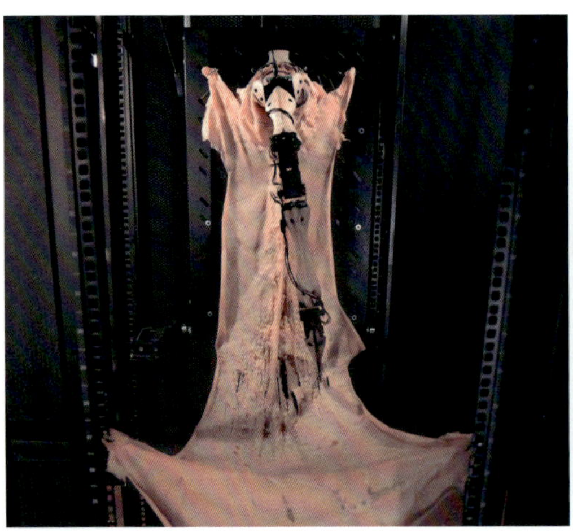

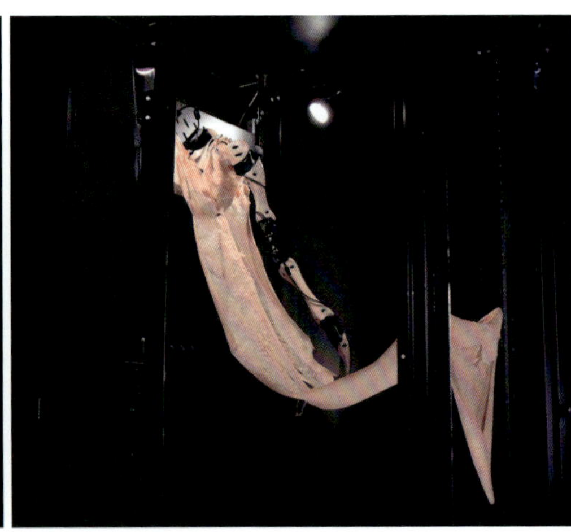

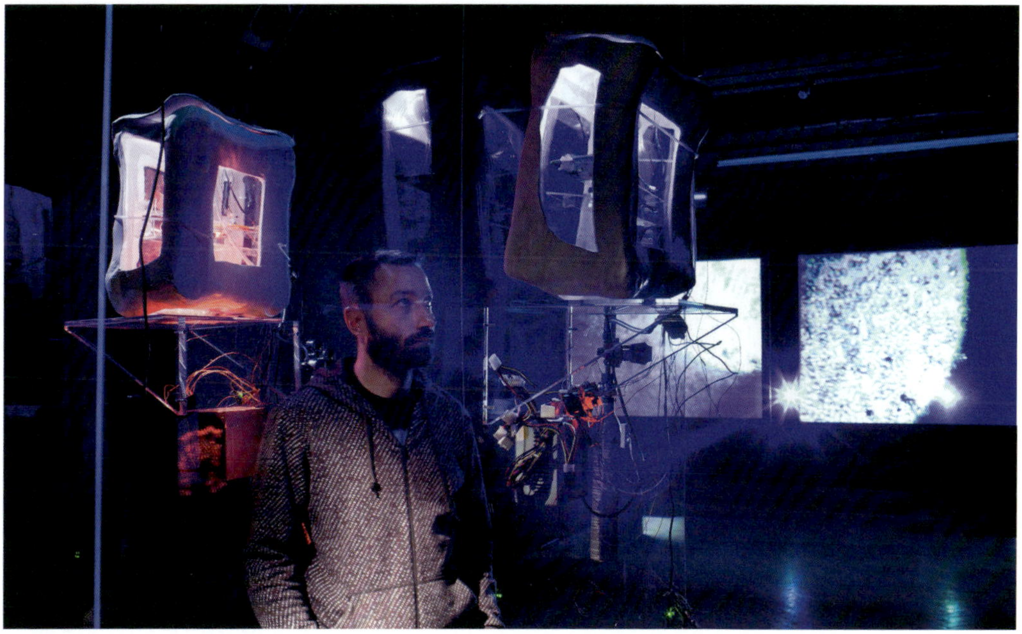

Miha Fras and Kapelica Gallery

Mojca Založnik (SI)

Infinite In-Between

In the project *Infinite In-Between* the author discusses the intermediate field between objectified medical diagnosis and anomalies that avoid this enforced objectivity. The viewer observes the culture of cancer cells and the artist's cells in incubators, which conventional diagnostic practice explores using standardized procedures, laid down by scientific protocols. On the basis of defined values, a grade is given to the aggressiveness of the cancer cells in the patient's sample, in the case of the non-cancerous cells, the sample is analyzed using a flow cytometer. The test environment of the artistic project of monitoring cell cultures is carried out intuitively, using sound and focused beams of light that make perceptible something that is otherwise undetectable or uncatchable. In quantum mechanics, it is accepted that measurement "does something" to the object being evaluated.

In the scientific experiment "double-split experiment," scientists confirmed that the atom acts differently if we view it. If the atom is not watched, it behaves as a wave and otherwise as a particle. This quantum paradox is also known in quantum biology, from which Založnik draws inspiration for the artistic layout, in which the visitor with his/her presence and perception impacts the development of cell culture.

Production: Kapelica Gallery / Kersnikova Institute
Technical solutions: Kristijan Tkalec
Programming: Blaž Berdnik, Matic Potočnik
Light design: Jure Sajovic
Acknowledgment: Maja Čemažar–Phd, Bsc Biol, Veronika Kloboves Prevodnik–Phd, MD, Mira Lavrič–lab technician, Simon Buček–Bsc Biochem
Support: The Ministry of Culture of the Republic of Slovenia, The Department of Culture of the Municipality of Ljubljana

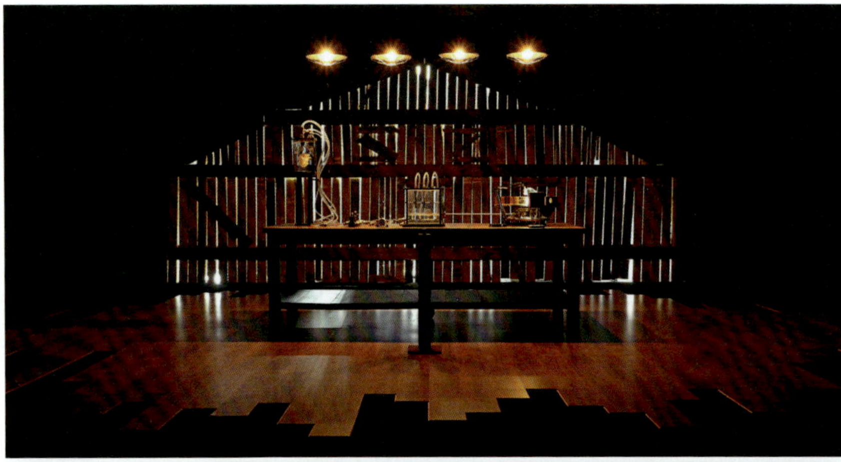

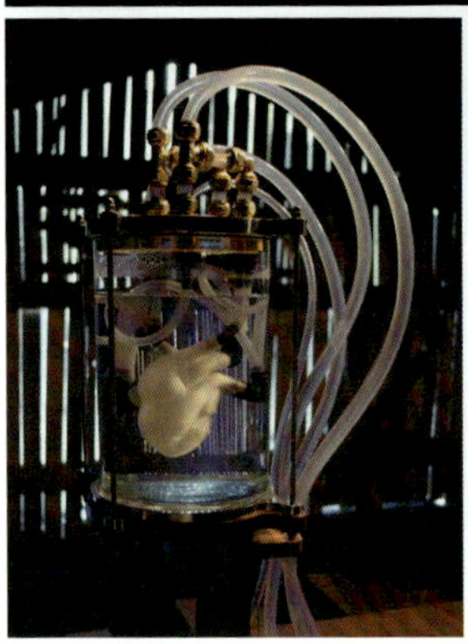

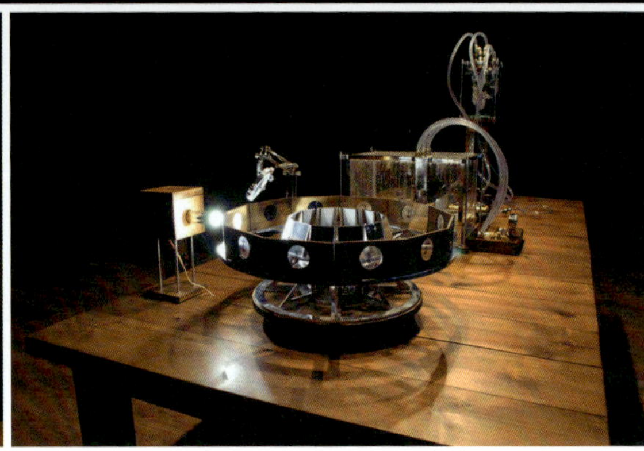

OPN Studio – Susana Ballesteros (ES), Jano Montañés (ES)

Give my creation...Life!

Give my creation... Life! is based on the generation of energy through the heartbeat, with the aim of granting autonomy to a machine. Issues such as extending a removed organ´s life, feeding it nutrients artificially, and using it as a source of natural energy were addressed during the research. The result of the study is a series of experiences materialized in an audiovisual with images taken from an experimental operative procedure carried out at CIBA (Centre of Biomedical Research of Aragón), reinforcing the project's narrative. The material conclusion of the process developed during the study embodies the goals, advances and outcome of the investigation. It is a heart rhythm simulation device, which captures and feeds on the generated energy to give life to a machine.

All this is accompanied by a more theoretical part, which introduces the project through stock footage concerning experiments and investigations in relation to extracorporeal circulation throughout history.

Supported by:
Etopía, Center for Art and Technology
Cooperation partners:
CIBA (Centre of Biomedical Research of Aragon)
BIFI (Institute for Biocomputation and Physics of Complex Systems)
EBERS Medical Technology
Zaragoza University

Haruo Usuda (JP), Matt Kemp (AU)

The Artificial Uterus

Life support platform for extremely preterm babies

Despite profound improvements in outcomes for preterm infants born at later gestational ages, there have been only modest improvements in outcomes for babies born below 24 weeks' gestation over the past two decades. If we are to improve outcomes for infants born at the border of viability, we must recognize that they are not "small babies;" rather, they are a unique patient demographic that requires an entirely different treatment approach from older preterm infants.

One opportunity to improve outcomes for these infants is a non-pulmonary life support system that will allow for the healthy survival of extremely preterm babies. In our *Artificial Uterus* platform, gas exchange is performed by a sophisticated artificial placenta connected directly to the fetal umbilical cord, with circuit perfusion driven solely by the preterm fetal heart. The present aim of this work is to bring to clinic a functioning life-support platform for infants born at the current border of viability.

We wish to gratefully acknowledge the generous support provided by the following collaborating institutions and funding partners:

Tohoku University *(http://www.tohoku.ac.jp/en/)*
Women and Infants Research Foundation *(https://wirf.com.au/)*
Nipro Corporation *(https://www.nipro.co.jp/en/)*
Government of Western Australia Department of Health (http://ww2.health.wa.gov.au/)
University of Western Australia *(https://www.uwa.edu.au/)*
Channel 7 Telethon Trust *(https://www.telethon7.com/)*

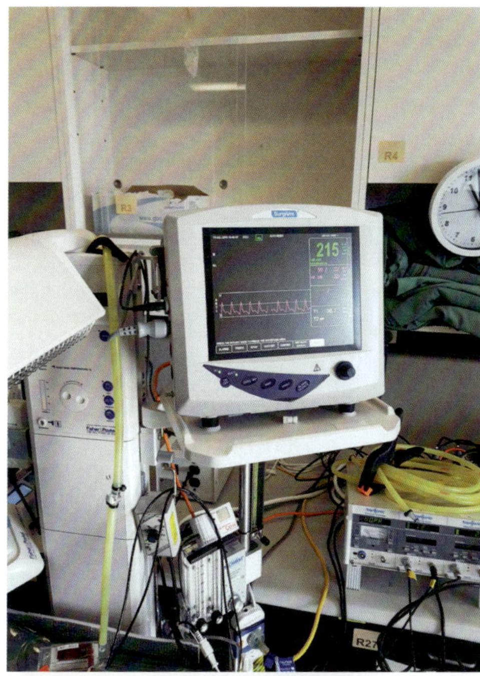

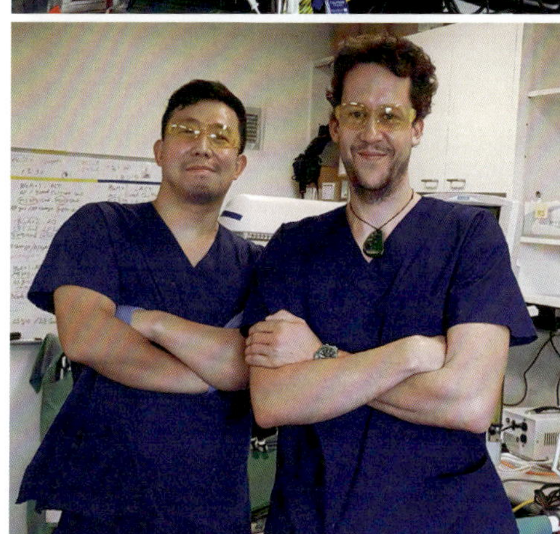

Yunchul Kim (KR)

Triaxial Pillars II and Argos

The mechanical and fluidic events in *Triaxial Pillars II* are implicated in cosmic events which come through *Argos*. *Triaxial Pillars II* is a fluid kinetic installation which consists of multiple planetary, helical structures with 23 axes, a double acrylic cylinder filled with nano-sized photonic crystals, and *Argos* is a cosmic ray detector which consists of 41 channels of Geiger Müller tubes. When muon particles collide with *Argos*, it flashes a light and transmits signals to trigger the algorithm of fluidic movements in *Triaxial Pillars II*. This magneto-hydrodynamic action causes the mass of moving tiny particles to scatter and converge, and black hollows and shiny structures of metallic filaments are formed. Particles, fluid, machines, other objects, and humans are interwoven in this work. It expands the spacetime of works, the reality of materials, materiality and even its meanings. Those material imaginations that come through the matter, or into matter, provide us with an affective material poetry that allows us to transcend the restrictions of language. It becomes reality through its materiality.

Argos was developed as part of the Collide International Award, a partnership programme between Arts at CERN and FACT and was co-produced by ScANNER.

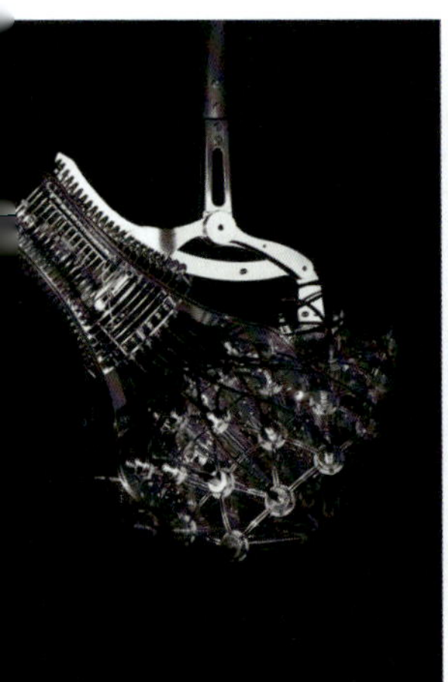

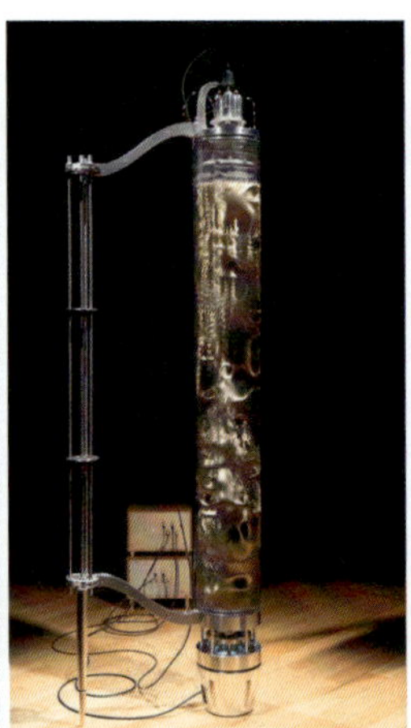

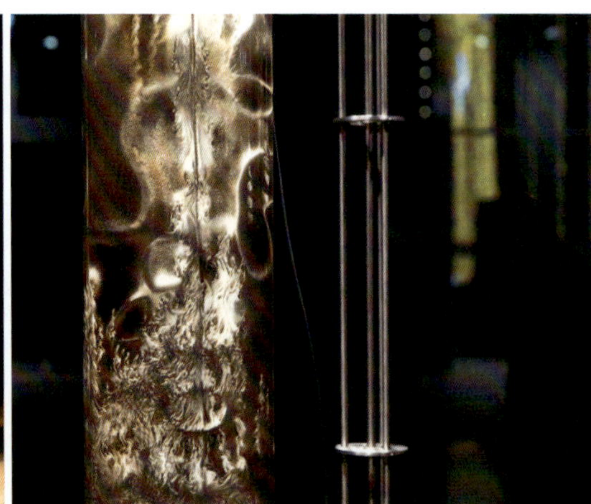

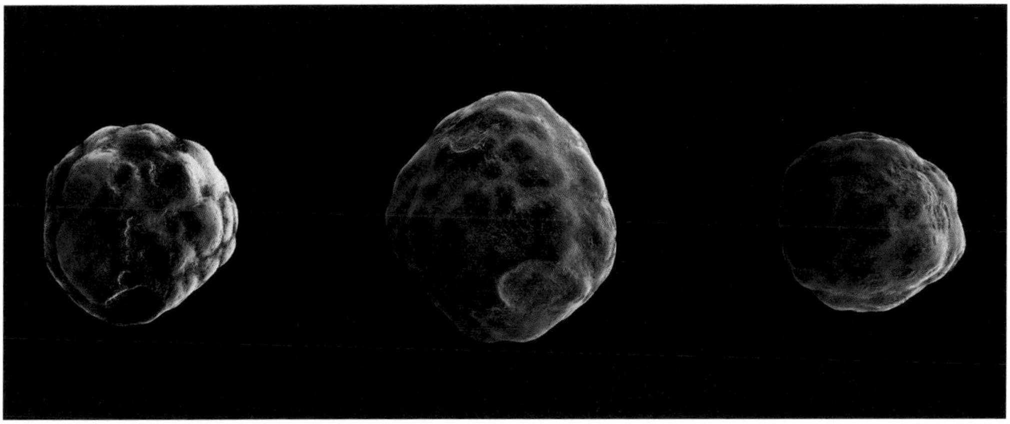

Bea Haines (UK)

Heavenly Bodies

Backlit Scanning Electron Micrographs of Human Gallstones

When my grandmother died, I inherited tiny gallstones removed from her body. Grotesque to the outsider, these nuggets of bodily imperfection became precious relics to me. Like the heavily adorned relics found in Catholic churches, my aim was to present the stones in a different light, replacing disgust with a sense of reverence. To visualize these minute forms, I chose to use science as an equivalent contemporary power to religion. *Heavenly Bodies* was created by using a Scanning Electron Microscope to visualize the detailed topography of the stones' surface, transmuting them into objects of celestial beauty. Body stones (or calculi) are formed by the build-up of mineral salts within the body. Like comets or pearls, they are accidental waste products of the body, or as nurses used to call them, "tombstones to dead bacteria".

With the support of Imperial College London

Cod.Act (CH)

πTon

πTon is an intriguing sound installation; it constitutes a new stage in the Cod.Act research into plastic and sound organicity. It results from an experiment on the relation between the distortion of an elastic structure and the synthetic transformation of the human voice. A long rubber tube, closed in a loop, is animated by contortions and undulations like an invertebrate body. Surrounded by a group of four dumb human beings equipped with strange vocal prostheses, the creature seems to try to release itself from this disturbing presence in vain. Its efforts and sufferings excite the curiosity of the four human beings and become the subject of primary and sophisticated polyphonic rituals only constituted by synthesized voices.

From raw materials and natural physical phenomena, *πTon* associates organic movement and vocal expression in their most primitive forms. The result is a striking sound and visual event that sends the spectator back to the origins of his behavior.

Cod.Act is Michel and André Décosterd (CH)

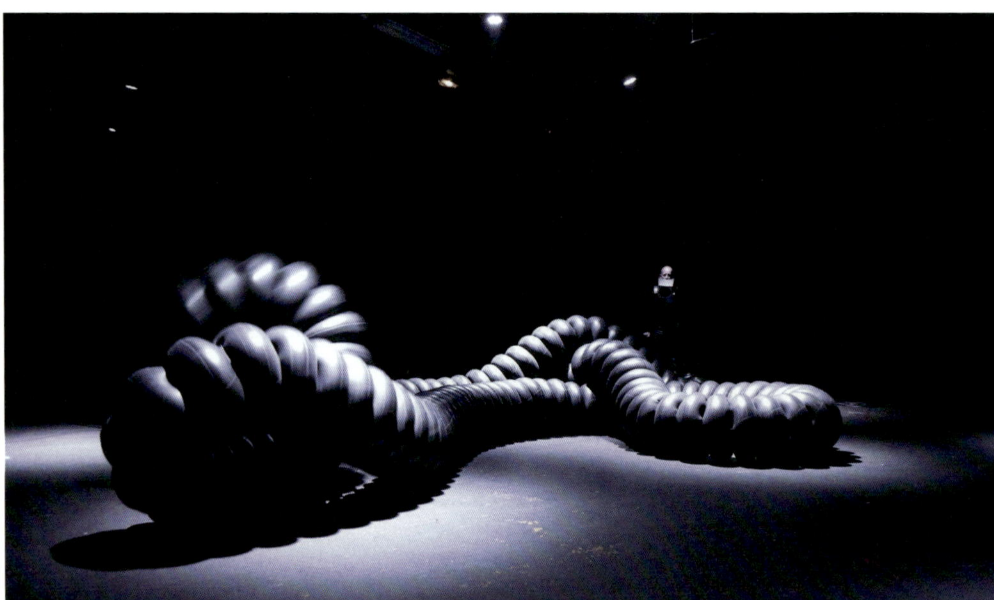

Xavier Voirol

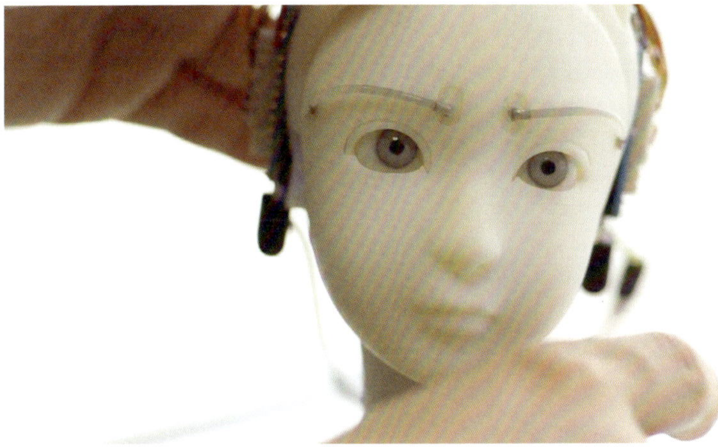

Takayuki Todo (JP)

SEER: Simulative Emotional Expression Robot

SEER is a compact humanoid robot developed as results of deep research on gaze and human facial expression. The robot is able to focus the gaze directions on a certain point, without being fooled by the movement of the neck. As a result, the robot seems as if it has its own intentions in following and paying attention to people and its environment.

Using a camera sensor and eye tracking, it has an interactive gaze. In addition, by drawing the curve of the eyebrow using soft elastic wire, I was able to enrich the emotional expression of the robot.

The purpose of my research and development is not to answer the philosophical question "Will a robot (or computer) obtain a mind or emotions like mankind?" but to portray the sense of conscious human emotion. I think it is possible to represent human-like communications by constructing an adequate interaction system between emotional sensing and expressions. If we understand and identify with robots that can learn the functions and usages of emotional expressions from interactions with people, get a good command of them accordingly with situations and context, could we distinguish them from those with real minds and emotions?

As the first step to realizing this, I thought two conditions were necessary: the eyes to detect the appearance of another's face, and the face to be made appealing to the viewer's eye.

Technically supported by Takanari Miisho, Yuki Koyama

Ben G. Fodor (HU/AT)

CARMINE

In a darkened exhibition space, a black cube equipped with special glass-optics projects lines and points of carmine-red laser light onto the space's walls. Several of the points oscillate and move about randomly due to a hanging chromium steel sheet that's spun by the air currents generated by installation visitors moving about the space. Fodor is also exhibiting images that were "painted" in his atelier with the laser tool used in the installation. The traces of the moving light were captured photographically through the use of time exposure. Fodor brilliantly employs this technique to create immaterial architectures consisting of lines and surfaces that open up a space of indeterminate dimensions. This cycle takes its place in Fodor's artistic investigation of utopias and dystopias. His works in recent years have revolved around the sociopolitical question of new "utopian" horizons – a reinvention of utopian thinking.

Ben G. Fodor, Sillyconductor, Marcello Farabegoli Projects & ARCC

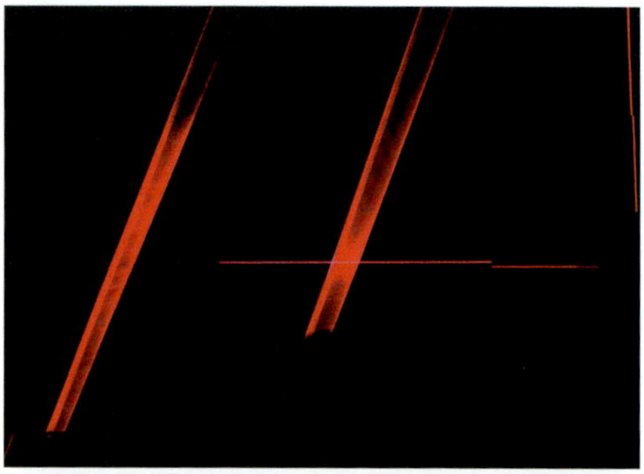

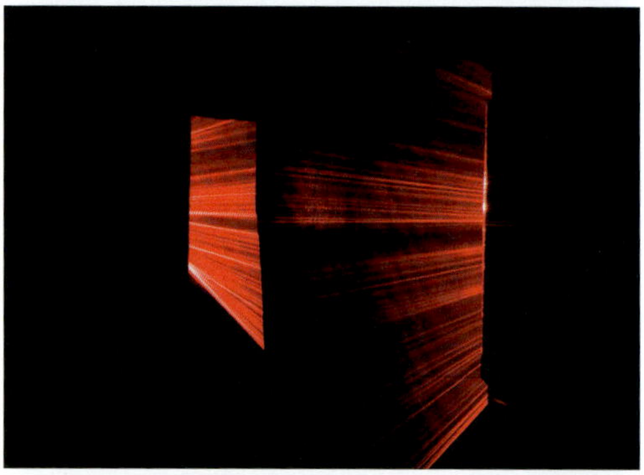

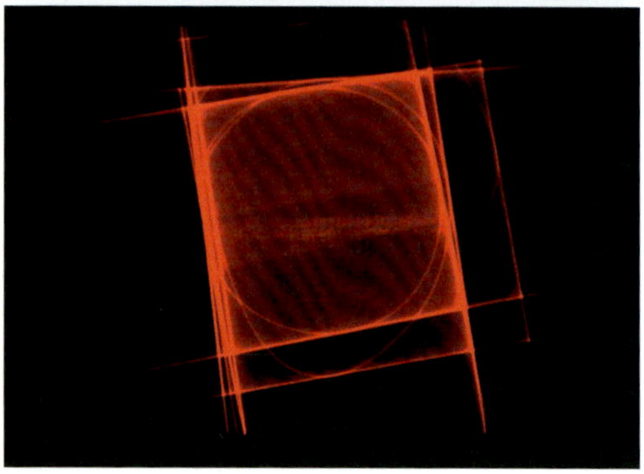

Yuki Anai (JP), Hideaki Takahashi (JP)

in the rain

Nature is a truth. It is often more than meets the eye because there are some things in nature that are invisible to the human eye. But that doesn't change the fact that they exist.
Rain has a variety of expressions: light rain, heavy rain, sun showers, etc. Sometimes, rain is revered as a blessing from heaven. Other times, it is disliked as a natural disaster. Rain is just rain, but its interpretation depends on the context. I believe rain is beautiful, but some people feel blue when it's raining. So, it is my desire to express how rain is beautiful and how nature has a message for us.

One day, I was at home watching heavy rainfall. At that moment, I remembered another rainy day and my mind recalled the view, the surroundings, my feelings and so on. I felt like I was communicating with my past self. Then, I realized that rain is a medium which is deeply linked to people's memories. Creating a rain-filled world, my "rainy world," may help open the door to people's pasts.

Yuki Anai: Concept, Visual Design, Programming
Hideaki Takahashi: Sound Design

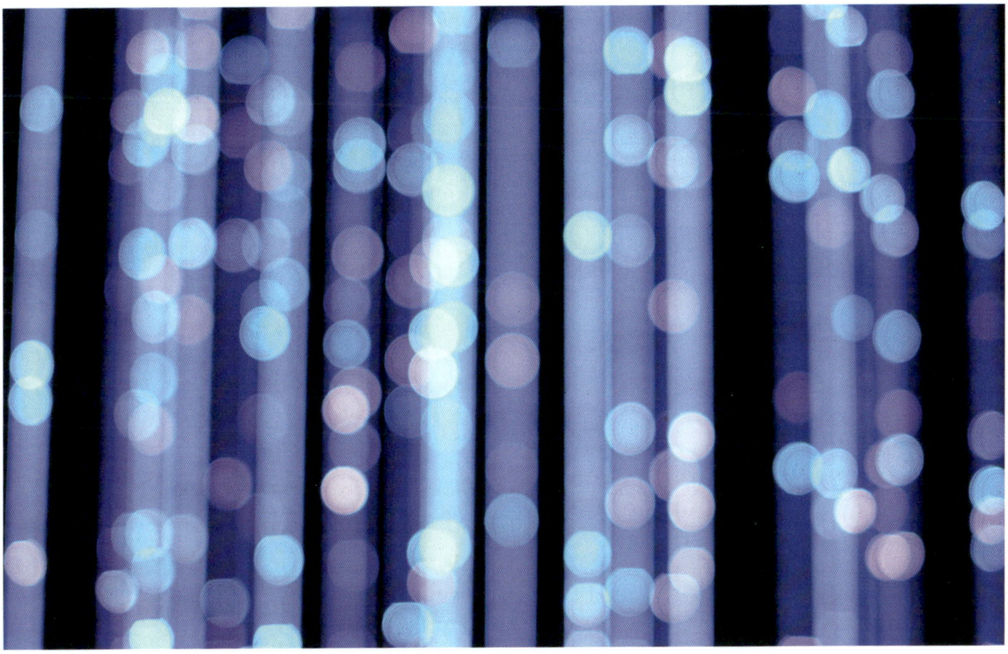

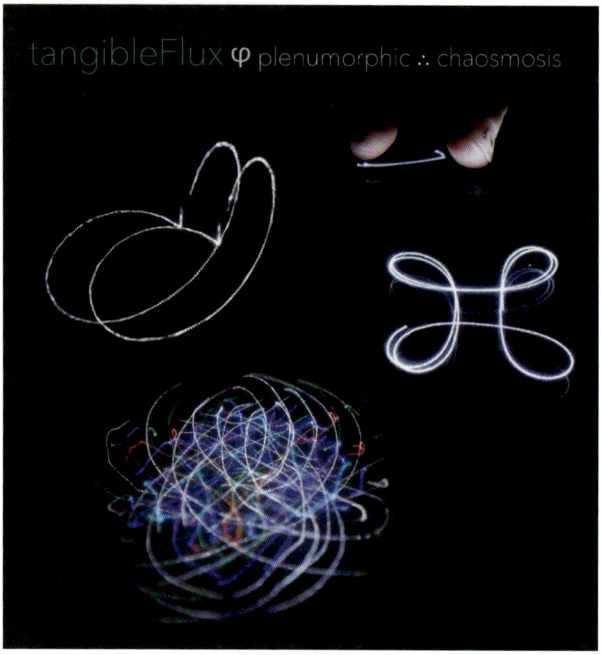

Navid Navab (CA/IR)

tangibleFlux φ plenumorphic ∴ chaosmosis

tangibleFlux φ plenumorphic ∴ chaosmosis is a sensorial installation inhabited by the playful ontogenesis of vibrant patterns. Participants encounter the organic formation of temporal textures, which emerge from continuous material-energy fields. In search for excitable mysteries of matter at the borders of tabletop astrophysics and natural fiction, three microcosms orchestrate states of sensory access to the phenomenological emergence of order out of chaos. This is physically investigated through collaboration with non-linear forces of complex harmonic motion. Each microcosm engages unique entanglements: between pattern and uncertainty, orbital resonance and topological semblance, percepts and the stuff of forms. Patterns of interaction between magnetism, gravity, matter, and light evoke the behavior of the cosmos at large, always spiraling into an improvisatory dance of hallucinatory forms. Eventually the kinetic event unveils messy realness: things doing... stuff seeking a minimum-energy state under magnetic flux, dissolving into upheavals at the edge of chaos. Sensually binding, spatially confusing, and temporally unsettling, *tangibleFlux* pulls everything into immediate vibrational relation with its vertiginous ritual.

Navid Navab: concept, direction, audiovisual design, composition, fabrication, programming, electronics, sculpture, technical design
Ceramic artist (microcosm no.3): Marie Côté
Programming assistance: Evan Montpellier
Fabrication collaboration: Studio Robocut (Montreal), SceneArt (Slovenia)
Design consultants: Garnet Willis, Tatev Yves
Research collaboration: Topological Media Lab
Special thanks: Michael Montanaro, Nima Navab, Peter van Haaften, Bruiser Bruce, Ashley Obscura, Robertina Šebjanič, Miha Colner, MGLC (Ljubljana), Lina Dib, Harry Smoke

Martin Hesselmeier (DE), Andreas Muxel (AT)

the weight of light

Light, as we usually interpret it, is an element without mass and gravity. For *the weight of light*, dots of light appear to have mass, momentum and kinetic energy on physical paths. One simulates the expected behavior while the counterpart inverts gravity, causing us to experience a fictional space of light. The installation plays with our sense of reality and emphasizes the materiality of immaterial elements in a revolving reality which overcomes the dualism of "virtual" and "real." Therefore *the weight of light* goes beyond what is expected; the matter of light traverses a re-interpretation of our known reality. Each moving light dot is underlined by an auditive representation in the low-frequency range. Phenomena of standing sound waves and frequency interferences support the shape waved structure. This creates an immersive space where the movement of light is also represented by vibrational patterns of oscillating sound waves.

the weight of light by Martin Hesselmeier, Andreas Muxel
Physics engine development: Micha Thies

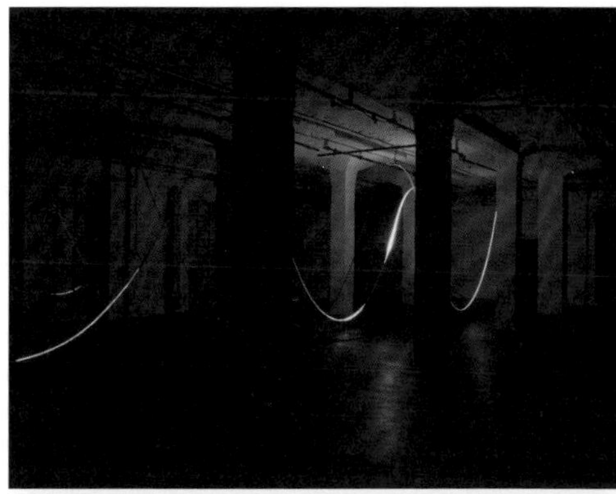

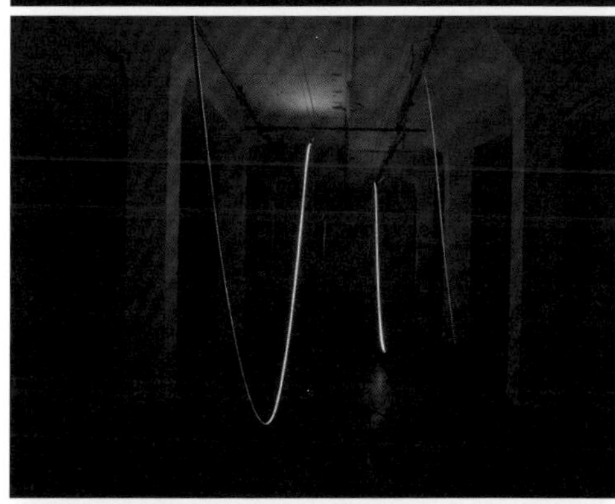

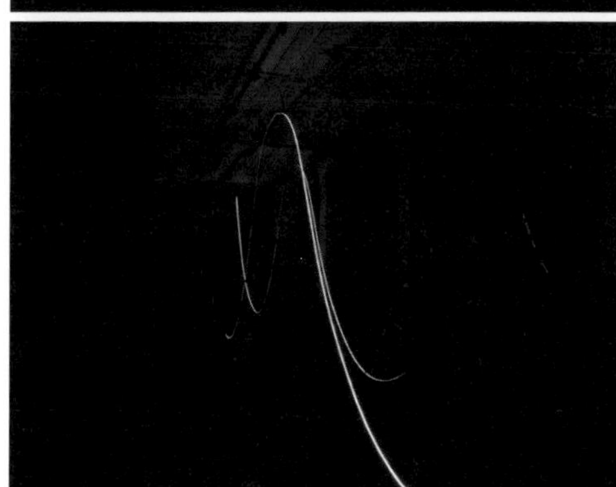

Stefan Mittlböck-Jungwirth-Fohringer (AT)

The Bien

Bees are fascinating beings. Bees have always been a motif and metaphor in artistic works. The colony offers a huge amount of visual impressions: bees at the entrance with their different tasks and activities; the comings and goings and flights of bees; a honeycomb full of nursing bees in brood care; the transfer of nectar or pollen from homecoming bees; the intriguing honeycomb construction and the aesthetics of the bee itself.

Due to the death of many bee colonies, the absence of the bee was the focus of the artistic research that manifested in the work *The Bien*. Dealing with the topic of transformation, the work is a representation of the progression from a living, animated colony to an empty, found hive, which is caused by "colony collapse disorder" (CCD).

Transformation and deconstruction are also made visible in the process, which the "supposedly" empty beehives are exposed to throughout the year. Through the description of the remaining environment of the organism of *The Bien*, a portrait of this fascinating being is created.

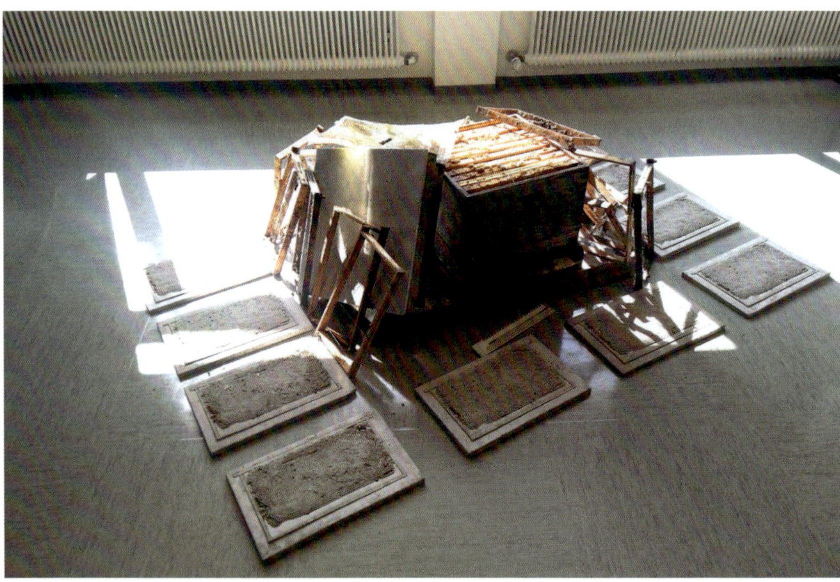

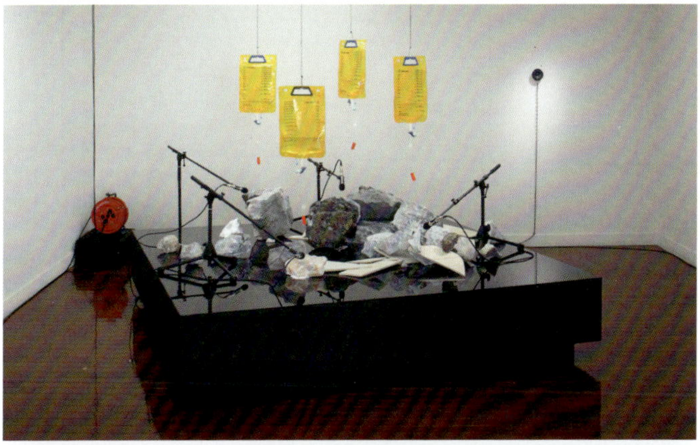

Keizou Kioku

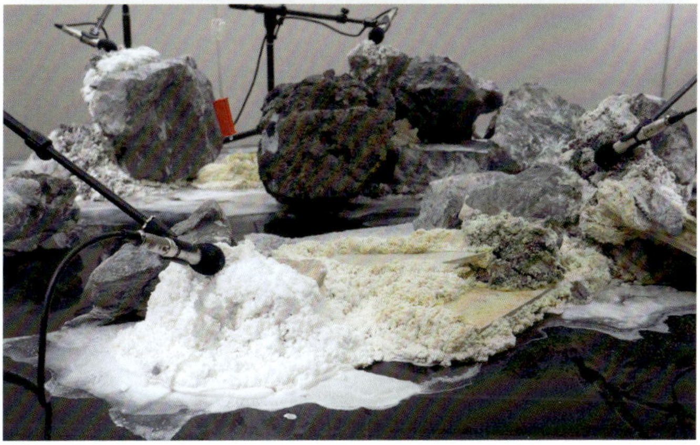

Ryohei Tomita

Shun Owada (JP)

unearth / Paleo-Pacific

unearth / Paleo-Pacific is a project that explores the fundamental elements of sound: the atmosphere and time. The work utilizes limestone produced through the fossilization of fusulina, small marine organisms that vanished during the greatest mass extinction 250 million years ago. Fusulina formed their bodies by biological fixation of CO_2 in the sea. Which means that CO_2 is stored in these fossils (limestones). Dripping dilute acid onto them causes the fossils to melt, and releases CO_2 into the space. Standing in front of the speaker, viewers can listen to the extremely subtle sound of fossils melting. The work questions how we can or cannot sense the aspects of time beyond the human scale by listening to sound.

Joana Moll (ES)

DEFOOOOOOOOOOOOOOOOOOOREST

DEFOOOOOOOOOOOOOREST shows the number of trees needed to absorb the amount of CO_2 generated by the global visits to google.com every second. Google is the most visited site on the Internet. The site has an average of 52,000 visits per second and weighs around 2 MB, resulting in an estimated amount of 500 kg of CO_2 emissions every second. On average, a tree can absorb 21.77 kg of CO_2 per year. Thus, in order to counteract the amount of CO_2 emissions caused by the global visits to google.com every second, we would need approximately 23 trees/second.

This project has been created with the aim of highlighting the massive environmental impact of ICT. In our contemporary algorithmic decision-making society, ecosystems are being increasingly considered as mere economic externalities, thus it is urgent to generate critical thought about the true nature of technology in order to imagine alternative techno-paradigms, which may coherently respond to our environmental and human conditions.

Cooperation partner: Ramin Soleymani

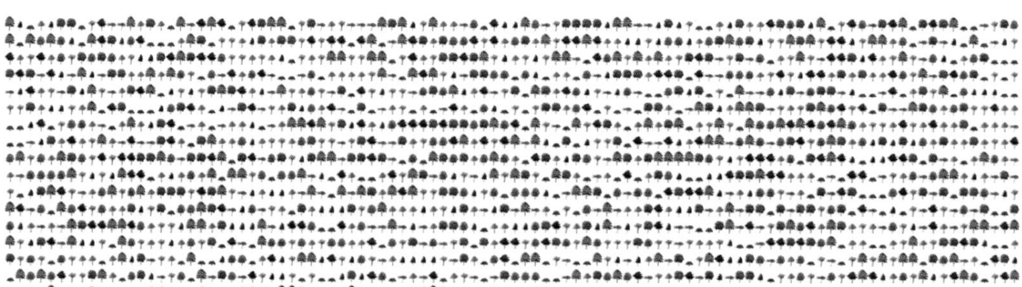

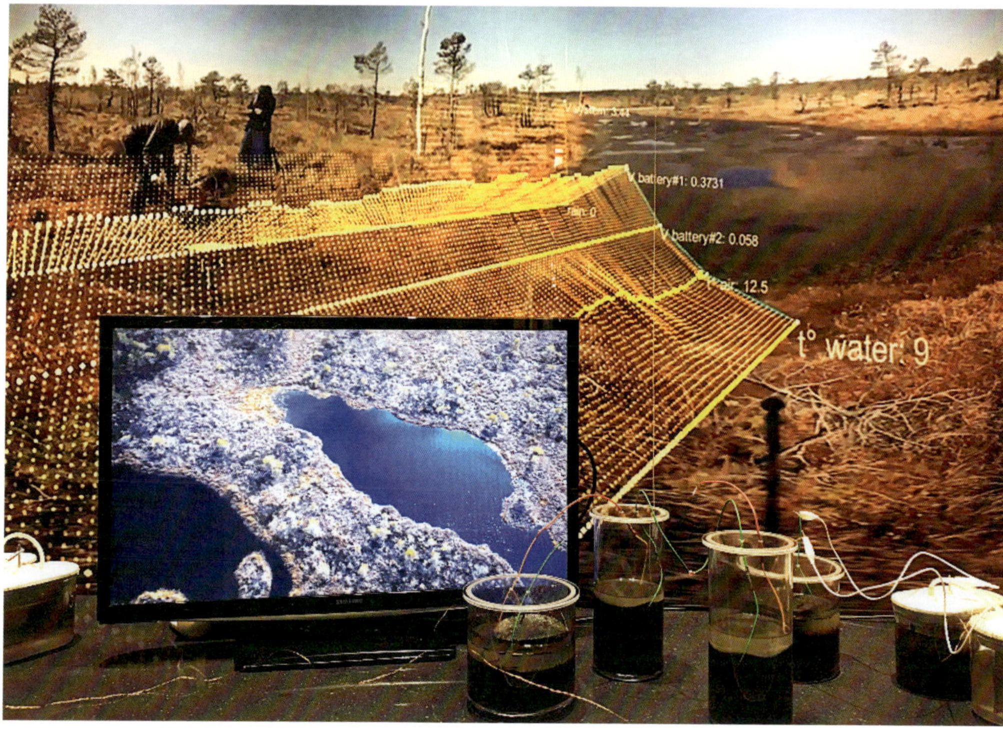

Rasa Smite (LV), Raitis Smits (LV), RIXC (LV), MIT ACT (US)

Swamp Radio

Swamp Radio takes on the challenge of giving a voice to many who are unheard and invisible. The swamps, these vast wetlands with their ancient ecosystems, are like time-capsules.

Yet they are also key players for providing a variety of ecological services for our modern society. Escaping from intensive agriculture, the swamps contain dormant resources, and myriad other species, with whom we share life on this planet.

Artists have been installing microbial fuel cells, environmental monitoring sensors and data transmission networks to explore the sonic environment of the various ecosystems—from bogs near Riga in Latvia, and marshlands of the Boston area in Massachusetts, to wetlands in Venice, Italy. Using LoRa radio technology, during the Ars Electronica festival, data from the local biotope and ecosystem monitorings on the roof of the festival building will be transmitted live in the exhibition space, creating real-time audio-visual interferences and immersive sound experiences.

Credits: Nicole L'Huillier / MIT Media Lab, Gary Zhexi Zhang / MIT ACT, Kristaps Biters / RIXC, Dr. Reinholds Zviedris / LU.
Support: RIXC, RISK CHANGE—Creative Europe's project, MIT ACT, Swamp School at the Venice Architecture Biennale 2018

Ryo Kishi (JP)

ObOrO

Since early times lights have attracted people, and people have gathered spontaneously around lights. Especially lights that have fluctuations, such as a sky lantern, the moonlight or the light of a flame, fascinate people. We feel compelled to put our hands above them. Fluctuations of light are not under our control, so they can express wonder in frailty and instability. *ObOrO* is a fragile kinetic light that has fluctuations and fascinates people with its frailty and instability. *ObOrO* is based on the Coandă effect. Levitating balls are suspended in midair, not with strings or wires, but only with an airstream. In midair, the levitating ball spins with instability.
Users can physically touch and feel zero-gravity materials, and change its motion by putting their hands above the airstream.

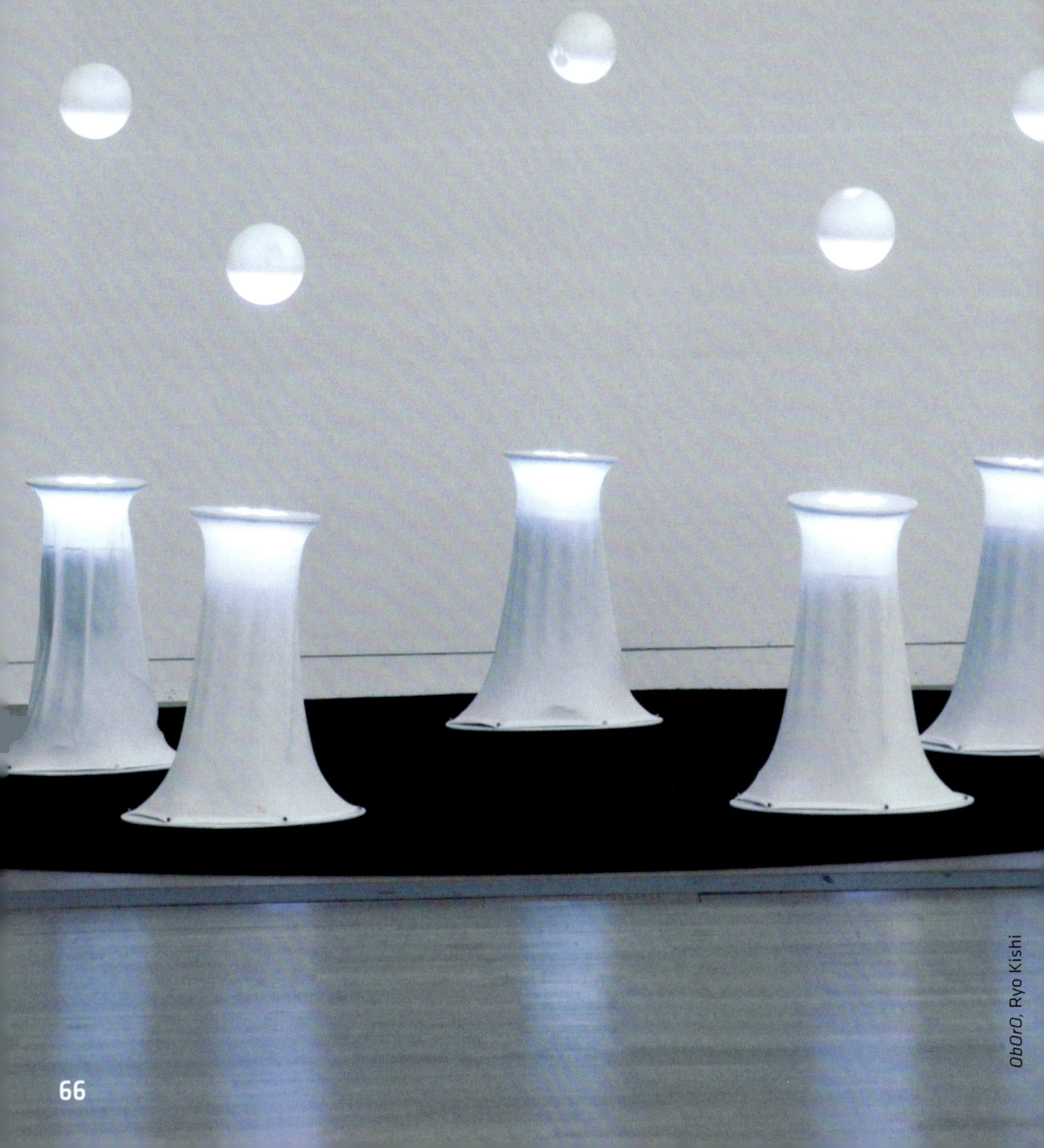

ObOrO, Ryo Kishi

Kathrin Stumreich (AT)

Sovereignty

robotic sound installation

The focal point is a translucent flag; its movement choreography is controlled by complex robotics. Beyond the flag´s operation range, a laser and a light sensor are positioned. When the laser and fabric collide, sound is created. This work's poetry lies in the apparently random combinations of floating silk and gritty sound. Underlying this is Stumreich's precise audiovisual composition. Every drape has been analyzed on the basis of its sound. According to these aesthetics, the robotics of the flag choreography *Sovereignty* was programmed to run in a 9-minute loop. Marked by sharp laser beams, elements associated with sovereignty (as the title implies) can be found, such as leeway, a change of direction, or a drop.

Robotics: Adam Donovan (AU)
Programming: Christoph Freidhöfer (AT/DE)
Exhibition Support: Bildrecht

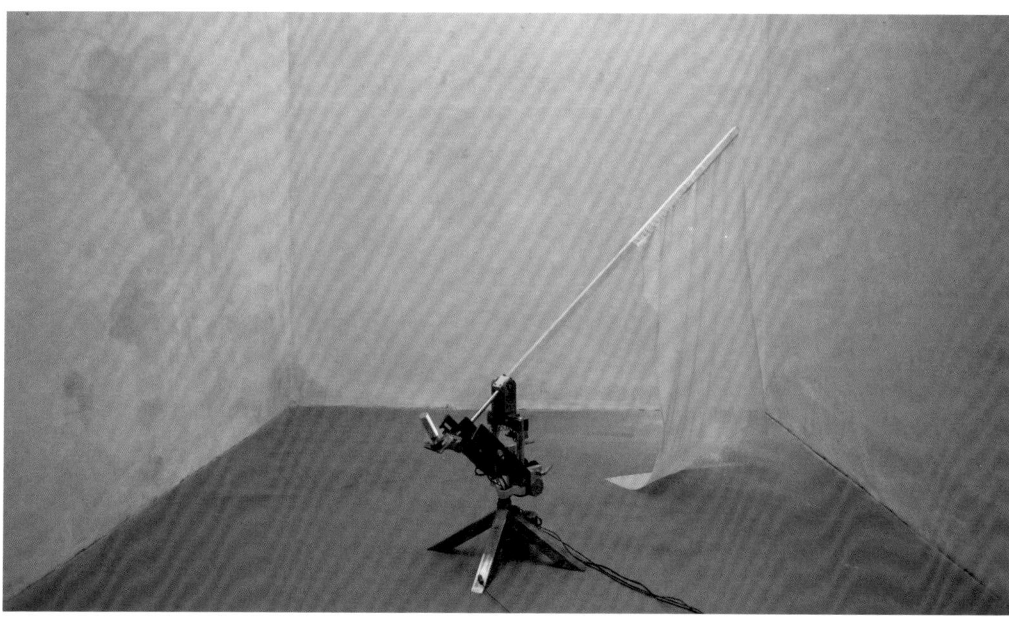

Mathieu Zurstrassen (BE)

I Love you / I Hate you _TDS*

Trump Derangement Syndrome / impartiality bot

In 1980, the self-proclaimed "Doctor" Masaru Emoto placed identical portions of cooked rice in two different containers. On the first one he wrote "thank you" and on the other "you idiot." He then asked school children to read the text on the jar's labels aloud every day as they passed by. After 30 days, the rice in the container with positive thoughts appeared barely changed, while the other became black. The automated installation tries to reproduce this famous rice experiment. A computer gathers all the tweets posted by the current President of the United States on one hand and by a "Love Quotes" tweeter account on the other. Every 25 seconds and by intermittence, a digital voice reads the tweets to the two beakers containing cooked rice. The installation addresses the theme of popular beliefs and the worrying success of unverified data available online as well as the manipulation of these pieces of information.

The work has received the support of "La Fédération Wallonie-Bruxelles"
http://www.federation-wallonie-bruxelles.be/
Programming of the RPI: Martin Pirson

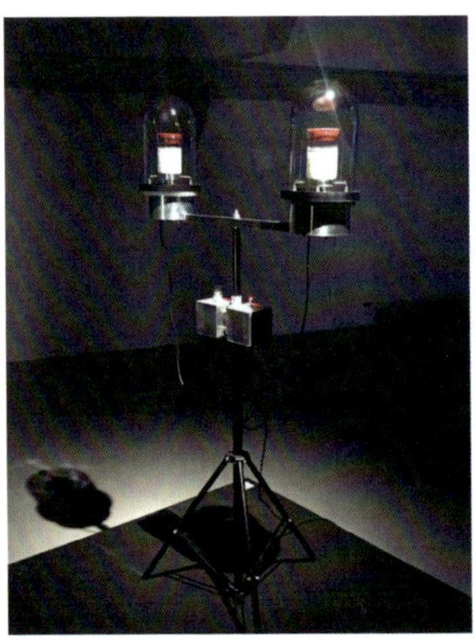

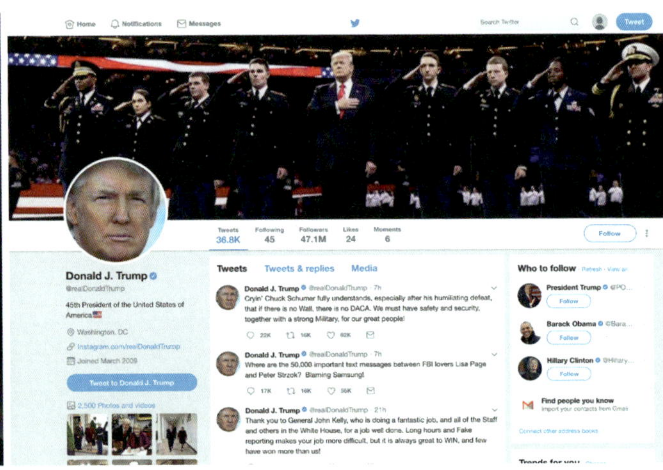

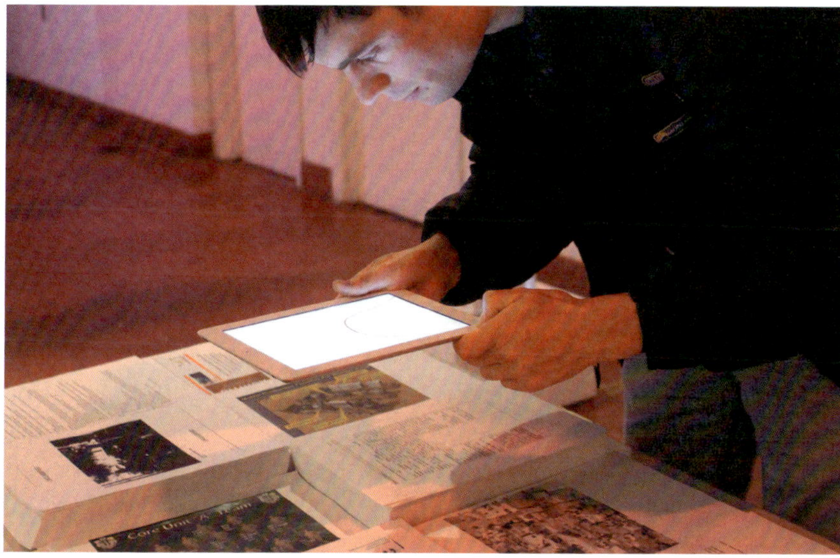

Daria Jelonek

Anna Ridler (UK)

Wikileaks: A Love Story

Wikileaks is a complex system trying to balance freedom of information against the right to know. This installation shows this collision of public and private that sits at the heart of the Wikileaks project. Stacks of printed paper – tens of thousands of emails and documents and photographs – show the macro political machinations that most people assume are being revealed by the database. However, when an iPad is placed above the top of the papers a hidden narrative is revealed to the viewer, a deeply personal love story of a couple who fell in love and then broke up, all found in the emails that were released as part of a data-dump, collapsing this collision of public and private.

Friedrich Boell (DE)

Dead Pixel

The installation consists of eight panels of acrylic glass which hang on a steel frame. 24 IPhone displays are glued onto each panel. Their backlights are controlled individually and cyclically play back an abstract light animation. Since the displays are based on the IPS technology, they would normally block all light when having an image signal. Therefore, I have destroyed the polarizing filters, causing the light to emit in uneven artefacts from the cell phone screens, creating irregular textures.

Friedrich Boell, KHM Academy of Media Arts Cologne.
Many thanks to Georg Trogemann, Martin Nawrath, Christian Faubel and Karin Lingnau.

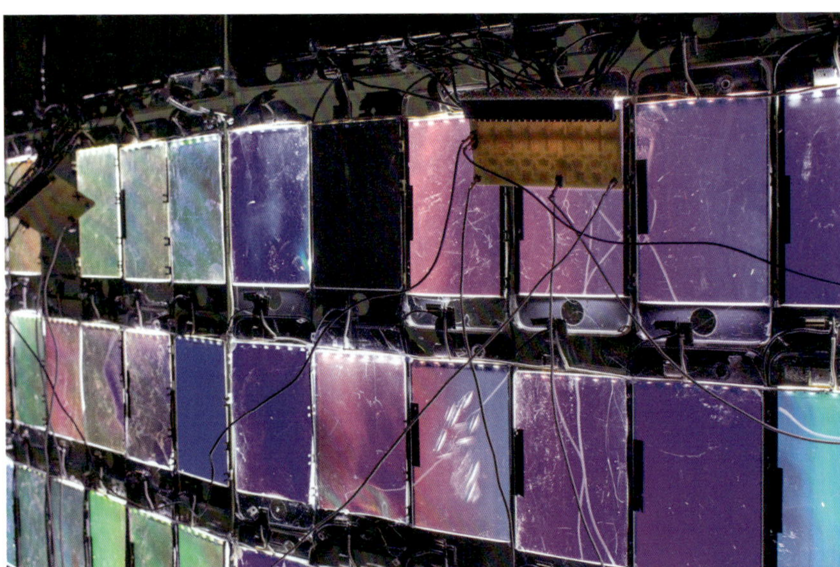

Frankie Macaulay, 2018

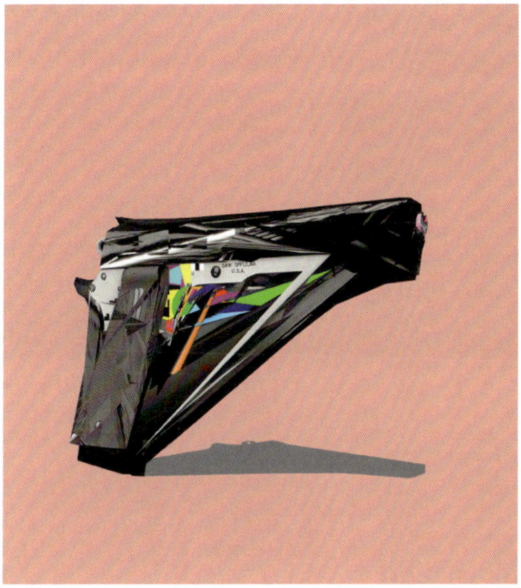

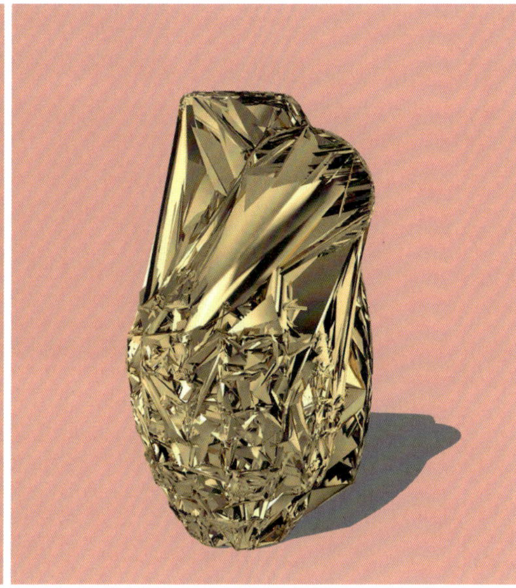

Alexandra Ehrlich Speiser (AT)

Useless Weapons Series

corrupting abstract violence

A glitch is an "error" in a system. Audio files with a broken tune, videos with corrupted images or sounds, colorful pixels which disguise and manifest themselves in the image. In our digital life we encounter such malfunctions on a daily basis.

Algorithms get confused. Computers seem to have grabbed ahold of their own life. Glitches arise, almost like the spirit of the machine, emerging from the depths just to say "hello" in their own form on the surface. "Glitch art" celebrates the aesthetics of the error and malfunctions. 3-D datasets are available in the dark web. Just 3-D-print your favorite weapon! Alarming fact: the ubiquity of destructive murder weapons will increase rapidly. Based on this thought, the artist had the inspiration to take over data sets of a S&W handgun, an AK47 and a grenade to glitch them and make them useless. She has given material form to her abstract sabotages and their ability to affect the equally ethereal file, and in turn the dangerous and threatening original.

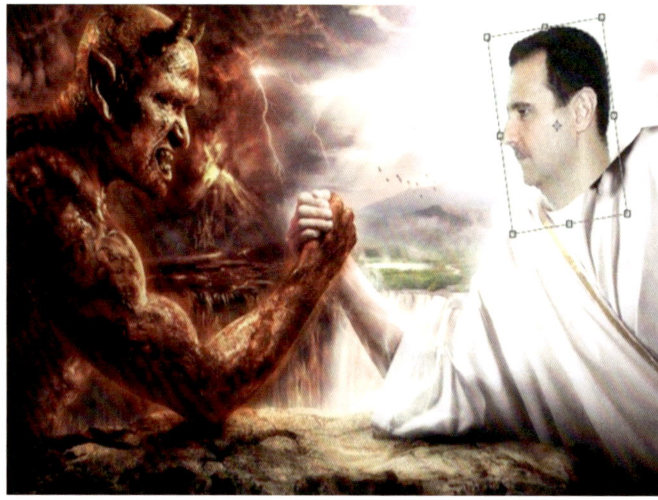

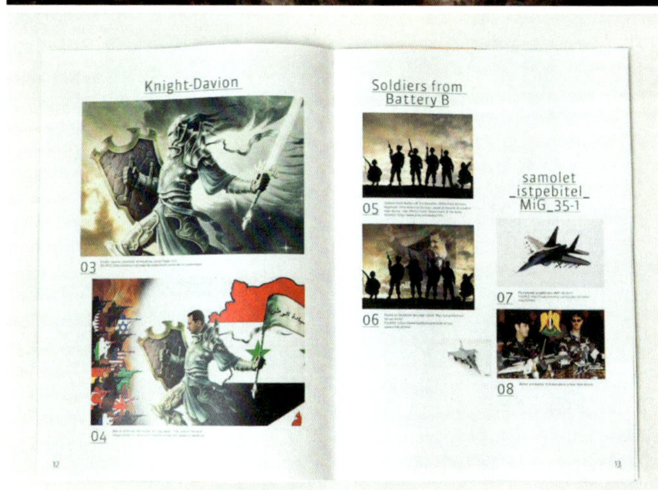

Foundland Collective – Lauren Alexander (ZA/NL), Ghalia Elsrakbi (SY/NL)

Real-time History

Foundland Collective reveals an on-going collection of social media evidence related to the conflict in Syria, sharing projects developed in 2012 and 2018. They focus on variable interpretations of media and generally assumed "truth" and how it is dependent on context, framing and intention. A recent case study deconstructs 2018 "chemical weapon attacks", as they are labeled in Western media, as being an important and complex example, since these events have been the grounds for recent Western intervention in the Syrian civil war, despite there being no confirmed proof. We live in an age where errors in news coverage can have huge consequences. These errors can be caused by deliberate manipulations, out of negligence, or because quality journalism and news reporting are under pressure. By using subjective methods of analysis, they slow down specific key images and details to allow for critical reflection.

Co-produced by the Impakt Festival, Utrecht

Mr. Erbil (IQ), Volker Kreidler (DE), WISP Collective – Nabeel (IQ), Nani Cooper (DE)

The Third Room

There is no region of the world from which we receive more media coverage than the Middle East. The images are of places completely destroyed by war. Some may still remember that the cradle of our culture lies between the Euphrates and Tigris. However, it is nearly impossible for most people to imagine that between bomb attacks and ruins there exists young, contemporary art which continues to develop, which both attaches importance to the preservation of its culture and seeks connection to the Western world. We move at the borders of cultures, the past and the future. What of humanity remains, what do people want to preserve or throw overboard? Is there an intelligent way of globalization in which the positive aspects of mixed cultures can be conserved and negative ones abolished? Do similarities lead to mutual understanding instead of differences to distrust? In pictures, in sound and video, with artists from Iraq, Kurdistan and Europe, we will tell a story.

Goethe Institut Erbil
Engineers without borders Austria
C.Rockefeller Center, Dresden
Galerie Genscher
Riso Club Leipzig
Nani Cooper

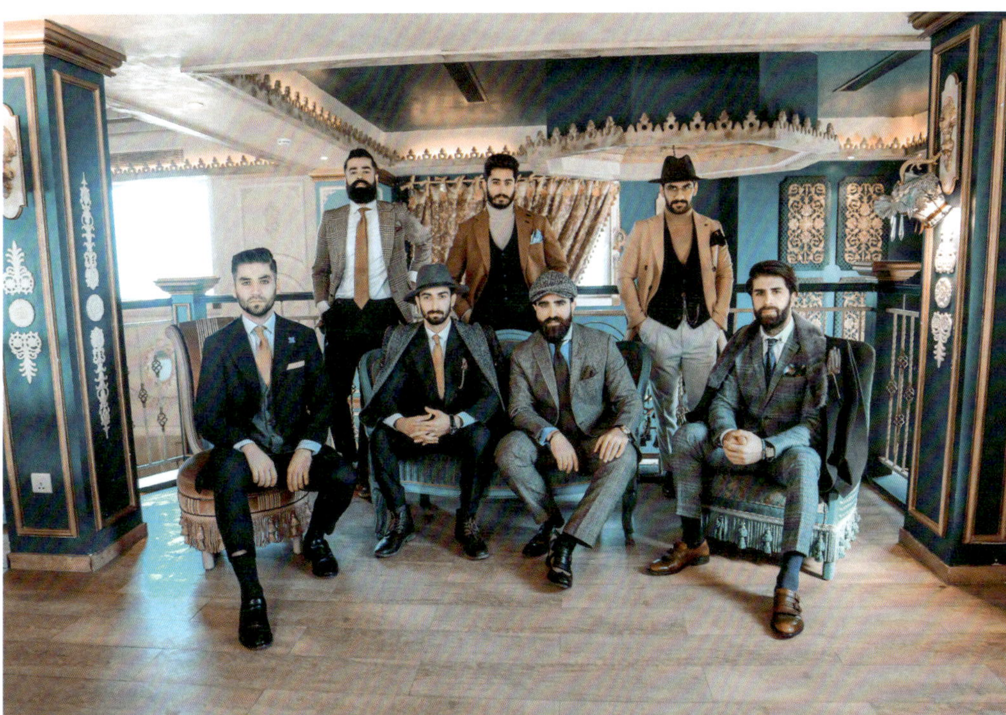

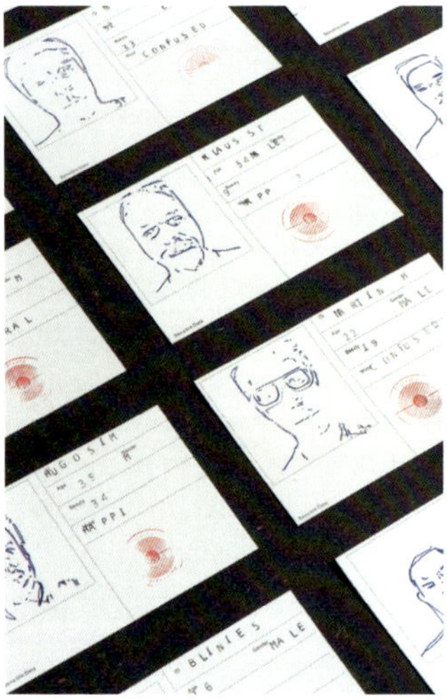

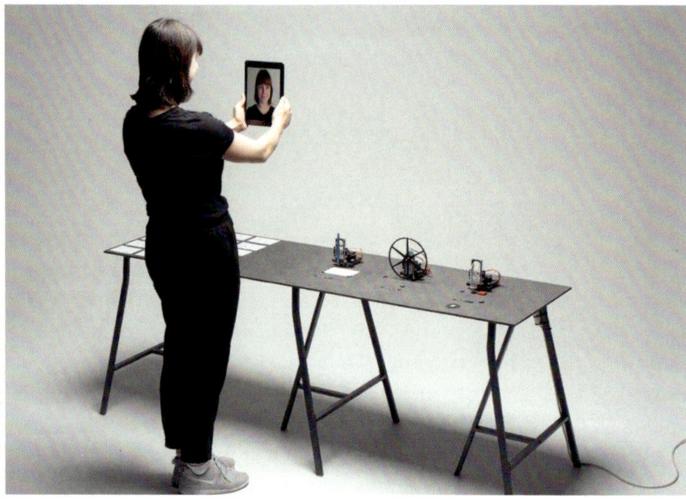

Martin Hertig (CH)

Sensible Data

Sensible Data is an interactive installation consisting of three machines that invite you to create a passport from your personal data. First, take a picture of yourself using a smartphone. A drawing machine will automatically sketch your portrait onto the card. Send an email to the machine that triggers the portrait to be processed on an online facial recognition service to judge your mood, age, gender and beauty. This information is then stamped onto the card letter by letter. By pressing a dubious button, you will get the finishing confirmation stamp. Thank you for your contribution! Offering fun in exchange for personal data, this project encourages us to explore questions about confidence in data collecting systems.

Made at écal

Alexander Peterhaensel (DE)

Smile to Vote—Political Physiognomy Analytics

Against the backdrop of currently trending AI-driven political campaigns, *Smile to Vote* escalates the latest research findings in the field of psychometrics and merges them with the worry-free large-scale implementation of facial recognition systems in lifestyle products and daily business processes. The conceptual art work, which spans multiple media, portrays the fictitious GovTech startup *Smile to Vote* and its cutting-edge product of the same name: an ultra-efficient e-voting booth. By means of AI-based facial scanning, the e-voting booth gauges the political conviction of any given person and emulates the process of digitally casting a vote at the upcoming 2019 EU Parliament elections by simply looking into a camera. The work confronts us with the implications for political processes as well as for our understanding of self-determination and freedom of will, once privacy is phased out for good and predictability of our behavior through IT systems becomes ubiquitous.

Artistic Direction: Alexander Peterhaensel
Executive production: audiovisual architectures lab
Software Development: Julian Netzer
Graphic Design: Christopher Höhn
Carpentry: Konstantin Hildebrandt
Cast: Anna Anders, Julian Netzer
Voiceover: Adam Gardiner
Support: University of the Arts Berlin

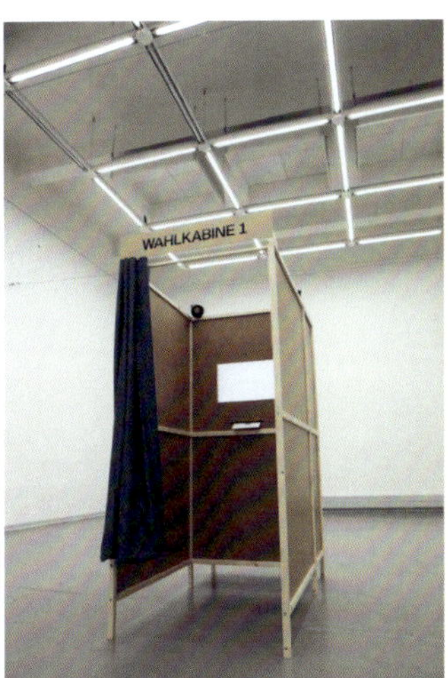

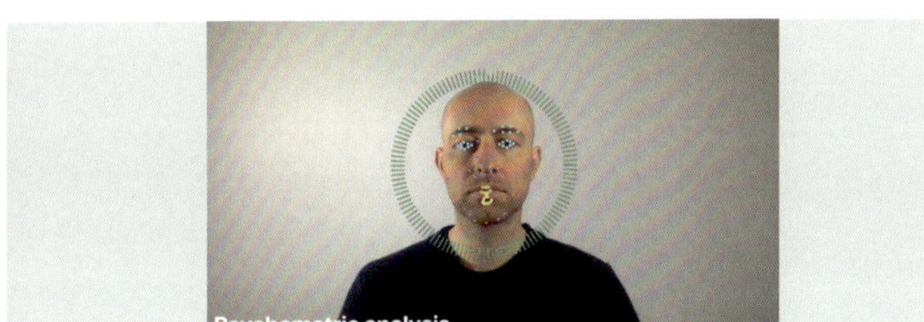

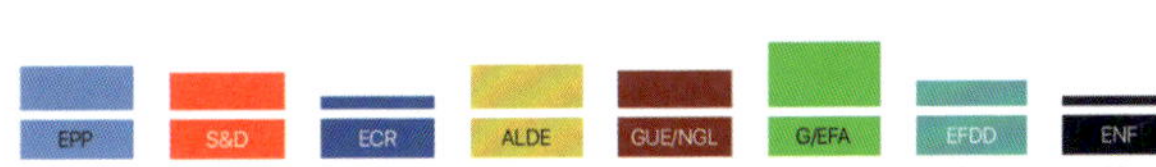

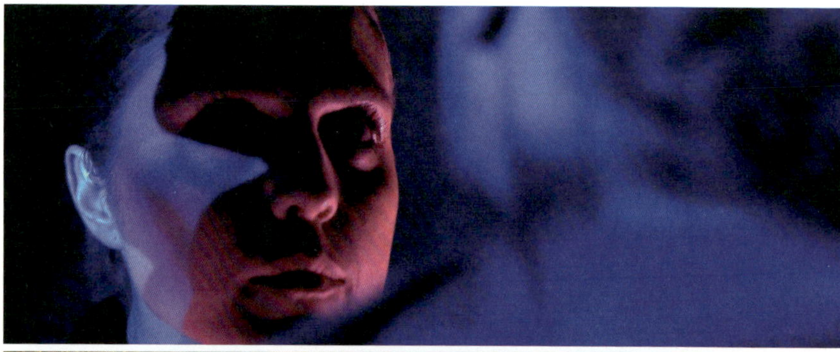

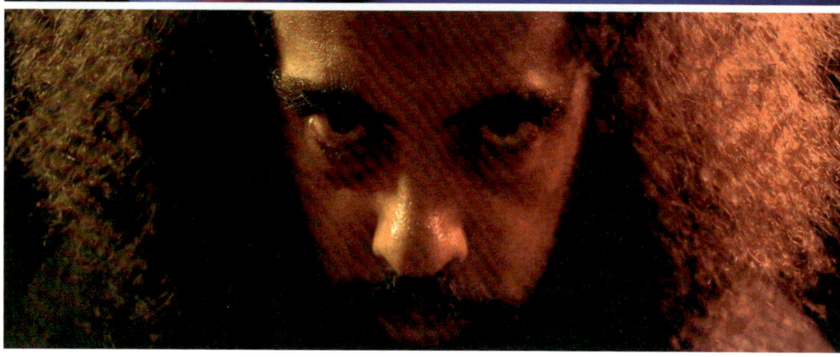

Rachel Ramchurn (UK), Richard Ramchurn (UK)

The MOMENT

A Brain Controlled Movie

The MOMENT is an interactive film which uses a Brain Computer Interface to collect attention data from the viewer that is recombined into a real-time narrative. Each time the film is watched the rhythms of the viewer's brain data create a new narrative combination; in total there are 18 billion possible combinations. *The MOMENT* is the second brain-controlled film from writer and director Richard Ramchurn.

University of Nottingham
Horizon Centre for Doctoral Training EP/G037574/1 EP/LG15463/1
Digital Economy
Arts Council England

Mario Klingemann (DE)

79530 Self Portraits

Neurographic Feedback Loop

This work explores how a feedback loop of generative adversarial neural networks (GANs) is transforming errors into meaning while at the same time confirming and emphasizing biases in the training data.
Three GANs and a camera form a closed chain in which they interpret and transform the input they receive amongst each other. Each model has been trained on thousands of painted "old masters" portraits harvested from various European collections. One model's purpose is to generate a biometric semantic map of all faces found in an image, another model translates these maps back into portraits, a third model adds details and texture to the result. A camera filming the artist is mixed into the loop in varying degrees, disturbing and changing the face markers.
Since all the models' knowledge consists of faces they will gradually transform any kind of incoming data, even noise, into their parameter space. The models' misinterpretations of small errors accumulate due to the closed loop, while at the same time the errors and imprecisions introduced during the generation process prevent the system from ending up in a static state and result in a fluid creative process that traverses the possibility space of the system.

http://quasimondo.com

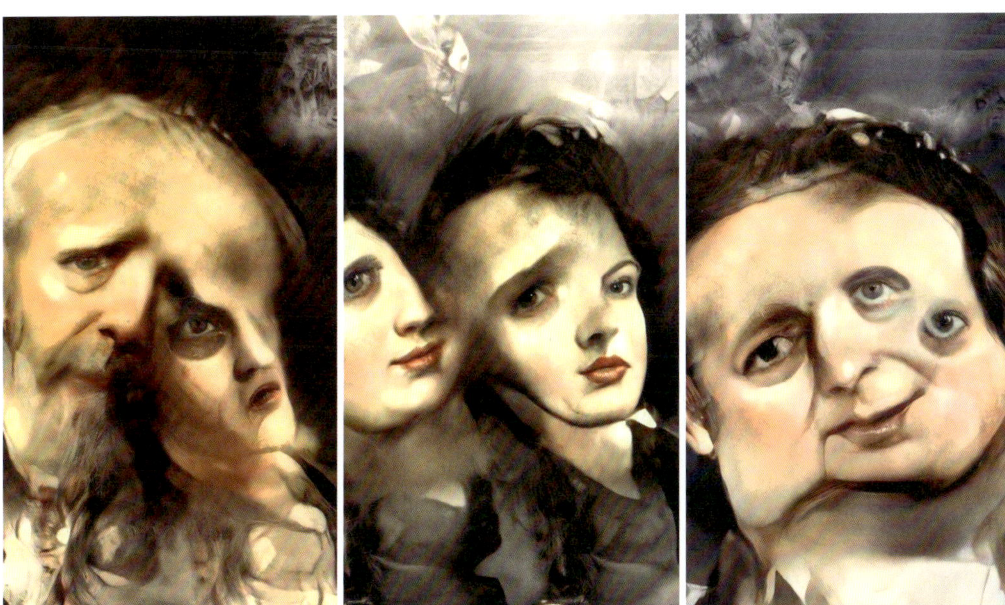

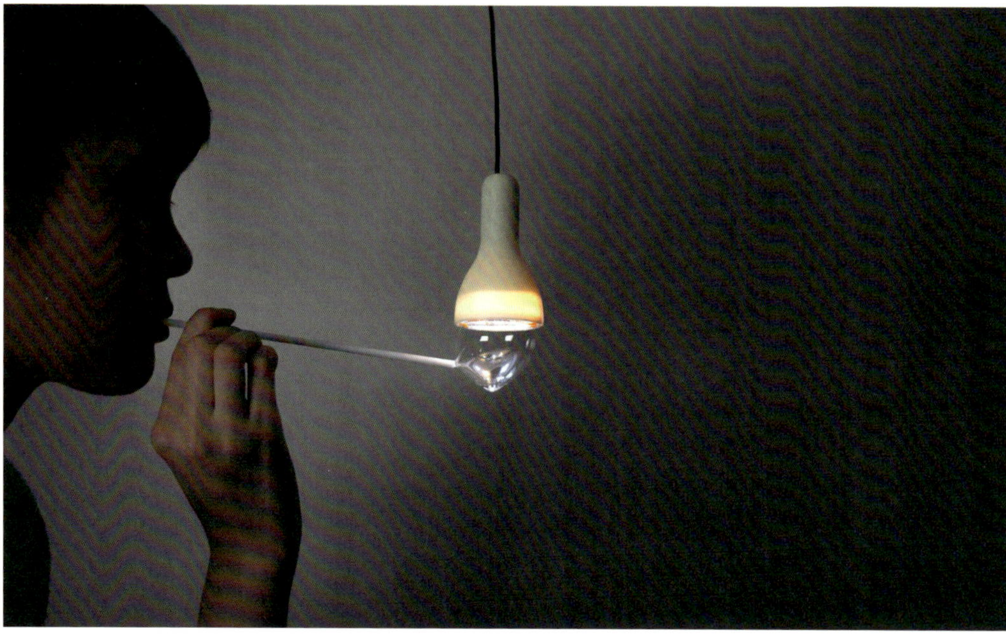

Risako Kawashima (JP), Yasuaki Kakehi (JP)

Anima

We often use the phrase "breathing life into." The target of this phrase can also be an artificial object like a digital device. In our work called *Anima*, we literally utilize our breath as an interface to activate a lighting system. In this installation, the participant creates a soap bubble using a straw. When he or she puts the bubble underneath a light socket, the light bulb starts to be activated. While the bubble lasts under the bulb, the light stays on. However, when it bursts, the light turns off just after it. Technically, this work utilizes the conductivity of soap bubble film to detect its existence on the light socket. The soap bubble works as a part of an electronic circuit in this system.

These interactions caused by the bubbles are based on an unstable, ephemeral phenomenon and look random. However, through the interactions derived from our breath and actions, we might be able to find some meaningful relations between the artificial objects and ourselves.

Direction: Yasuaki Kakehi
Device: Risako Kawashima
Research: Risako Kawashima, Shiho Hirayama and Yasuaki Kakehi

The Agency for Cultural Affairs, Government of Japan organizes participation in various overseas media arts festivals etc. The aim is to introduce outstanding works in such fields as media art, video, websites, videogames, animation and comics. Exhibitions, screenings, presentations and so on are arranged at overseas media arts festivals and other facilities with their focus on award-winning works from the Japan Media Arts Festival.

Tomoya Sasaki (JP), MHD Yamen Saraiji (SY)

MetaLimbs

21st Japan Media Arts Festival, award winner Entertainment Division

MetaLimbs adds "new arms" that can be freely manipulated by strapping on two robotic arms. The robotic arms are worn on the shoulders, and sensors placed on the top of the feet and the knees of both right and left legs track the movements of the legs to move the arms. Moving the toes moves the robot hands, which are equipped with tactile sensors, and the sensations of the robotic hands are fed back to the legs. These robotic arms can provide functions that surpass human physical abilities by replacing the end of the robotic arms with tools such as a soldering iron. As trends in technology transition from prosthetic to human enhancement technology, the question is raised as to how technology will change human physical sensation.

Industrial JP (JP)

INDUSTRIAL JP

21st Japan Media Arts Festival, award winner Entertainment Division

INDUSTRIAL JP is a music label that creates and distributes musical compositions and music videos by re-editing video samplings of machine tools in operation and field recordings of sound in small factories located here and there in Japan. Numerous musicians and factories producing springs, screws, and the like collaborate to produce club music out of analog sounds of machine tools at work. Seven works have been released as of March 2018. The reverberating sounds of operating machines and gleaming, well-oiled machinery in motion continuously construct a constant rhythm that is presented in beautiful sound and video. Establishment of the label was motivated by a strong sense of the impact of dwindling domestic industry caused by globalization. The label aims to help boost domestic manufacturing by communicating the excitement of small factories in Japan. The website features interviews with the various factories and has become a platform for sharing the advanced technological capabilities of small factories and the appeal of state-of-the-art products crafted by that technological power.

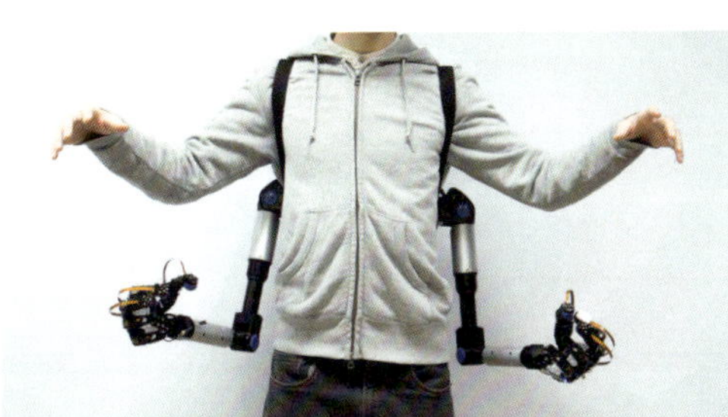

MetaLimbs

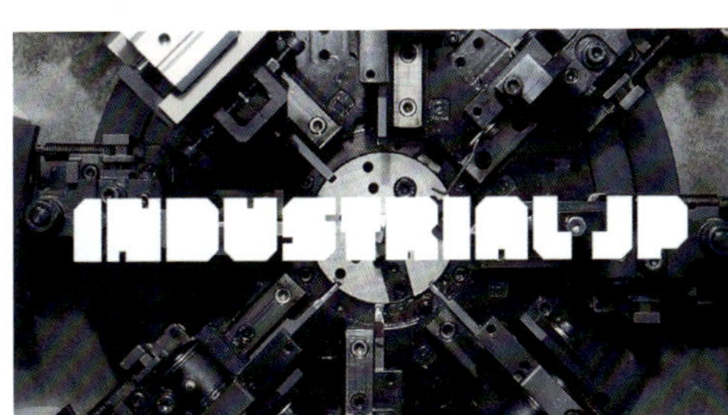

INDUSTRIAL JP

Chen Chen (CN), Qifeng Chen (CN), Vladlen Koltun (US)

Learning to See in the Dark

Imaging in low light is challenging due to low photon count and low SNR. Short-exposure images suffer from noise, while long exposure can lead to blurry images and is often impractical. A variety of denoising, deblurring, and enhancement techniques have been proposed, but their effectiveness is limited in extreme conditions, such as video-rate imaging at night. To support the development of learning-based pipelines for low-light image processing, we introduce a dataset of raw short-exposure nighttime images, with corresponding long-exposure reference images. Using the presented dataset, we develop a pipeline for processing low-light images, based on end-to-end training of a fully convolutional network. The network operates directly on raw sensor data and replaces much of the traditional image-processing pipeline, which tends to perform poorly on such data. We report promising results on the new dataset, analyze factors that affect performance, and highlight opportunities for future work.

University of Illinois at Urbana, Champaign Intel Labs

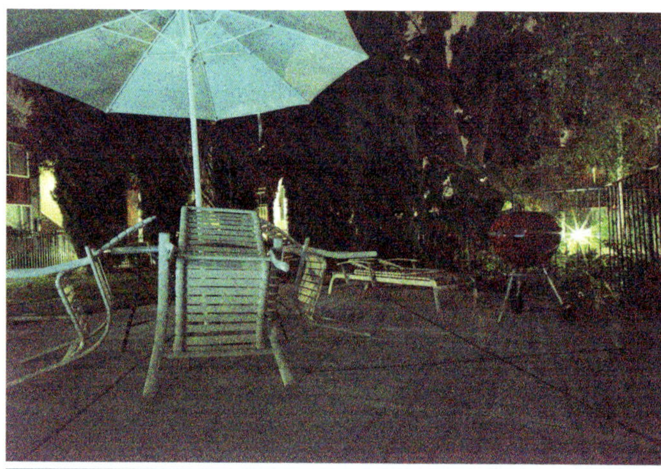

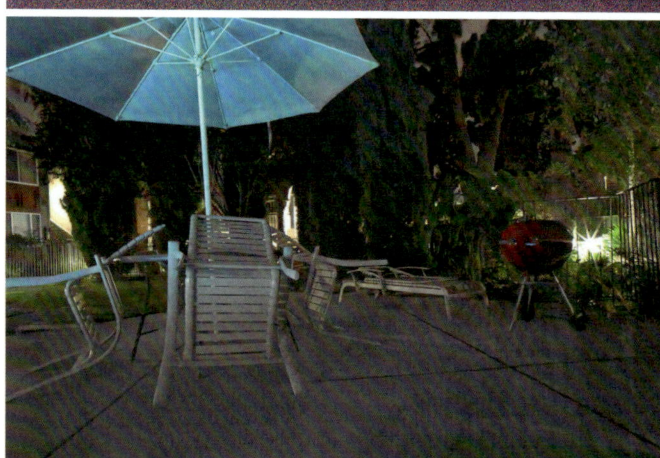

Peter van Haaften (CA), Michael Montanaro (CA)

Spiel | an in-situ performance for prepared mouth

While absorbed in conversation you notice a stranger approaching. With a curious instrument affixed to their face the visitor leans in, and listens. The mouth opens, patterns of rhythm and sound emanate from within: voices recognizable as your own. Spun out of focus, words reveal their ingrained subtleties as the collector of conversation captures the sentence but not the sentiment. Vocal exchanges are recalled and reflected. Voices are transformed by physical formant inflections, while acoustic hallucinations seem to reference what might have been said. An étude on hearing lips and seeing voices, the performer's mechanically augmented vocal tract reshapes and filters conversational spectra into new modes of mis-communication. *Spiel* physically unravels the tenuous synesthetic relationship between what is seen, heard and understood.

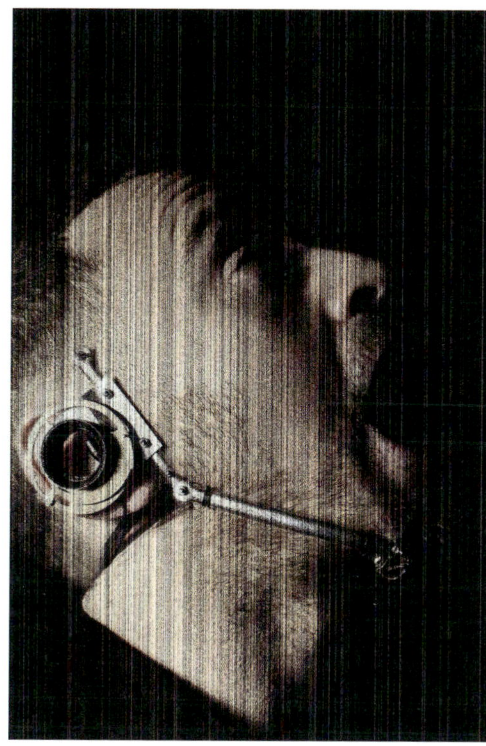

Haadi Jamali

Creative direction, composition, sound, interactive design: Peter van Haaften
Original concept, creative direction, visual design: Michael Montanaro
Visual design, fabrication: Tatev Yesayan
Performer: Nien Tzu Weng
Research collaboration: Topological Media Lab, Concordia University [an "in time" FRQSC funded project]

Tommy Pallotta (US), Femke Wolting (NL)

More Human Than Human

More Human Than Human explores the rise of artificial intelligence (AI) and its effects on our lives. Artificial Intelligence was once the realm of science fiction and futurist visions, but it is part of our current reality. Self-driving cars, autonomous drones seeking their next target or government agencies mining our personal data: we are surrounded by smart machines already. Increasingly, we expect machines to know what we want and understand us when we talk – just ask Siri. As we take a look at the history and current state of AI, filmmaker Tommy Pallotta takes us a step further as he builds his own robot to see if it can replace him as a filmmaker. He and his team design, build, and program the robot to think autonomously and test if it can direct and interview him. Leading them to question "Can AI replace us as filmmakers and storytellers?" The complexity of tasks that smart machines can perform is increasing at an exponential rate. Where will this ultimately take us? If a robot can learn to fold a towel on its own, will it someday be able to cook you dinner, perform surgery, and even conduct a war? *More Human Than Human* instigates this debate between futurists and sceptics, about the potential of Artificial Intelligence. The filmmakers take us on a quest of understanding these innovations, examining opposing views and challenging our aspirations for the future of man and machine.
More than just an exposé on the pros and cons of this new technology, the filmmakers' journey takes them to the world's leading AI experts and robot pioneers, confronting them with the existential questions that artificial intelligence begs us to ask such as: What are the values and ideas these scientists built into their creations? Has the artificial intelligence revolution taken us to the verge of witnessing the birth of a new species? Why are humans so driven to create a new species that might make themselves obsolete? How are our lives changing by these smart machines, what do they give us and what do we lose?

Produced by Submarine Amsterdam
In co-production with SubLA, Savage Film, VPRO
Written & directed by Tommy Pallotta, Femke Wolting
Producers: Femke Wolting, Bruno Felix, Tommy Pallotta
Co-producer: Bart Van Langendonck
Director of Photography: Guido van Gennep NSC
Sound: Eddy de Cloe
Editor: Chris van Oers

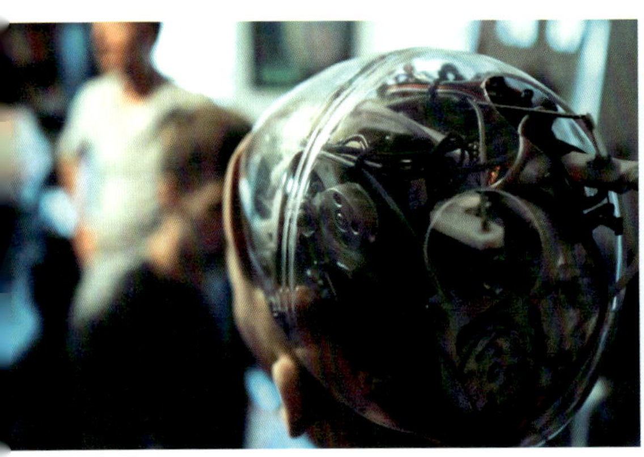

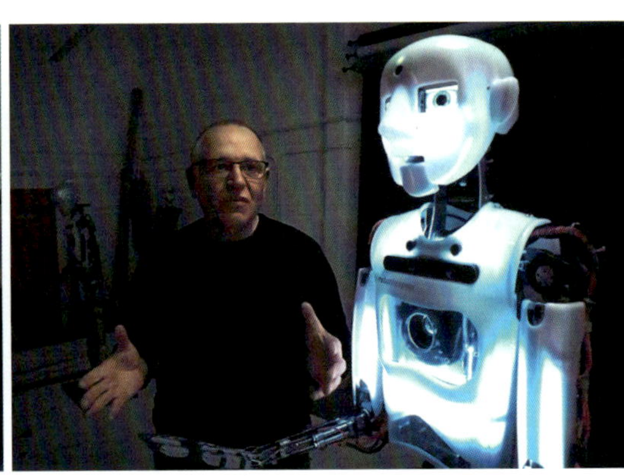

Submarine Amsterdam, 2018

FAIRY BOT – Sandra Trostel (DE), Thies Mynther (DE)

All Creatures Welcome

A contemporary documantary

All Creatures Welcome sketches a utopian image of society in the digital era. Accompanied by the appeal to "use hacking as a mindset," the viewers immerse themselves, together with the filmmaker, in a documentary adventure game and explore the world of digital communities at the events held by the Chaos Computer Club; a real-world reflection of the virtual spectrum. *All Creatures Welcome* is not merely a documentary: it's a transmedia project, complemented by modular additions that enhance the classic cinematic framework according to the themes it deals with:

- The *Create Your Own Culture* platform
- The music project *Thies Mynther's Fairy Bot Orchestra*
- A cultural-political debate on the topic of *Public Money = Public Culture*

The entire venture aims to forge ahead with the discourse on the free availability of culture and knowledge and transpose it into a concrete example—"Sharing something can make more out of it!"

Director/ Camera/ Editor/ Producer: Sandra Trostel
Co-Producer/Music: Thies Mynther
Animation and Computer Interface Voice: Jon Frickey
Camera @32c3: Lilli Thalgott
Sound Mixer: Jonas Hummel
Production company: FAIRY BOT
Supported by *https://ipredator.se/* and approximately 1000 crowdfunding backers.

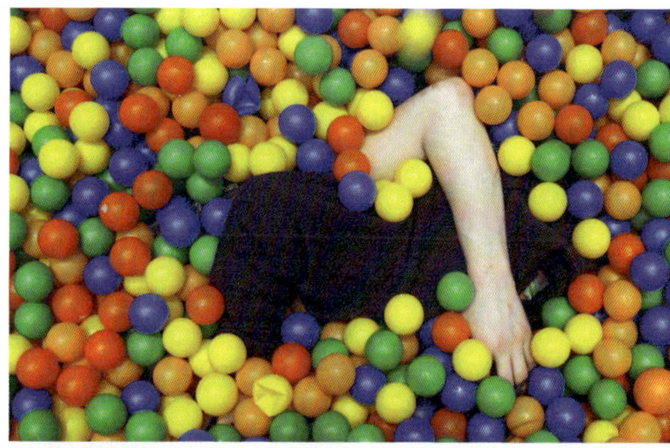

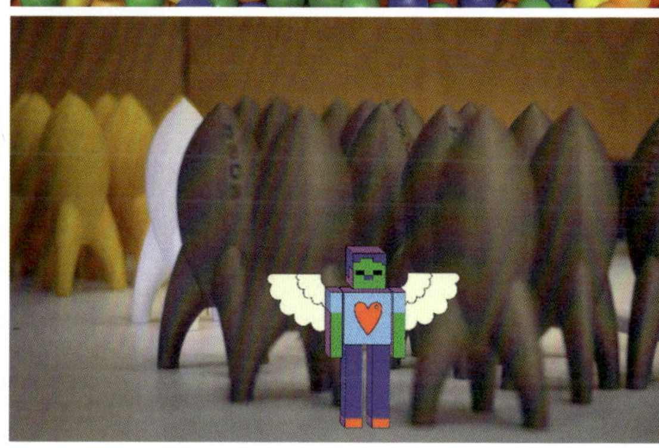

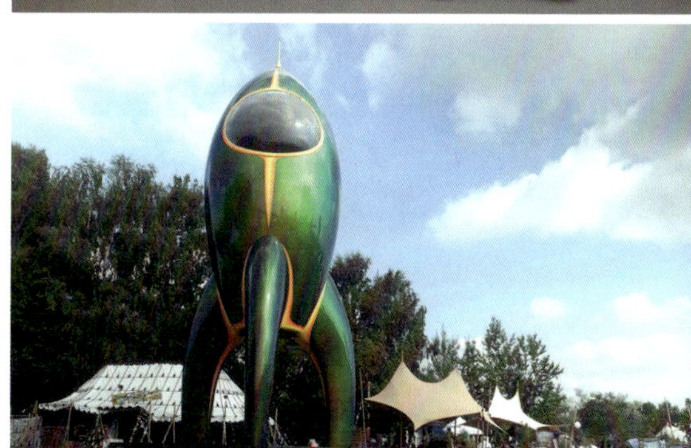

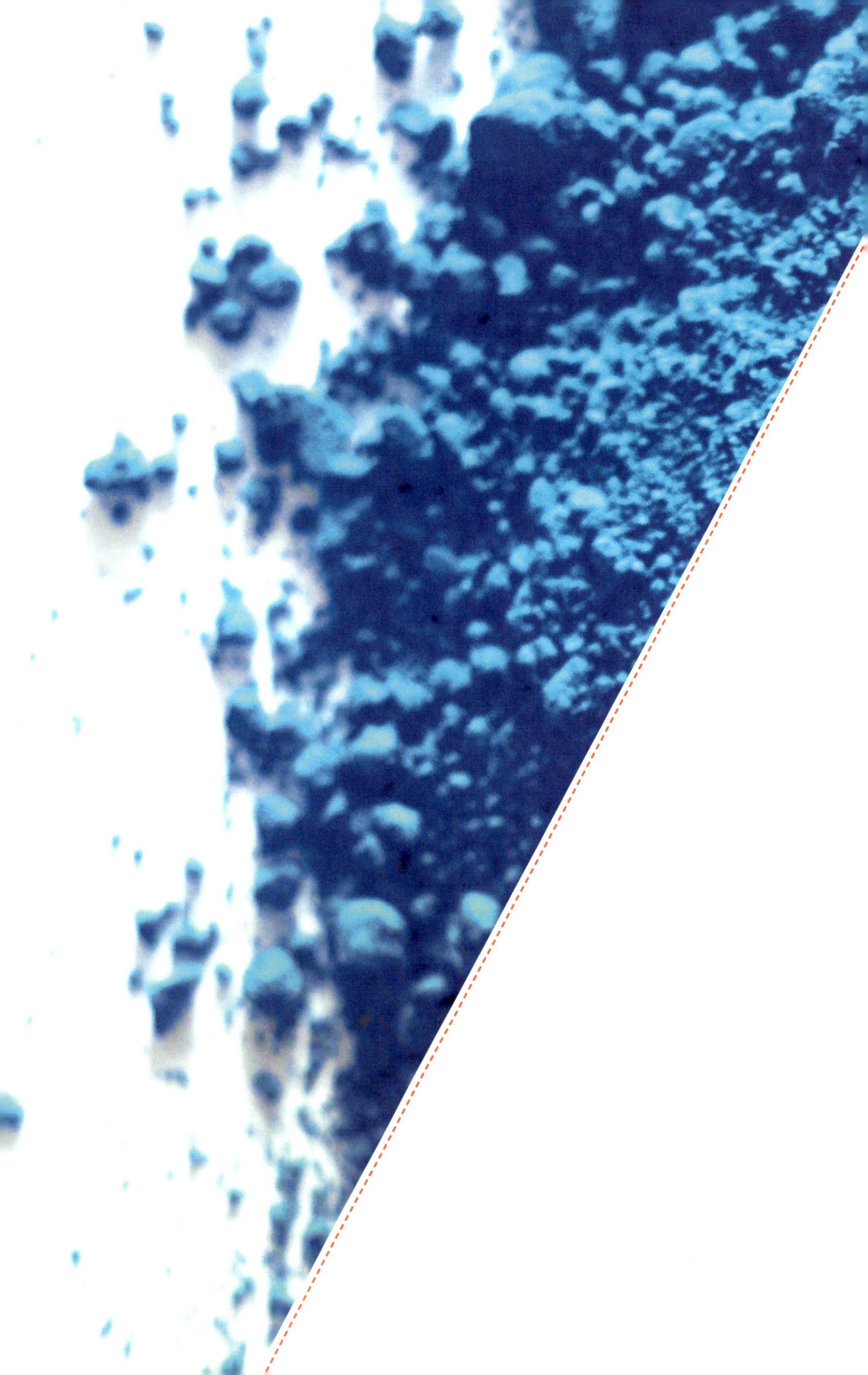

THE PRACTICE OF ART AND SCIENCE

The rapprochement, as it were, of art and science, the artistic exploration of new applications, is a key factor in the increasingly social dimension of new technologies in order to comprehend how reciprocal human-machine relationships, interactions among individuals and globally networked systems can not only be better understood but, above all, better designed.

Since the inception of the Festival in 1979 by artist Hubert Bognermayr, scientist Herbert W. Franke and journalist Hannes Leopoldseder, art and science has always been a key aspect of Ars Electronica. In 1996, the increasing interest in collaborations among artists and scientists resulted in the foundation of two pillars of learning, research and presentation: Ars Electronica Center as "Museum of the Future" and Ars Electronica Futurelab as "Laboratory for Future Innovations." As sort of a melting pot, where different cultures of knowledge can—or should—mingle and discover their synergies, Ars Electronica has been a place for artists, scientists, researchers, designers and engineers to cooperate on multifarious projects for years. Based on a collaboration with CERN, which started in 2011, the plan for a network for art and science residencies on a European scale evolved. Ever since its initiation, the European Digital Art and Science Network with the partners ESA, ESO and Fraunhofer MEVIS sparked interest from artists as well as institutions and continued to grow.

This year, various international crews of artists and scientists have also taken on the task of exploring art and science further, and now present their work in this exhibition space.

University of Graz – Artificial Life Lab (Project coordination) (AT), Université Paris Diderot – LIED (FR), École Polytechnique Fédérale de Lausanne – LSRO (CH), University of Zagreb – LARICS (HR), FCiências.ID – Associação para a Investigação e Desen, Cybertronica Research (DE)

ASSISIbf–Symbiotic computation of bio-hybrid systems

Bees, fish and robots solve difficult network problems together

Networks are ubiquitous in our world today and are growing and complexifying at a global scale, from traffic and logistics to social networks and the "Internet of Things."

Optimal design of such networks is a difficult task, exploding in complexity with network size and connectivity. Thus, smart heuristics are a key technology –using randomized guessing as a form of provoked error is central to their functionality. In *ASSISIbf* we aim for a radically different approach towards network optimization of this kind: We use "fuzzy" swarms of honeybees and fish, in association with autonomous robot swarms. These novel bio-hybrid "computational symbioses" efficiently search for optimal, or near-optimal, network configurations. With large populations of animals we hope that future bio-hybrid applications will be able to solve global-scale network problems in a shorter time than today's computer algorithms, exploiting the parallelism of organismic swarm computation.

EU-FP7 Project no. 601074. Objective ICT-2011.9.10: Fundamentals of Collective Adaptive Systems – FoCAS, Duration: 5.5 years; Start: 1st of February 2013, Budget: 6 Million Euro,
Project website: *http://assisi-project.eu/*

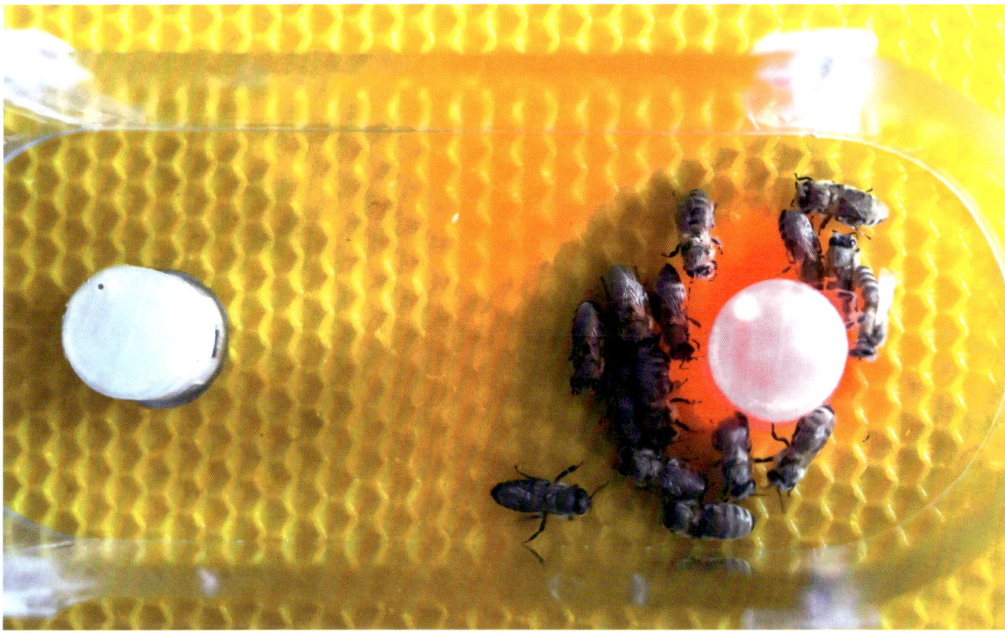

Artificial Life Lab Graz

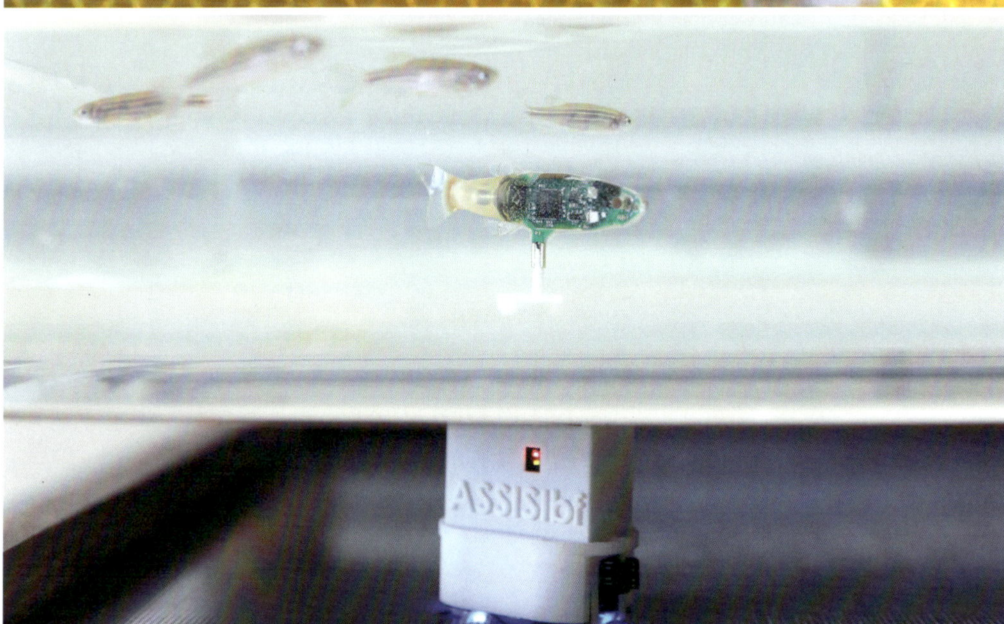

EPFL

g.tec medical engineering GmbH (AT)

BR41N.IO Hackathon

The Brain-Computer Interface Designers Hackathon

A BCI provides a direct link between the brain and an external device. 20 years ago, BCIs could only spell or move computer cursors. Today, BCIs are being used in many different fields of neuroscience, such as motor rehabilitation for stroke patients, assessment of and communication with coma patients, control of devices for disabled people, cognitive training or neuromarketing. Machine learning, dry electrodes, wireless electrode caps, and other technologies are making BCIs more powerful and practical for a growing number of users.

The *BR41N.IO* Brain-Computer Interface Designers Hackathon Series has been created to show these current and future developments, and the unlimited possibilities of BCIs in creative or scientific fields, and brings together engineers, programmers, designers and artists. Each team must design and build a wearable BCI-headpiece that can measure brain activity in real-time to create any sort of interaction. The hacking projects use EEG electrodes and amplifiers, and challenge programmers to code an interface that enables them to control devices, robots or applications, post messages on social media, draw paintings, or a myriad of other applications by using their thoughts only. *BR41N.IO* also challenges creative minds to design a BCI headset with 3-D printers, handcrafted materials and sewing machines.

Children are invited to create their own Brain-Computer Interfaces and can take their handicraft with them. *BR41N.IO* aims to promote awareness of artificial intelligence, life science, art and technology, and how these can merge into innovative and exceptional BCI systems.

BR41N.IO is organized by g.tec neurotechnology GmbH
BR41N.IO is sponsored by the IEEE Brain Initiative

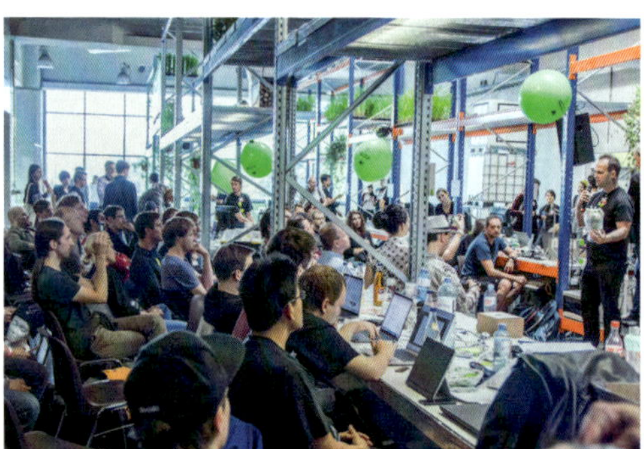

g.tec neurotechnology GmbH

BR41N.IO Linz 2017

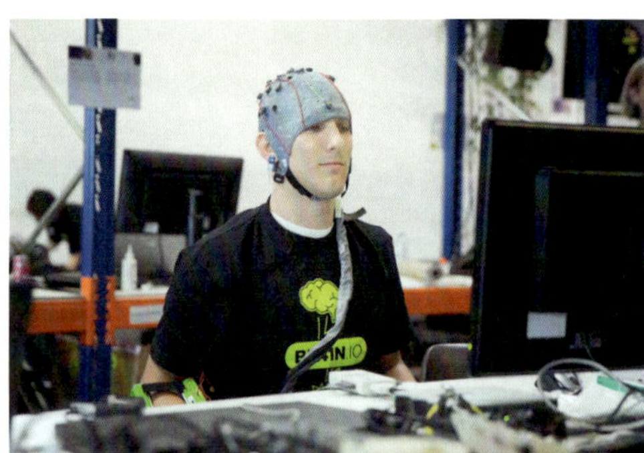

Tom Mesic

BR41N.IO Orthosis Control at Ars Electronica 2017

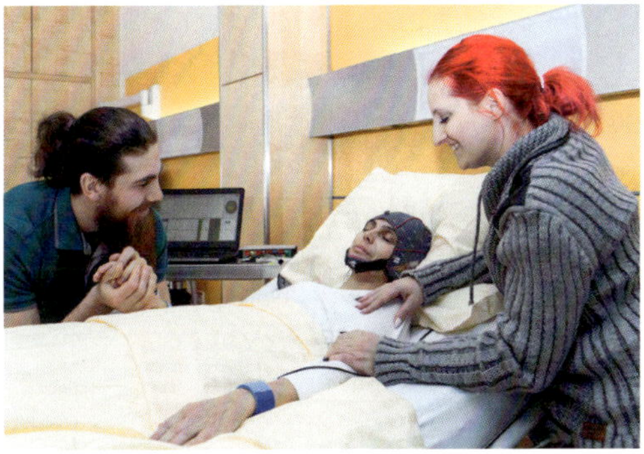

Florian Voggeneder

mindBEAGLE © g.tec medical engineering GmbH

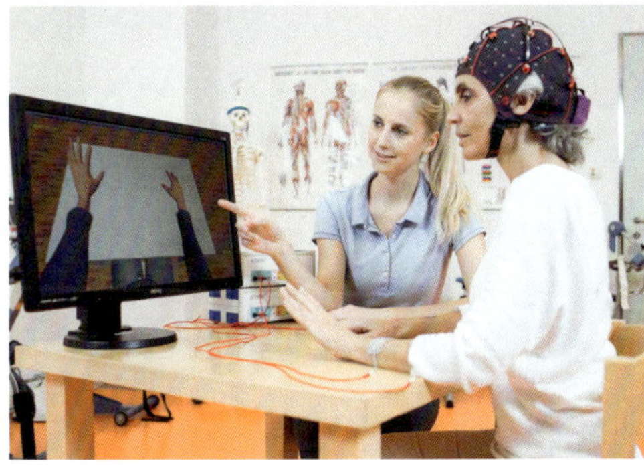

Florian Voggeneder

recoveriX © g.tec medical engineering GmbH

g.tec medical engineering GmbH (AT)

Bugfix the Brain

Rehab Technologies for Medical Professionals and Patients

Imagine being able to think, hear, and feel – but not to move or communicate. The exhibition *Bugfix the Brain* focuses on patients who suffer from motor disabilities due to stroke or disorders of consciousness and shows state-of-the-art rehabilitation and assessment tools based on Brain-Computer Interfaces. Neurologists, physical therapists, caregivers and patients are welcome to test these brain rehabilitation technologies on themselves.

recoveriX is the first rehabilitation system for stroke patients that pairs mental activities with motor functions. For example, if a stroke patient imagines a hand movement, the BCI system gives a visual feedback through Virtual Reality and feedback through muscle stimulation at the same time. This fosters brain plasticity and patients learn to move hands again.

mindBEAGLE is a pioneering method that can be used with patients who are living with coma, unresponsive wakefulness syndrome, minimal consciousness or locked-in syndrome. *mindBEAGLE* uses auditory and vibrotactile stimulation to assess whether a patient is conscious. If patients are able to perceive stimuli given by the *mindBEAGLE* system, they might be able to communicate and answer Yes/No questions! The *Unicorn-Speller* is an affordable Brain-Computer Interface that allows patients to write words and sentences just by thinking. This is an important tool for completely paralyzed people to establish communication.

Marjan Colletti (IT/AT/UK), Daniela Mitterberger (AT), Tiziano Derme (IT/AT), Georg Grasser (AT)

FrAgile 6 – Pahoehoe Beauty

The trans-disciplinary and post-digital series entitled *FrAgility* addresses the fragility/agility of design, technology, materiality, environment, nature and architecture in the 21st century, and results from the collaboration of a plethora of disciplines (industry, craftsmen, material scientists, biologists, couturiers, artists, engineers...).

Fully robotically 3-D-printed, *Pahoehoe* is a landscape, a garden—a chimera of both natural and man-made materials, objects and species—yet with antithetical techno-organic aesthetics. The project reflects with particular attention on the often troubled, in some cases incompatible relationship of the built environment with the natural environment. It creates a hybrid domain beyond the oppressive binaries of western modernity, defined by an ecosystem of specious symbionts and mistaken materials.

With technology disappearing into the background, visitors focus more on the value of design experimentation, on the spatial experience of hybrid artefacts, and on the potentialities of architecture as cultural catalyst.

Supported by the University of Innsbruck (Faculty of Architecture, Institut für experimentelle Architektur.Hochbau, REX|LAB); and by UCL (The Bartlett School of Architecture)

Collaborations:
Dr. Aurelien Forget; Jan Contala (Rex-Lab); Jonathan Raphael Hanny; Moritz Riedl; Michael Schneider (Tyroler Glückspilze); Philipp Schwaderer (Rex-Lab); Lukas Vorreiter

Thanks to Pavlos Fereos and all Hochbau E3 students: Claire Hentgen; Tobias Hinterschwepfinger; Laura Schwarz; David Haslgruber; Florian Heinrich; Tobias Sam; Kristan Walder; Sandra Al jbali; Cendrine Peters and Kilian Bauer; Marina Niederleitner

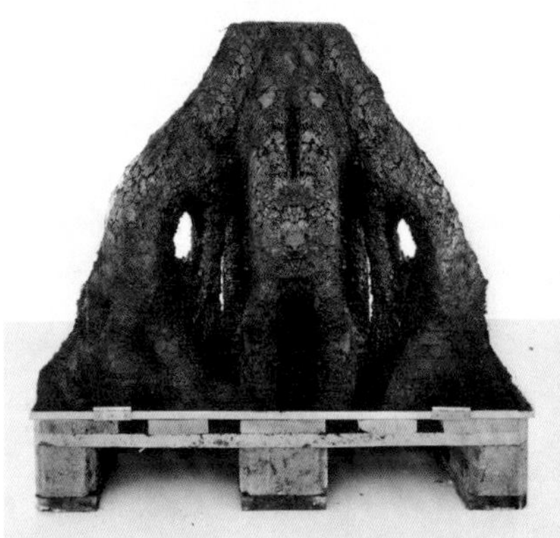

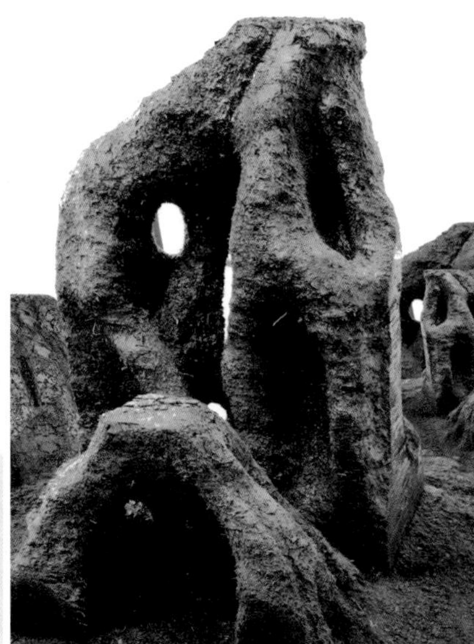

maeid copy mc

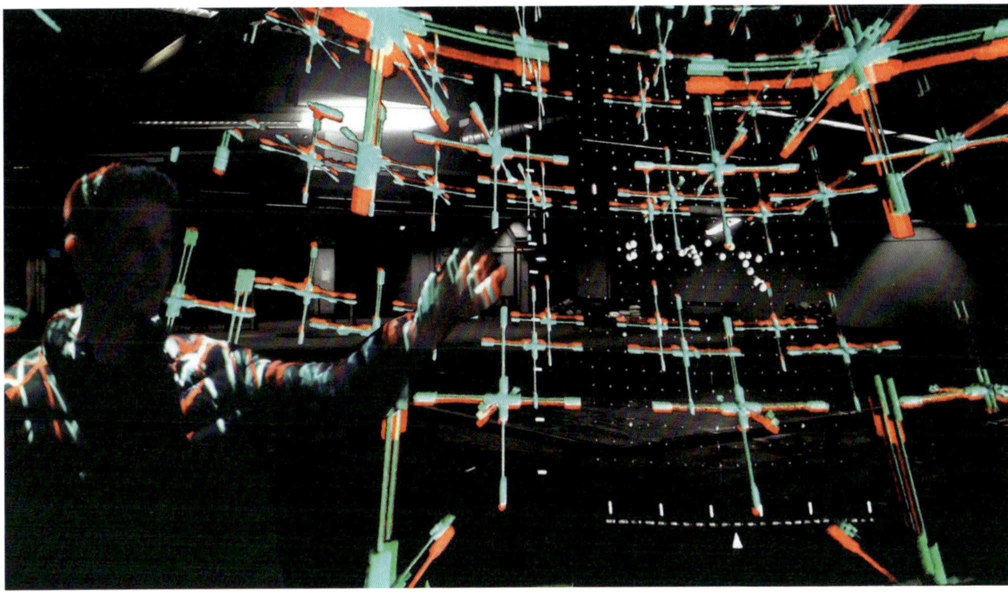

arc/sec Lab – Uwe Rieger (DE/NZ), Yinan Liu (NZ), in collaboration with the Augmented Human Lab

LightTank

LightTank is a mixed-reality installation that augments a space frame structure with holographic line drawings. Using the anaglyph (red/cyan) stereo projection technique on transparent screens, the haptic-digital construction merges with its surrounding and blurs the boundaries between tactile and virtual. *LightTank*creates a communal immersive experience with an intimate responsiveness to its audience. The project was conceived by the arc/sec Lab for Digital Spatial Operations at the School of Architecture and Planning at the University of Auckland. Headed by Assoc. Prof. Uwe Rieger, the Lab explores real time Reactive Architecture as a fusion of physical materiality and digital information. *LightTank*'s interface was developed in collaboration with the Augmented Human Lab at the Bio Engineering Institute. Under the lead of Assoc. Prof. Suranga Nanayakkara, the AHLab focuses on creating enabling human-computer interfaces as natural extensions of our body, mind and behavior.

www.arc-sec.com; www.ahlab.org

Project concept and development by arc/sec Lab: Uwe Rieger and Yinan Liu, the University of Auckland

Sensing solution by Augmented Human Lab: Roger Boldu, Heetesh Alwani, Haimo Zhang, Suranga Nanayakkara, the University of Auckland

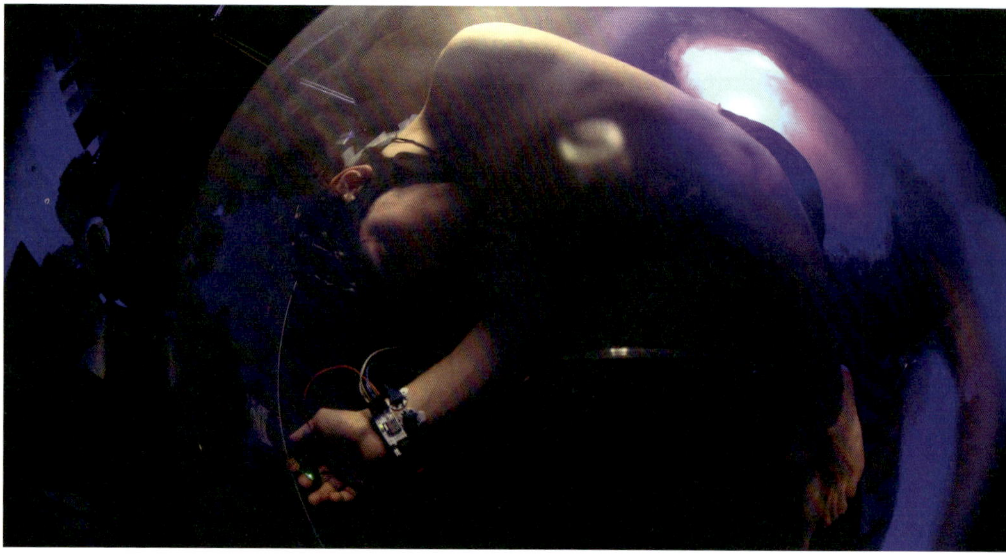

MIT Media Lab, Fluid Interfaces Group (US)

Cocoon

Interfacing with the Sleeping Mind

Cocoon transplants ongoing research at the MIT Media Lab's Dream Lab into the electronic arts context. We will perform live experiments that focus on tracking, influencing and extracting content from dreams. We present three custom-built wearable electronics for sleep science which make this interface across levels of consciousness possible: Masca, Essence and Dormio. Our wearables gather eye, heart, brain, breath, muscle and skin biosignals for sleep staging and in turn output smell, audio and electricity to manipulate the dreaming brain. The intersection between signals and stories where we work, between concrete inputs and the algorithms which abstract them in the mind, offer a unique meeting point for scientific regiment and experiential art. The nightly dissolution of our perceptually grounded experience into dreams, and the clear signals associated with them, link science and the imaginary.

The MIT Media Lab is an antidisciplinary research laboratory at the Massachusetts Institute of Technology. Its research is not restricted to fixed academic disciplines, but draws from technology, media, science, art and design. The Dream Lab is an initiative from the Fluid Interfaces Group at the MIT Media Lab. The group designs systems for cognitive enhancement building upon insights from neuroscience and psychology. These systems help us exploit the untapped powers of our minds and seamlessly supplement our natural abilities to support attention, memory, emotional regulation, creativity, learning, decision making and more.

Pierre Cutellic (FR/CH), Maria Smigielska (PL/CH)

Proteus 2.0

Modulating Matter with Mixed Intelligence

Proteus 2.0 is an installation that modulates ferrofluid patterns, with both human and machine intelligences, in a closed loop. Through an individual and prolonged visual experience, it immerses the visitor in an implicit interaction with the material through a brain-computer interface. A pre-trained dedicated machine learning model is informed by real-time neural signals, produced by the visitor's gaze while being exposed to the rapid serial change of patterns without any explicit instructions to follow. Over the time of the gaze experience, visitors may witness a certain stabilization of their own modulated picture of the material. This model not only modulates the ferrofluid display over time but also slowly builds a digital motion picture of its behavior. All the recorded sequences are then displayed separately to exhibit the variety of patterns collected throughout the period of the festival. The project relies on digitally encoded material properties with the use of a computational mixture of both human and machine vision models.

Pierre Cutellic *(compmonks.com)*, Maria Smigielska *(mariasni.com, creativerobotics.at)*
Supported by :
Creative Robotics UfG Linz, Austria *http://creativerobotics.at/*
Chair for Computer-Aided Architectural Design (CAAD), Institute of Technology for Architecture (ITA), ETH Zürich, Switzerland *http://www.caad.arch.ethz.ch/*
Sponsors : SMC Corporation *(smc.at)*, supermagnete

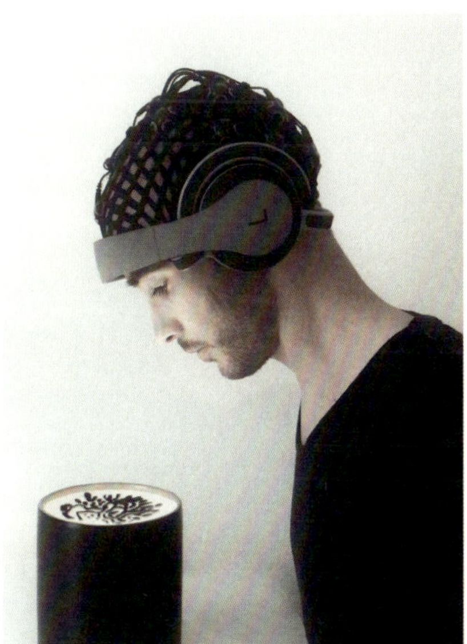

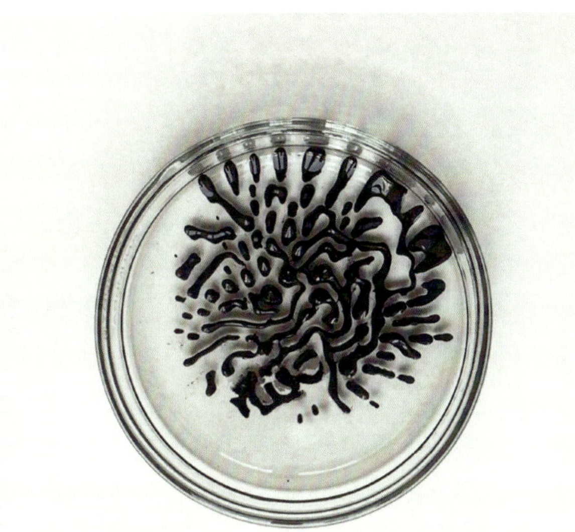

Maria Smigielska

Seoul LiDARs – Hyun Parke (KR/US), Jinoon Choi (KR), Sookyun Yang (KR)

Volumetric Data Collector

Seoul LiDARs focuses on the non-intended use of industrial technology through various artistic approaches. *Volumetric Data Collector* is built on the idea of mobilizing a LiDAR sensor, a 3-D-laser sensor commonly used on autonomous vehicles, as an extended sense organ of a human body. *Volumetric Data Collector* is our first experiment translating between a LiDAR sensor and a human sensor, and from a 3-D-point cloud to visual data. We packed a LiDAR, a display monitor and peripheral devices in a portable unit, to be carried around Seoul by walking to places that are otherwise inaccessible. This (in) efficient device is combined with human mobility, performing a simple task of collecting a 3-D-point cloud of the carrier's surroundings. For three days, we collected volumetric data of Seoul's historic places such as a high security inner-city mountain, alleys of an old electronics market and a now-defunct prison. This seemingly parasitic device visualizes three-dimensional spatial information onto the monitor on the carrier's back.

This experiment shifts LiDAR senses into a completely different perspective by de/re-purposing its intended function and integrating human mobility. It is an examination of how spatial sensing technology, a non-human sense that could either be a potent counterpart or an expansion of human senses, could define city space, thus influencing human perception.

Hyun Parke, Jinoon Choi, Sookyun Yang
Supported by ZERO1NE

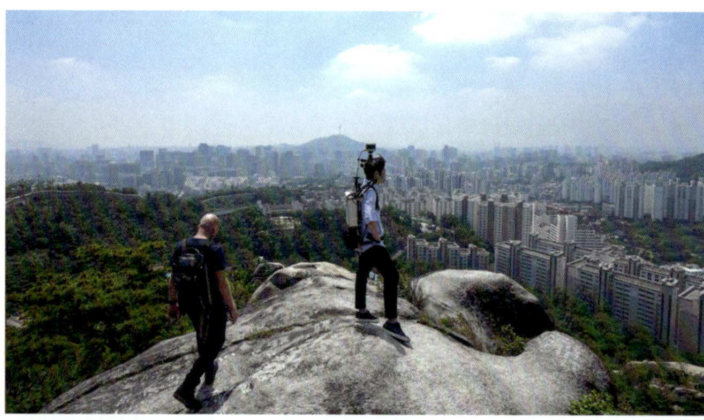

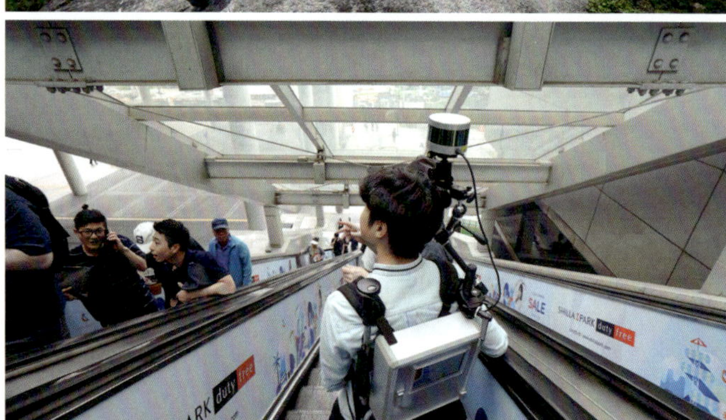

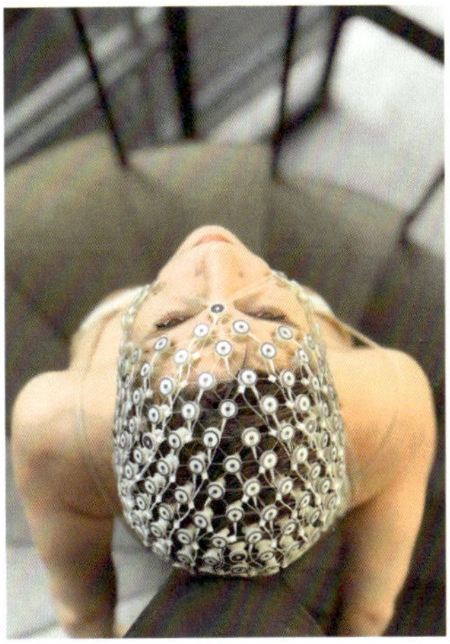

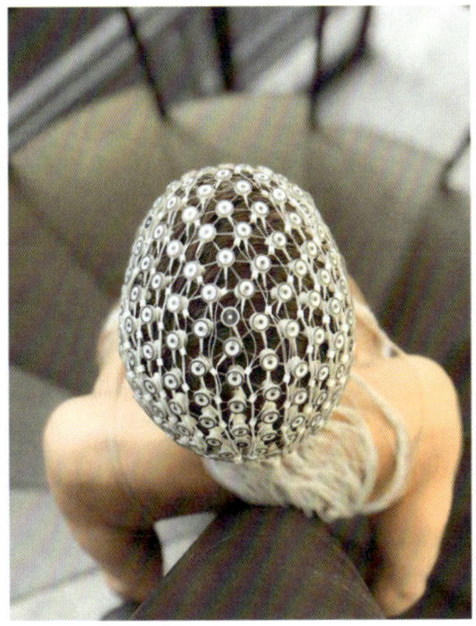

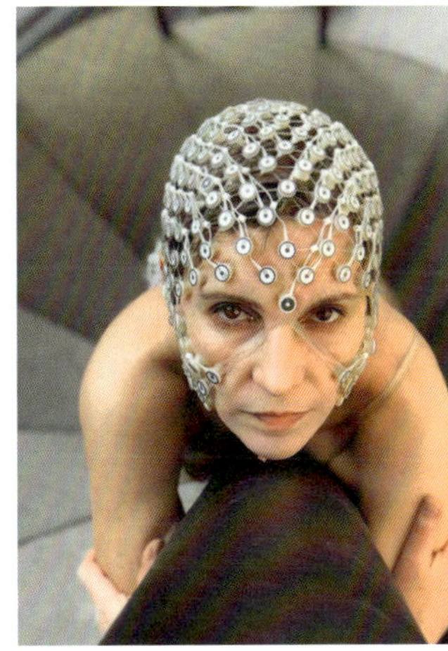

Nathalie Regard (FR/MX), Guillaume Dumas (FR), Roberto Toro (FR/CL)

The 101-nights

Dream Sessions developed an experimental protocol for sleep monitoring. Integrating Nathalie's body into a study, we reconstructed the environment of dream emergence upon the permanent exchange between the internal perception and the environment. Adding an exceptional length of time in comparison to common studies, in *The 101-Nights* we perform the recording of her sleep with 256 EEG channels, a performance in which her dreams were electronically recorded and stimulated with audios. These stimuli occasionally demonstrate their influence in the account of her dreams. The Sleeper proposes to reproduce the landscape between the brain activity made from encephalographic data and the subjective experience of dream, recorded during the performance, where the viewer is invited to interact and look for possible correspondences between the plot of her dreams and the auditory influences triggered by the computer system. The protocoled writing becomes sound and the stories solved through voices tell us strange facts. The invisible, silent and impersonal EEG data are carried over into a visual projection, regaining reality in a new way.

EGI Electrical Geodesics Inc. / cooperation partners /

Copenhagen Institute of Interaction Design, IT University Copenhagen (DK)

VIRT-EU

Tools for Ethical Thinking While Creating Connected Devices

We are exposed daily to creepy news stories about smart objects gathering people's data, yet such an unethical practice is widely left unaddressed for the sake of technology-led economic growth. No one yet knows how to solve the challenges of ensuring ethical data practices in the way new technologies are designed. The EU has made significant legislative and advocacy efforts to address societal concerns brought about by technological innovation in data processing. Nevertheless, we lack practical guidelines and assessment procedures to embed ethical, social and data protection values in the design and development of data-intensive technologies and services.
VIRT-EU applies an interdisciplinary research approach to generating new knowledge and methods for countering the unproved assumption that technological development leaves no room for ethical and moral reasoning.
VIRT-EU aims to intervene at the point of design to foster ethical thinking among developers of IoT solutions. In fact, addressing social concerns in new technologies not only impacts changes in regulatory regimes, but also influences the process of imagining and developing the next generation of digital technologies within European clusters of creative innovation.

Copenhagen Institute of Interaction Design, IT University Copenhagen, Open Rights Group, London School of Economics, Uppsala University, Politecnico di Torino

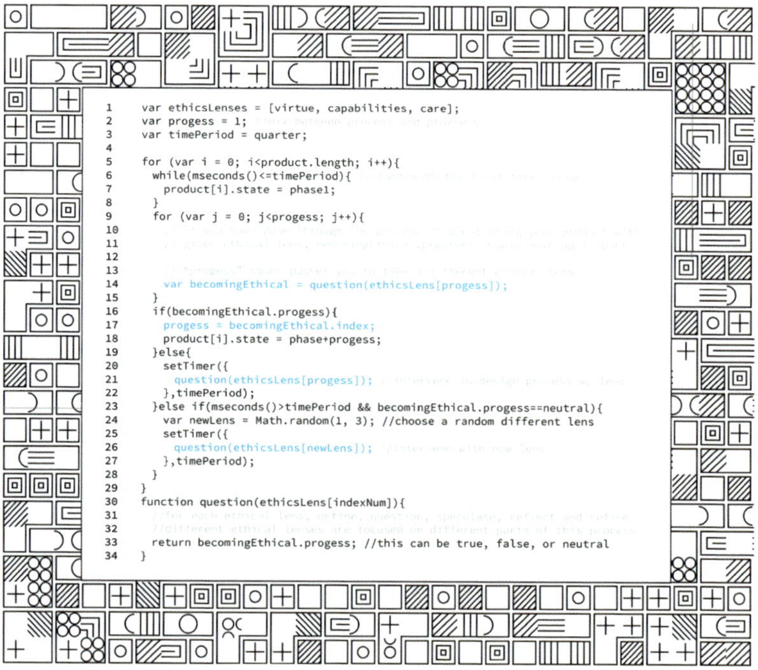

Greiner Group (AT)

Plastics for Life

At Greiner, every employee has the opportunity to present and discuss ideas online. So far, hundreds of ideas have been posted and developed, innovations have been promoted and creative employees connected. The *Plastics for Life* campaign is about sustainability and plastics. Particularly attractive are those two ideas that have the goal of the entire value-added cycle:

At *Enzymatic Recycling*, the ability to break down plastics, derived from specialized bacteria, is studied and improved. The resulting decomposition products can be used as raw material for the production of plastic granules.

The *Life Capsule* aims to replenish fallow land. The degradable plastic containers contain plant seeds, nutrients and water and are dumped over affected areas. The capsule is made of 100% biodegradable plastic and protects the seedling in the initial stage.

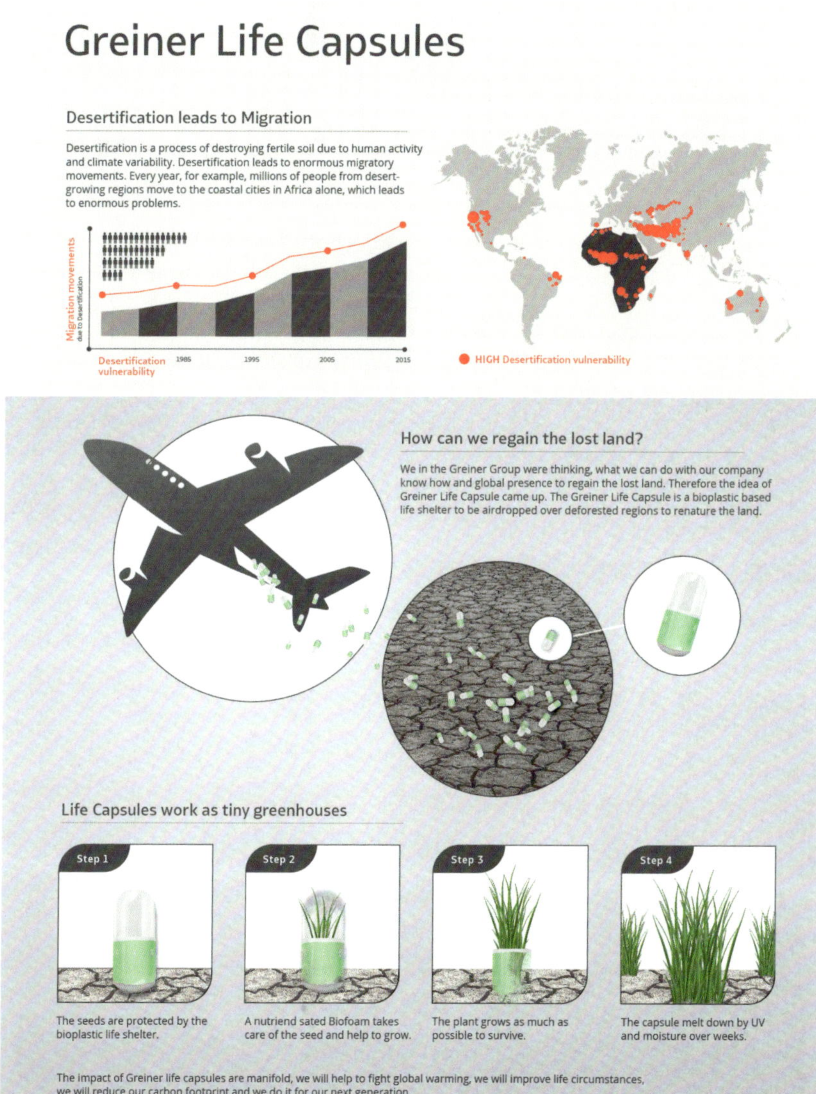

Mozilla (US), The Tech Museum of Innovation (US)

▸ Reality Redrawn Challenge

The *Reality Redrawn Challenge*, launched by Mozilla, is intended to tackle the current topic of misinformation with the help of mixed reality and other art media. Artists and developers were invited to conceive of projects that reveal and illustrate processes of misinformation as well as their potential societal impact. The five winners' works were presented at The Tech Museum of Innovation in San Jose, California. Two of these projects—*Bubble Chaos Echo Chamber* and *Faking News* —are exhibited at the Ars Electronica Festival 2018.

Stu Campbell (AU)

▸ Faking News

If the front-page news could be broken down and analyzed in real-time, what would it look like? Using the *EyeJack* Augmented Reality App, you can see the front page of the *New York Times* come to life with animation and sound as the stories are deconstructed. In this context the *EyeJack* app allows the user to perform a news autopsy—a post-print examination that explores the climate of events surrounding the news story. In an era where the term "fake news" is wielded frequently, the legitimacy of news is under threat and therefore it is important for people to question the authenticity of news. The *EyeJack* app is a time-lens where multiple points of view can be presented simultaneously.

The juxtaposition of stories and events presents context that may explain why influential leaders and institutions choose to omit or distort facts or fabricate news altogether. This piece was first seen at Mozilla's Reality Redrawn event at The Tech Museum of Innovation in San Jose.

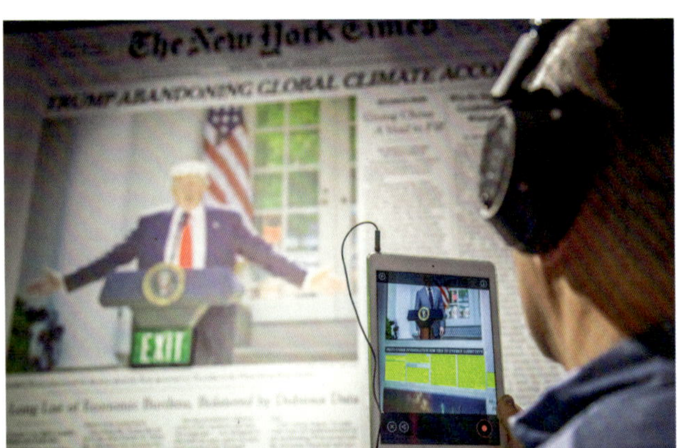

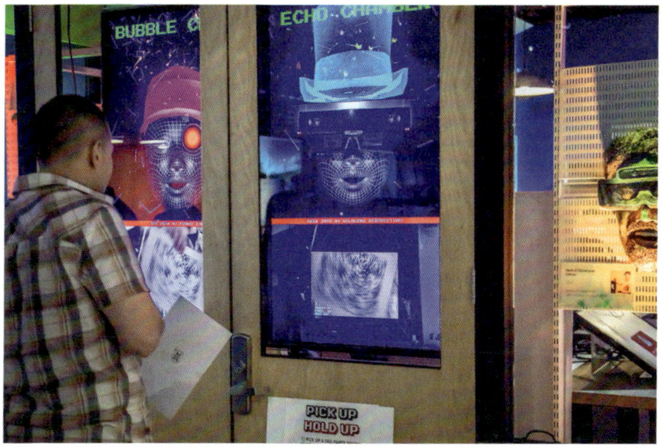

Nick Leoni

Yosun Chang (US)

▶ Bubble Chaos Echo Chamber

Bubble Chaos Echo Chamber is the experiential embodiment of the visitor-viewer peering into the viewpoint of someone ("It") stuck in an echo chamber bounded by a filtered reality.

"It" sees the world through a head-mounted display. But, we see it is actually a digitally processed version of the world—call it reality-replacement lenses or augmented reality (AR) taken to extremes. "It" is at the mercy of the system, and we see in this case, its face detector has been hacked to identify people incorrectly. As a pun on "filter bubble" and AR "filters" made famous by Snapchat, the visitor-viewer is invited to pick up a "bubble marker" to hold in front of "It". These AR markers can augment and trigger video filters—and can even evoke different facial emotional responses from "It." If the visitor-viewer chooses to hold their "bubble marker" in front of their face, they could mask themselves, and effectively inject their own "filter bubble" as perceived by the system into "Its" echo chamber, continuing the bubble chaos loop.

Mozilla and The Tech Museum of Innovation

Rosenbauer International AG (AT)

CHALLENGE: Emergency Error Battle

The Future of Helping

In a future with a growing world population, with worsening climate change accompanied by a whole series of novel challenges, with so-called smart homes and other new technologies in use in everyday life, certain aspects of emergency service organizations' mission and tasks will change accordingly. Rosenbauer, the world's leading manufacturer of systems for firefighting and disaster protection, is facing these future issues now. The company presents newly developed technologies that will shape the future of firefighting and ultimately make it easier. An example is the concept study of a self-driving fire truck of the future, Concept Fire Truck (CFT), which will be presented. The development engineers focused on how the architecture of the vehicle is determined by future requirements for fire departments.

These developments also take social structures into consideration. What errors could possibly creep into the use of this technology in an emergency situation? Can natural human intuition be replaced by artificial intelligence? How can new technologies deliver support to existing, well-functioning team structures?

What this future will look like is now assuming increasingly clear contours. In cooperation with the Upper Austrian Regional Fire Brigade and g.tec Corporation, Ars Electronica and Rosenbauer invite festivalgoers to experience this future right up close, right now.

The big event highlighting the presentation will be a special CHALLENGE for VOLUNTEER FIRE COMPANIES in Upper Austria. POSTCITY LINZ's bank of spiral packet chutes will serve as the setting for this "Emergency of the Future."

In Cooperation with OÖ Landes-Feuerwehrverband and g.tec

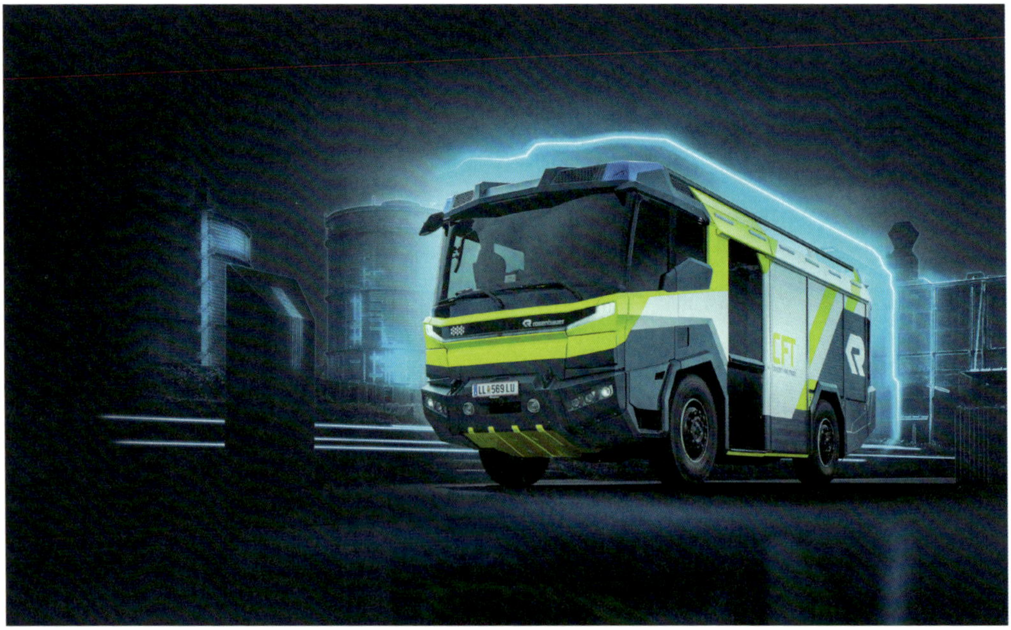

Rosenbauer

Artshare (PT), Ecole Polytechnique Fédérale de Lausanne (CH), Fraunhofer Group Information and Communication Technologies (DE), French Tech Culture (FR), Inova+ (PT), IRCAM (FR), Libelium (ES)

STARTS Residencies

STARTS Residencies form one of the STARTS Pillars. This program is supporting and funding 45 artistic residencies that bring original artistic contributions to technology-based research projects between 2017 and 2020. During each STARTS Residency, a tech project collaborates with an artist, leading to the creation of an original artwork, and to the development of the innovative aspects of the tech research.

A grant (up to 30,000€) is awarded to the artist of each STARTS Residency as a contribution to their involvement in the residencies programme. Additional support to the residency can also be brought by a producer.

With the support of the European Commission's DG Connect in the framework of STARTS initiative (Science, Technology and the ARTS).

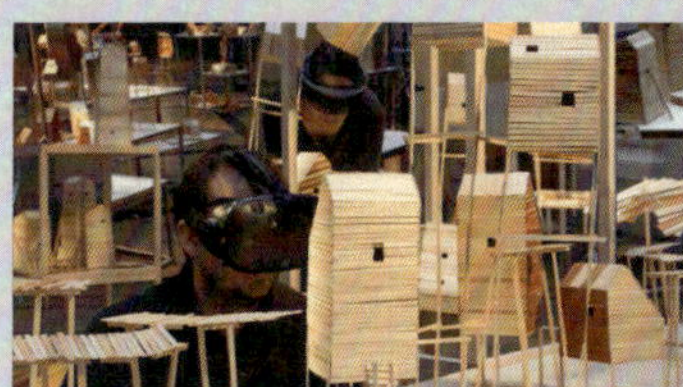

Atlas

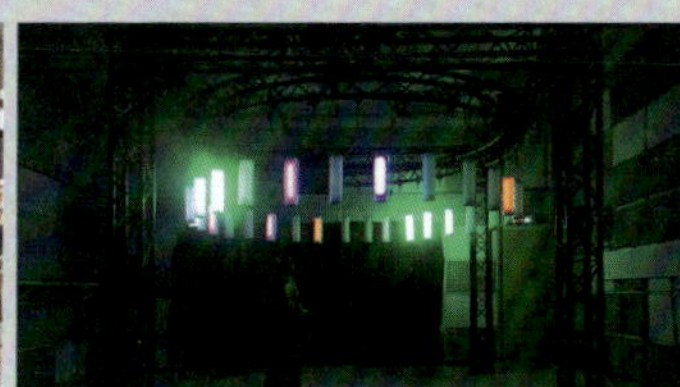

SMART>SOS

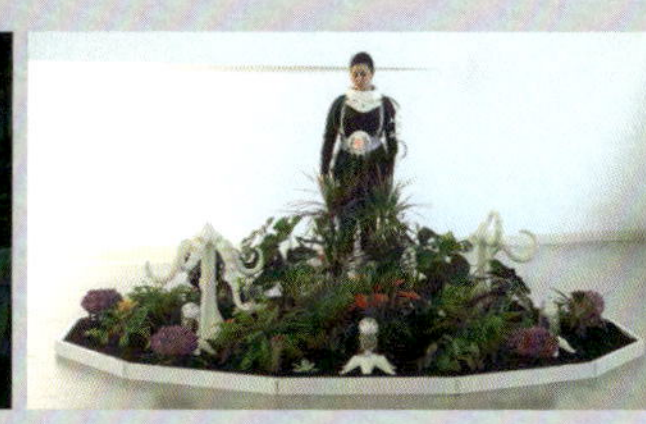

The Plants Sense

Yann Deval (FR), Marie G. Losseau (BE)

Atlas

Atlas is a work between digital arts and visual arts, in the form of an interactive and scenographic exhibition (a mix between real models and a VR/AR experience). After being plunged in an archipelago of poetical islands, spectators are invited to build virtual cities using a "seed launcher". Each launch causes a house to grow. The growing houses follow some urbanistic rules, adapting to their environment. There are cities in the clouds, uprooted cities, cities on stilts, flying cities...
Spectators create empty cities, without inhabitants, given free rein to imagine what happens inside these houses. The cities take on a life of their own, with or without the interactions of users, just like living organisms...
The work allows you to create huge cities, in which you can wander and lose yourself. It provokes a reflection on urbanism, architecture, and their influence on our lifestyles. It gives life to inanimate things...

An experience by Yann Deval & Marie G. Losseau
With the support of VERTIGO STARTS program of the European Commission with IRCAM-Centre Pompidou and EPFL Museum Lab, Fédération Wallonie-Bruxelles, Cocof, Maison des Cultures et de la Cohésion Sociale de Molenbeek Saint-Jean, Fablab'ke, Wallonie-Bruxelles International

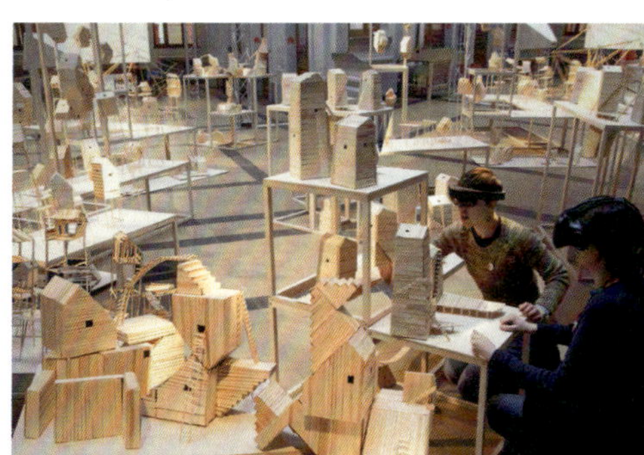

Fashion

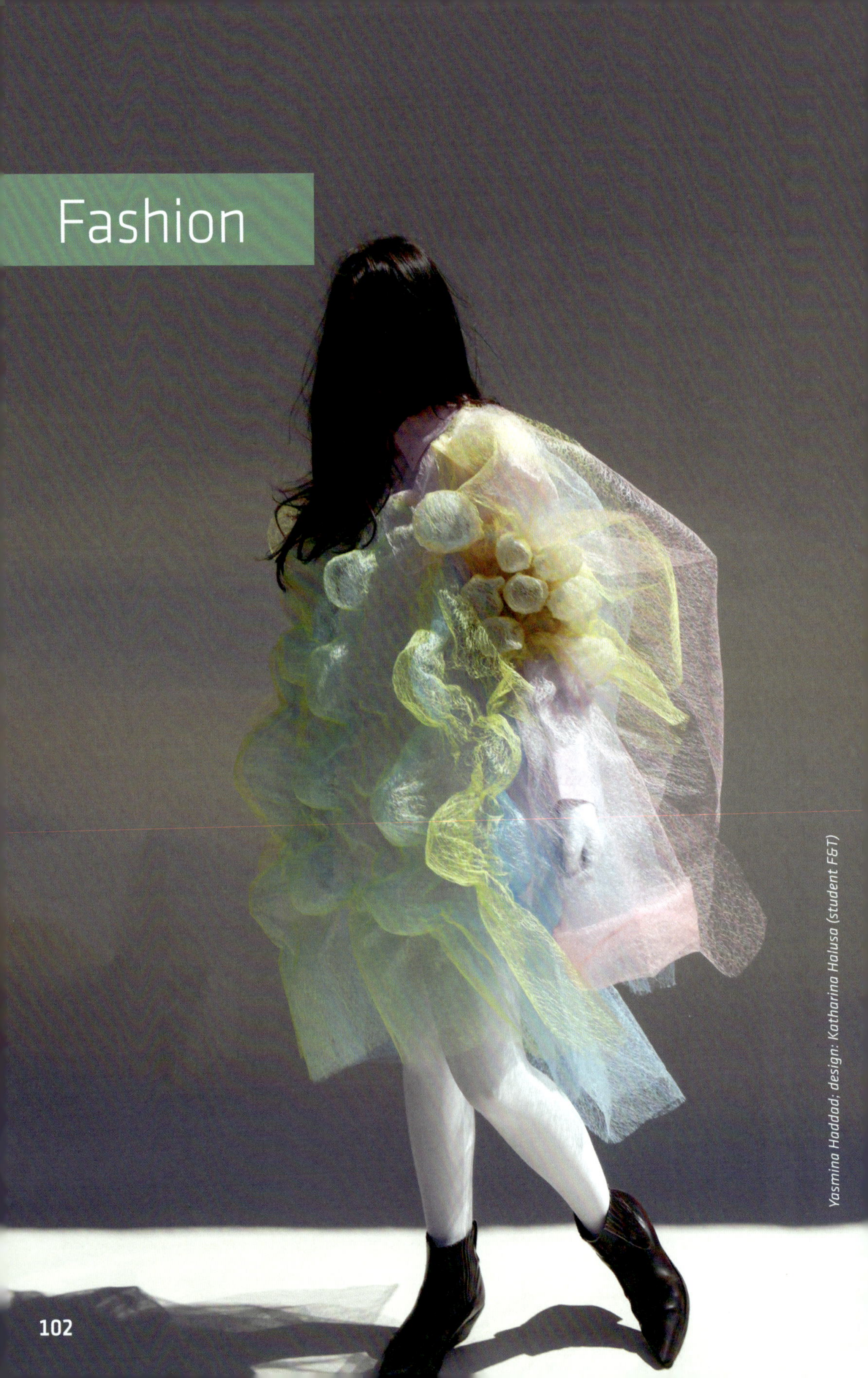

Yasmina Haddad; design: Katharina Halusa (student F&T)

Fashion & Technology (FAT)

University of Art and Design Linz

Curators: Veronika Krenn, Christiane Luible, Ute Ploier, Nina Wenhart

The Fashion & Technology program at Linz Art University was initiated about three years ago. It commenced as a bachelor's degree program in autumn 2015; an internationally oriented master's program will be offered beginning in March 2019. F&T is presenting a small selection of the first bachelor's degree projects at the 2018 Ars Electronica Festival.

The interplay of fashion and technology is nothing new per se. One such example was of fundamental importance for the invention of the computer—the Jacquard loom. Its principle of perforated cards was the basis of the programming of Charles Babbage's unfinished analytical engine. This means of storing and retrieving data was used until the 1960s. Today, the pairing of fashion and technology is most often associated with smart textiles and soft circuits, wearables and 3-D printers.

But Fashion & Technology is much more than this. The program is governed by an integrative philosophy. Technology and fashion don't just encounter one another; they blend. This isn't simply the consolidation of two separate spheres; their merger brings forth something totally new.

The Fashion & Technology program aims to impart a critical way of dealing with the social relevance of fashion as an instrument of differentiation, as an emancipatory means of reflection on the conditions of design and production as well as on the relationship between the body and its media-pervaded environment. This means not only deploying technology but also critically reflecting on technology by means of fashion. Fashion & Technology sees fashion as a complex fabric in which many different threads of discourse run together and materialize.

From its very inception, Fashion & Technology has experimented with alternative show formats. In 2016, the Ars Electronica Center's Deep Space 8K hosted Anatomies of Fashion in which state-of-the-art CT software delivered a walk-through 3-D experience. In 2017, students' designs were shown in stereoscopic selfie videos featuring industrial robots instead of human models. This year, a round-table discussion and keynotes are launching THE POLITICS OF FASHION: FASHION AS SOCIAL BOT, a new platform event designed to rethink and aesthetically form the phenomenon of fashion at the nexus of art and technology.

Mirela Ionica (RO)

▸ The Weakest Force

The project *The Weakest Force* explores simple ways of using and simulating the effects of natural forces and phenomena in the design process in order to create structures, textures or prints. Gravity is one of the fundamental forces in the universe which can either be viewed as a strong, binding force on a large scale or as a weak force at particle level. The project starts with simple silhouettes and shapes, which are then distorted in an analog and digital way. Distortions are caused by various gravity forces: the stronger the gravity, the stronger the distortion.

Sara Kickmayr (AT)

▸ Visual Invisibility

Inspired by biological phenomena, this project deals with molecular technology and its aesthetic. In nature, the beauty of lucent objects lies within the microscopic structure. It is possible to imitate and recreate these effects with additive nanoimprint technology. Various textiles have been treated applying this technique. Results are materials with a constant moving visual aesthetic that plays with light-reflecting effects.

VISUAL INVISIBILITY has been developed in cooperation with Profactor GmbH.

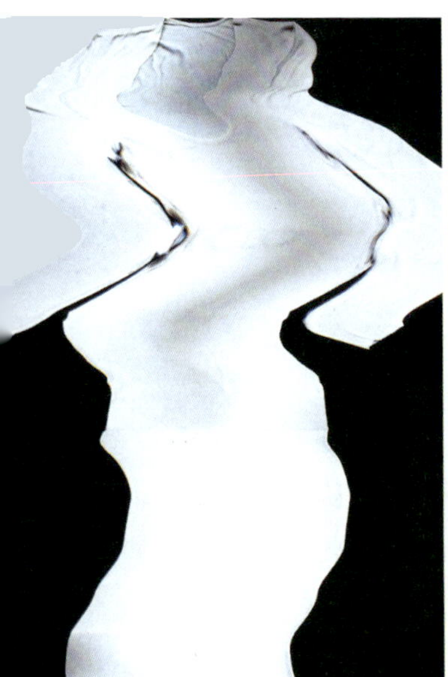

The Weakest Force

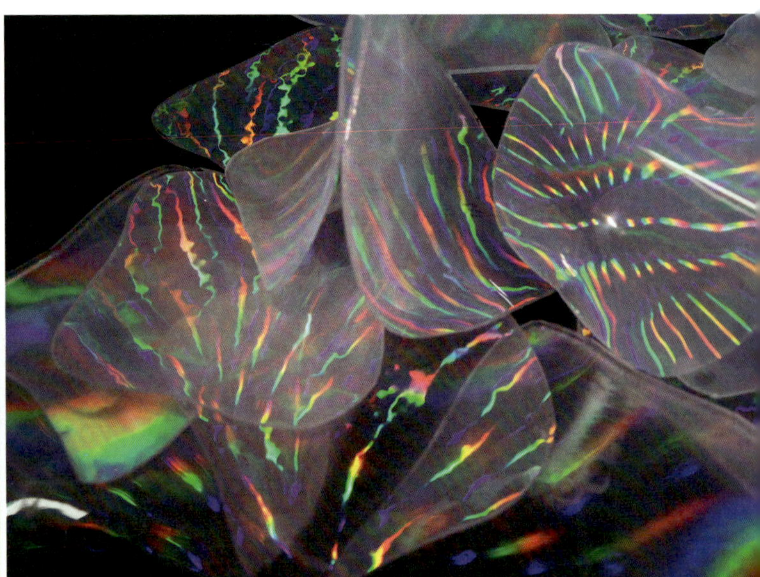

Visual Invisibility

Nina Krainer (AT)

Sculpting Identity

Identity can be explored, questioned and translated into an aesthetic language or form. For this project, an individual wardrobe serves as a "database." Each piece of clothing is disrobed, photographed and analogously converted into an icon that constitutes the starting point of a new pattern-development process. At a first glimpse the resulting sculptural objects look like clean, digitally developed objects. The fact that everything is developed within an analogue process values craftsmanship and traditional production methods in a digitalized world. Furthermore, external and internal data sharing and profiling on the Internet are critically examined.

Michael Wieser (AT), Simon Hochleitner (AT), Viktor Weichselbaumer (AT)

Generative Rendered Design

Generative Rendered Design describes an open process, which aims to come up with a new method to generate and present fashion, or garments in general. 3-D data is the basis for a digital process chain, which rethinks the relationship between body and garment.

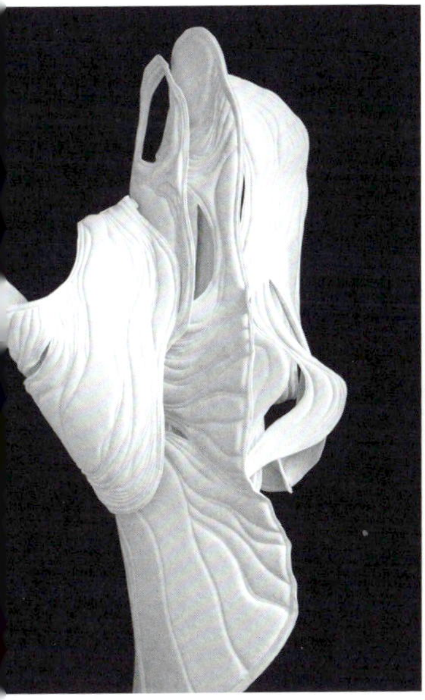

Sculpting Identity

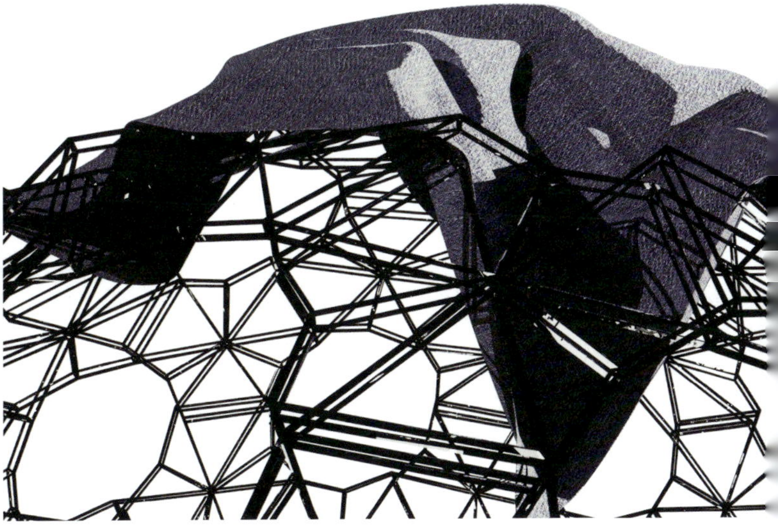

Generative Rendered Design

Yasuaki Kakehi Laboratory (JP), HOSOO (JP), YCAM (JP)

Heteroweave

The work aims to create new functional woven fabrics that can undergo dynamic changes in their characteristics by combining the structure of Japanese traditional Nishijin weave with new materials, which are not commonly used in that way. *Heteroweave 001*: We wove foils coated with an ink that changes color when it reaches a certain temperature and produced a dynamic color-changing woven cloth. We explore the possibility of developing it as a cloth with a rewritable pattern or as a cloth that responds to the surrounding environment. *Heteroweave 002*: The whole cloth works as a computer interface, which visualizes the cloth structure on the screen. We created a woven cloth that has foils on which position and ID data are printed as fine dots. By sliding the cloth on a table, the data can be acquired in real time by a sensor embedded in the table. *Heteroweave 003*: Synthetic seam leather, which is a highly absorbent material, was cut and woven into filaments like a foil. This material becomes soft when it absorbs water and hardens when dried. Utilizing the hardness during drying, we explore the possibility of application to three-dimensional structures that can be "reshaped."

Co-researched & developed with YCAM, HOSOO, and Yasuaki Kakehi Laboratory
Project Members:
Direction: Yasuaki Kakehi and Masataka Hosoo
Material Research & Design: Yumi Nishihara, Satoshi Nakamaru, Juri Fujii, and Shingo Maeda
Interface Design & Programming: Shohei Takei and Hiroki Kaji
Supported by JST ERATO, Japan.

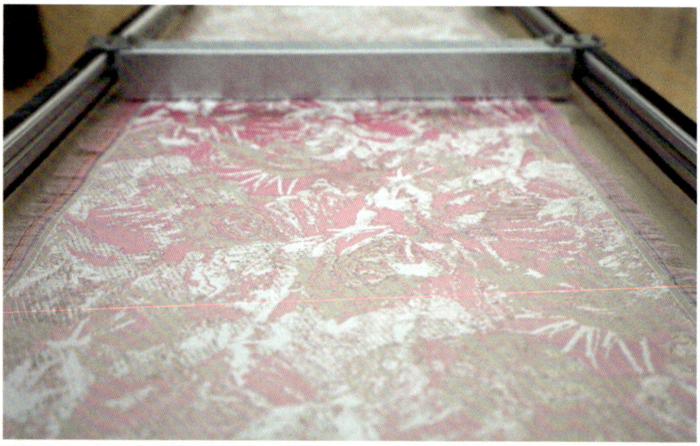

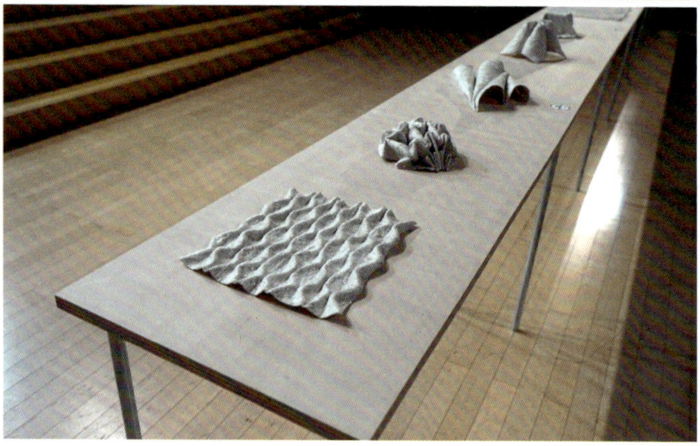

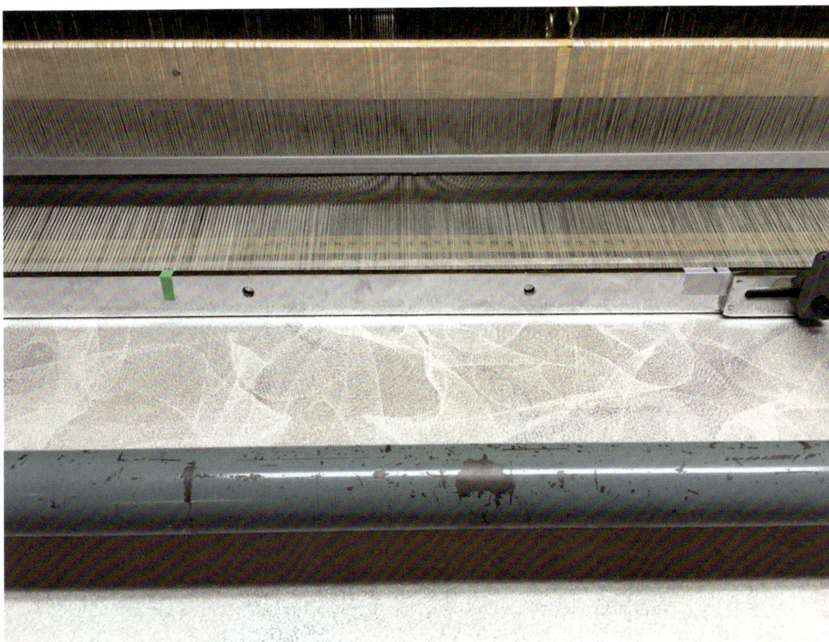

Ken Furudate (JP)

Complex Order

Complex Order is a research and development project between the artist/programmer Ken Furudate and the company HOSOO which is a Kyoto-based textile company founded in 1688 specializing in superior quality fabrics for high-end interior design and fashion. In this project, Ken Furudate has made the software to create weaving patterns by generative algorithms for jacquard loom, specially focusing on the structures of the fabric. He uses the computer program to reconstruct fabrics from their most primitive element and realizes a complex construction which is difficult with human handicraft.

Co-researched & developed with YCAM, HOSOO Co.,Ltd. and Ken Furudate

Bart Hess (NL), in collaboration with Maria Dada (UK) Marco Coluccia (IT)

Life Instinct

PVC worms, linked together with hairy cords, squirm subtly. The movement intensifies when one comes closer. Awkward feelings arise when you see this work by Bart Hess. It is an intense physical experience that comes with a little discomfort, but at the same time fascinates immensely.

This three-part installation from artist Bart Hess, in collaboration with Maria Dada from the Digital Anthropology Lab at the London College of Fashion, provokes and humanizes our idea of programmable materials. Bart Hess applies his exploratory approach to materials, creating interactive silicon (with nitinol), animating robotic worms and giving personality and character to magnets, blurring the boundaries between the physical and the emotional. The installation space is filled with dark, spectral soundscapes created in collaboration with Treale aka Marco Coluccia, imagined to draw the audience deeper into *Life Instinct*.

Created by Bart Hess
www.barthess.nl
In collaboration with Maria Dada
www.digital-anthropology-lab.com
www.mariadada.com
Sound design: Treale
www.marcocoluccia.com
Additional support: Wojtek Tusz
www.studio-wt.com

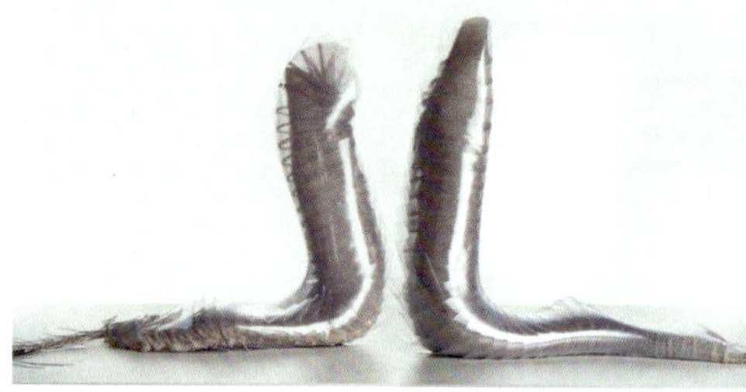

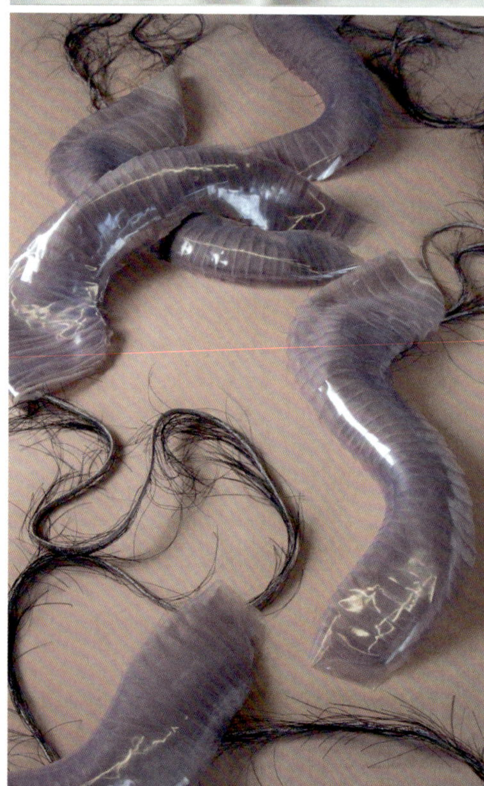

Bart Hess Studio, info@barthess.nl

Kasia Molga (UK)

Human Sensor LDN

Pollution, lack of greenery, and congestion are key environmental factors that impact negatively on the day-to-day existence of city dwellers and their long-term health. A recent study by King's College London revealed that nearly 9,500 people die a year as a result of air pollution in London alone. *Human Sensor* is a large-scale digital and performance artwork from award-winning artist and creative technologist Kasia Molga. This exhibit demonstrates the creative technology used in the recently premiered *Human Sensor LDN* project exploring London's inhabitants' daily exposure to air pollution through a combination of cutting-edge wearable technology and immersive theater. Wearables created by Kasia Molga and her team (Ricardo O'Nascimento and Erik Overmeire). Performance directed by Kasia Molga in collaboration with Ruth Jones and air pollution scientist Professor Frank Kelly. *Human Sensor LDN* was commissioned by Invisible Dust and Euston Town BID (Business Improvement District).

Human Sensor LDN was commissioned and produced by Invisible Dust in partnership with Euston Town BID as part of "Under Her Eye"
Invisible Dust, Euston BID (Business Improvement District)

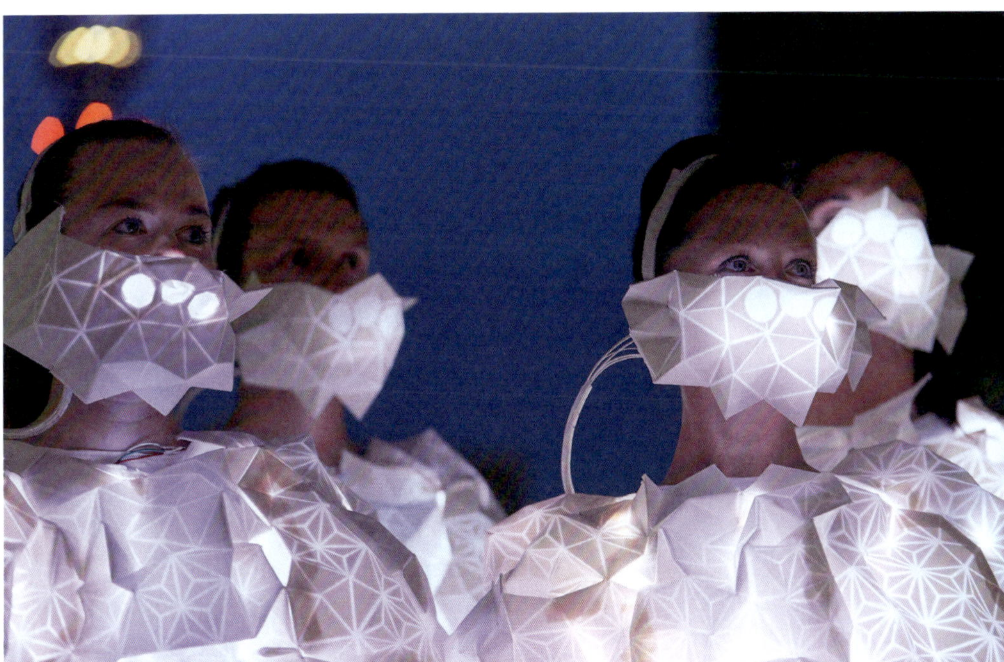

Angela Dennis

Wearable technologists Engage with Artists for Responsible innovation (WEAR)

WEAR Sustain

The STARTS Prototyping project WEAR Sustain is a two-year and three-million-Euro project funded by the European Commission Horizon 2020 research and innovation initiative.

It aims to engage artists and designers to work with technology and engineering experts, to shift the development of the wearables and smart/e-textile landscape and industry towards a more sustainable and ethical approach to development, from design to end-of-life and waste management. 2.4 million euros have been awarded to 46 teams for 6+ months, each with €50,000 to support further development of each prototype, with input from mentors, experts and hubs across Europe. The project runs between January 2017 and December 2018.

WEAR's aims are to:

- Develop a sustainable European network of stakeholders and hubs, to connect and push the boundaries in the design and development of ethical and sustainable wearables and smart/e-textiles;

- Encourage cross-border and cross-sector collaboration between artists and technologists to design and develop a new generation of ethical and sustainable wearable technologies and smart/e-textiles;

- Develop a framework within which future prototypes can be developed to become the next generation of what ethical and aesthetic wearables should be;

- Lead the emergence of innovative approaches to design, production, manufacturing and business models for wearable technologies and smart/e-textiles;

- Make citizens, industry, and other stakeholders more aware of and responsible for ethical, sustainable and aesthetic issues in the design, development and use of wearable technologies and smart/e-textiles.

https://wearsustain.eu
https://network.wearsustain.eu/dashboards/home
The project is managed by a consortium of 7 organizations across 5 EU countries:
IMEC (Brussels, BE)
University for the Creative Arts (Epsom, UK)
Queen Mary University of London (UK)
Berlin University of the Arts, UdK (Berlin, DE)
Blumine (Milan, IT)
Digital Spaces Living Lab (Sofia, BG)
DataScouts (Brussels, BE)

Lara Grant (US), Anna Blumenkranz (DE/UA), Adrian Freed (UK)

FlexAbility

FlexAbility is a modular kit prototype created from ethically and sustainability vetted components facilitating the creation of made-to-measure wearables for people with physical disabilities. This open-source kit allows us to build customized wearable solutions for accessing computer and other devices, and will be made available to a network of designers, fabricators and community centers through the web portal *flexability.org*, along with informational and support services. This project is rooted between San Francisco, US, and Munich, Germany, where we are collaborating with consultants at the community centers Ability Now Centre and Stiftung Pfennigparade to conduct living lab research.

Supported by WEAR Sustain–EU Horizon 2020, and Hans-Sauer-Stiftung.
Cooperation partners: Pfennigparade München, Ability Now Centre Bay Area, Maker Space UnternehmerTUM

by-wire.net (NL) – Marina Toeters (NL), Holst Center (NL) – Margreet de Kok (NL), Katoenenzo (NL) – Melissa Bonvie (NL)

Closed Loop Smart Athleisure Fashion

Intelligent Fashion, containing technology that can measure your health, and is still sustainable. My shirt keeps track of my heartbeat and respiration. So I can be more effective. I know there is a lot of technology in there, but I don't feel it. The sensors are seamlessly laminated onto the fabric. It's very comfortable. It's beneficial to wear these clothes on a daily basis, because they check my well-being and stress level. The lease and recycle system is a closed loop. I don't have to worry about what to do when the garment has no use for me anymore. The technology will be de-laminated from the shirt, and recycled. It sounds like a dream of the future but it's right here and we are wearing it. The smart shirts continuously measures the ladies' key vital signals based on Holst Centre's advanced printed sensor technologies on flexible substrates for textile integration. The laminated sensors are truly wearable, comfortable, robust, invisible during use, washable up to 25 cycles, and designed for unobtrusive integration in conventional fashion production.

We would like to acknowledge all project partners and contributors listed here: *http://www.by-wire.net/clsaf/*. The WEAR Sustain project has received funding from the European Union's Horizon 2020 research and innovation program under grant agreement no. 732098. This publication reflects only the author's views. The sole responsibility of this publication lies with the WEAR Sustain Consortium. The European Commission is not responsible for any use that may be made of the information contained therein.

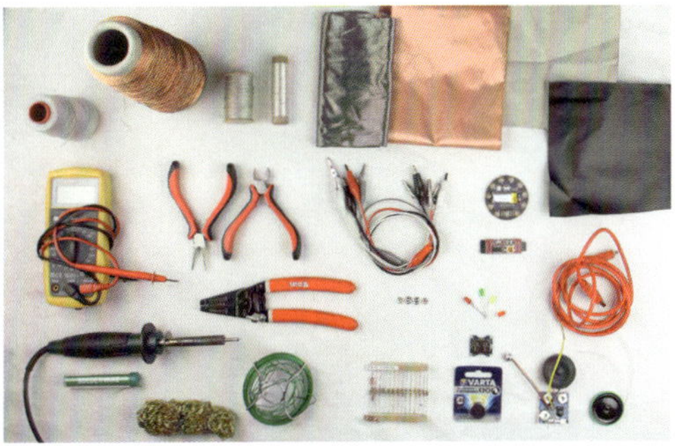

FlexAbility

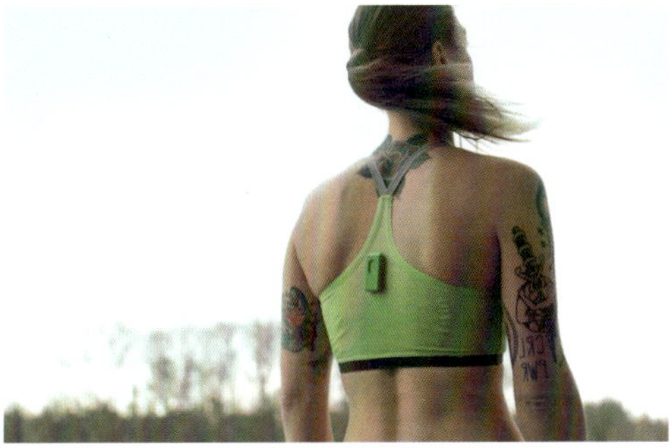

Closed Loop Smart Athleisure Fashion

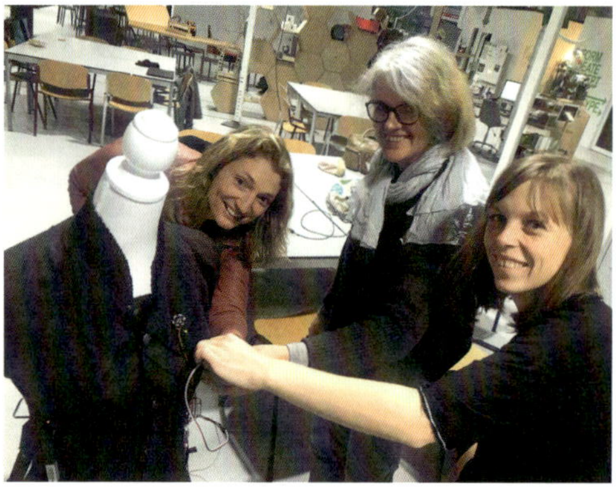

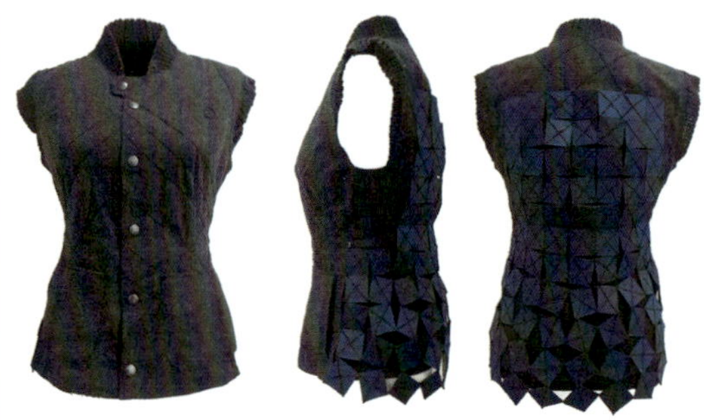

Hellen van Rees (NL), University of Twente (NL) – Angelika Mader (DE), Geke Ludden (NL)

Textile Reflexes

This project aims to further develop the shape-changing textile developed by Hellen van Rees. The flexible textile is made of squares of post-consumer textile waste.

Through a unique approach, the different squares can move and respond to each other, creating a flexible, expanding & contracting surface. Through this effect a dynamic and playful texture emerges that easily adapts, leaving continuously changing open gaps between the squares. Shape-changing sustainable textiles give feedback to users and can be applied in coaching situations. The first working prototype of this project is a posture correction coach. The vest can be worn over regular garments and it measures when the user has an incorrect posture. It then pulls the flexible panel of squares at the back to subtly remind the user to sit back upright.

Supported by WEAR Sustain, Horizon 2020, Designlab, University of Twente, Hellen van Rees

University of Siena (IT) – Patrizia Marti (IT), Matteo Sirizzotti (IT), Pietro Rustici (IT), Simone Guercio (IT), Glitch Factory s.r.l. (IT) – Michele Tittarelli (IT), Iolanda Iacono (IT), T4All (IT) – Gianluca Daino (IT), Riccardo Zambon (IT)

Quietude

Quietude is a collection of interactive jewels that enhance the experience of sound for deaf women. They detect sounds and translate them into vibrations, light and shape changes. The jewelry collection is completed by an app for smartphone allowing personalization of both input and output, and the construction of a personal library of sounds that can be monitored for and replayed on demand through the jewels. The project has been developed in co-design sessions with deaf women to enable the development of accessories that respond to the emergent discoveries and desires.

www.quietude.it
Quietude has been designed and developed by a team composed by Santa Chiara Lab – University of Siena (coordinator), Glitch Factory and T4All.
Siena Art Institute and University of Southern Denmark collaborated as service providers.
Authors of the project are: Patrizia Marti (University of Siena), Matteo Sirizzotti (University of Siena), Simone Guercio (University of Siena), Pietro Rustici (University of Siena), Michele Tittarelli (Glitch Factory s.r.l.,*www.glitchfactory.it*), Iolanda Iacono (Glitch Factory s.r.l., *www.glitchfactory.it*), Gianluca Daino (T4All s.r.l.,*www.t4all.it*), Riccardo Zambon (T4All s.r.l.,*www.t4all.it*).
Supported by WEAR Sustain

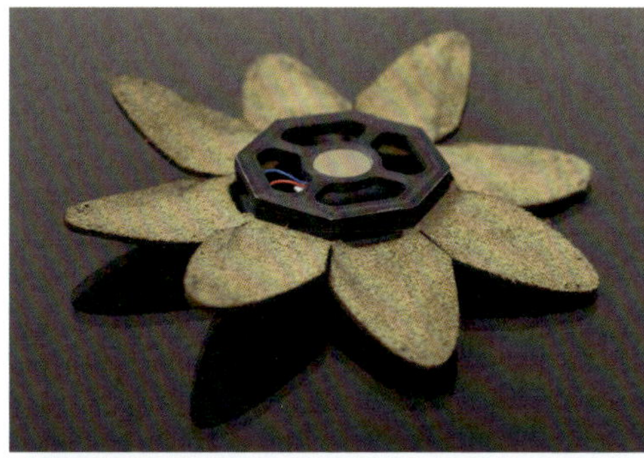

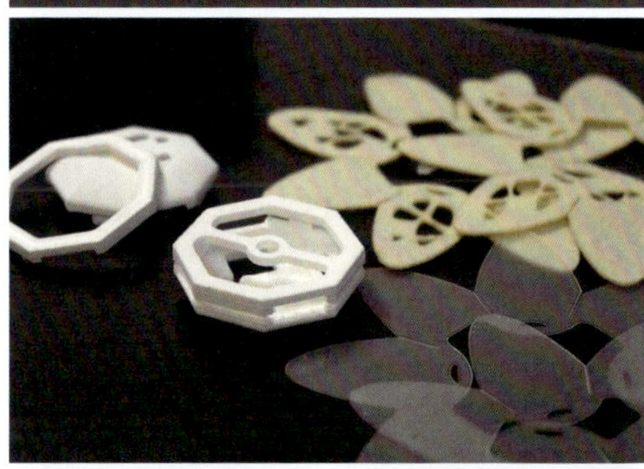

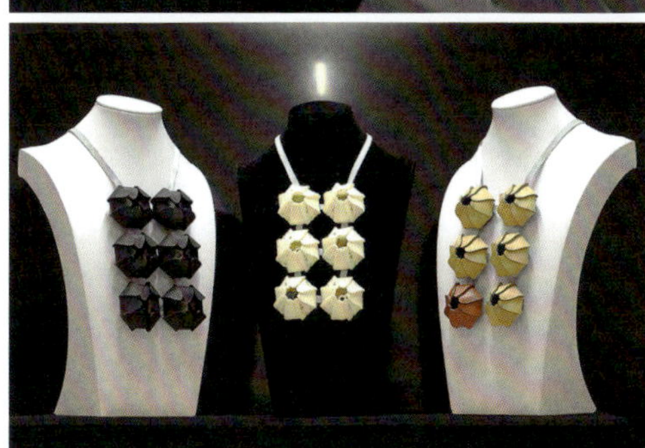

Quietude Core Team

Mogu S.r.l., Maurizio Montalti (IT/NL)

MOGU Leather

MOGU Leather is a revolutionary line, comprising a range of materials and products exclusively consisting of fungal mycelium. *MOGU Leather* looks at a future where animals are not killed for human's material desires. As opposed to traditional animal leather, *MOGU Leather* can be grown in rather short time-frames, with a limited amount of resources, and its production process is not wasteful. Some of the outstanding qualities of *MOGU Leather* include: strength, flexibility, breathability, water resistance, compostability, customizability and cost-competitiveness. *MOGU Leather* is to be looked at as an alternative to traditional animal leather, though different as grown by fungi and by means of mycelium technology; a novel natural material with comparable experiential/tactile qualities to those of animal leather, promising to bring about a great material revolution, leading towards the establishment of a more ethically and environmentally responsible fashion industry.

Mogu S.r.l.
Designer / Researcher / R&D Director: Maurizio Montalti
CEO: Stefano Babbini
in close collaboration with Utrecht University / Microbiology Department (Prof. Han Wösten)
supported by WEAR Sustain

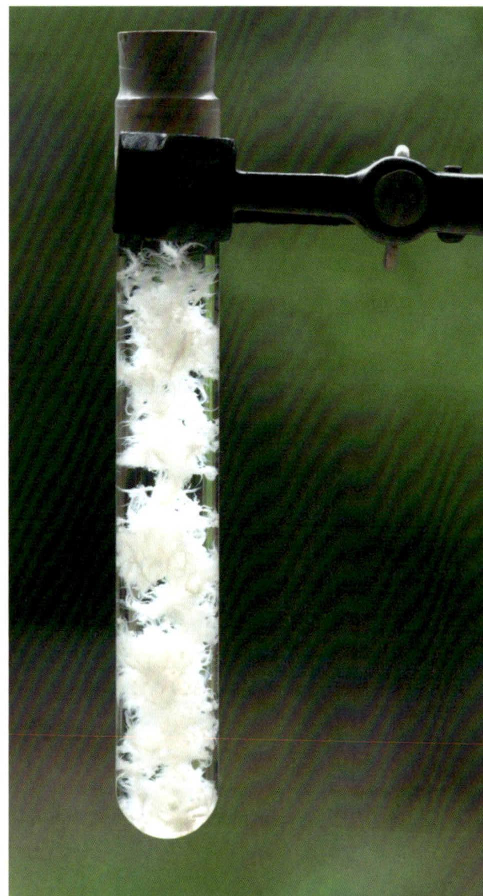

datable (US) – Elizabeth Esther Bigger (US), Luis Edgardo Fraguada (PR)

GenCloth

GenCloth is a 3-D toolkit and data set archive for generative garment design. The ambition of the project is to empower local fashion designers with 3-D software toolkits for rapid client customization in order to grow local talent, local economies, and local demand for bespoke clothing. This project began as an idea in 2013 after 3-D-scanning the bodies of 150 people in Barcelona and discovering the wonderful and rich diversity in each of those bodies. The challenge was to design a simple software system that would encompass the process to create a bespoke garment from the initial client measurements or body scan to the final flat garment pattern pieces. After utilizing this process in practice and with real clients for several, the project won a sub grant from the WEAR Sustain EU H2020 funded project for further development. The main objective of the project development during the grant is to lower the barrier to entry by creating a simple toolkit, create a community around the tools, and create open data sets to enrich the process of generative garment design.

The GenCloth Project has received funding from the European Union's Horizon 2020 research and innovation program under WEAR grant agreement no. 732098 *https://wearsustain.eu/*

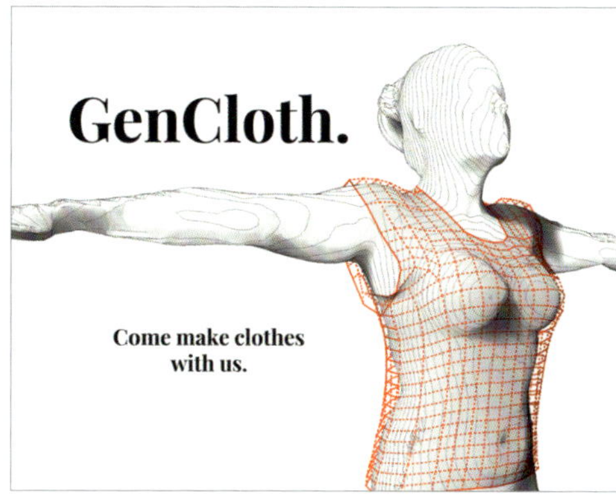

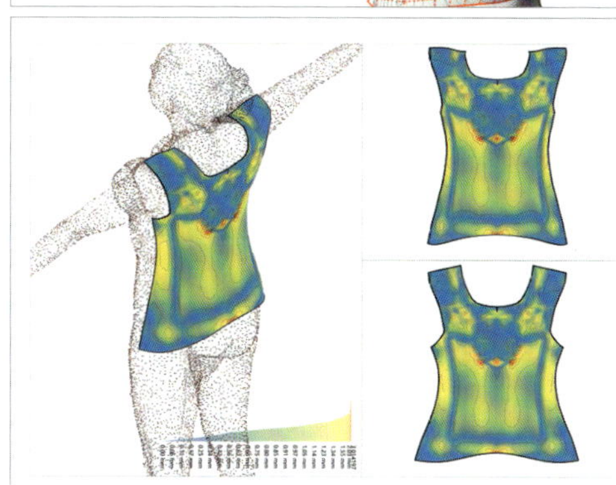

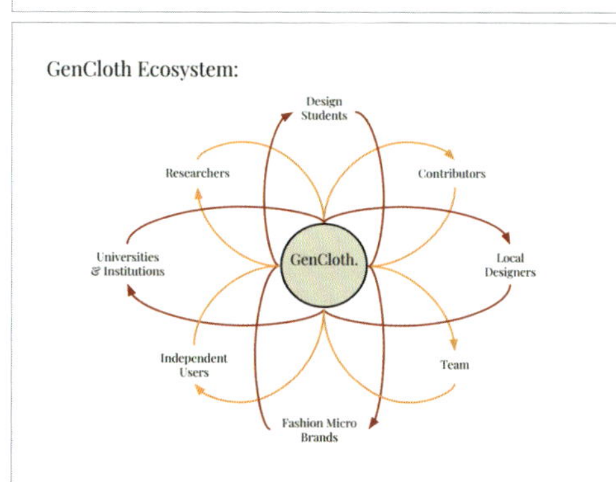

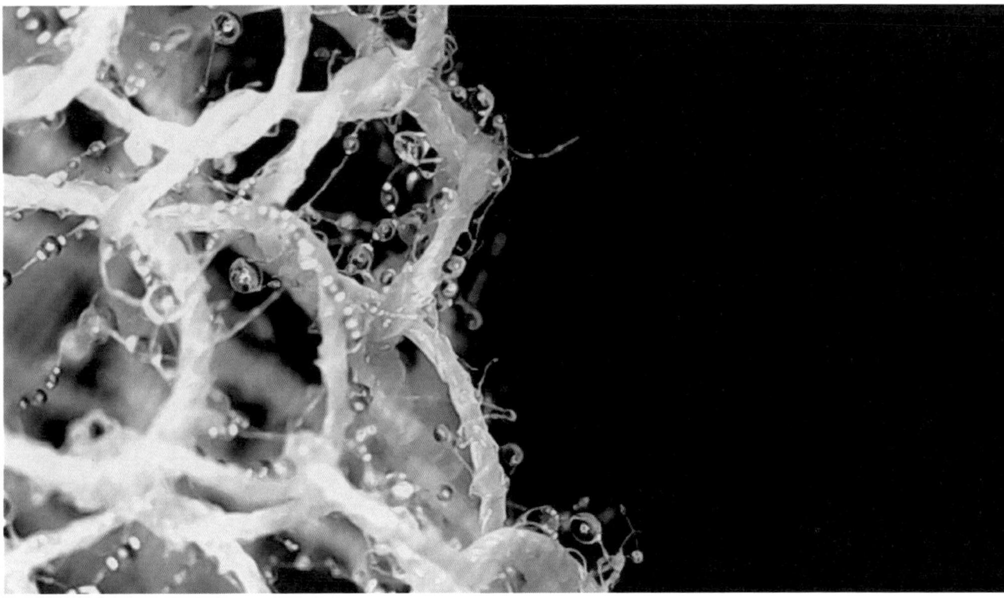

Francesca Perona (IT/UK), Jane Scott (UK), Tiffany Wood (UK)

Biocoatile

What if we could harness natural properties of biological materials to enhance the performance of textiles used in our daily life?

Biocoatile brings together soft matter physics, biotechnology and programmable knitting to envision sustainable, chemical-free, smarter textiles.

Chemical finishes are commonly used by the textile industry to provide comfort and adaptation to varied environmental conditions. However, these treatments contain toxic chemicals and nano-particles that disperse in water streams and bioaccumulate in our bodies. A team of scientists and designers is working with a newly discovered functional protein produced by harmless bacteria, to create more sustainable alternatives in line with the growing bioeconomy. They are testing how properties of common fibers are enhanced by protein-based coatings. Using these materials together with biomimicry-inspired knitting construction, they are creating adaptive textiles that respond to environmental humidity and skin moisture.

This project has received funding from the European Union's Horizon 2020 research and innovation program under WEAR Sustain grant agreement no. 732098.

Ilian Milinov (BG), Evgeniya Tsankova (BG), Yana Tankovska (BG), Georgi Chervendinev (BG), Petar Zhivkov (BG)

Airlief

Airlief is an effective and innovative air-pollution mask. It is comfortable and looks more like a fashion accessory than a complicated air-filtering device. The 3-D-knitting technology for the front layer also allows artists and designers to customize the mask and to apply different patterns as desired by the user. The filters are 99 % effective against PM2.5 (biggest killer in cities) and change based on how often the mask is worn and how polluted the environment is. That's why *Airlief* is developing a smart device and a mobile app to track when you wear the mask and how much air pollution there is. It will also be able to notify you when to change the filter and when the outdoor air is highly polluted. The mask has a smart ventilator that improves its comfort. This is a big advantage because harder breathing, higher humidity, and fogginess are the biggest problems for all masks.

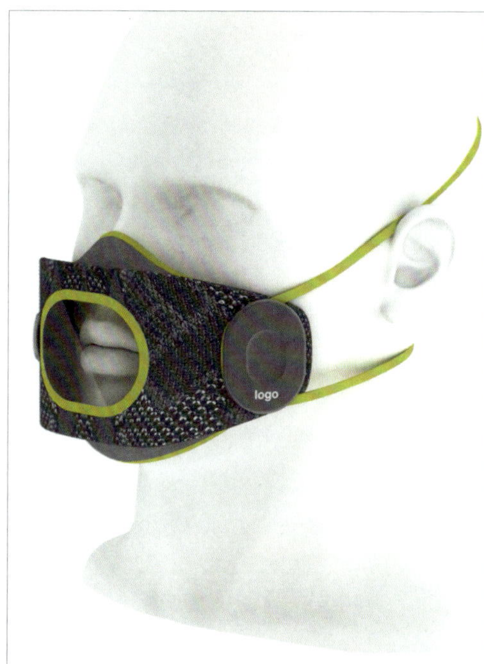

Supported by wearsustain.eu
It is a project that aims to engage art, design and creative industries to work closer with technology and engineering industries, to shift the development of the wearables and e-textile landscape towards a more sustainable and ethical approach.

Space Art

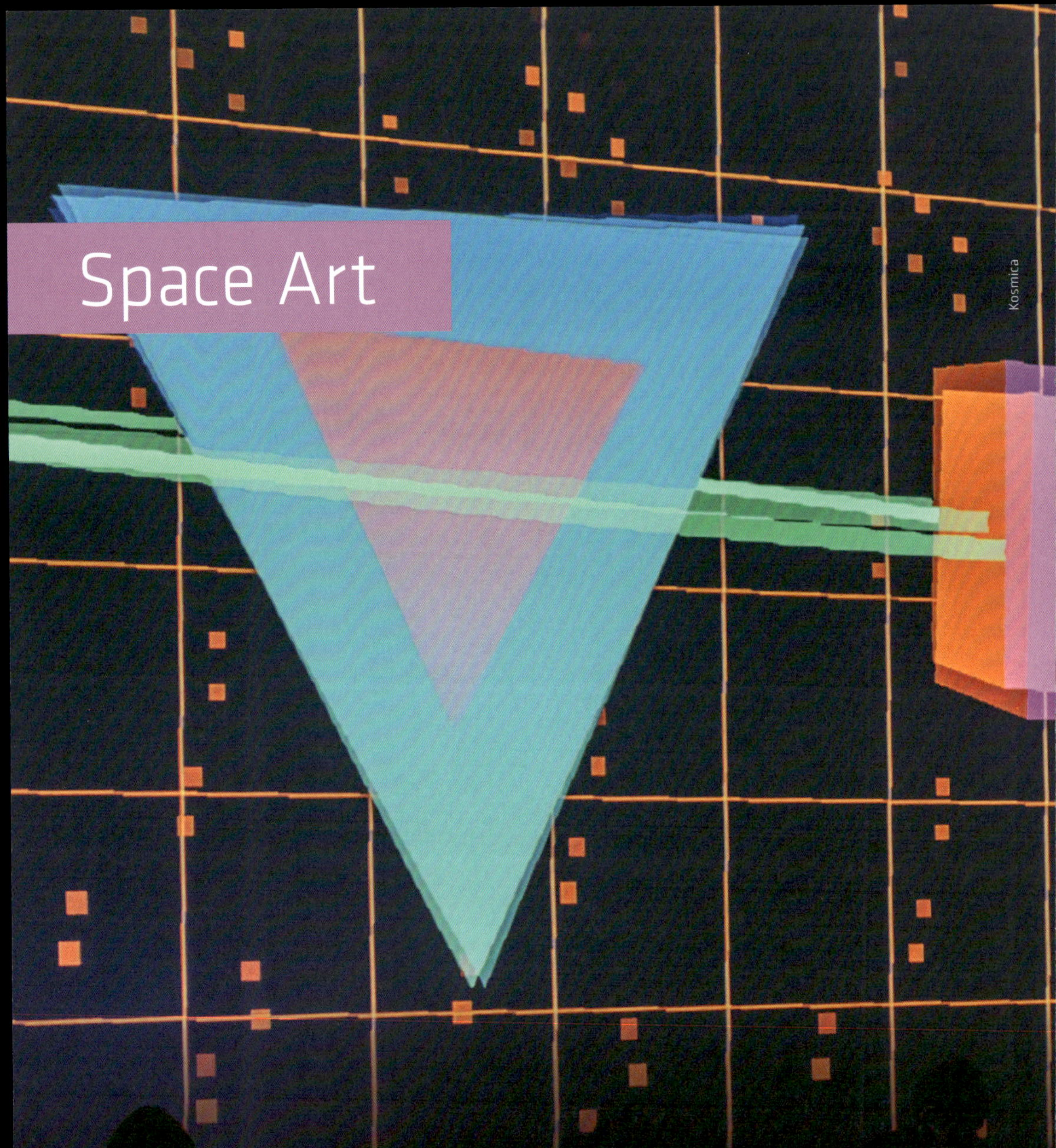

Kosmica

Art might not be the first category that comes to mind when thinking of actual space travel and exploration; the technical aspect – rockets, spacesuits, space platforms – would probably be more dominant. Nevertheless, space travel has always been a topic for artists to unleash their creativity, which has resulted in a multitude of genres. And everybody who has seen a sci-fi movie knows: there is no wiggle room for errors in space. So there is art where the error can be an integral part of creation, and there is space travelling, where a minor error could end deadly – at first glance, it almost seems impossible to combine these two. Yet, exploring is a vital part of both domains: searching for the formerly unknown or unconscious or, at least, looking at the dark corners of our universe. Maybe this is one facet of why space travel and art have crossed paths before and their combination sparks the interest of researchers, scientists and artists: from sending a couple of oil paintings to space in 1986, to Arthur Woods' *Cosmic Dancer*, a sculpture which was sent to Mir station in 1993, to the foundation of *KOSMICA* in 2011, an institution that entirely focuses on Space Art, to the projects exhibited at Ars Electronica Festival 2018.

People are already envisioning space as a travel destination for civilians or even as a human living environment (although this may take a while…). Why shouldn't we start establishing a cultural and artistic discourse about it right now?

MIT Media Lab, Space Exploration Initiative (US)

▸ A glitch in the stars – Space Exploration Initiative Exhibition

In 1990, from 6 million kilometers away, Voyager 1 took a snapshot of our existence in the universe: a pale blue dot. In it, we saw the loneliness and impermanence of our species, a realization that continues to sustain a thriving, resonating call for the future. However, space is not for humans. We are never meant to be there, an error in the wild. The isolation, lack of gravity, radiation and all the risks there can kill us in minutes. What is human experience beyond the earthbound? Here, six projects form the Space Exploration Initiative of MIT Media Lab are asking the same question and bringing possibilities to the toughest, impossible space:
A musical instrument that only plays in zero-gravity, pneumatic surface that morphs to embrace the human body in zero-g, self-assembly infrastructure for the next generation of zero gravity habitats, spider-like performance with the three-dimensional movements of a weightless body, scents that capture the memories of our homeland and a grappler for landing foundational infrastructure on an asteroid.
All the projects were successfully deployed and performed in a zero-gravity parabolic flight last year. They are hopes beyond solutions, imaginations more than facts. Just like generations of observers, they see our future in the stars.

All the projects are supported by the Space Exploration Initiative of MIT Media Lab.

Nicole L'Huillier (CL), Sands Fish (US)

▶ Telemetron

The *Telemetron* is a musical instrument designed explicitly for performance in the zero gravity environment of space. Leveraging gyroscopes and wireless data transmission, the instrument transforms the poetic motion of the internal chimes into musical notes. The performance is a dance between human and non-human bodies and explores a new body language for music.

Engineering Assistance: Thomas Sanchez Lengeling
MIT Media Lab, Opera of the Future

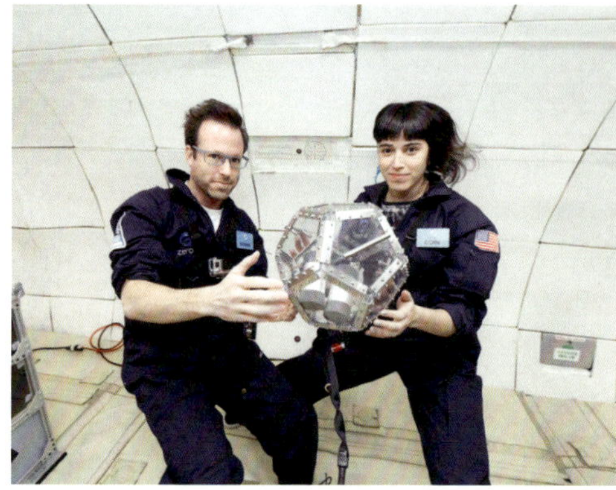

Steve Boxall

Xin Liu (CN)

▶ Orbit Weaver

Is the weightless state a moment of true autonomy or does the ungrounded body simply lose control? Inspired by the three-dimensional mobility of arachnids, *Orbit Weaver* uses a hand-held device to regain control of her body and move freely through weightlessness. The device shoots strings out and rewinds to drag the weightless weaver, transforming her into a "spider woman," weaving her web in space.

Costume design: Andrea Lauer

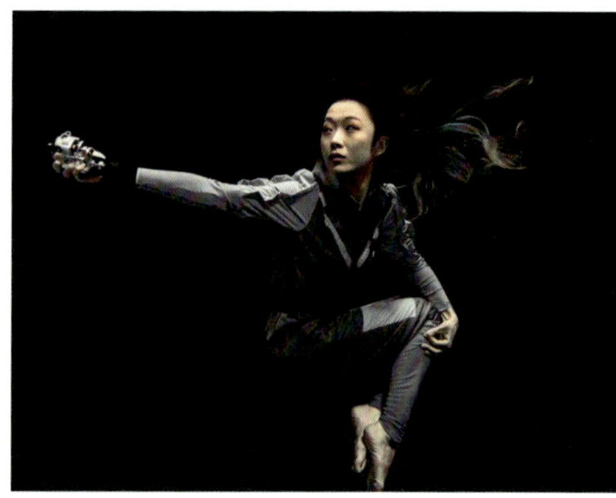

Rob Chron

Ariel Ekblaw (US)

▶ TESSERAE

Self Assembling Space Architecture

TESSERAE demonstrates a self-assembling geodesic dome structure for future space habitats in orbit. The "Tessellated Electromagnetic Space Structures for the Exploration of Reconfigurable, Adaptive Environments" (*TESSERAE*) prototype has been tested in zero gravity, and explores the use of natively embedded sensor networks and magnetic jointing for in-space, floating assembly.

MIT Media Lab, Responsive Environments, PI: Joseph Paradiso

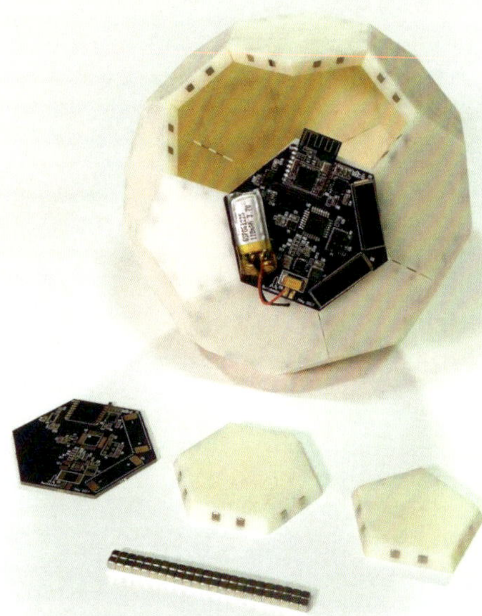

Chrisoula Kapelonis (US), Carson Smuts (ZA/US)

▶ Spatial Flux

Spatial Flux is a seamless pneumatic surface that morphs to embrace the human body in zero gravity. This project redefines our relationships with surface and volume in architecture. The surface is in constant flux—allowing for new temporal possibilities between the body and the architecture.

MIT Media Lab, City Science Group

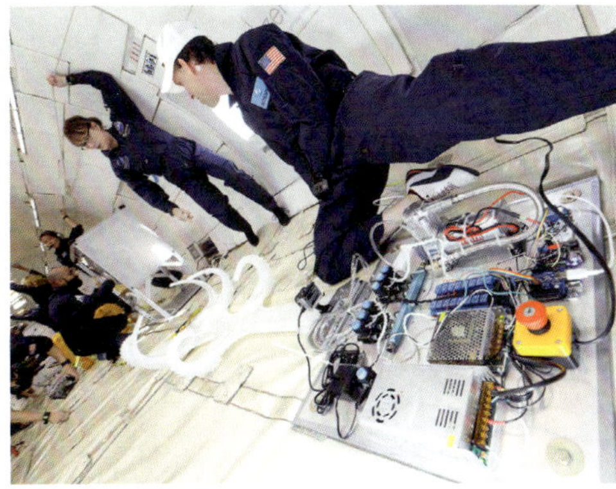

Steve Boxall

Ani Liu (US)

▶ Smells for Space

Olfactory Time Capsule for Earthly Memories

This is a set of olfactive tokens containing precious smells of Earth for future cosmonauts. Speculating on a future where some might embark on a one-way trip into space, this project investigates the sensory modalities of memory beyond the digital. In addition to terabytes of data, what other forms of communication and connection might we invent for an extraterrestrial future?

International Flavors and Fragrances Inc.

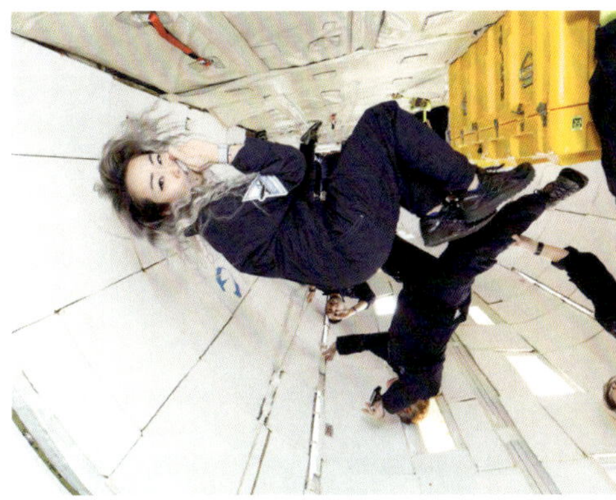

Steve Boxall

Juliana Cherston (US)

▶ Grappler

Arrays of bistable elements for landing distributed sensor networks on low gravity bodies

Can a modified snap bracelet be used to land infrastructure on an asteroid? *Grappler* is part of the mission concept in which a rope or a net is used to grapple onto a low-gravity body of interest.
The net doubles as infrastructure for a network of tiny crawlers that move across the net's surface, primarily for applications in in-situ distributed sensing.

MIT Media Lab, Responsive Environments, PI: Joseph Paradiso

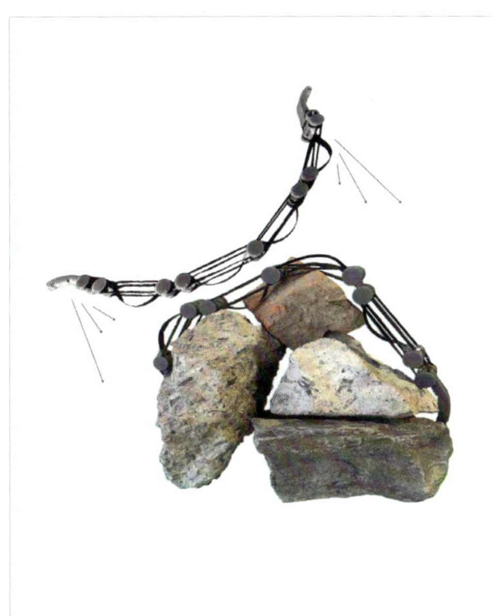

Florian Voggeneder (AT)

The Kepler Station

A space station on Mars, habitat of space sojourners, who, far from their home planet, live isolated from the hostile environment. Extravehicular activities behind the mountain ridges at the horizon. The series *The Kepler Station* depicts scenes of a Mars simulation that serves to gain insights for future flights to the neighbor planet and documents the endeavor of volunteers and researchers to grasp another planet.

Following the original idea of a speculative, alpine space program, this work is the preliminary result of a three-year investigation into astronautic space travel. To be admitted into the one-month AMADEE-18 analogue Mars Mission in Oman, Florian Voggeneder had to complete a multi-month educational and physical program at the Austrian Space Forum. As a field crew member, he participated in research studies and conducted various experiments.

With the support of Austrian Space Forum

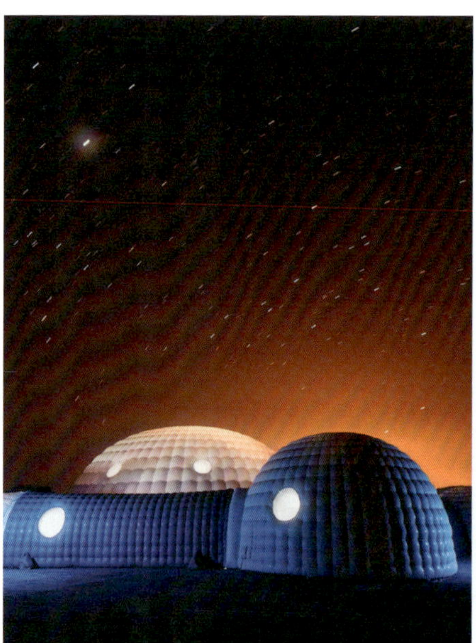

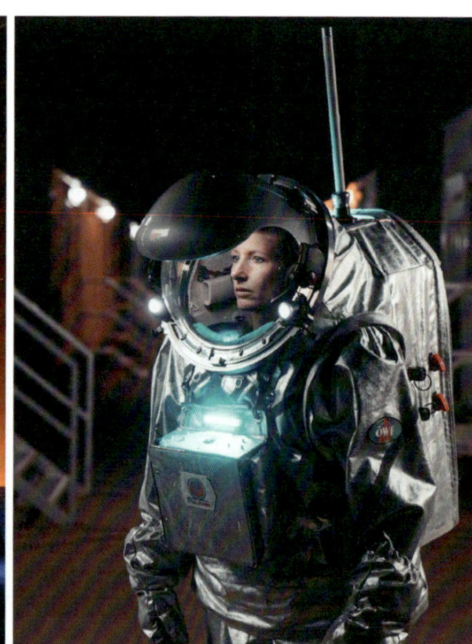

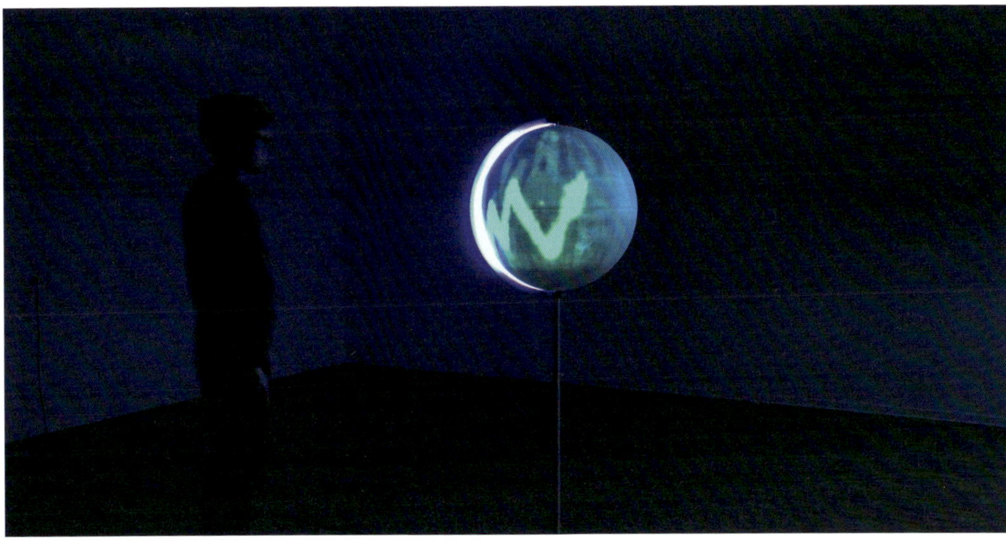

Gregor Göttfert (AT), Florian Kofler (IT)

SPHERE

The installation *SPHERE* is comprised of a 65 cm (25.5 inch) orb. The shimmering sphere seems to hover in the darkness of the room. Optical sensors slowly circle the object, scanning and monitoring the environment like radar and projecting a transient image of time and space onto the phosphorescent surface of the globe. Visitors and spectators become recognizable on the globe and with them the changed dimensionality of *SPHERE*: no longer is it just an image of purely spatial dimensions, but rather the factor of time, which manifests itself temporarily on the surface of the object through optical distortions and compressions. The slowly fading images offer a visual reflection on our very existence at the surface of a sphere.

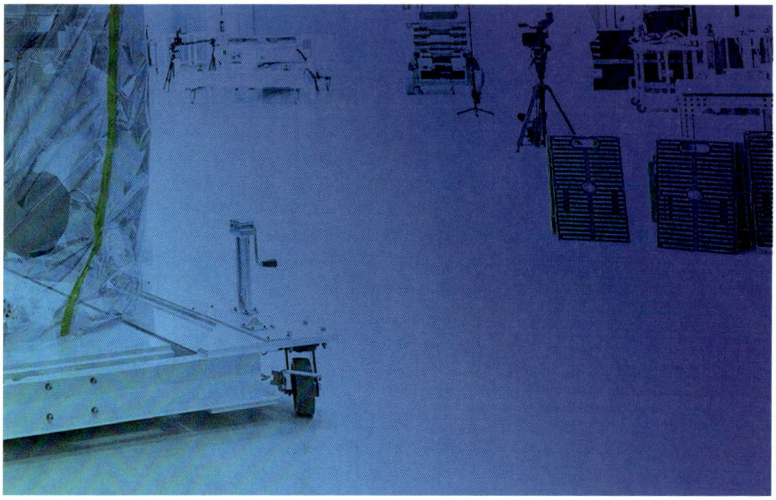

NASA

Miha Turšič (SI/NL)

KOSMICA Parliament

KOSMICA Parliament is a series of performative events featuring the artistic community of Ars Electronica and their views about human activities in outer space. Inside a space capsule, artists will be welcome to give a performative statement about the errors of space exploration. This kaleidoscope of critical views will create the arena to collectively reflect on these issues and inspire new modes of human becoming in outer space. "Failure is not an option" is the famous quote from Apollo 13; however, space activities have always been plagued by errors and mistakes. *KOSMICA Parliament* will invite the Ars Electronica community to actively reflect on these issues in order to envision novel ways of exploring and inhabiting space through emerging understanding of relations between humanity, non-terrestrial environments and technologies.

Artistic director: Miha Turšič (SI, NL)
Production: Kosmica Institute, Waag, KSEVT

Nahum (MX/DE)

The Contour of Presence

On June 29, 2018, a SpaceX Falcon 9 rocket launched from Kennedy Space Center—half artwork and half ethereal presence. *The Contour of Presence* makes use of an object called Pulse that is currently orbiting Earth in an elusive time and place onboard the International Space Station, where it was brought by the Falcon 9 rocket. Through a series of interactive video performances, this ethereal presence will tell a story in real time to explore the meaning of presence and the politics of existence. Audiences will find themselves immersed in an intimate narrative unfolding simultaneously on Earth and somewhere in outer space. The resulting encounter will become the stage to reflect on how existence is not an individual affair and how we are closer to each other beyond distance and borders.

This project is a collaboration between Nahum Studios and the International Space University. The artwork will travel to the International Space Station with the new ICE Cubes Service by Space Applications Services in conjunction with the European Space Agency.

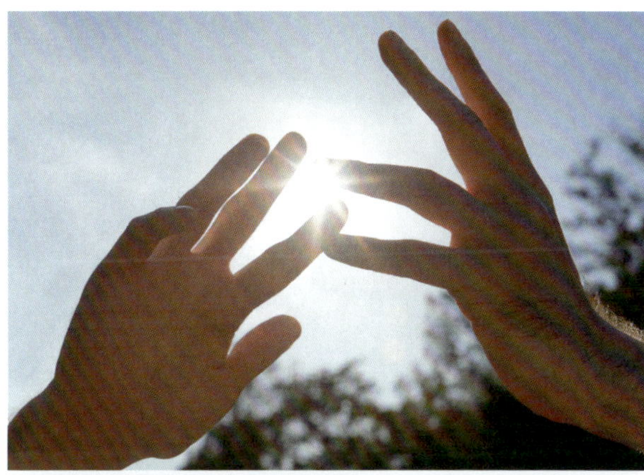

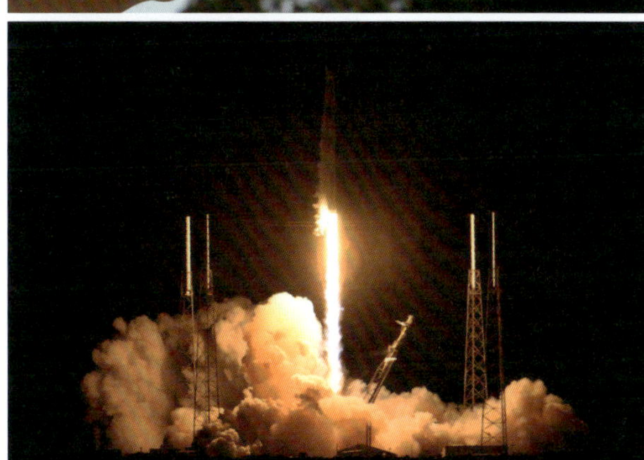

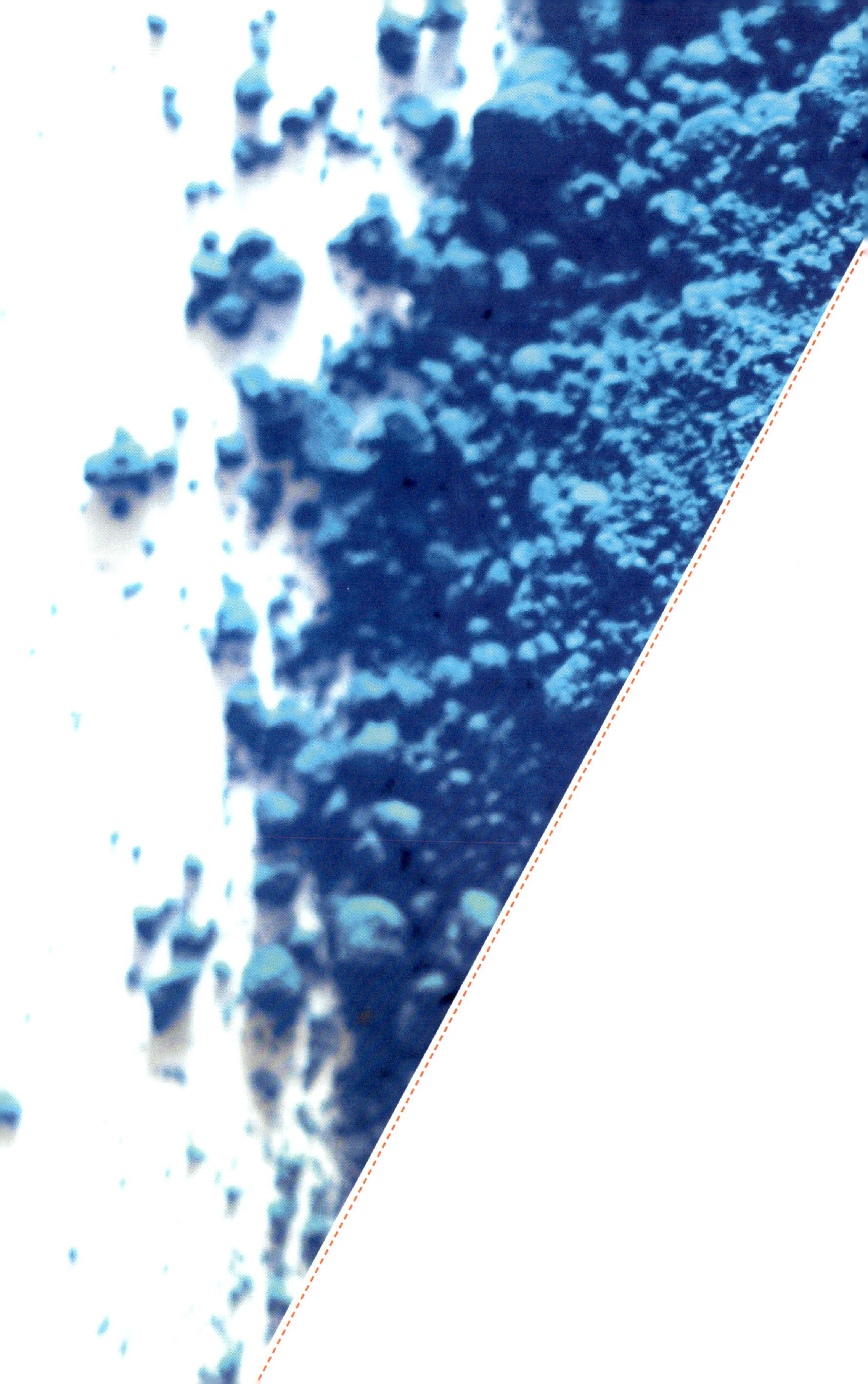

HIMATSUBUSHI TRAIL

For the first time the Ars Electronica Festival is climbing the heights of the roof of POSTCITY. Remote locations directly under the roof of the postal machine also offer exciting views and insights. Himatsubushi promises time to stroll, discover and waste time. The *Himatsubushi Trail* consists of three vital parts—Lab, Trail and Lounge. In the lab, visitors experience prototypes from five international research and design teams. The trail is dedicated to distraction and contemplation, lined with works of art on the subject of time and productivity for man and machine of the future. The lounge entices with distraction and relaxation, with a view similar to a desired summit.

Himatsubushi

The Art of Time Killing

Humanity has become richer than ever before due to creative energy and rational/efficient systems. Continuous trial and error have driven us to be more efficient and productive, to the extent where we are overwhelmed by the furious speed of progress. We feel as though we are left behind by the system we've generated through history.
Behind this was a big dream. A dream to be freer from labor and stress, a craving for culture and leisure to reclaim our life on a human scale – to become more human. On the other hand, as we are coming closer to realizing this dream, calls are arising for a sense of purpose in life through labor and rationality, also in order to become more human. We are going back and forth, not knowing how to balance the two ambivalent ways of being authentically human.
There are many discussions on labor, productivity, efficiency, optimization, rationalization, for these are the overt characteristics and values of our current society, and topics that are already widely discussed. But is the future of humanity only about such things? How can we highlight the other side of human activities, moments where humans are sensual, relaxed and irrational "beings," in order to update our discussions?

暇つぶし; Himatsubushi – A Japanese term for a playful time-killing.

Hima emerges when a meeting is canceled, waiting for a delayed train, waiting for the app to be downloaded – these are wasted times, empty times, free times, void times; where there is nothing required to be done, times that need not be spent efficiently or used productively, moments of mind wandering, moments of finding new ways to play, moments of departing from everyday life whilst still being in it.

To have another understanding of *Himatsubushi* and what lies within this moment can bring us new insights into our perspective on the future.

> Is himatsubushi a beautiful luxury to retain your true self?
>
> Is himatsubushi something that should be removed because it's a waste for productivity?
>
> Is himatsubushi a tool for unleashing more productivity and creativity?
>
> Is himatsubushi a way to hack the blind belief in optimization or does it assist it instead?

The Himatsubushi Research Team, which consists of several groups of research labs, is throwing open the rooftop area of Post City in the context of Ars Electronica Festival 2018.
Welcome to the Himatsubushi Zone & Himatsubushi Trail, where we explore the art of time killing.

Text by Himatsubushi Research Team

Yasuhiro Suzuki (JP)

Aerial Being

Aerial Being has been produced in varying sizes ranging from 1 meter to 30 meters, inflated using air from numerous locations around the world. From something small on a molecular level to a vast size that encompasses the air within the atmosphere – it envisions as an ideal the manifestation of beings that are omnipresent and transcend scales perceivable by human beings. *Kuuki* which is the word for "air" in Japanese, harbors meanings such as the emotions of the people who are there as well as the atmosphere. While elusive and ephemeral, it is an ambiguous concept that at times can exert a highly dominant sense of power within society. *Aerial Being* is perhaps a mediating presence that while being something familiar can lead us to consider things we often are not particularly aware of. One realizes in the first place, that it is only natural for a balloon to inflate with air that is invisible to the eye – so much so, that no one today thinks of it as anything out of the ordinary.
Maybe *Aerial Being* is like an afterimage, or traces of a presence that comes to our minds after leaving the place we had once seen with our eyes. It harbors the atmosphere of the place and invites the people who have gathered there to participate in its landscape. Blending into its surrounding environment due to its transparency, it serves to transform familiar places into a fresh scenery that extends beyond context and meaning.

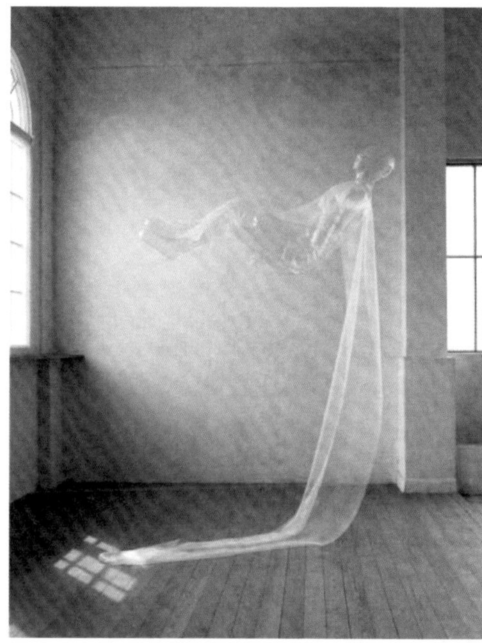

Yasuhiro Suzuki (JP)

Balance Scale for Lightness

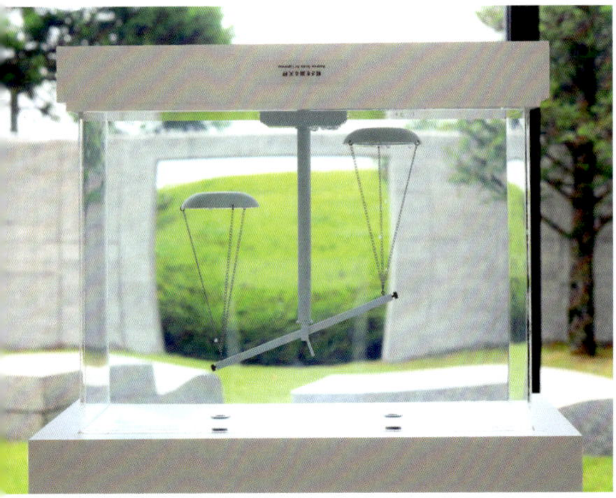

Air bubbles gather slowly beneath a pair of plates that float upside down in water. An attempt to reverse the relationship between "heaviness" and "lightness" under water had incidentally resulted in an interesting aspect of the accumulated air overflowing at unexpected times.

The gaze of viewers became fixated on the swelling surface tension and the state of balance in which the bubbles barely maintain their steadiness, seemingly able to overflow at any moment. Upon repeated observation, it becomes possible to recognize the signs of change. Nevertheless, one cannot determine when the bubbles will actually overflow. The work transcended the intentions of its production to give rise to a space that endlessly continued to draw the attention of the viewer.

Yasuhiro Suzuki (JP)

Nature's Time Metronome

1 second, 1 minute, 1 hour, 1 week, 1 month, 1 year, 10 years, 100 years, 1000 years, 10000 years. The interval of the metronome can be adjusted accordingly. The way we look at various elements such as people, the trees, ocean, mountains, and the sky, changes in correspondence to this metronome. The work considers the means by which we engage with the time inherent within things that appear unmoving, things where we cannot measure their change.

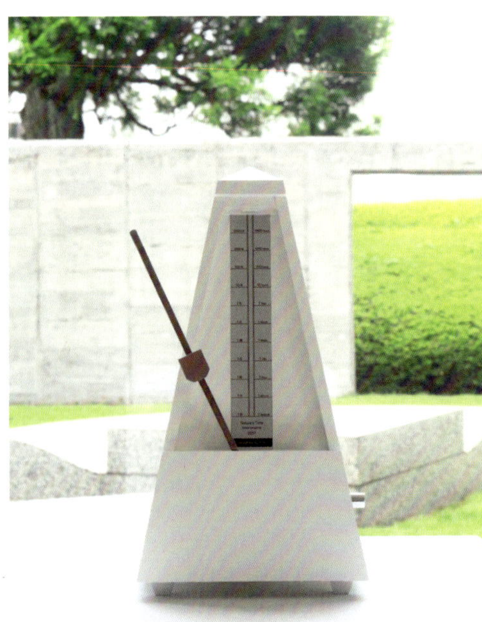

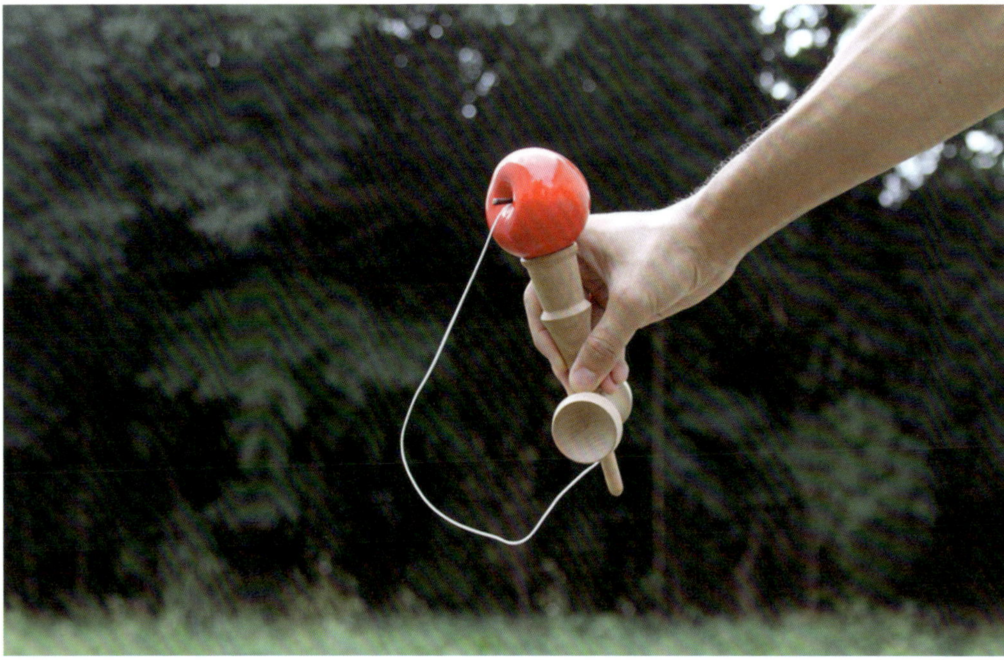

Yasuhiro Suzuki (JP)

Apple Kendama

A Kendama is a toy with which one plays on a worldly scale, against the very gravity of the earth. In extending one's awareness towards the "earth" that exists beyond the Kendama as a medium, the red ball was remade in the shape of an apple in keeping with Newton's law of gravity. The Japanese notion of "association" that connects two things such as the "moon" and an "apple" that are superficially quite unalike one another, also bears links to a scientific image of the world founded upon the continued discovery of perspectives that transcend human perception and imagination. By making the red ball into an apple, one came to consider how perhaps Kendama's roots could be traced back to the game of capturing an apple that could fall at any time from the tree on the tip of a stick, leading perhaps to contemplating this risky game of challenging an unpredictable entity that exists between oneself and nature.

Perhaps mankind created Kendama in various parts of the world and has continued to practice playing with it as a means to confront the unpredictable phenomena of the future?

Erwin Wurm (AT)

ONE MINUTE SCULPTURES

As an artist Erwin Wurm devotes himself to the encyclopedic enlargement of the concept of sculptures. He has been evolving *ONE MINUTE SCULPTURES* since 1988 where he deals with the connection between sculpture and everyday life, the border between sculpture and action, time and duration, as well as the reproduction of sculpture and its interaction with the public. They are characterized by their absurd and paradox subtle tensions, and represent succinctly drawn instructions for changing protagonists to utilize a number of different utensils in the creation of temporary sculpture.

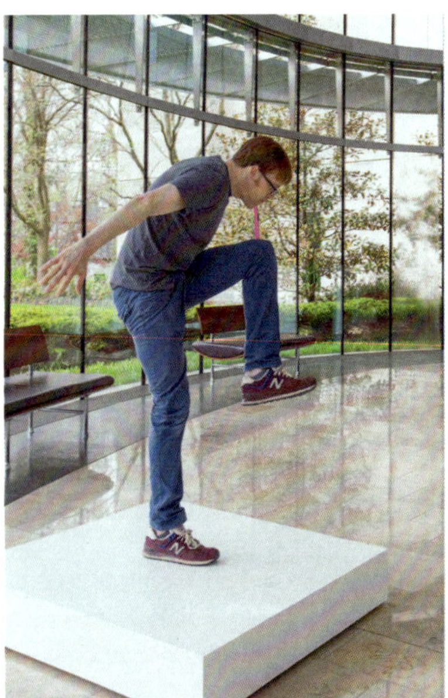

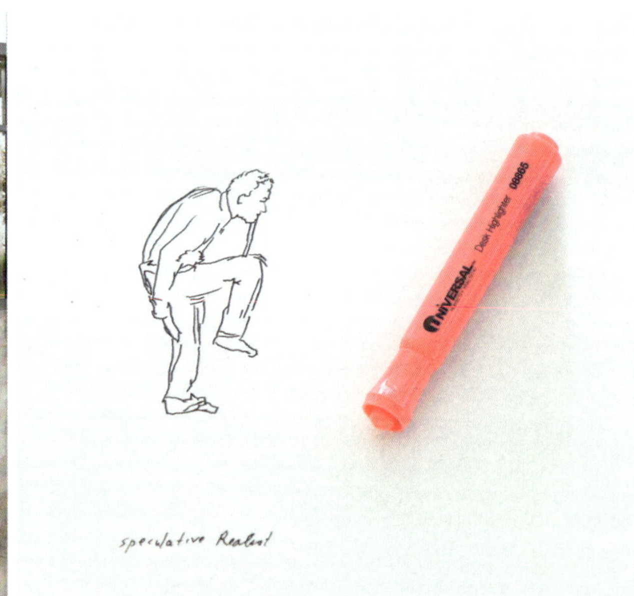

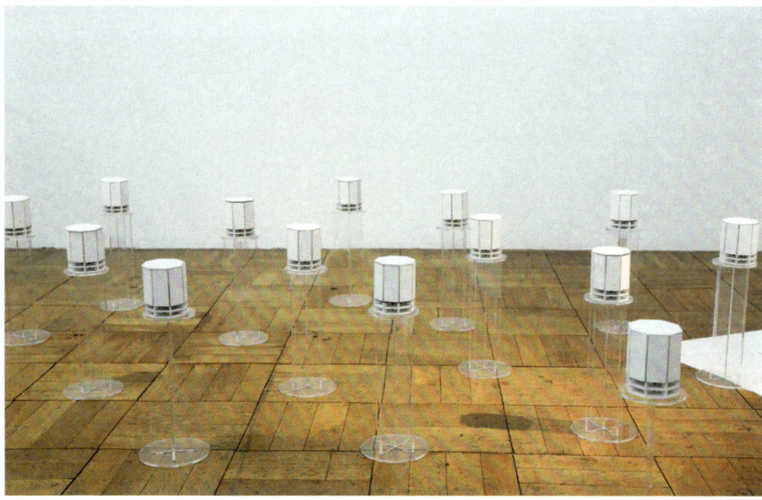

Shintaro Wada

Jun Fujiki (JP), Katsuhiko Tabei (JP), Tomihiro Akagawa (JP)

etheroid

An etheroid is a device working as a medium to embody the existence of an invisible "something" in space. Each of these etheroids, these objects, will propagate its movement to one of the surrounding etheroids repeatedly. There is no God who orders the etheroids; each of them acts autonomously. From the movement of all of the etheroids together, a "something" that can't be seen, appears to be jumping from one etheroid to another–thus coming into existence in its invisibility.

We decided to use an infrared LED that illuminates light invisible within a certain range and a general-purpose communication module called XBee. First, the "current" etheroid uses XBee to transmit information to all of the etheroids and to emit infrared LEDs. Next, the etheroid which received the infrared LED transmits its information back to the current etheroid so it is able to build a list of information for the surrounding etheroids.

This information list corresponds to a temporary storage area. The system randomly selects one from the list and sends the information to the chosen etheroid. Since the list is constantly updated, the operation will continue even if an etheroid is removed.

University of Tsukuba, Ochiai Laboratory, Digital Nature Group (JP)

Naturalization of Technology Towards Japanese Aesthetics

Digital Nature is a concept that underlies our work. We were given two further themes to incorporate into this exhibition: "Beauty of Natural Resolution" and "End to End Transformation of Material Things". Based on this combination, we described the exhibition as "an attempt to update our aesthetics to the extent, where we sense something as being natural in our industrialized times."

An important theme for this exhibition is the vernacular industrial society. To people living in advanced technological societies, "nature" is no longer the same as it was, when humans first emerged back in prehistory. We consist of over 40 people including students, researchers and professors who are interested in Wave Engineering / Machine Learning / Material Research. Our intent is to promote research and development not only for academic reasons but within the broader society. From the viewpoint of computer science research, we are prototyping systems combining wave engineering, bio- and meta-materials of digital fabrication and deep learning in order to explore new ecosystems in the digital age.

These prototypes are different from modern standardized social forms, modern mass production formats or mass communication. They define their view of the world as computationally incubated diversity, and by tackling the expansion of the body as well as the production process, audio-visual communication by holographic wave engineering for individual communication, and machine intelligence.

We are trying to figure out the digital-age ecosystem by these emerging technologies. This is what we always keep in mind in the process of combining art, science and technology, thereby trying to solve real social problems by these technologies.

The Digital Nature concepts are combined with the ancient Japanese aethetics "wabi-sabi", defining it as "stabilization of the instability between complexity and simplicity through the iterations towards Nature".

Yoichi Ochiai (JP)

Structure of Industrial Nature

Since the 19th century trains have traveled on railroads which enable them to move with less friction and these structures are a component of dynamic cities. If a city is regarded as a box garden, the railroad is one of the dynamic components of the garden – like a fountain or a windmill. This installation fuses Japanese garden style and industrial structures to create dynamics and wabi-sabi.

Yoichi Ochiai (JP)

▸ Wavefront of Life: Kandagawa

What color do you imagine on the surface of monochrome water?
The colorization is a subjective process in our mind. Yet, image coloring by AI is becoming common. However, there is a limit to the accuracy of coloring by only learning generic datasets. Will the neural network behave subjectively, if trained based on personalized datasets? What colors will a neural network as an "externalized" artist use/see?
The project explores the external-subjective process using a neural network that only learned from Yoichi Ochiai's photo dataset.

Yoichi Ochiai (JP)

▸ Morphing Iterations of Vase and Flower @Ippei Suzuki (Pixie Dust Technologies, Inc.)

What color you imagine on the surface of monochrome water? The colorization is a subjective process in our mind. Yet, image coloring by AI is getting common. However, there is a limit to the accuracy of coloring by only learning generic datasets. Will the neural network behave subjectively, if trained based on personalized datasets? What colors will a neural network as an "externalized" artist use/see? The project explores the external-subjective process using a neural network that only learned from Yoichi Ochiai's photo dataset.

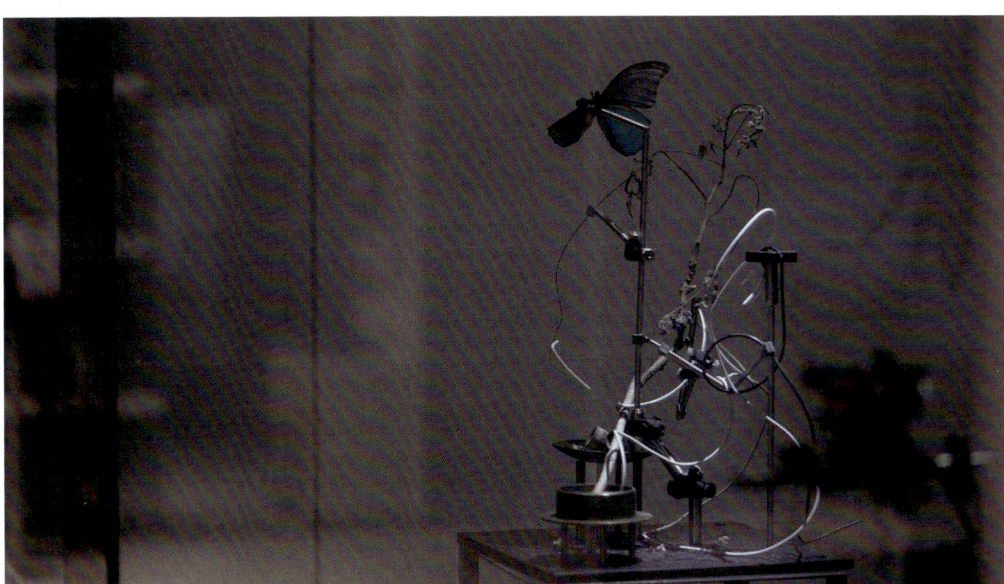

Yoichi Ochiai (JP)

▶ Wavefront, Reflection, Pointillism of Sunlight and Sea waves@IppeiSuzuki (Pixie Dust Technologies, Inc.)

Mackerels' body patterns are optically imitating the sea and the sun. Here, the GAN process can be seen in their natural texture, a natural survival strategy. The canvas in this project is covered with stretched silver and painted with high resolution textures of mackerels. This optical mimicry of mackerels is a naturally generated scenery of horizons. It shows us "wabi-sabi," the stabilization of instability between complexity and simplicity through the iterations towards nature.

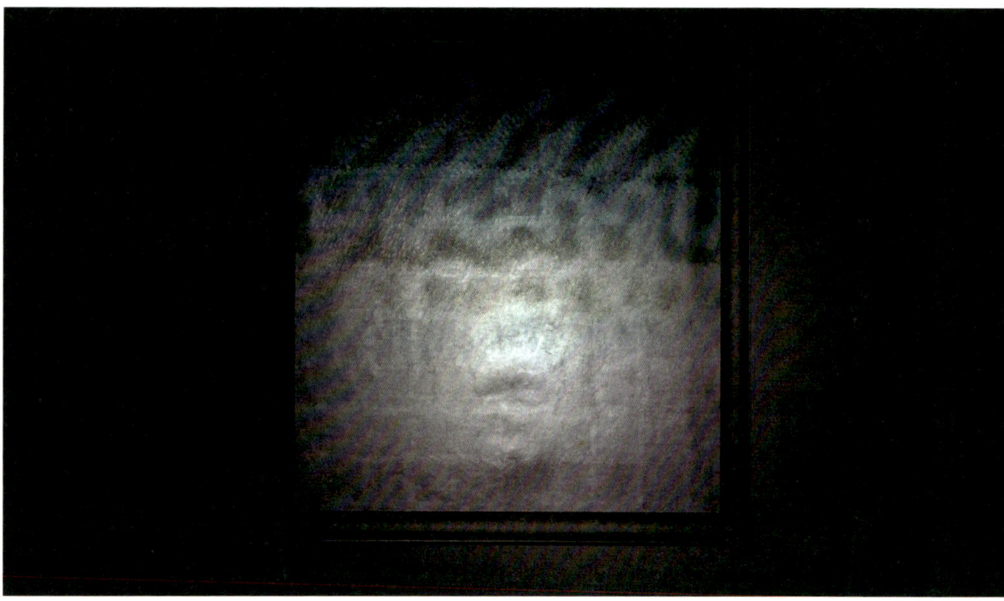

Kohei Ogawa (JP), Yuta Sato (JP), Atsushi Shinoda (JP), Yoichi Ochiai (JP)

▶ Chrome

Chrome is a black Chinese calligraphy which is drawn by femtosecond laser. It has a meta-material structure that looks super black as it absorbs more than 99% of the light.
Chinese calligraphy is commonly referred to as being three-dimensional art by two-dimensional shades of Chinese ink. "黒無" expresses Japanese "書" by using optical characteristics and making it a perfect 2-D representation. Moreover, the black of this work cannot be reproduced with natural objects.

Riku Iwasaki (JP), Yoichi Ochiai (JP)

▸ Impulsed Air Chime: Insect / Impulsed Air Chime: Fish@IppeiSuzuki (Pixie Dust Technologies, Inc.)

Wind chimes are traditional Japanese furniture. Here is the plasma chime, which, activated by high voltage, produces the impulse-sound in the glass tubes with metallic structures. This impulse-plasma-chime mimics the sounds of insects or fish. The generated plasma shakes the atmosphere, and its digitally generated randomness makes the sound of natural phenomena. The sound from the industrial structure of glass and machine permeates into the perspective of Digital Nature.

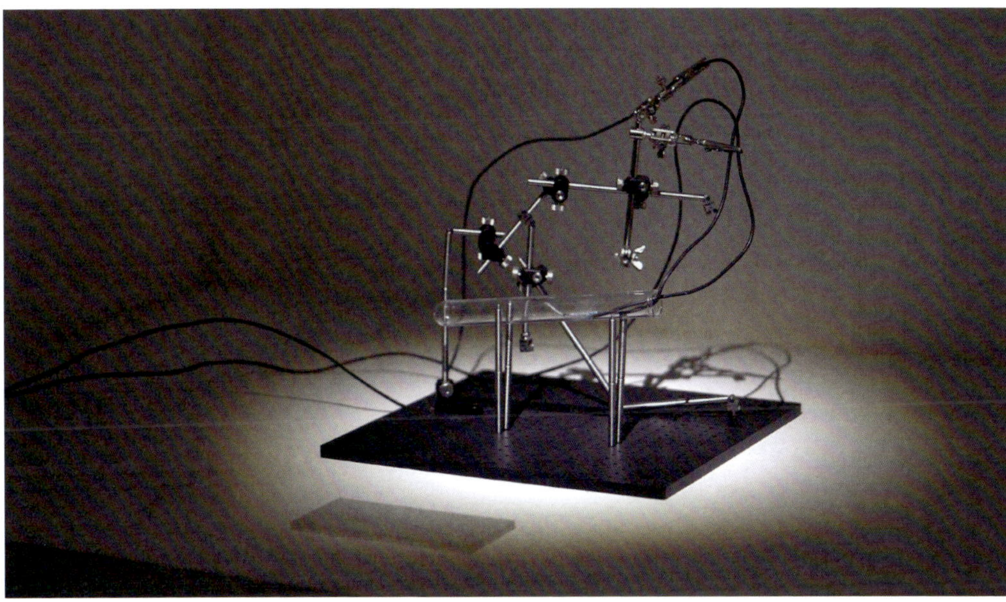

Chun Wei Ooi (JP), Yoichi Ochiai (JP)

▸ Beyond the Iris

Beyond the Iris is a visualization of the light merged by our eye lens. The light undergoes a Fourier transformation, where the retina senses it and decodes it into the world as we know it. This work brings new perspectives of the raw information our eyes receive into a dimension that we can process.

Yuzuha Ito (JP), Riku Iwasaki (JP), Yoichi Ochiai (JP)

▸ "Kesa" for lantern

"Kesa" is the costume that a Buddhist monk wears. As Buddhist monks could not possess private property, they made "Kesa" to cover themselves by joining together torn pieces of cloth. This work is the "Kesa" for lantern, which was made using three-dimensionally shaped silk fabric produced by silkworms. The cloth lights in computationally fabricated organic shapes. This artwork inherits in its visual representation a classical Japanese way of lighting.

Shinnosuke Ando (JP), Yoichi Ochiai (JP)

▸ 砂注像

In Zen there is the word "雨滴聲". This word means "if you look at things without being trapped by bias, everything is one and only." In the anecdote behind this word the teacher preached this lesson through perception of the sound of rain. This work provides the reliving of this word through perception of the image projected on the falling beads.

Kohei Ogawa (JP), Tatsuya Minagawa (JP), Yoichi Ochiai (JP)

▸ 山川草木

山川草木 is a 3-D printed vase that interacts with water. This vase expresses flowing water by capillary action. This property is provided by the capillary effect. We created a structure with a lot of thin tubes inside the 3-D model, which causes the capillary effect. The fresh color expression is realized by capillary action of water colored with edible pigment inside the 3-D print model made of transparent material.

Dmitry Morozov / ::vtol:: (RU)

teleknitting

The concept of the project is to transform the television signal into a multicolored thread that wraps around objects installed on a rotating table. The installation picks up a TV signal in real-time mode and scales it down to a single-pixel image. A special program gradually lowers the digital image resolution. Each time the number of pixels is cut by half, until the image becomes a single pixel, the color of which is the one dominating in each specific frame. The program recognizes and interprets the color, giving the device commands to activate arms. The arms are rigged with ink-ingrained sponges; the ink can be of 5 basic colors: red, yellow, green, blue and black. The sponges press the white thread in the special thread-stretching device and color it. The concept of the project is to ironically transform and reduce the trivial and annoying data stream into a creatively different kind of interpretation. A one-dimensional materialized lo-fi stream covers and mantles familiar objects.

::vtol:: 2016
Special for Polytech Fest, Moscow, 2016

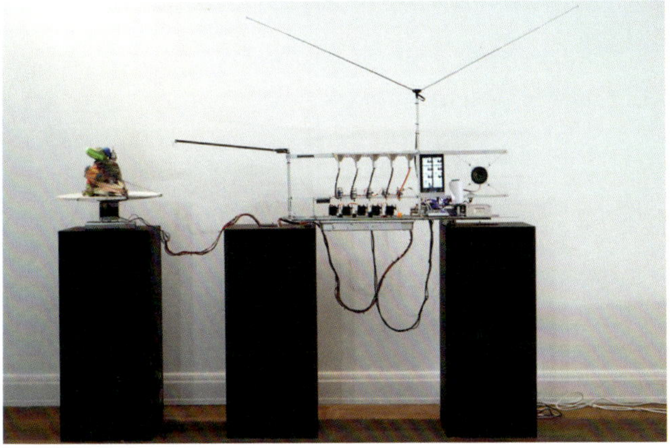

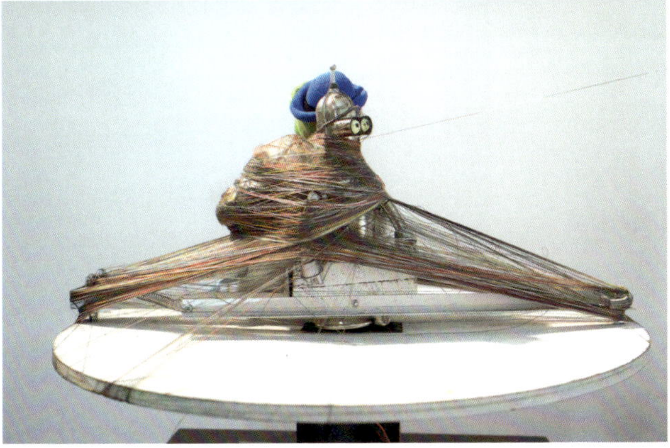

Jiwon Woo (KR)

Mother's Hand Taste (Son-mat)

Mother's Hand Taste explores the complex relationships between intangible cultural heritage, microbiology, immigration, and notions of a "transient self". Woo experimented with the control and development of generational inheritance by visualizing and then bio-fabricating the hand yeast of multi-generational family members across four global locations, to examine its effects on the taste of fermented food. By Investigating "hand taste" through artistic and scientific means, it reflects critically on the origins, authenticity, and preservation of cultural heritage. *Mother's Hand Taste* involves social research, laboratory work, computational design, and additive manufacturing technology. Its main outcome is the creation of a conceptual and mechanical object aimed at capturing, storing, and growing one's own son-mat–specifically, to be used in the brewing of the traditional Korean fermented rice wine, Makgeolli.

With the support of Han Wösten / Utrecht University / Bio Art and Design Award

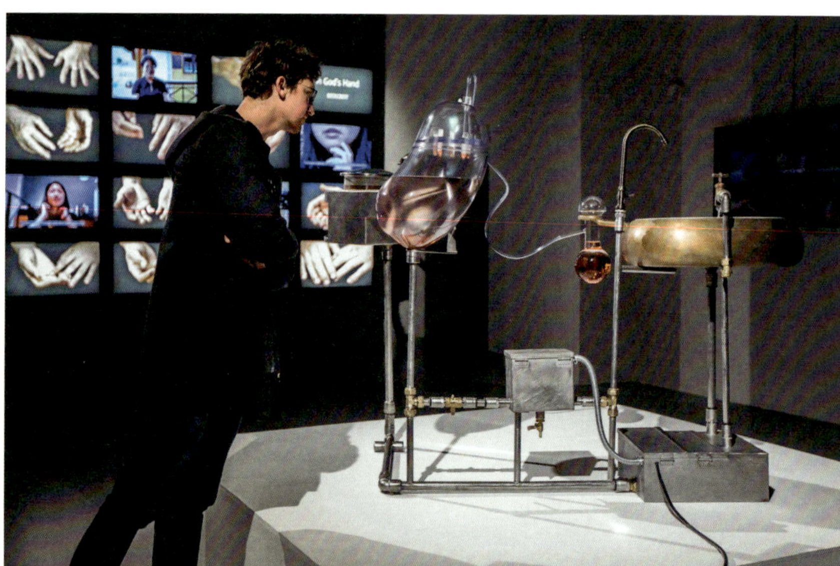

Henneke Wetzer

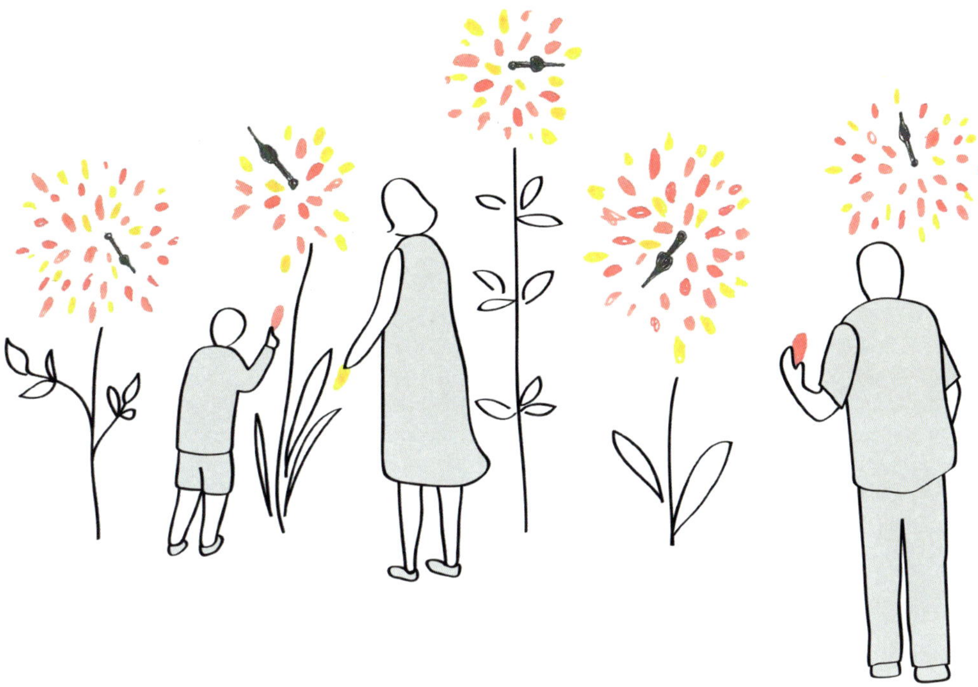

Kyoko Kunoh (JP/AT), Hideaki Ogawa (JP/AT)

Flower of Time

The *Flower of Time* is a participatory installation to create flowers and collect ideas of "rest" for different scales of time. Clock hands with different rotation speeds, such as 1 minute, 1 hour, 1 day, are lined up on the wall. Around the axis, many petals are posted by visitors to express the idea of resting.
Depending on the situations and individuals, the idea of time is elastic. This project aims to visualize the sense of time.

Ars Electronica Futurelab (Kyoko Kunoh, Hideaki Ogawa)

Lisa Aoyama (JP)

The Dive

The Dive is a series of tanks filled with blue, viscous liquid, with bubbles being discharged from the bottom. By gazing at the organic movement of the bubbles, thought gradually ceases and gives way to a deeper dive into the mind, inducing a connection to the subconscious and allowing viewers to enter a meditative trance. I believe that this peculiar state of mind is precisely the "error" that separates us from machines; that makes humans such complex beings. It is perhaps the necessary—and ultimately effective—escape from the monotonous routines of daily life, in which our minds are constantly saturated with information.

Lisa Aoyama, Tokyo University of the Arts

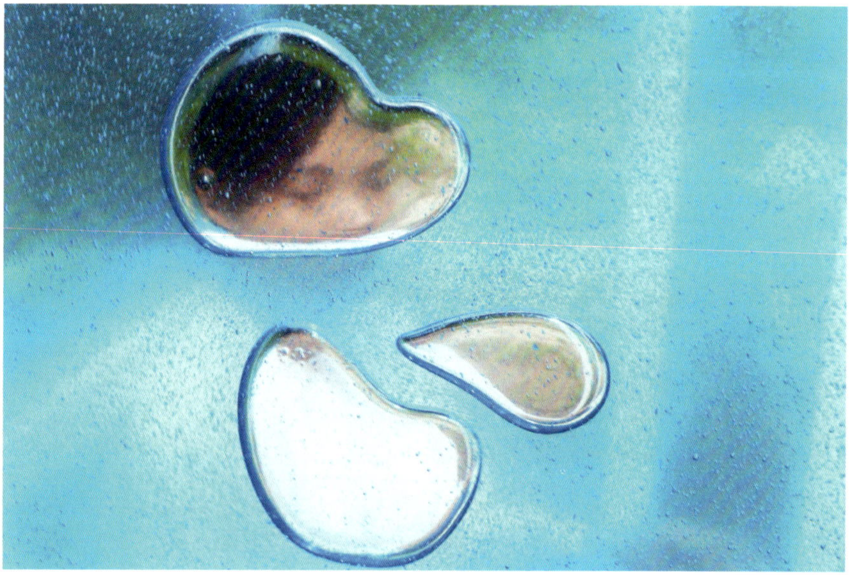

Risako Aoyama

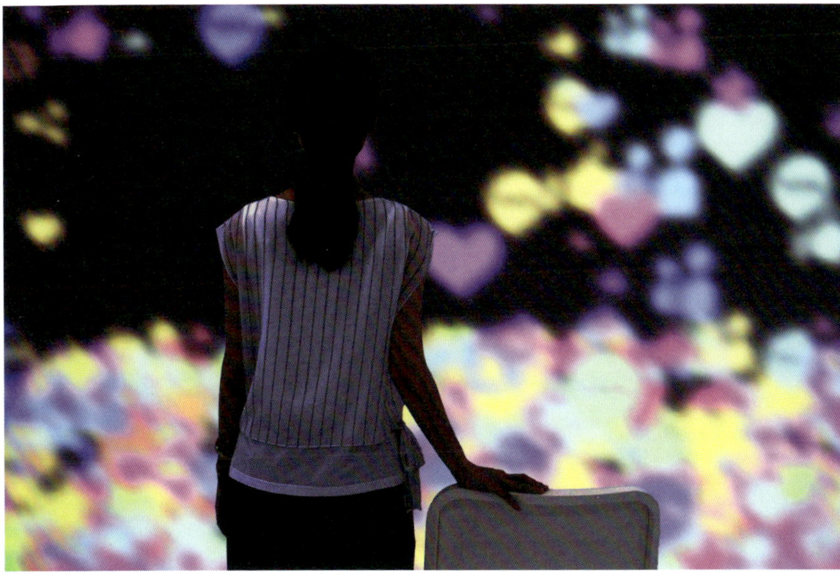

Bokashi B.

MiND X (JP)

Comfort ZONE

The act of praising someone was once delivered through a weighty process like letters, sentences, dialogues...In recent years, social networks have penetrated the lives of people and made this action a daily phenomenon.

This is an era when praise for individuals has become visible, not in terms of depth or weight but in terms of numerical values. This has created a lifestyle that could also be called "approval supremacy"; people create experiences to earn praises, rather than acting because they want to experience things.

But we may think that this behavior of seeking approval has been something we were waiting for, the act itself possibly might have been the hidden value we had craved. Given the choice, would you rather obtain satisfaction through the approval of others, or refuse approval and prioritize personal contemplation? It is for us and the people of the future to decide whether this is an error or not. How should we face the evolved modern society and how should we live?

Ryo Hashimoto (SG)

Slow

"Hima" is a Japanese word for "time when there is nothing to do."
That kind of time is often considered negative and dull. However, most people in today's society are swamped with tasks, therefore their minds are full of their own duties and there is less time to think about what individuals want to think about. In other words, there is less "hima" in individuals' minds. "Hima" may seem to be an unproductive time, but if we interpret it in a positive way, "Hima" is a time when people could freely think about what they want to think about, and that is very human. I propose a new type of pen, a daily tool in which positive "hima" is embedded. Pens are one of the tools that most people have experience using and are often used in the process of completing daily tasks. This is a proposal for tools that allow you to obtain "hima" and an opportunity where each and everyone's mind can freely wander in everyday life.

Ryo Hashimoto, Tokyo University of the Arts

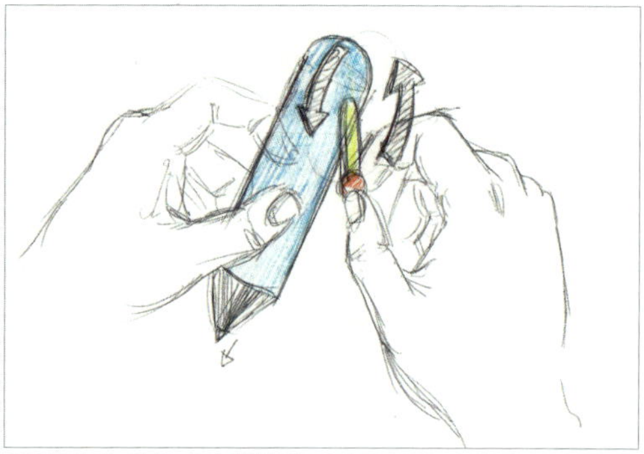

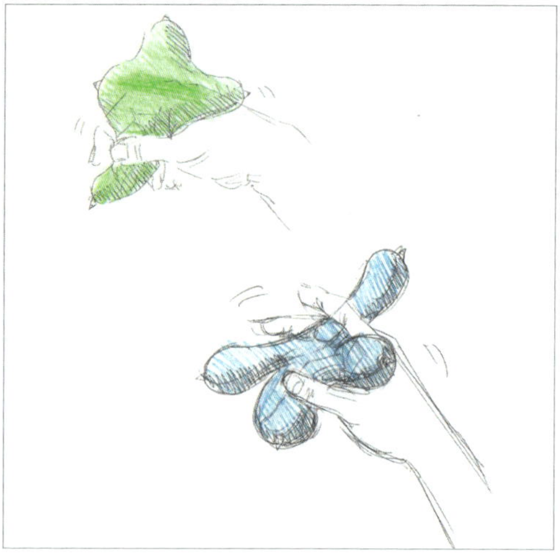

Noga Sapir (IL)

Reflect

REFLECT is a series of smart knit orbs that combine design and technology to create a unique calming experience. The orbs detect stress and help the user practice relaxation techniques to reduce anxiety. The orbs sense how stressed the user is by measuring the skin conductivity level, a physiological parameter that correlates to stress, and respond with soft, changing LED lights.
The resulting experience aids relaxation through the therapeutic process of biofeedback, in which the user is made aware of his/her physiological parameters. By acknowledging them the user learns to gain control over them and affect his/her health and well-being. The main body of the orbs is knit with hand-dyed Tencel yarn that is made in an environmentally friendly process. LED lights shine in soft pulses, creating a quiet and pleasant experience that enhances the intimate relationship between the user and the orb.

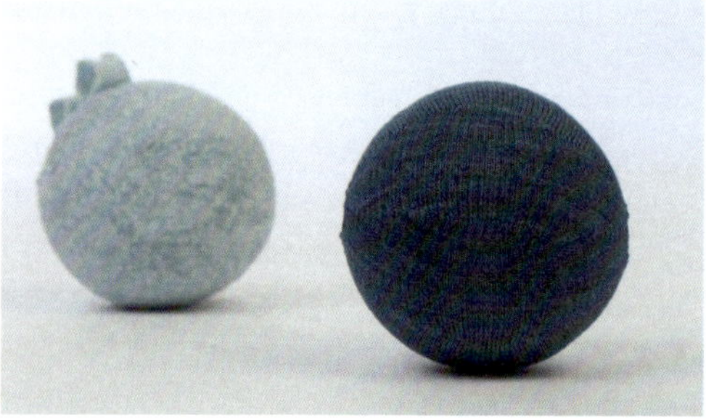

Ahikam Ben Yosef

Yihyun Lim (US), Kacper Pietrzykowski (PL), Andrea Piccolo (IT), Patrik Dolo (IT), Stella Kim (US), Minh Dinh (VN)

Break Time Experience

The Delightful Moment of Time Killing

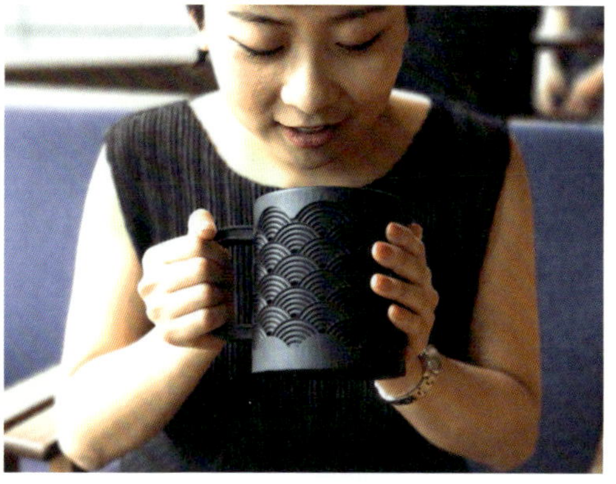

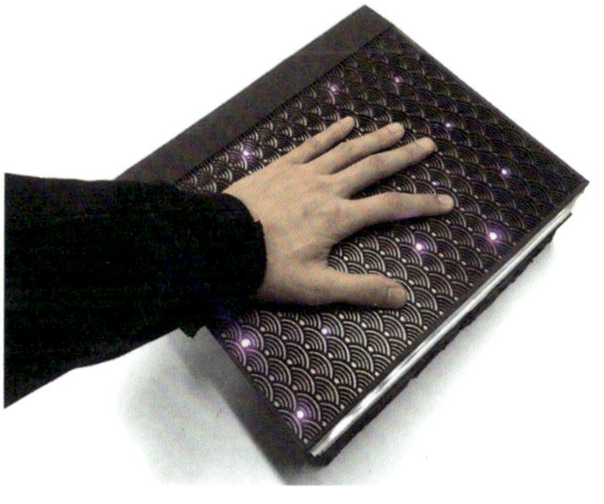

Take a break with us. Stare out the window, look into yourself, talk to strangers, and listen to bits of stories left behind. We are asked to spend our time efficiently, productively and wisely. But perhaps, creativity and human experiences come from in-between moments, in our "break-times", where it's ok to "kill-time".
We want to re-imagine our break experience. What if we could have a break that heals you emotionally, or connects you to unexpected acquaintances? With inspiration from our day-to-day breaks such as smoking, coffee, and meditational breaks, we augment an ordinary break into a sensory, emotional, surreal and social experience. The installation brings the experience of the daily "object" to the forefront. We want to focus on the sensory, tactile, portable, and sharable qualities of our break-mate, which bring social and personal experiences to people's daily lives in a very personal and unexpected way. We have embodied the break experiences in to objects of everyday: the Cup, the Book, the Card, and the Window. The installation reflects some of the fundamental human values and interactions which are embedded into everyday objects. These objects are designed to provide, ultimately, a better human experience. Imagine you're taking a break with these ordinary yet smart objects. We hope this installation gets you to think about what taking a break means and how we can create this experience in the future. Walk into our cafe, where extraordinary experiences are embedded into ordinary objects.

Yihyun Lim, Kacper Pietrzykowski, Stella Kim, Minh Dinh,
Andrea Piccolo, Patrik Dolo
And IXD–Innovation by Design

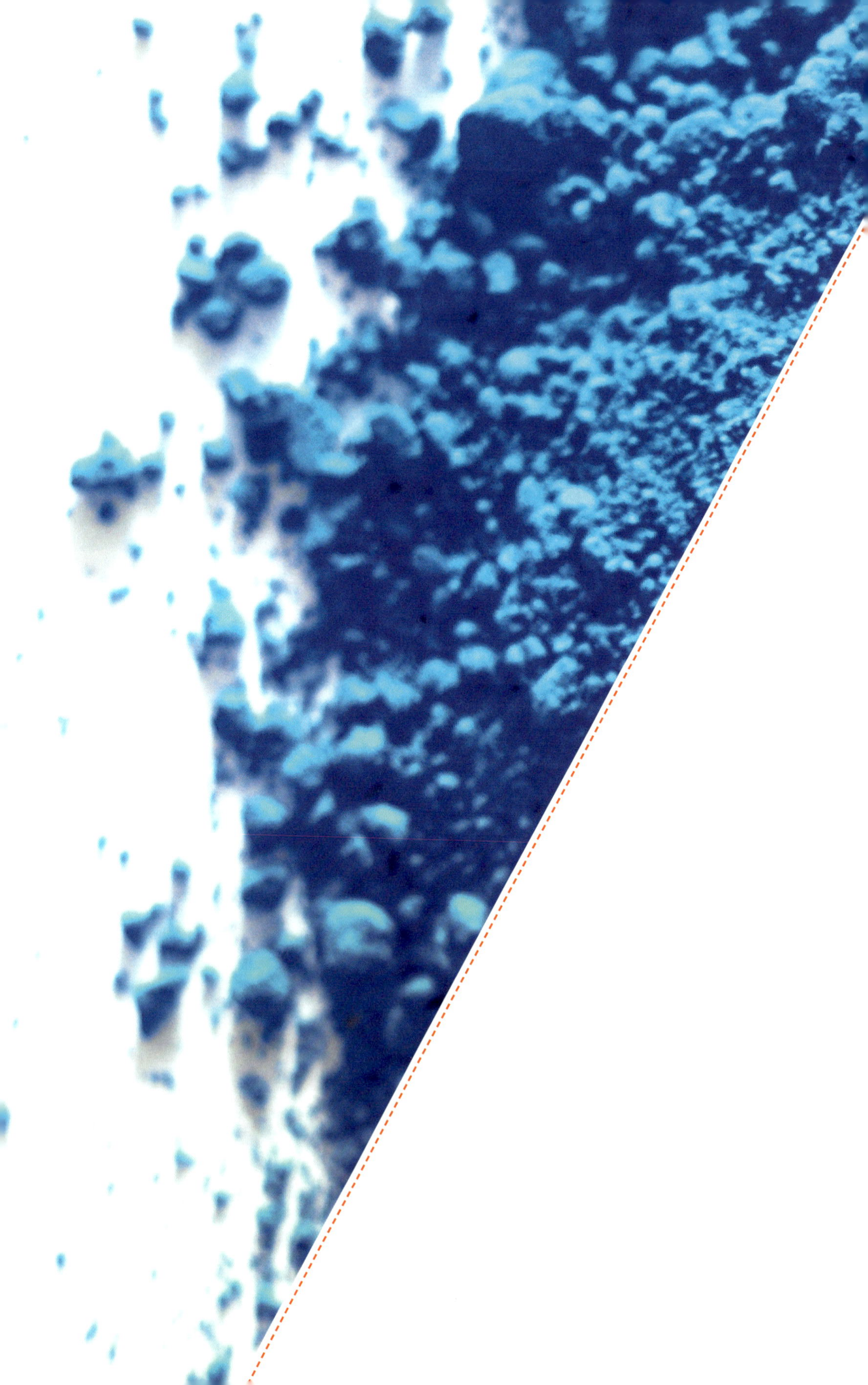

DEEP SPACE 8K

The Ars Electronica Center offers its visitors something that can not be found anywhere else in the world: 16 x 9 meters of wall and another 16 x 9 meters of floor projection, laser tracking and 3-D-animations make the Deep Space 8K something very special indeed. Furthermore, Deep Space 8K presents challenging infrastructure to media artists. As they go about adapting existing works and, above all, designing installations custom-made for this space, they're entering artistic terra incognita. The position of visitors amidst the projection surface and participation by them call for a well-thought-out aesthetic composition and concepts for the resulting dynamics.

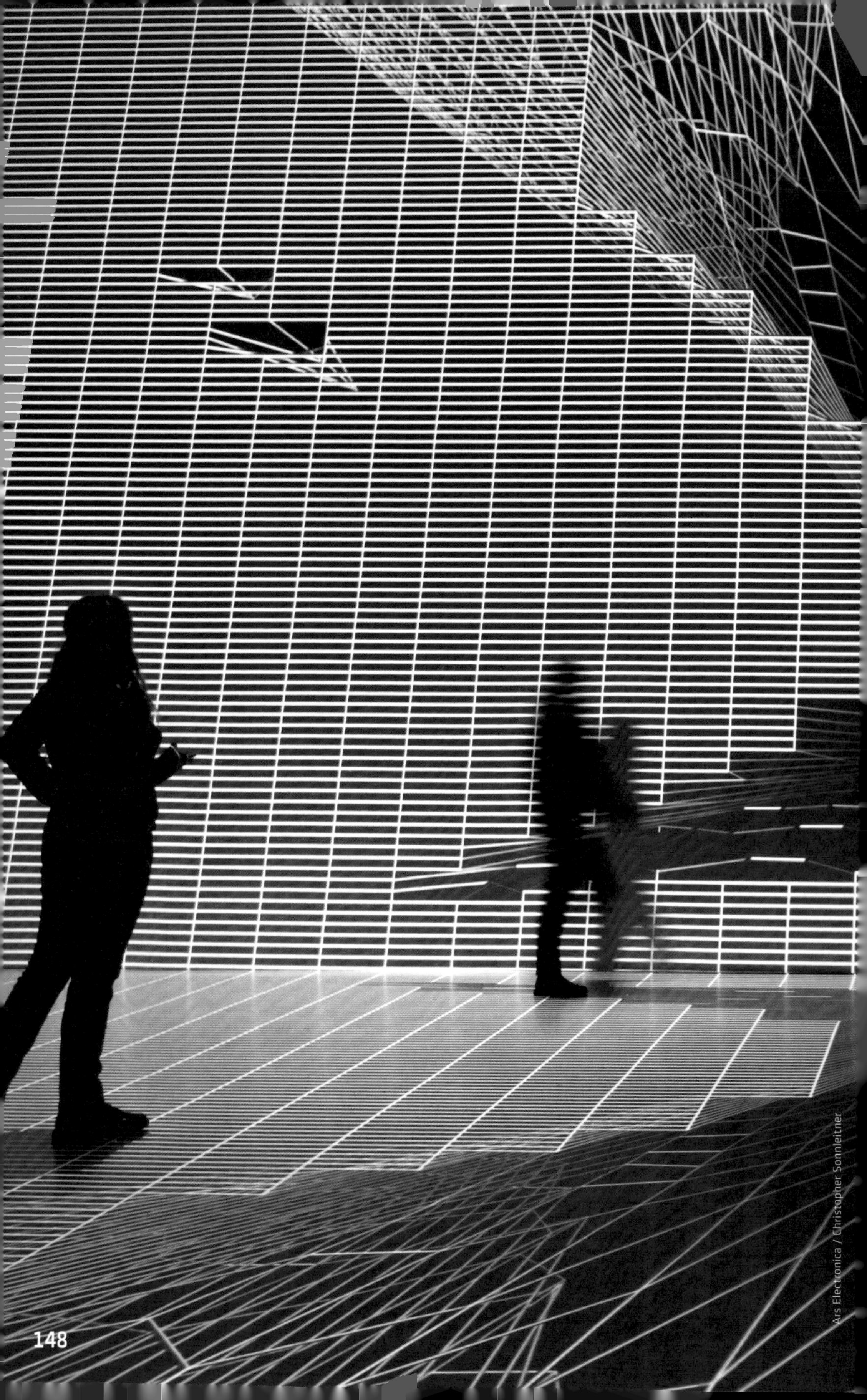

Ars Electronica / Christopher Sonnleitner

Memo Akten (TR)

WAVES

Inspired by the well-established history of both artistic and scientific study on oceans and oceanic waves, *WAVES* is a data dramatization of complex ocean simulations, distilled and re-imagined in the form of abstract visuals and sounds. An ongoing series of studies that investigate the tension and delicate balance between the immense power and incredible fragility of the oceans; simultaneously mighty yet delicate; calming yet terrifying; graceful yet violent; a symbol of fear, danger and death as well as hope, freedom and life.

Memo Akten
Earlier WAVES studies: WAVES 2014 series.
Original score by Rutger Zuydervelt (Machinefabriek)

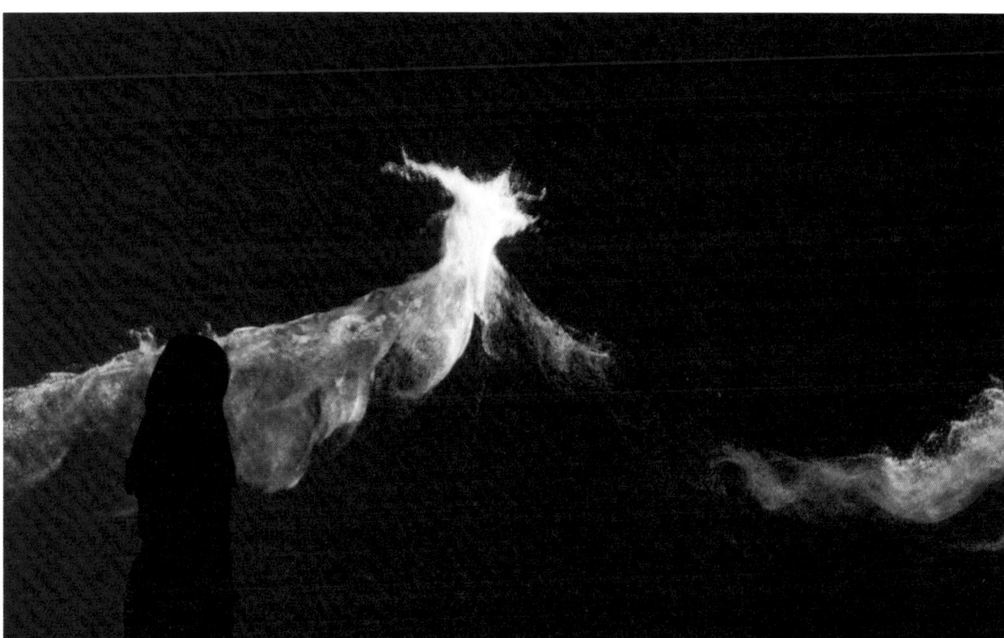

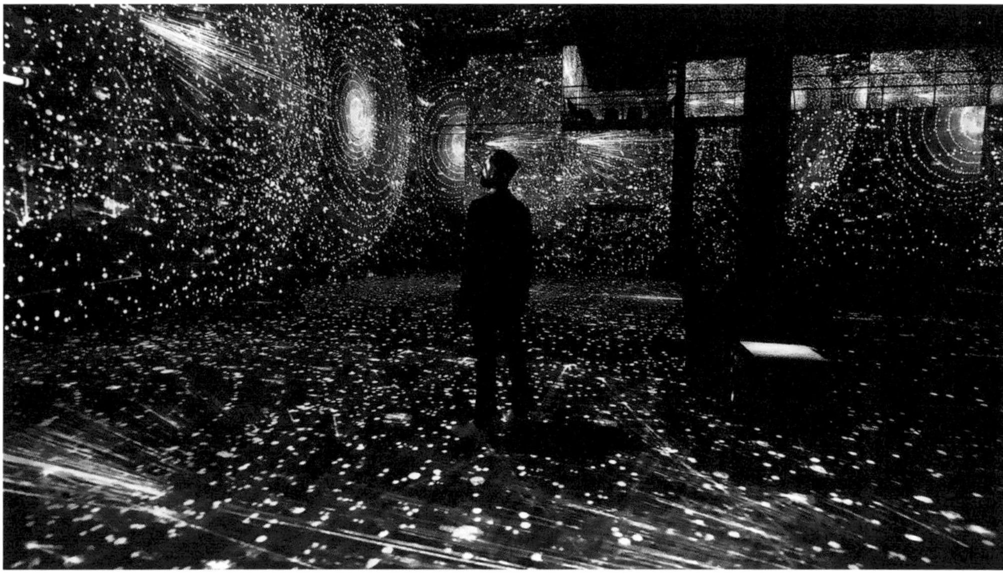

Ouchhh (TR)

POETIC AI _ ERROR

Are you ready to experience the art of imperfection, which is made with the poetic refraction of AI? These refraction errors are created by using machine learning and AI algorithms. We create a "scientific conscious" Poetic Refraction of AI reality—the errors it learns are from millions of lines of theory, articles and books about light, physics and space-time, written by scientists, who change the destiny of the world and write history.

Direction_Design_Animation: Ouchhh (ouchhh.tv)
Sound Design: AudioFil

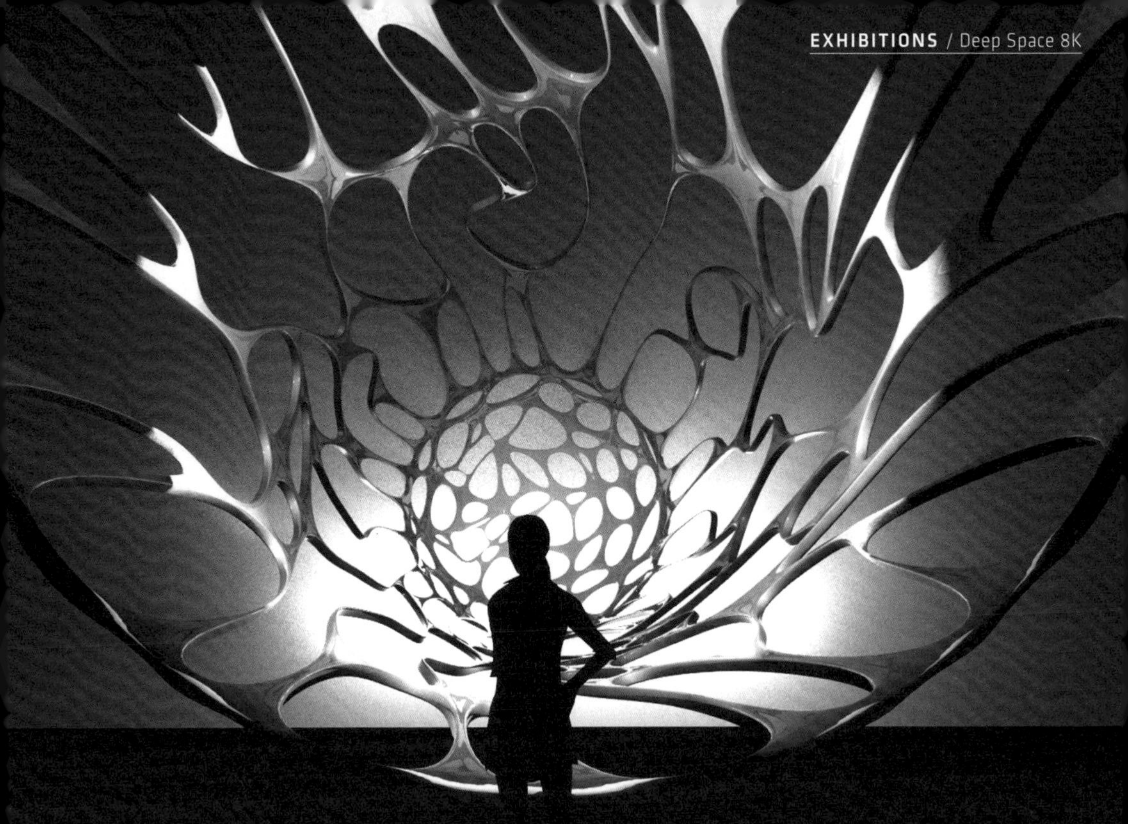

dastrio (DE/KR) & Ouchhh (TR)

"SAY_SUPERSTRINGS"

The concert is dedicated to Turkish composer Fazil Say and played by the South Korean/German trio. The visualization is provided by the new media studio OUCHHH and especially created for the Deep Space 8K at Ars Electronica Center in Linz.
What if matter is nothing but notes coming out of a vibrating string?
According to superstring theory, all matter in the world is made up of one thing: vibrating thin strings. These strings, which vibrate at different resonances, bring everything into existence in the known universe. Matter consists of small strings. When these strings are pulled in a certain way – just like a violin or guitar string – they create a frequency. Therefore, the pitches/notes occur. We have become aware of the existence of the little notes that these superstrings create, and we realize that the universe is a symphony, and all the physical laws of the universe fit these superstrings. *Ouchhh* will take inspiration from the notes that exist in the universe while micro-strings vibrate (Subatomic Particles) in real time and define the melodies created by the notes as "Matter" and symphonies of these melodies as "Universe". With *dastrio*, *Ouchhh* will take 11 dimensions in abstract directions in super grade gravity theory and move them beyond space in real time. The dimensions captured intuitively in living space will constantly change and turn into reality.

Fazil Say:
"Space Jump" for piano trio
Sonata for Violin and Piano
dastrio:
Bernhard Metz – Violin
Manuel Von Der Nahmer – Violoncello
Suyang Kim – Piano
OUCHHH:
Ferdi Alıcı
Eylül Duranağaç Alıcı
Atay İlgün

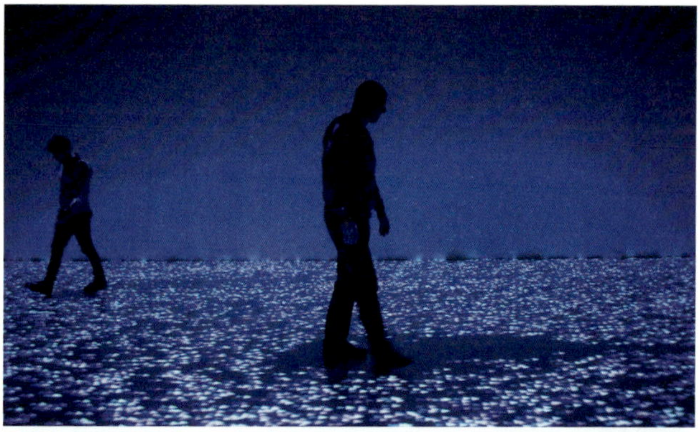

Şehmus Poyraz Birusk

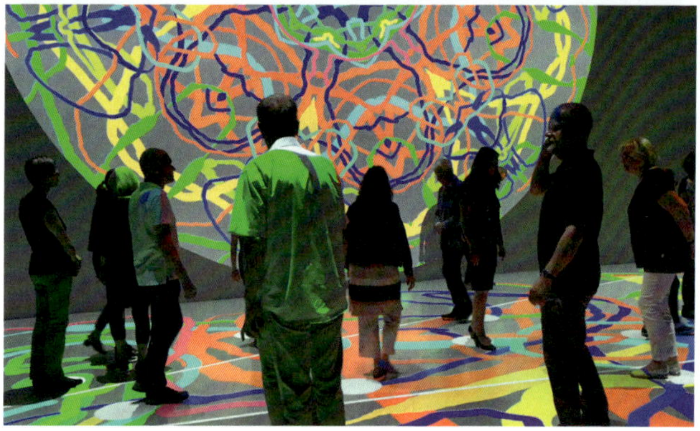

Kaleidoscope, Gerhard Funk

Gerhard Funk (AT), Christian Berger (AT), Şehmus Poyraz Birusk (TR), Clemens Niel (AT), Fabian Terler (AT)

Cooperative Aesthetics – Next Edition

The main idea behind the concept of *Cooperative Aesthetics* is to build an immersive, interactive space in which the users can produce a collaborative audiovisual aesthetic experience. Using the whole floor as an interface, people can influence and animate the audiovisual outcome by their positions and movements. An essential aspect is the interaction between all users, their communication and collaboration. So the concept of *Cooperative Aesthetics* has a strong social component to bring people together. *Cooperative Aesthetics* is the result of a close long-term cooperation between Ars Electronica and the study program "Time-based and Interactive Media" at the Art University of Linz, where Prof. Gerhard Funk and his students have the opportunity to use the Deep Space 8K with its laser tracking system. For this year's festival, new selected works from this series will be shown.

Kaleidoscope: Gerhard Funk (AT)
Critical Mass: Fabian Terler (AT)
Particle Collage: Christian Berger (AT)
Sound Towers: Sound: Gerhard Funk (AT), Clemens Niel (AT)
Particle Flow: Şehmus Poyraz Birusk (TR)

Sophie Amelin (AT), Jeremiah Diephuis (US), Georgi Kostov (BG), Marlene Mayr (AT), Tobias Sichmann (AT), Eric Thalhammer (AT)

City Lights

After a power outage, the city of New Orleans is in complete darkness. Players have to collectively solve puzzles and cleverly guide and manipulate rays of light with prisms, lenses and mirrors to bring the city back to life.

The interaction with the game world is made possible via a laser tracking system, which tracks the current positions of the players. The life-sized stereoscopic visualization of Bourbon Street on the immersive 16-by-9 meter wall projection makes players and spectators feel as if they were actually looking down the street.

City Lights was created in the course of a semester project by Media Technology and Design students from the Hagenberg Campus of the University of Applied Sciences Upper Austria in cooperation with the research group Playful Interactive Environments and was specifically developed for Deep Space 8K at the Ars Electronica Center.

Playful Interactive Environments / FH OÖ Campus Hagenberg

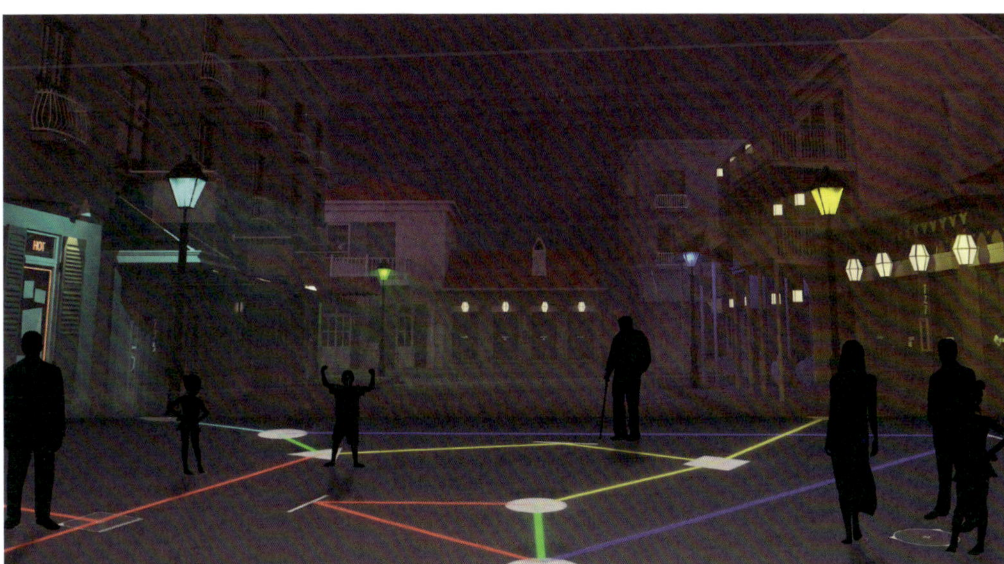

Jürgen Hagler (AT), Andrea Aschauer (AT), Jeremiah Diephuis (AT)

House of Medusa

The *Virtual House of Medusa (VHM)* is a supervised, seated virtual-reality (VR) installation, including one VR player and up to four fellow co-players. By putting on a head-mounted display (HMD), the visitor enters the installation and can switch between the four virtual stations via gaze interaction. Using a VR controller, the player can interact with the virtual world and grab digitized fragments of the fresco to slip into the role of the restorer. An additional perspective is provided by a public display that allows spectators to watch the virtual experience.

A majority of VR installations are designed as single-user experiences, and do not typically facilitate communication or interaction between multiple participants. In contrast, the VHM is constructed as a cooperative VR Experience between the VR player and the audience. Audience members have the ability to interact directly with the VR player and control their own individual point of view of the virtual world. Using tablets combined with positional trackers, the co-players can look into the VR world, and via touch-based interaction, give the VR player hints. This allows co-players to move freely in the physical, tracked space and view the VR world from any direction, regardless of the perspective of the VR player.

Team: Andrea Aschauer; Jürgen Hagler; Michael Lankes; Jeremiah Diephuis; Wolfgang Hochleitner; Georgi Kostov;

Project Partners: Federal Monuments Authority Austria; Kunsthistorisches Museum Wien

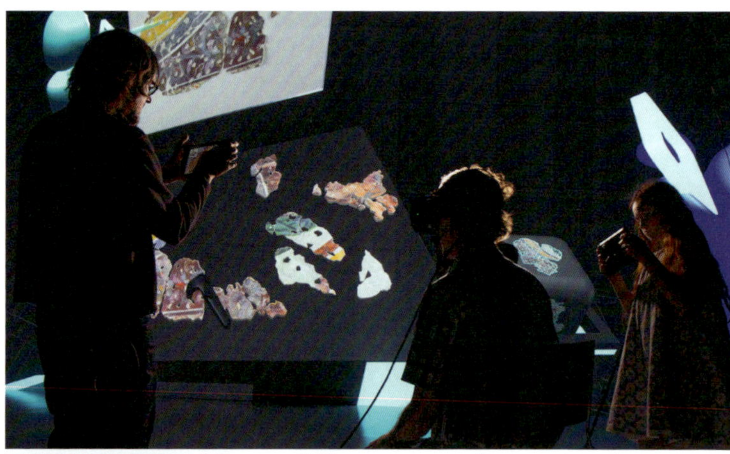

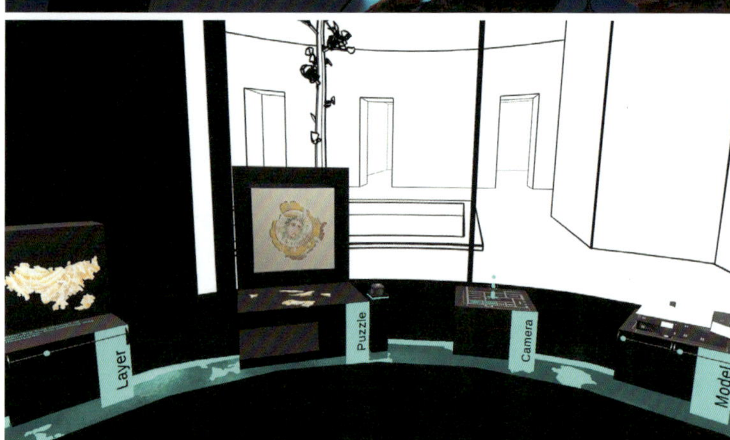

FH Hagenberg

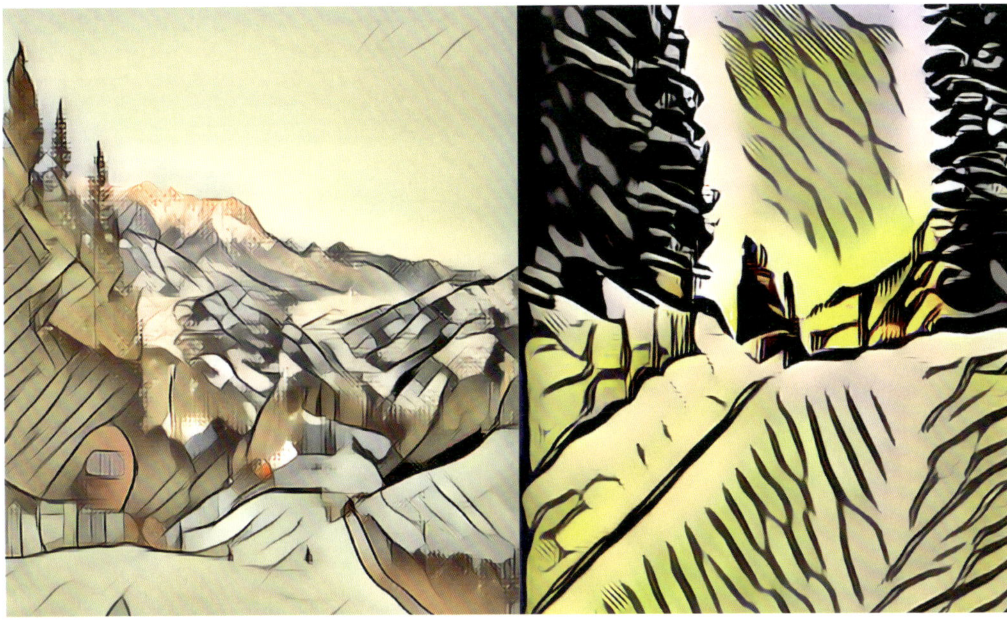

Ars Electronica Center

Ali Nikrang (AT)

Creative IA – Style Transfer

StyleTransfer is a technique from the field of artificial intelligence (Deep Learning) with which everyday images can be transformed into an artistic and painterly style. This project uses a stereoscopic video from the Streif in Kitzbühel. This is one of the most difficult and dangerous downhill slopes in the world. As far as we know, this is the first time this technique has been applied to a stereoscopic video.

Another special feature of this project is the 8K resolution: For this purpose, too, a technique from the field of Deep Learning is used, which quadruples the resolution of the original video.

The result immerses the audience in the forms and colors of the artists Wassily Kandinsky and Roy Lichtenstein and lets them experience the Streif as a three-dimensional painting in Deep Space 8K.

Ali Nikrang (Ars Electronica Futurelab)
Juliane Leitner (Ars Electronica Center)
Thomas Kollmann (Ars Electronica Center)

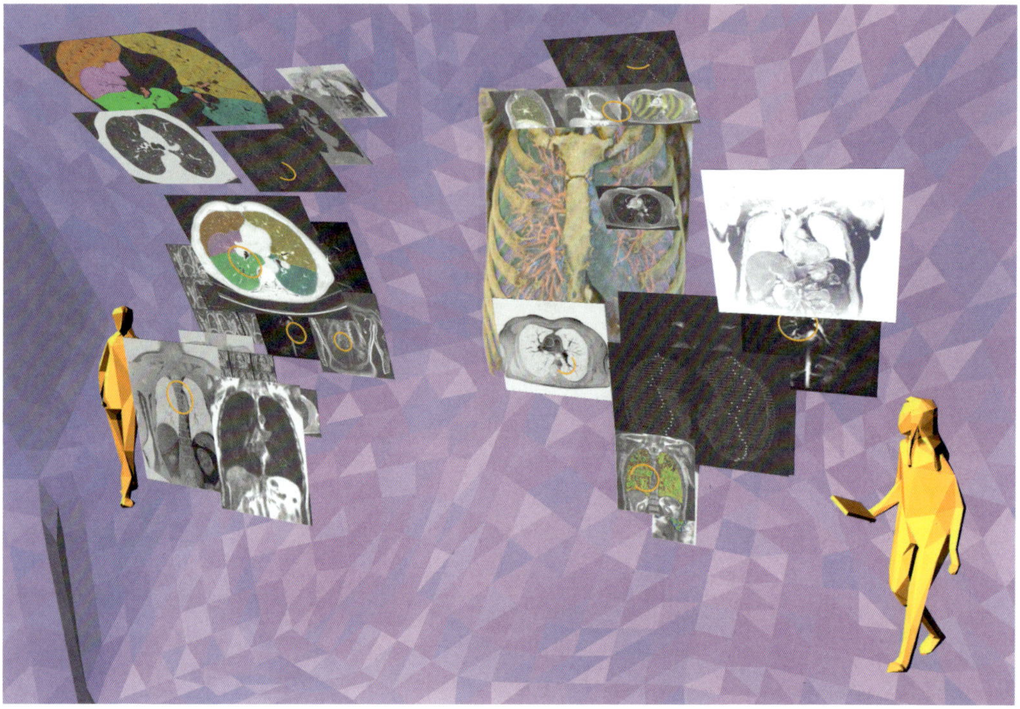

Fraunhofer MEVIS

Fraunhofer MEVIS (DE)

Physicians' Colleague, Patients' Helper: The Cognitive Computer

Mr. Dr Wenzel from Fraunhofer MEVIS: Institute for Medical Image Computing introduces how deep neural networks contribute to the evaluation of medical data and how this also changes the role of physicians and patients.

Fraunhofer MEVIS: David Black (US), Alexander Köhn (DE), Bianka Hofmann (DE), Mathias Neugebauer (DE), Markus Wenzel (DE)

Fraunhofer MEVIS (DE), Media Art Nexus NTU Singapore (SG)

Before Us Lies Eternerdy

The short movie shows different scales of the human body, from digitized microscopic lymphoma tissue examined with the molecular cytogenetic technique Fluorescent in situ hybridization (FISH) to detect abnormal changes in DNA, to 3-D reconstructions of a liver as well as a whole-body MRI. At a simultaneous launch at Ars Electronica's Festival and Media Art Nexus NTU Singapore, we will talk with Ina Conradi and Mark Chavez in a live video call about how art can contribute not only artistry and a sense of awe to removing barriers for engagement with severe health topics, but also as a transdisciplinary approach to innovation in digital medicine. The movie marks the beginning of the cooperation of Media Art Nexus NTU Singapore and Fraunhofer MEVIS, which will provide techniques, tools, scientific expertise and an educational environment for students from NTU Singapore in Bremen, Germany.

Fraunhofer MEVIS; FISH data kindly provided by ZytoVision
Fraunhofer MEVIS (DE): David Black (US), Henning Höfener (DE), Bianka Hofmann (DE), Andre Homeyer (DE), Alexander Köhn (DE), Mathias Neugebauer (DE)
In cooperation with Media Art Nexus NTU Singapore (SG):
Ina Conradi (US, SG) & Mark Chavez (US, SG)
Project integration Singapore: Media Art Nexus, Nanyang Technological University Singapore, Ina Conradi and Mark Chavez

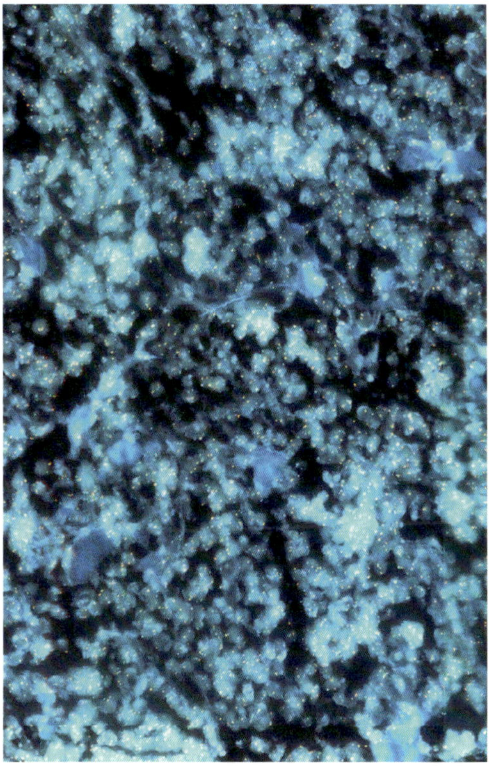

ZytoVision

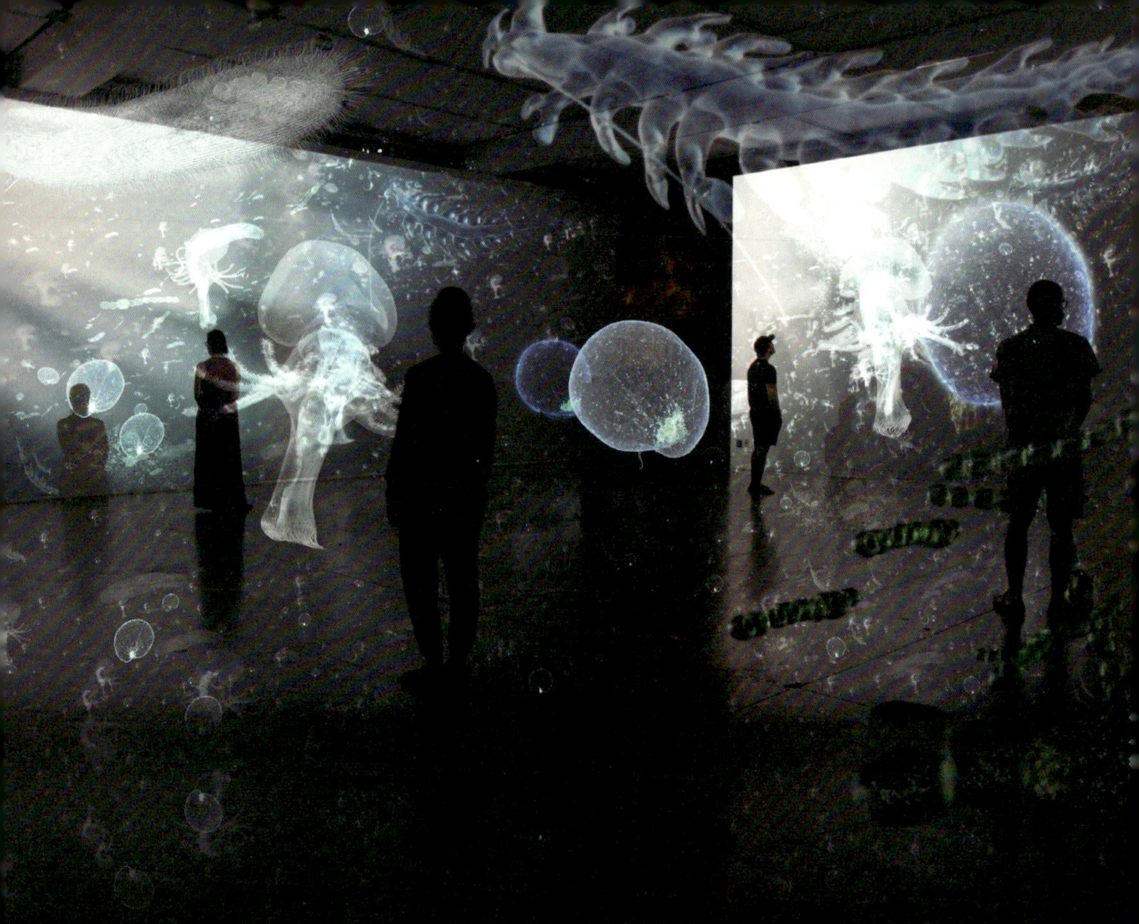

Victoria Vesna (US), Alfred Vendl (AT), Martina Fröschl (AT)

Noise Aquarium

Oceans and seas should never be considered as flat blue interfaces or dumping holes for our anthropogenic remains – vast amounts of diverse organisms live down there, adversely affected by our waste and noise pollution. Current scientific studies have demonstrated how noise sources such as sonar and fracking influence large marine life with shocking examples such as stranded whales and dolphins. However, almost nothing is known about the possible impact on marvelous microscopic organisms such as the plankton, and with the entanglement of micro plastics, the ecological balance is further compromised. *Noise Aquarium* utilizes 3-D-scans of these micro creatures obtained with unique scientific imaging techniques and immerses the audience in the 3-D "aquarium" of diverse planktons projected as large as whales. With their presence alone, participants create destructive visual and audio noises, demonstrating how we are all implicated by our inaction.

Artist: Victoria Vesna, UCLA Art Sci center
Director: Alfred Vendl, Science Visualization Lab at Department DIGITALEKUNST/Ruth Schnell, University of Applied Arts Vienna
Animation: Martina Fröschl, Science Visualization Lab at Department DIGITALEKUNST/Ruth Schnell, University of Applied Arts Vienna
Programming: Glenn Bristol, United Motion Labs
Audio mixing: Paul Geluso, NYU Steinhardt
Scientific imaging: Stephan Handschuh, University of Veterinary Medicine, Vienna
Scientific imaging: Thomas Schwaha, Department of Integrative Zoology, University of Vienna
Sponsors: Gerald Bast Rektor, Ruth Schnell, Digital Art, University of Applied Arts Vienna; UCLA faculty research, UCLA Art Sci center; Harvestworks.

Hans-Peter Bunge (DE), Bernhard Schuberth (DE), Markus Wiedemann (DE), Dieter Kranzlmüller (DE)

A Journey into the Earth

Hans-Peter Bunge, professor for geophysics at the Ludwig-Maximilians-Universtät (LMU), Munich, demonstrates computer simulations which explain the convection processes inside the earth's mantle. Spectacular animations show that some cold plates within the earth move down into the core, while other warm areas rise to the boundaries of the earth's mantle. Professor Bunge will explain the physical processes in more detail, as well as their effects on the geological evolution of our planet. The simulations were performed on the supercomputer SuperMUC of the Leibniz Supercomputing Centre (LRZ), led by Prof. Dieter Kranzlmüller, and visualized by Markus Wiedemann at the Center for Virtual Reality and Visualization (V2C).

Ludwig-Maximilians-Universität München (LMU) / Leibniz-Rechenzentrum (LRZ) der Bayerischen Akademie der Wissenschaften

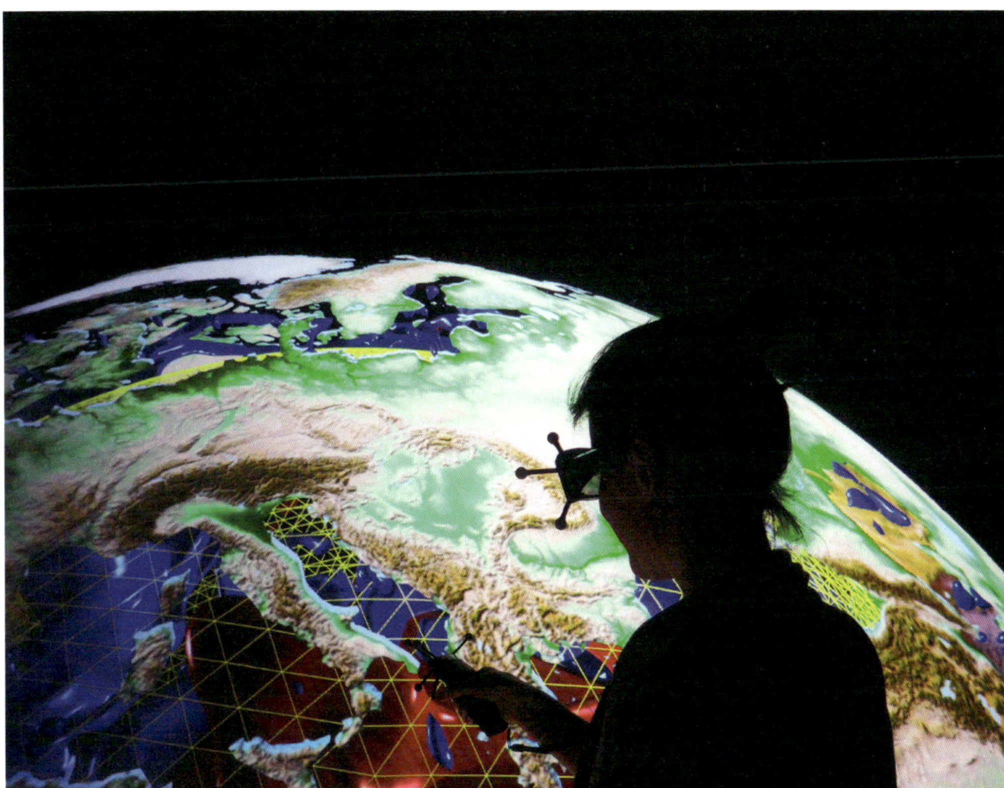

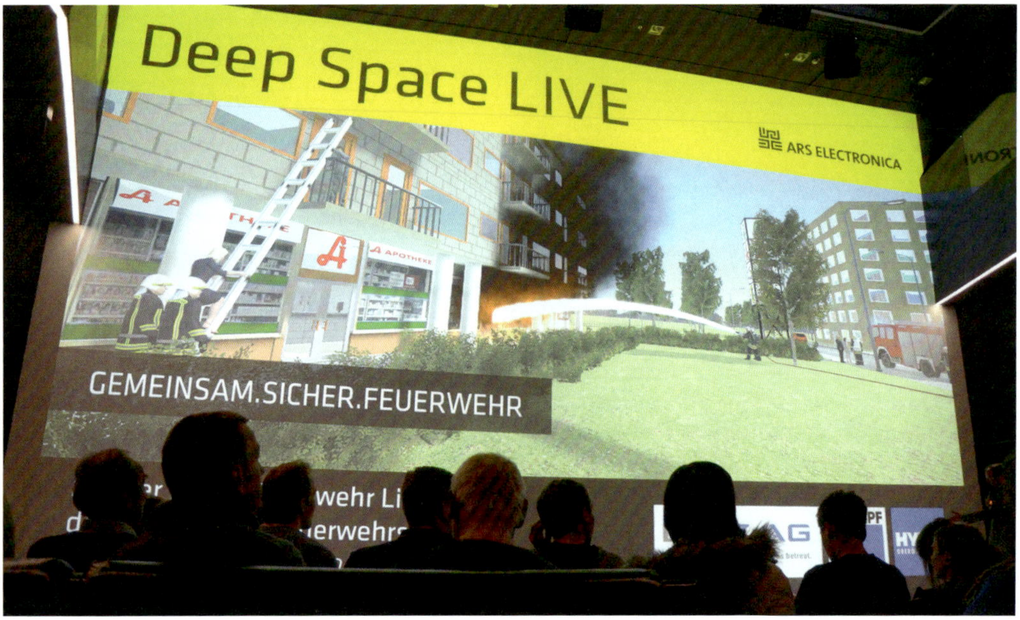

Ars Electronica/ Magdalena Sick-Leitner

Stefan Krausbar (AT), Michael Buchbauer (AT)

GEMEINSAM.SICHER.FEUERWEHR

With the education initiative *GEMEINSAM.SICHER. FEUERWEHR* the Austrian fire brigades would like to make an important and essential contribution to fire and disaster protection education in kindergartens and schools. The pupils receive important information and educational download materials for training. The Linz Fire Brigade and the Upper Austrian Fire Brigade School will present a virtual fire brigade mission in Deep Space 8K, offering a new and modern way of training. An exciting firefighting event is imminent!

DEEP SPACE LIVE

High-resolution image worlds in the format of 16 x 9 metres meet expert commentary. Deep Space LIVE stands for insightful entertainment amidst impressive pictures. Further dates can be found at *ars.electronica. art/dsl*

Berufsfeuerwehr Linz und OÖ Landes-Feuerwehrschule

Beyond the Frame: 8K Future Research Project

Japan's largest public broadcasting organization NHK and Ars Electronica Futurelab joined forces in 2017 to examine how 8K, the next generation of ultra-high definition TV technology, can be integrated into daily life. The innovative broadcasting giant had already started creating and using 8K footage on a trial basis a year earlier, providing a resolution 16 times higher than Hi-Vision to its audience. *NHK meets Deep Space 8K* offers a glimpse into the joint research project between the Ars Electronica Futurelab and NHK (for more information see p. 368).

project by NHK (Japan Broadcasting Corporation) + Ars Electronica Futurelab
Author: Vanessa Graf

IMMERSIFY

Virtual reality and other forms of immersive media have the potential to disrupt the entire media industry with new user experiences that are more immersive and interactive compared to current video, cinema and TV. In order to reach a mature state beyond the current niche markets, the quality of experience of VR media has to be improved in several ways. *IMMERSIFY* is developing key tools for allowing the next generation of immersive media applications. *IMMERSIFY* is a European R&D consortium funded by the EU's Horizon 2020 program and Ars Electronica Futurelab is one of the partners (for more information see p. 364). During this year's festival, the ongoing progress of R&D is presented in the form of demos in the Deep Space 8K. Two time-lapse videos, one of Los Angeles – Pano LA 10K – by Joe Capra (Scientifantastic) and the other of Atacama Desert in Chile by Martin Heck (Timestorm Films) are shown. Further Prima Materia from NOHlab, a stereoscopic piece taking the audience on an audiovisual journey, can now be enjoyed in 8K.

PSNC – Poznan Supercomputing and Networking Center; Spin Digital Video Technologies GmbH; Ars Electronica Futurelab; Marché du Film – Festival de Cannes; Visualization Center C

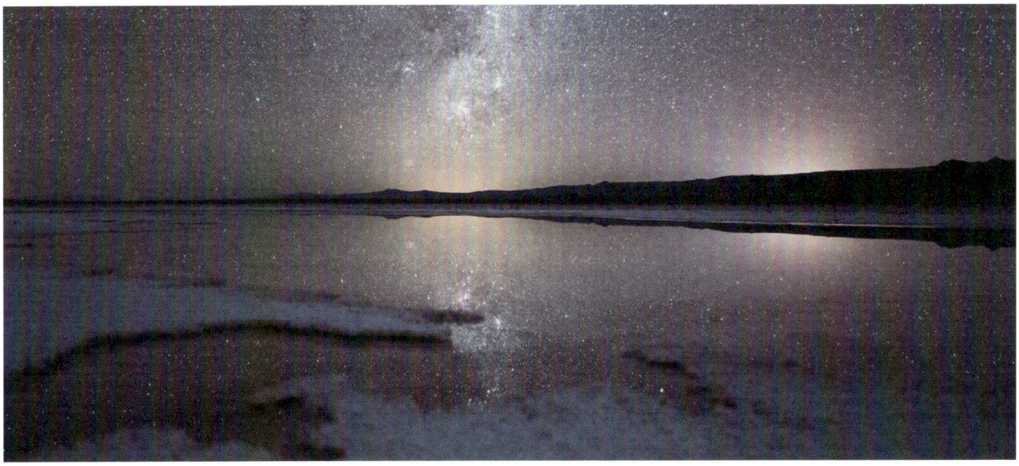

Martin Heck

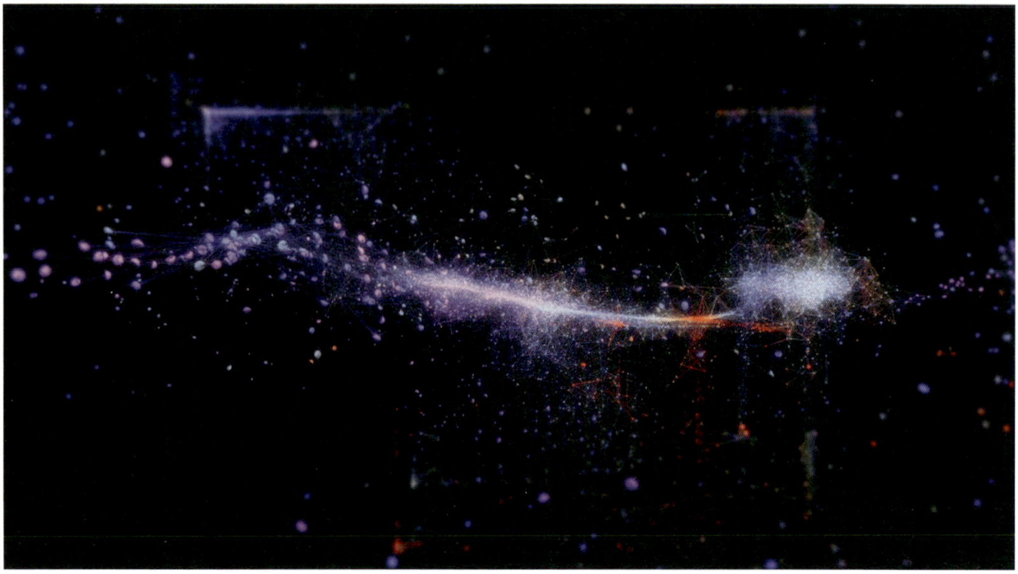

Tadej Droljc (SI)

Singing Sand

Singing Sand was inspired by the sonic potential of abstract 3-D computer graphics. It explores how various particle fluctuations, tensions or shape morphing sound, the internal rhythms they create and how that in turn affects our perception of the visuals.
Grains of sand resembling visual particles are sonified by mapping their individual velocities to various parameters of individual grains inside a custom-made granular synthesizer. Velocities also determine the color of the particles, which together create a spectrum of colors and sounds.
The aim of the piece was to explore how such audiovisual material could function in a free audiovisual paradigm and how to meaningfully force it onto the grid of harmony and metric rhythm. In the latter case the material created liquid grooves on top of dub-tech influenced fixed rhythmic elements, as well as ever-changing spectral swirls that emerged from cross-breeding an original noisy and non-stable sound source with a pitch-based material.

Singing Sand represents a part of my PhD composition portfolio
Special Thanks: Dr Alex Harker and Prof. Pierre Alexandre Tremblay (PhD mentors), Centre for Research in New Music (CeReNeM), University of Huddersfield (Dennis Smalley scholarship in electroacoustic music), Ministry of Culture Slovenia (Scholarship for post graduate studies abroad)
Mastering: Gregor Zemljič

Bela Usabaev (DE), Dawid Liftinger (AT), Florian Liesenfeld (DE)

Old News From New Friends

A video and an interactive sound installation share the Deep Space 8K. While the video is decomposing cinematic story-telling, the space is divided into acoustically shared rooms. The gaze and the ear are led down different paths for a moment and space for interaction is created. Spontaneous collective experiences become possible. The experience-building process is not guided by any rules, but only by the structure of the installation. However, not everybody shares the same experience, and the amount of collective exchange is decided individually.

7062 Lines, Dawid Liftinger, Audiovisual Installation
Audio: Sebastian Jazura
Time, space, acceleration, experience. An examination of horizontal polychromatic lines as a perception distorting experiment.
metadigma, Florian Liesenfeld, Interactive Installation
How individual is a world that one shares? How do you sensitize yourself to the perspectives of others? Metadigma is an interactive installation for the cooperative consideration of one's own environment.
Academy of Media Arts Cologne

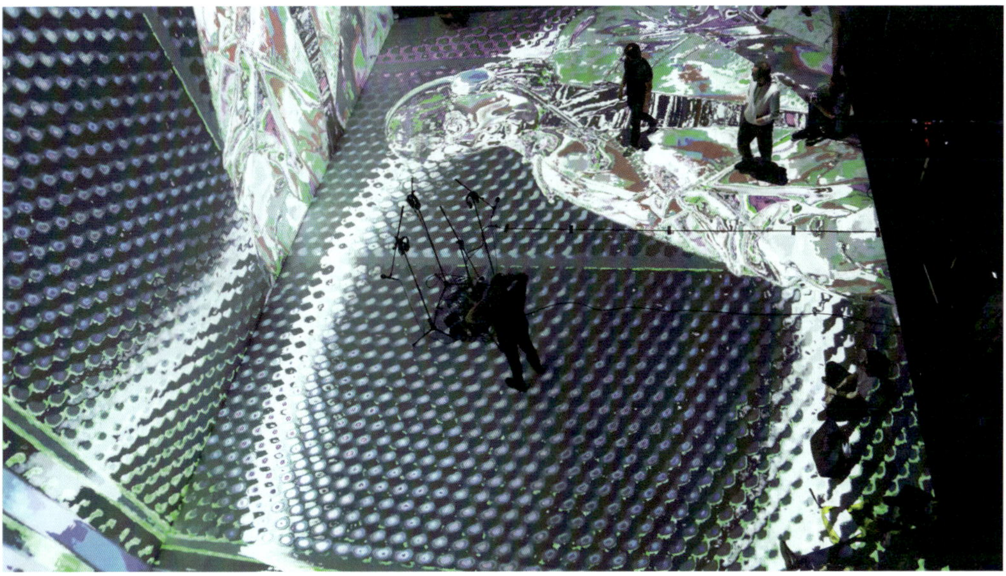

Bruno Manguen

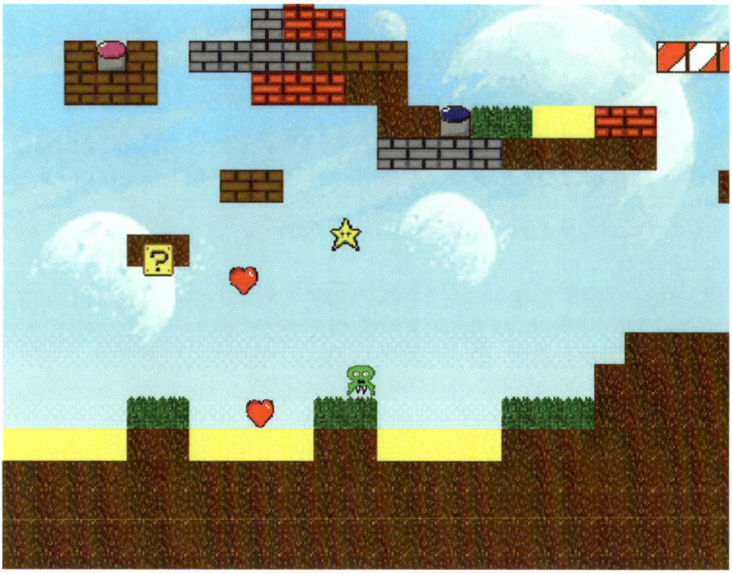

Marko Ciciliani (HR/DE), Christof Ressi (AT), Barbara Lüneburg (DE)

GAPPP – Gamified Audiovisual Performance and Performance Practice

GAPPP: Gamified Audiovisual Performance and Performance Practice is an arts-based research project conceived and run by composer and audiovisual artist Marko Ciciliani. Since 2016 Marko Ciciliani, violinist and artistic researcher Barbara Lüneburg, and musicologist Andreas Pirchner have conducted this artistic research project at the IEM, the Institute of Electronic Music and Acoustics of the Kunstuniversität Graz, where they explored various possibilities for utilizing elements from computer games in the context of audiovisual works. Altogether a dozen new works have been created by six artists, in which the combination of game strategies and game-driven performer interactions are explored for their artistic potential. In the event in Deep Space 8K M. Ciciliani, B. Lüneburg and C. Ressi will perform two works that emerged from the project, preceded by an introduction to the research design.

Partner: IEM (Institute of Electronic Music and Acoustics) of the Kunstuniversität Graz/University of Music and Performing Arts Graz
GAPPP is funded by the Austrian Science Fund FWF as part of the PEEK program for artistic research. *http://gappp.net*
Supported by: FWF/Austrian Science Fund as AR364-G24

Benjamin Weber (AT)

IPv4 Cube Anomalies

An online error is usually an accident. Conversely, in China, error is often intentional. Censors decide what information will be displayed or not. In the latter case, a message would alert the netizen: "Page not found." This error message is an intentional anomaly that appears by the design of The Great Firewall of China. *IPv4 Cube: Anomalies* visualizes and sonifies the entire IPv4 address space (Internet) and indicates the Chinese cyberspace within—shielded by the Firewall. Often described as a "giant cage," the Firewall here is neither fixed nor stable, but fluid and dynamic. The IPv4 address space is traced and visualized as a cube. Each cube situated inside the larger IPv4 Cube represents 256 individual IP numbers. A total of 4.3 billion addresses are visualized. The addresses pertaining to Chinese cyberspace are rendered in red. *IPv4 Cube: Anomalies* is based on Benjamin Weber's IPv4 Cube (2012), which was remastered and extended for this year's festival.

A special thanks to cyber security researcher Valentin Weber (University of Oxford), for his insightful advice and research during this project. I am also very grateful to the Center for Applied Internet Data Analysis (University of California, San Diego) for providing the data that made this artwork possible.
Benjamin Weber, digital artist and researcher living in Vienna, Austria
Valentin Weber, cyber security researcher at the University of Oxford
CAIDA UCSD Anonymized Internet Traces 2018

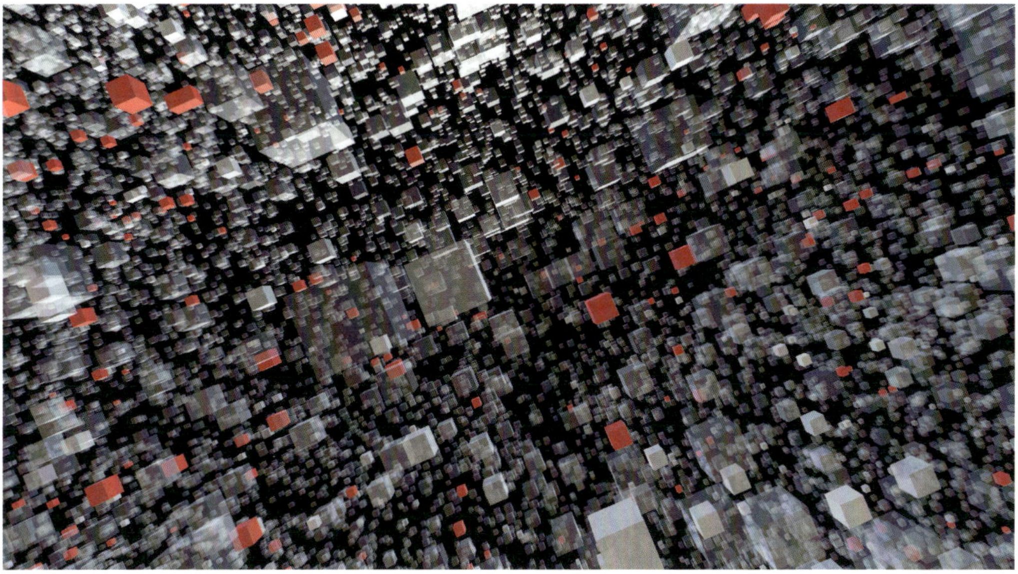

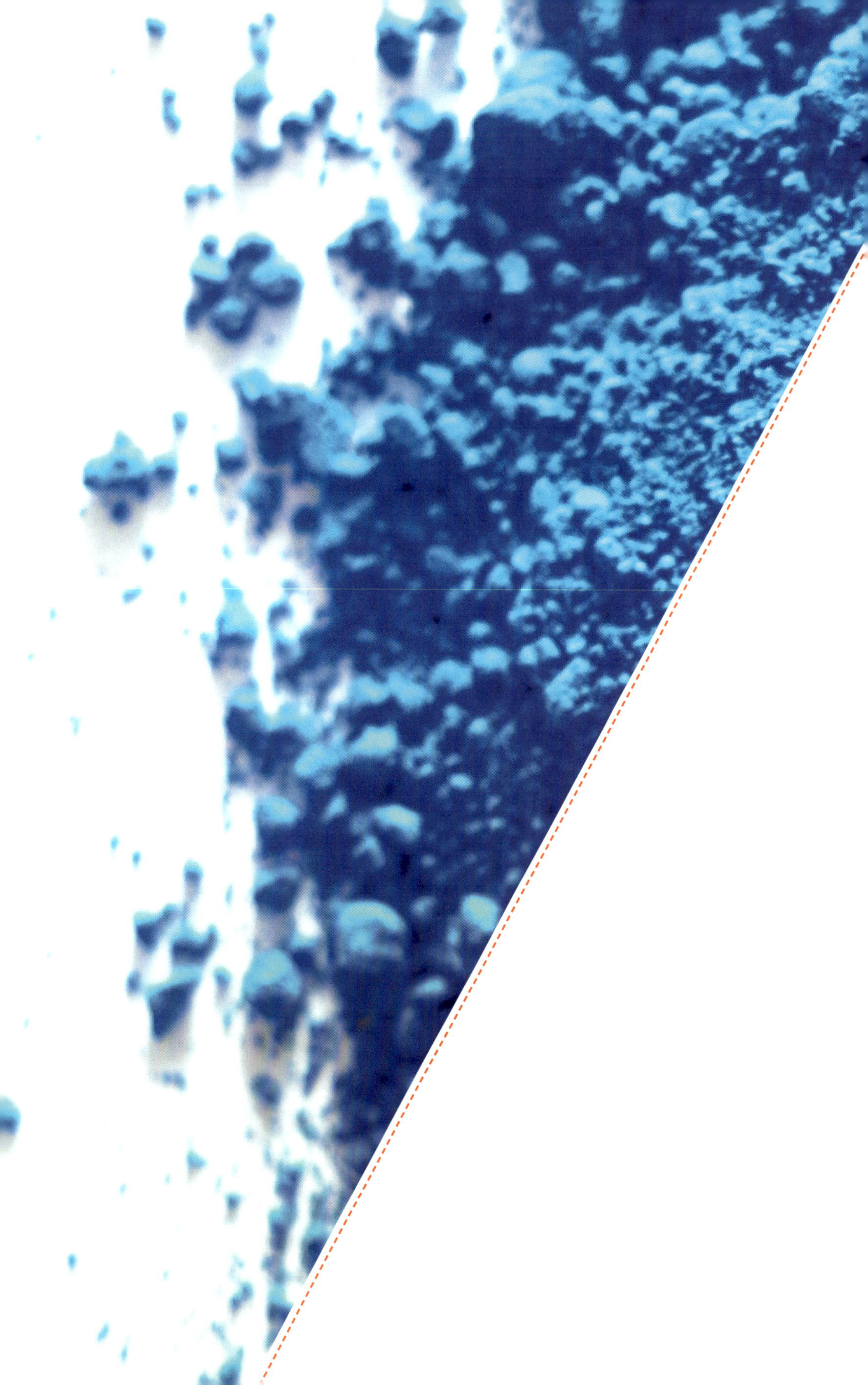

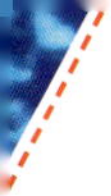

FEATURED ARTIST

Elisabeth Schimana (AT)

Hidden Alliances–Elisabeth Schimana and the IMA*fiction* series

In Search of Female Forebears

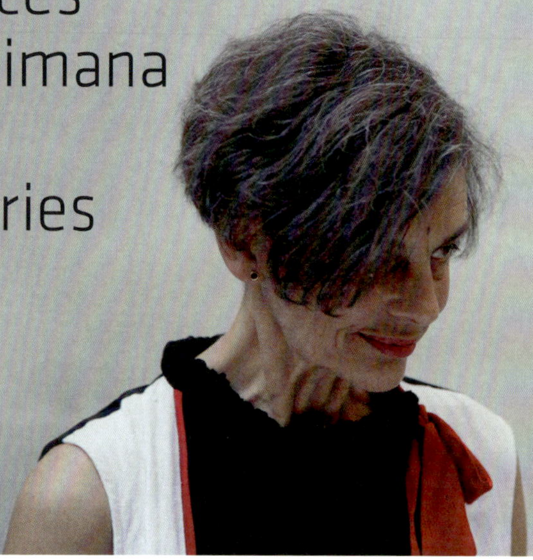

1998

After a decade of work in the electronic music/art field, I began to inquire into another story. In all the years of education and work in this male-dominated genre (with respect to both its teachers and practitioners), almost nothing had been imparted to me about women in the history of electronic music/art. And so I got started weaving new threads into the narratives.

I began with Italian Futurism, which was founded by Filippo Tommaso Marinetti. The noisy music favored by the Futurists, the music of machines, can possibly be considered the starting point of experimental electronic music. Luigi Russolo wrote about this in a manifesto dated March 11, 1913, *The Art of Noises*. In search of the female Futurist, I discovered ideological principles in Italian Futurism, especially with respect to women, that made it seem highly improbable to come upon a fragment of female Futurism: "9. We will glorify war–the world's only hygiene–militarism, patriotism, the destructive gesture of freedom-bringers, beautiful ideas worth dying for, and scorn for woman." [1] But in an out-of-print book, *Geschichte des Futurismus* by Christa Baumgarth, I nevertheless found what I was looking for.

A TRACE

Futurist painters Umberto Boccioni, Carlo Carra, Luigi Russolo and Gino Severini staged their first exhibition in Paris in 1912. Their sojourn there provided the painters with an opportunity to get acquainted with artists of the Parisian avant-garde. Valentine de Saint-Point, the niece of Alphonse de Lamartines and an artist in this scene, hosted an evening of music and literature, after which she joined the Futurists and authored the *Manifesto of Futurist Woman*[2].

"The women: the Erinyes, the Amazons; the SEMIRAMIS, JEANNE D'ARC, JEANNE HACHETTE; JUDITH and CHARLOTTE CORDAY; (...) May the next wars bring forth heroines like CATERINA SFORZA. During the siege of her father-city, she looked out from the ramparts and saw how the enemy threatened her son to force her to surrender. But she revealed her gender in heroic fashion and cried out: 'Kill him; I'm fertile enough to give birth to others!' " [3]

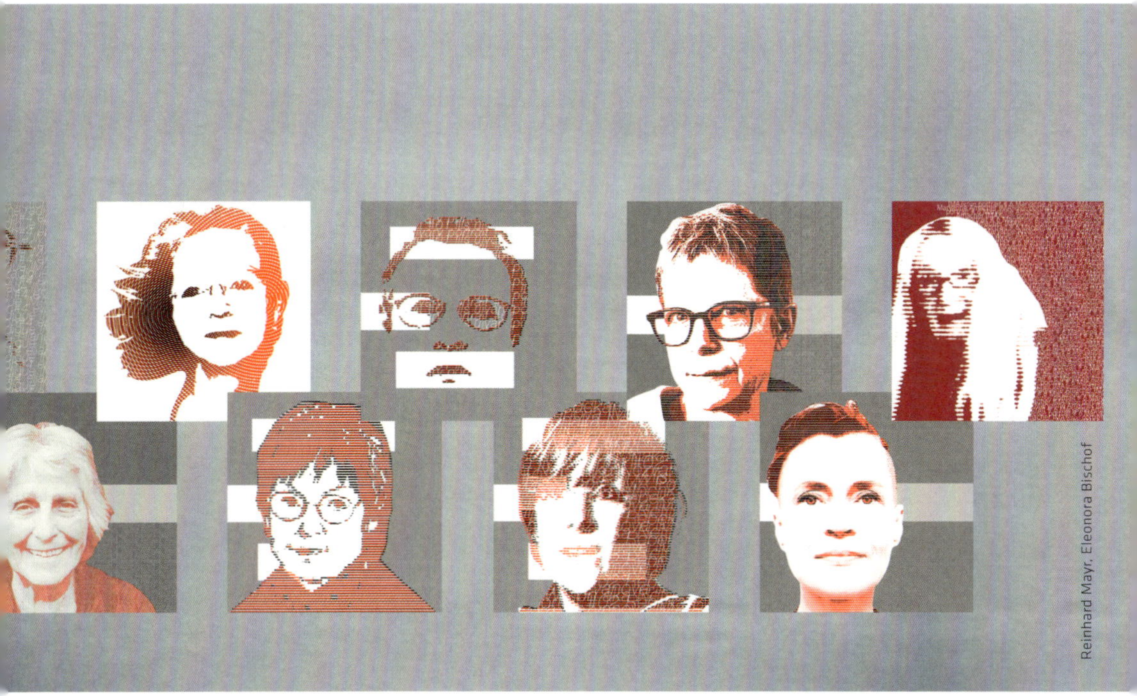

Reinhard Mayr, Eleonora Bischof

2002-03

This research manifested itself artistically in a performance entitled *Portrait 01–Die Futuristin*. I interwove my voice and face with those of Andrea Sodomka (fellow-traveler right from the start), Rebekah Wilson (aka Netochka Nezvanova / with NATO.0+55 software) and Tatjana Komarova (director of the electronic studio at the Yekaterinburg Conservatory) to form a fictional portrait.
http://elise.at/projekt/Portrait-01-Die-Futuristin

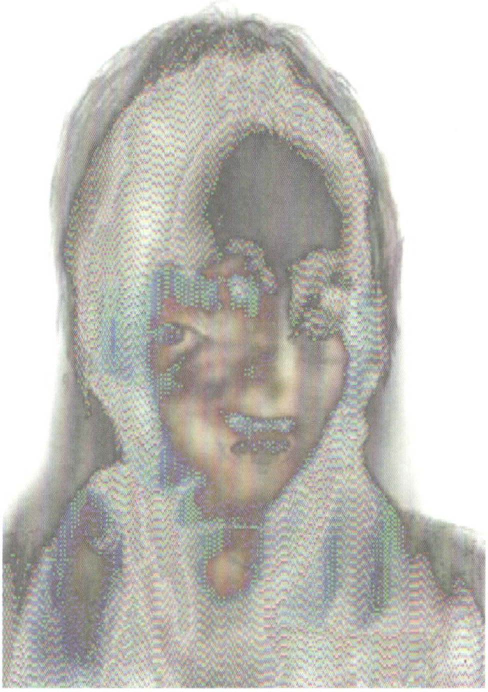

2005

This was the year in which Andrea Sodomka and I jointly established the IMA Institute of Media Archeology. For IMA, technology and art by women has a clearly positive interrelationship. Whether this art is now produced by means of processing via devices or programming computers is beside the point. The primary consideration is an active process of coming to terms with technological developments.
IMA*fiction* is the Institute's portrait series dedicated to women doing electronic art with a focus on sound.

What have appeared are video portraits of Liesl Ujvary (AT), Rebekah Wilson aka Netotschka Nezvanova (NZ), Heidi Grundmann (AT), Eliane Radigue (FR), Andrea Sodomka (AT), Maryanne Amacher (US), Anne La Berge (NL), Electric Indigo (AT) and Beatriz Ferreyra (AR, FR); the tenth portrait, my own, brings this series of five Austrian and five international artists to a preliminary close. Most of the artists made their own choice as to whom they wanted to be portrayed by. The results are homages that are aesthetically diverse, intimate, and not in enforced conformity with any conventional film format. *https://ima.or.at/imafiction/*

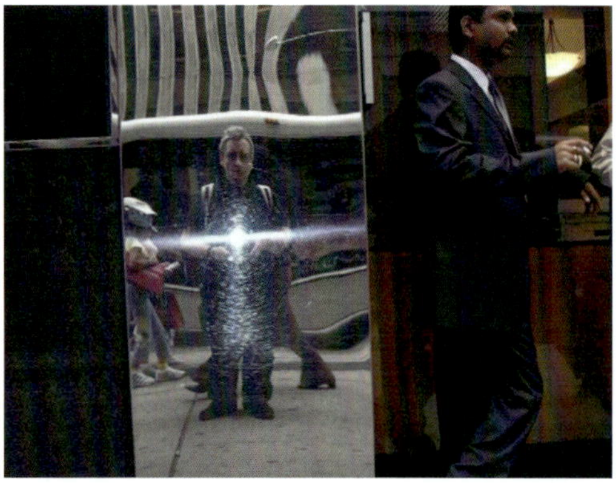

Liesl Ujvary, portrait #01

2018

What has changed since 1998? The intensive encounter with the world of ideas, the working methods and the music/art of female contemporaries and forerunners has enriched me and supplemented the overall account. Much still lies fallow—and thus a call goes out to musicologists and curators to cultivate this unworked soil. It's the purposely focused gaze that brings different things to light.

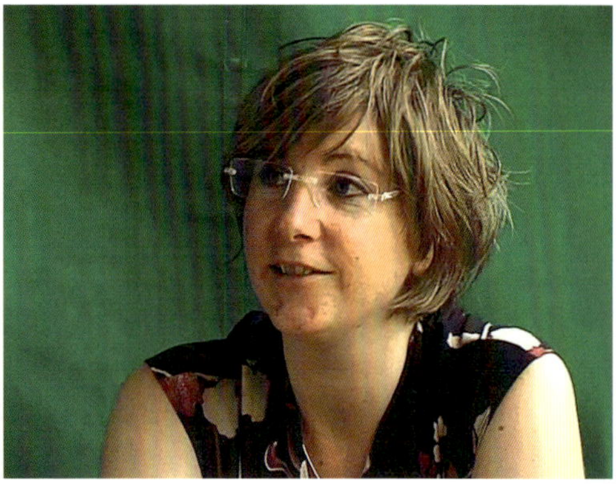

Rebekah Wilson, portrait #02

In the exhibition entitled *Hidden Alliances*, 10 female artists, pioneers of their time, tell rather a different story and together weave a fascinating network of interrelationships.

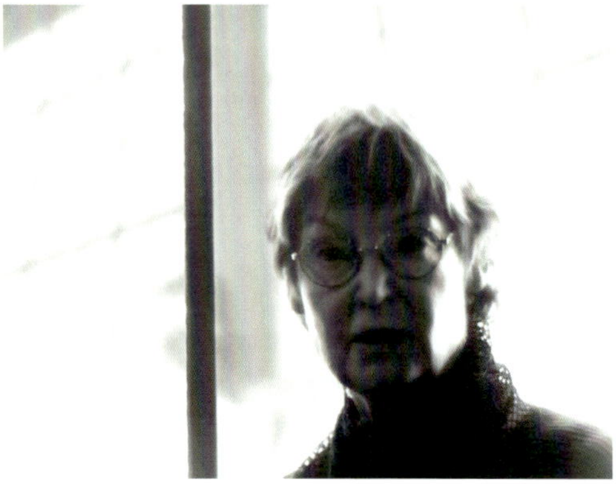

Heidi Grundmann, portrait #03

www.ima.or.at
www.elise.at

1 Marinetti, Filippo Tommaso, Manifest des Futurismus, 1909 (read in: Geschichte des Futurismus, Baumgarth Christa, 1966, p. 26)

2 Valentine de Saint-Point, *Manifeste de la Femme Futuriste*, 1912

3 de Saint-Point Valentine, Manifest der Futuristischen Frau, 1912, (read in: Geschichte des Futurismus, Baumgarth Christa, 1966, p. 237, translated by Jean-Jaques)

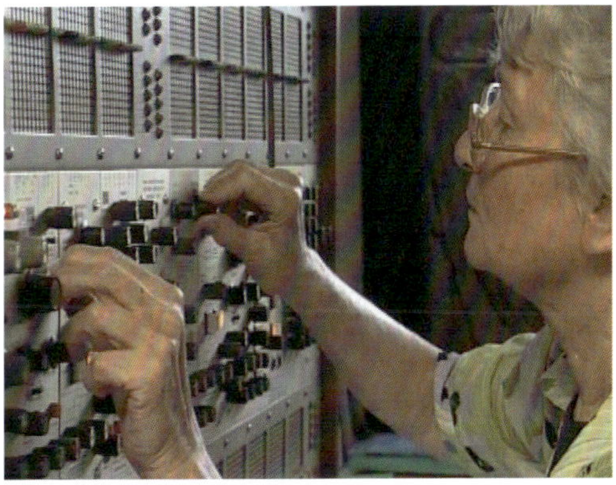
Elaine Radigue, portrait #04

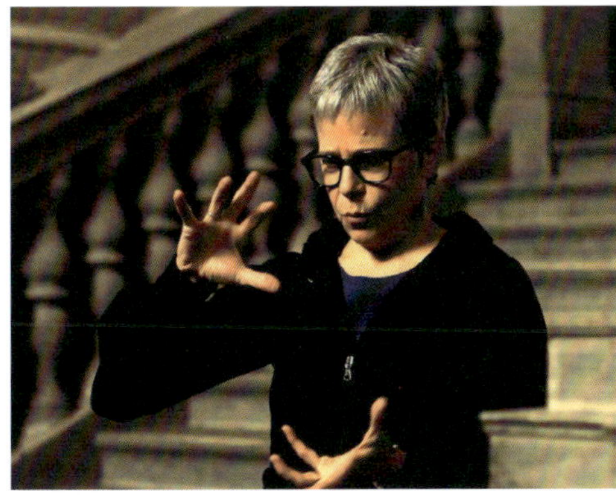
Anne La Berge, portrait #07

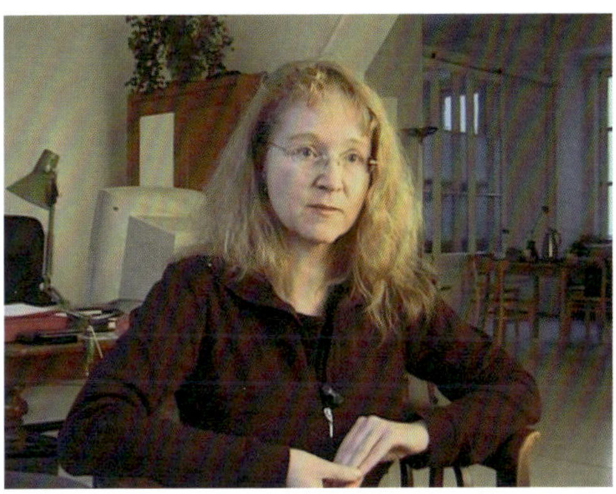
Andrea Sodomka, portrait #05

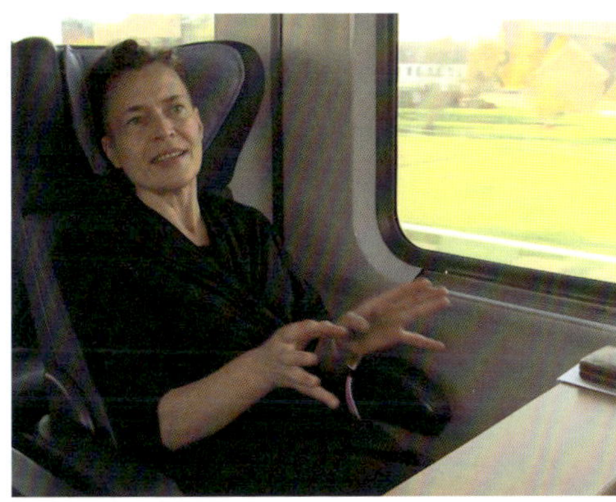
Electric Indigo, portrait #08

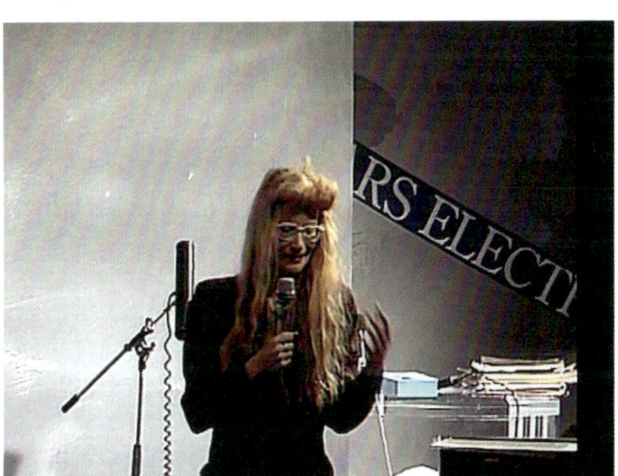

Maryanne Amacher, portrait #06

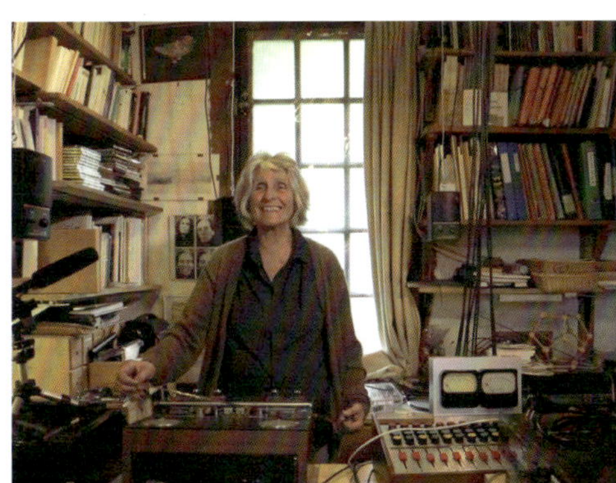
Beatriz Ferreyra, portrait #09

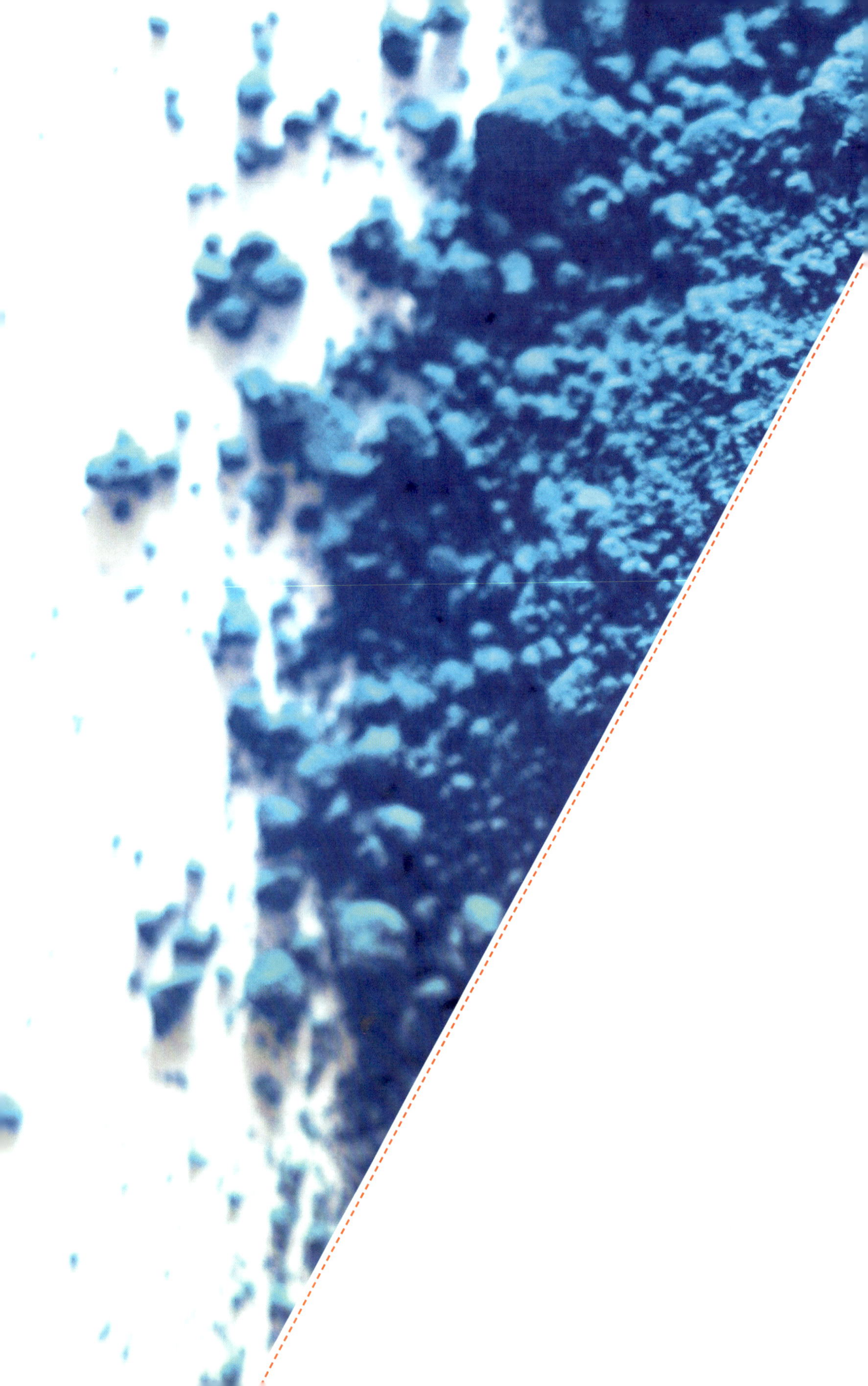

GALLERY SPACES

Gallery Spaces

Since its successful launch last year, the Gallery Spaces Program, deliberately written in the plural, has brought a large number of international galleries and collections with their different positions on digital art to the Ars Electronica Festival.
But there is more to it than showing digital artists represented by galleries - it is above all about the changing conditions of creating and marketing art under the impact of digitalization.
New possibilities and business models for positioning oneself in the digital online marketplace are discussed as well as digital strategies for museums and established art collections.
These workshops and roundtables also discuss how digital art can be preserved in the long term and how the many works created since the 1960s can be restored and preserved for the future.
With its continuous work and experience in the production and presentation of media art and digital art since 1979, as well as the remarkable exhibition areas of "Postcity," Ars Electronica is the ideal environment for this exchange.

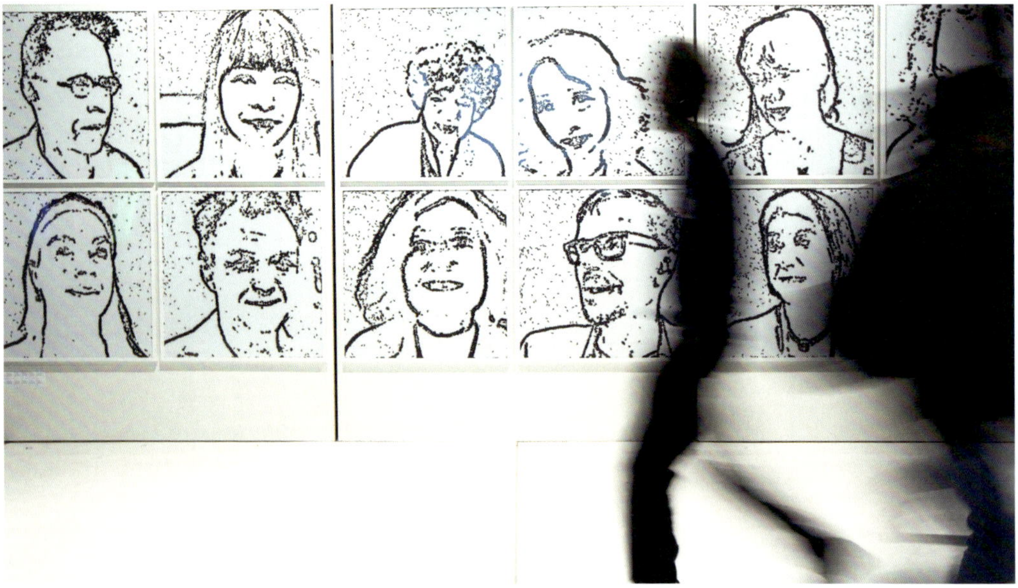

Tom Mesic

Portrait on the Fly, Laurent Mignonneau & Christa Sommerer (AT), Galerie Charlot, Paris/Tel Aviv-Yafo

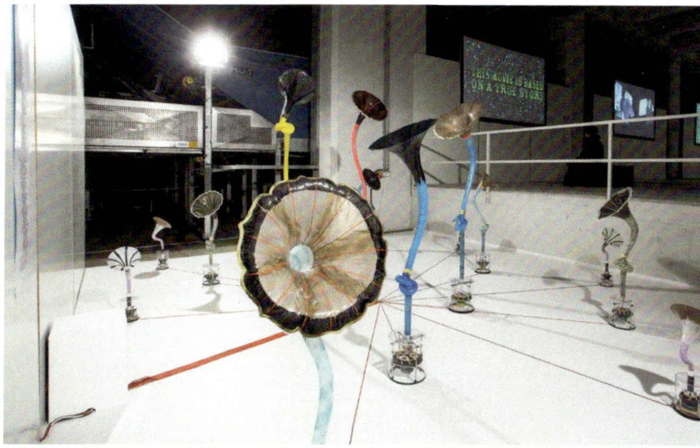

Tom Mesic

Summer of Love, Douglas Henderson

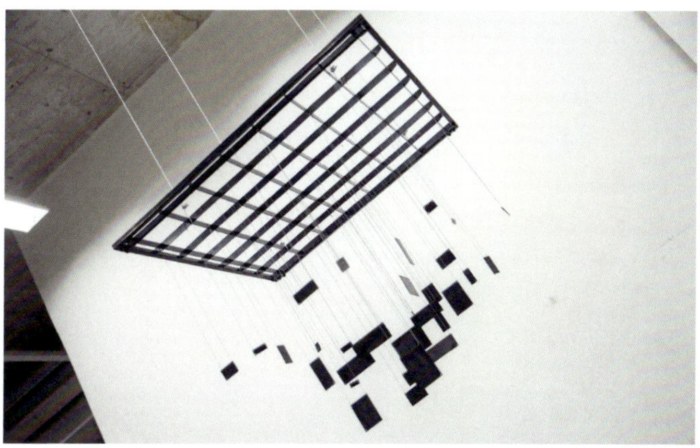

Tom Mesic

Untitled (Attractors Series), Voldemars Johansons (LV)

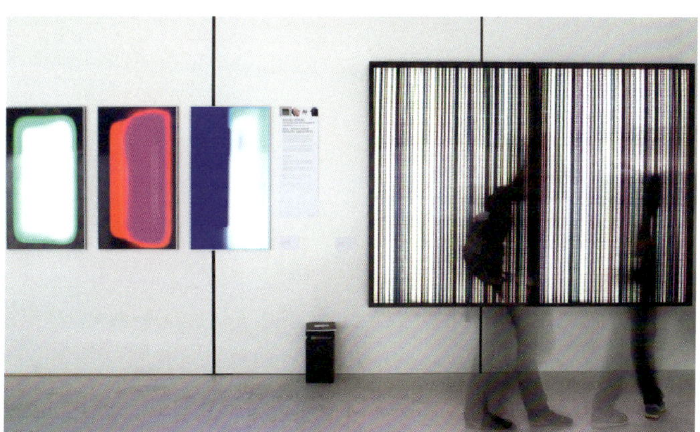

Christopher Sonnleitner

AROTIN & SERGHEI – contemporary art research & creation in cooperation with W&K – Wienerroither & Kohlbacher, Gallery Vienna

ERROR

the Art of Imperfection

EVENTS, CONCERTS & PERFORMANCES

Ars Electronica Opening

As every year, the Ars Electronica Opening will bring compelling and thought-provoking acts to the stages of PostCity. The night begins with the contorting and undulating *πTon* installation by Cod.Act. Following this otherworldly experience is Italian artist Gabriele Marangoni, who will send visitors on a journey to the limits of perception with his project *Silent*, a collaborative piece between a deaf performer and a vocal soloist.

Chris Ziegler's presentation, on the other hand, focuses on the limits of visual perception. *Corpus* sparks the visitor's imagination with an abstraction of a moving body creating visual and sonic poetry. *Interférences* by Alexis Langevin-Tétrault however relates to the topic combining gestural interaction with an unique device on the main stage. In the following performance *Field* of Martin Messier, residual electric signals, imperceptible to eyes and ears, are picked up by electromagnetic transducer microphones and become the material of the work.

Continuing the night are Chicks on Speed's Alexandra Murray-Leslie and Melissa E. Logan, live composing on self-made electroacoustic objects; a High Heeled Shoe Guitar, self-contained amplified hats and Theremin Tapestry culminating in an audiovisual performance art spectacle. Kutin/Kohlberger's audiovisual interplay of the ancient, resonating sound of a prepared hurdy-gurdy with sonified results of machine learning algorithms of the 21st century will be the last performance before Dino Spiluttini's drones and ambient soundscapes send the visitors off to another musical dimension.

The program will be rounded off by Welia who will play one of her fantastic DJ-sets in the courtyard.

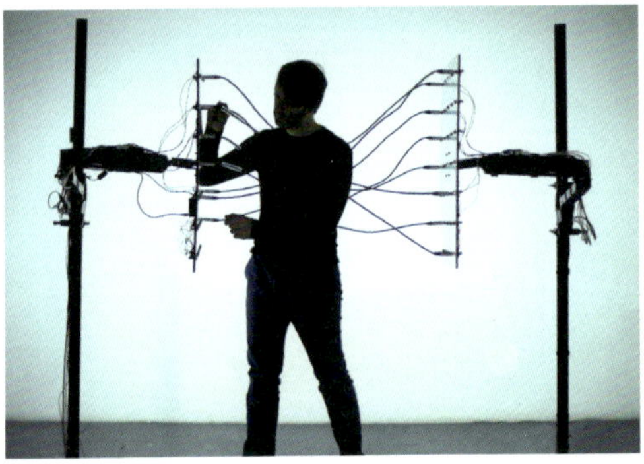

Field, Martin Messier

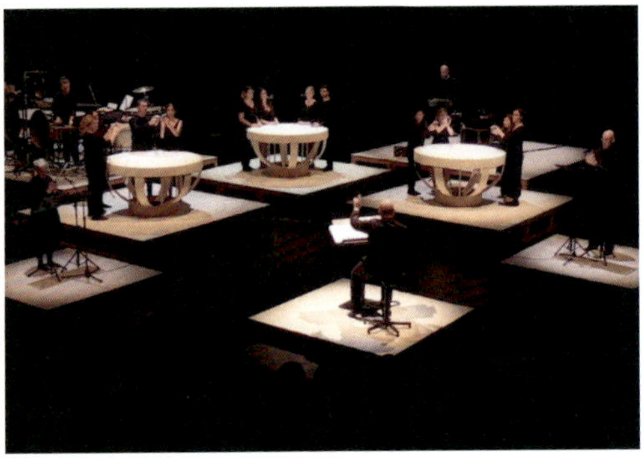

Silent, Gabriele Marangoni

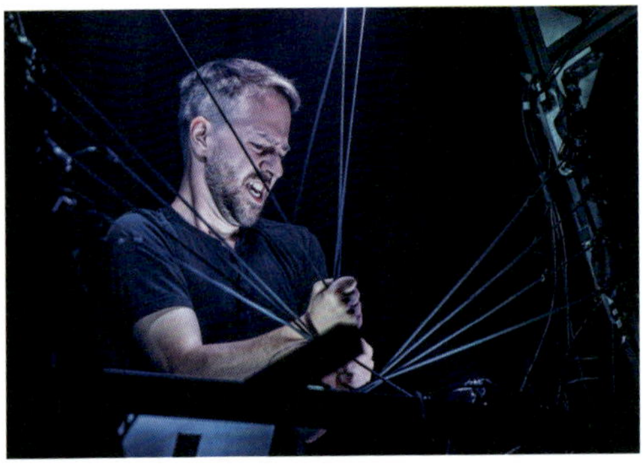

Interférences, © Trung Dung Nguyen

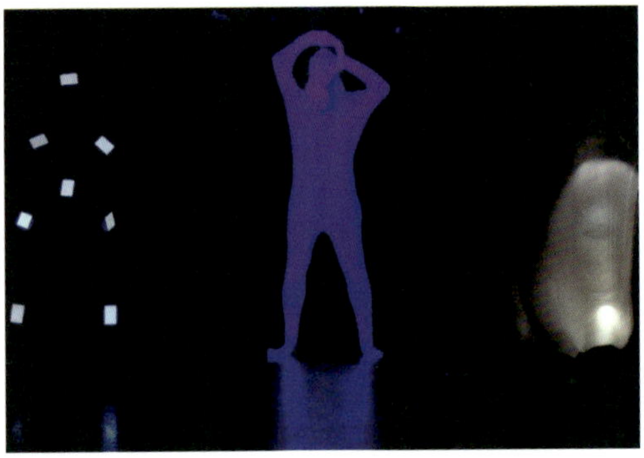

Corpus

Ars Electronica Nightline

Once again the Ars Electronica Nightline will be showcasing contemporary electronic music by international and local performers. B.Visible, the maestro of broken beats and inventive samples accompanied by his live-band, kick off the evening. Following up is Catnapp (Monkeytown Records), singer/rapper and striking performer from Berlin, who takes whatever she needs from all genres of electronic as well as pop and rap music, combining it into an energetic and eclectic live performance. Group A will be delivering a hard break from pop-based structures, playing with noise, experimental electronics and voice, combining it all into one beautifully chaotic experience.

Iglooghost, the UK-based electronic producer and newest signee to Flying Lotus' Brainfeeder label, invites us to live it up on the dancefloor with hyper-speed booms, fizzes, squeezes and stretches. Visionist continues on the main stage with detailed visceral soundscapes to encapsulate the experiential onslaught of an anxiety attack, accompanied by striking visuals by filmmaker Pedro Maia. Closing the mainstage will be Golin whose performances are full of improvisation, with a strong focus on the body and its movements, as well as an instant spontaneity that turns her shows into perpetually surprising cinematic experiences.

Parallel to the main stage, a program of anarchic high-energy non-conformism confronts the visitors on the Salon Stage. The show opens with Linz-based accordion noise performer Stefan Mittlböck-Jungwirth-Fohringer supported with visuals by Peter Freudling. After this, the Viennese duo Dvrst plays a show reflecting on the unity between club culture and experimental arts.

The tracks of the Polish-Austrian duo Mermaid & Seafruit inhale the deep, corporeal bass of Grime, the snottiness of Hip Hop, the frenzy of Hardstyle, the confrontational dynamics of Noise as well as the patient elegance of R'n'B and theatrical Spoken Word passages. Closing Friday night is the famous part-human part-meme, flute-dropping rule-breaker DJ Detweiler. He will give people a reason to dance to the weirdest mix of musical genres well into the morning hours.

For everybody who needs to step outside for a minute and take a break, Linz' all-time favourite Andaka plays in the courtyard in front of the Train Hall.

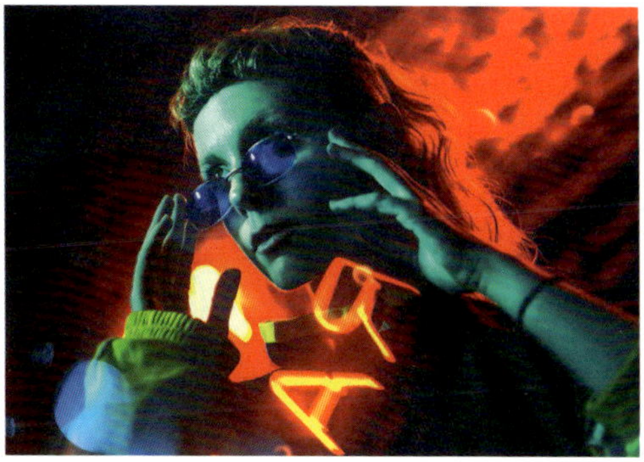

Catnapp, © *Calcio*

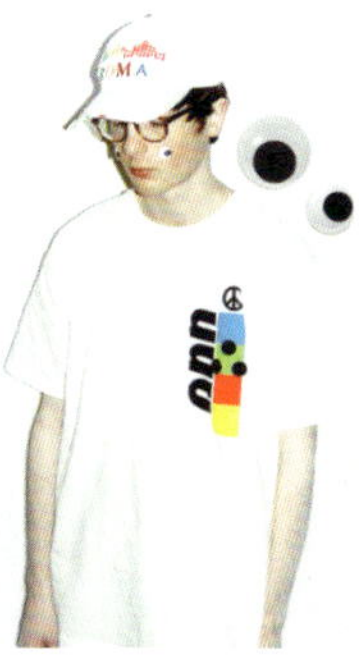

Igloohost, © Tim Twiss

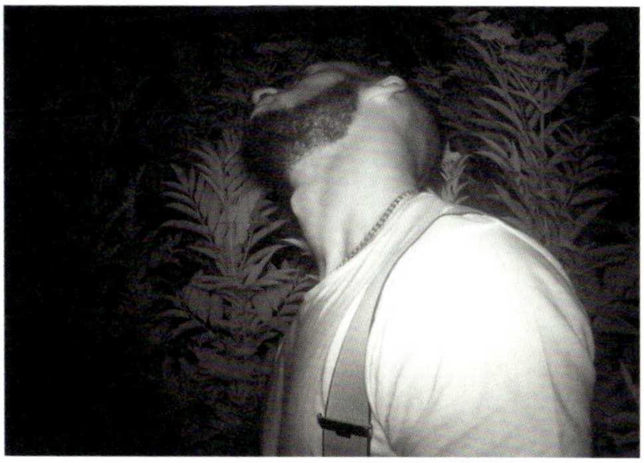

Visionist, © Daniel Sannwald

Alexandra Murray-Leslie (AU/ES), Melissa E. Logan (DE/US)

Chicks on Speed

Chicks on Speed began to make *OBJECTINSTRUMENTS* (self-made post-digital musical instruments) out of necessity in 2003. The art group began with a minimal stage set up and felt there was something missing on the big stages. It all started with the *OBJECT-INSTRUMENTS* when Alex and Melissa performed a show with F.M. Einheit. Alex was impressed by Mufti's custom-made metal instruments and tools—it was loud, messy and dusty. Backstage Alex asked Mufti which technical set-up he uses. After that, Melissa started experimenting with contact microphones and other materials. The experimentation of building and performing *OBJECTINSTRUMENTS* continued with iconic body-centric pieces like the *High Heeled Shoe Guitar*, *E-Shoe* and *Theremin Tapestry*. The group now possesses an ever-expanding *OBJECTINSTRUMENTS* archive of around 30 works. The risk-taking involved and the experimental nature of performing with self-made musical instruments has been an integral part of the artistic development of *Chicks on Speed* live performances. The group isn't interested in perfection, repeatability or virtuosity, but in exposing experimental processes live on stage, sometimes at the risk of jeopardising the perfect pop show.

Chicks on Speed's Alexandra Murray-Leslie & Melissa E. Logan perform OBJECTINSTRUMENTS in collaboration with technologist Dr. Sam Ferguson & costumes by Nina Kraine, Julio Escudero, Department of Fashion&Technology, The University of the Art and Design Linz. The Chicks on Speed archive of OBJECTINSTRUMENTS commissioned by: ZKM Centre for Art and Media, Karlsruhe, Artspace, Sydney, Craft Victoria Melbourne, NTU Centre for Contemporary Art Singapore, Dundee Contemporary Arts, Institute of Modern Art Brisbane and Milani Gallery Brisbane. *Computer Enhanced Footwear prototypes 1 & 3* by Alexandra Murray-Leslie made at Autodesk Pier 9 Technology Centre, San Francisco and Creativity and Cognition Studios, The University of Technology Sydney.
Tina Frank (AT), live video

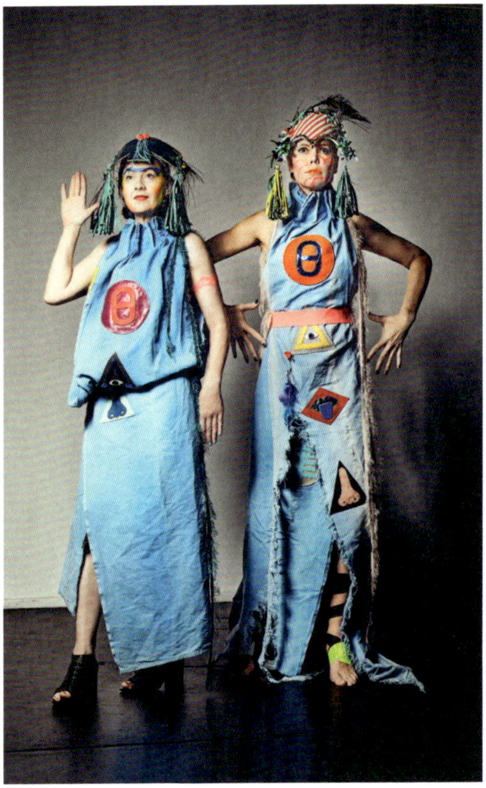

Alexandra Murray-Leslie & Melissa Logan (Chicks on Speed)

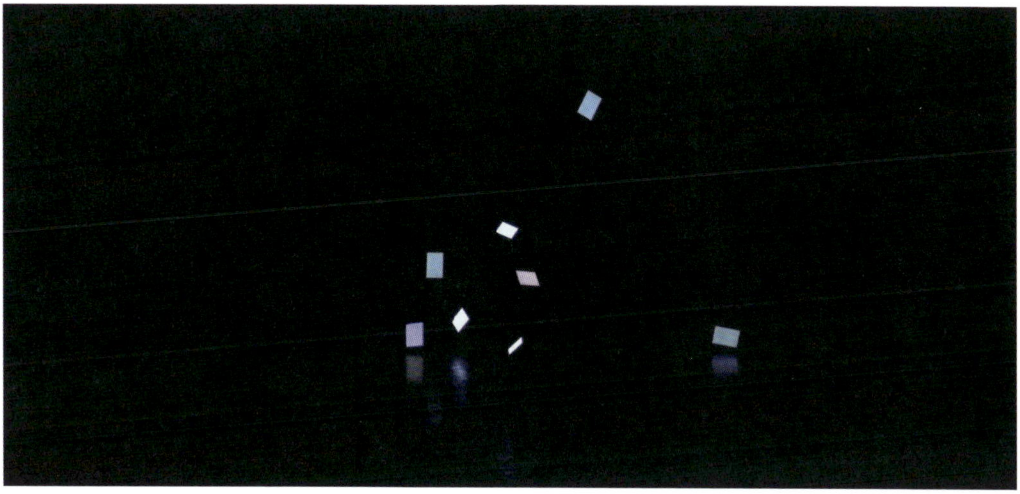

Chris Ziegler (DE)

CORPUS (2011/2017)

Ovid's "Pygmalion" redefined—a body in morphogenesis. A solo performance that uses digital, visual, sensory and material interactions to create a body of images and sounds. *CORPUS* articulates a process of interaction between the artist and the artifact in a feedback loop. Where is the border of analog to digital, where are the glitches in the conversion? Where "the apparatus" first needs the visible body in full flesh to generate images, *CORPUS* renders the perception of an animated figurine first, before revealing the body. *CORPUS* begins with an evolution of a single light cell organism to a human figurine. Flesh fills another body shell, creating a sculpture in a semi-transparent cocoon . In "Dance" the body can be perceived both as a subject and as an object of an artistic process. *CORPUS* explores the moving body and its frictions with a visual and acoustic "interaction body." The disguised moving flesh in a digital costume gives the viewer room for imagination.

Direction: Chris Ziegler
Dance: Unita Gaye Galiluyo
Music: Hugo Paquete
Costumes: Ismenia Keck
Software: Martin Bellardi, Nikolaus Völzow, Chris Ziegler

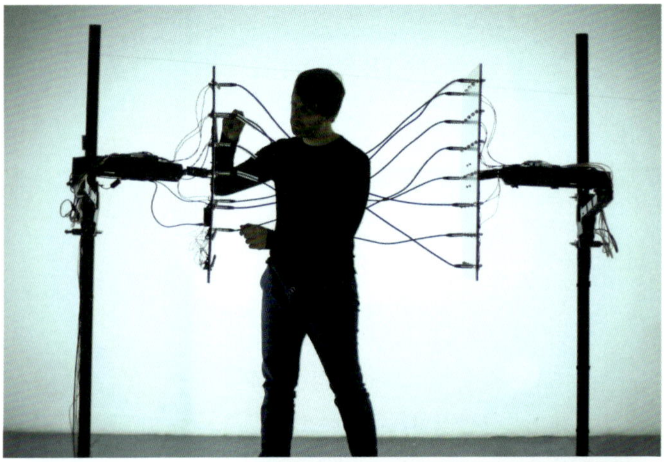

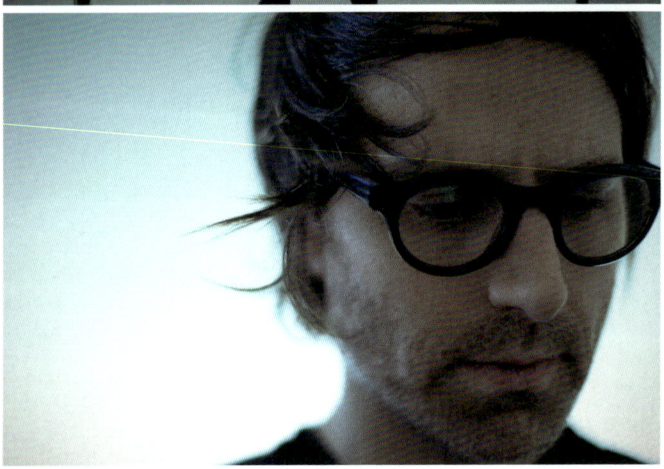

Martin Messier (CA)

FIELD

With *FIELD*, Messier gives material form to this otherwise inaudible, invisible flow. He becomes the operator through whom the work is activated and brought into the real world. A veritable mimesis of electromagnetic current, the visual aspect can plunge the audience into a hypnotic state: the omnipresent imperceptible power that surrounds us is stripped of its mystery here and finally seems accessible to us. *FIELD* speaks to the invisible forces around us: their ascendancy and their interdependence. Although they interact at an absolutely indiscernible level, they underlie our gestures and movements and carry us, in a way. In this sense, the work and the artist arrive together at an exchange that determines the conditions of the visual and sound elements of the performance: blinding lightning and electricity conduction.

Concept, audiovisual composition, programming and performance: Martin Messier
Interface: Thomas Payette
Technical design: Thomas Payette, Maxime Bouchard, Frédérique Folly
Production: 14 lieux

Alexis Langevin-Tétrault (CA)

Interférences (String Network)

Interférences (String Network) is a 20-minute audio-visual performance that explores the possibilities of embodying an electroacoustic work in real time through gestural interaction with an unique device. On stage, an audioreactive play of light unfolds gradually: Alexis Langevin-Tétrault builds a network of strings and captors with which he interacts to generate lighting and create a sound universe between industrial noise, electronica and acousmatic music. Through the staging of corporality and the dialectical relationship between human and machine, *Interférences (String Network)* presents an allegory of the globalized and interconnected modern world in which the individual seeks to derive meaning from his experience and attempts to preserve his freedom of action.

Composer and musician from the post-rock, acousmatic and electronic scenes, Alexis Langevin-Tétrault proposes a singular brutalist universe and infuses a dynamic of live music to a musical genre rarely incarnated by performance.

This project was made possible by the support of: Conseil des Arts et Lettres du Québec (CA), Château Éphémère - Fabrique sonore et numérique (FR), Université de Montréal (CA), the Social Sciences and Humanities Research Council of Canada (CA), Fonds de Recherche Science et Culture du Québec (CA), Exhibitronic Festival (FR) and Centre national de création musicale Césaré (FR).
Concept, device manufacturing, programming audio and lights, composition and performance: Alexis Langevin-Tétrault
Help with the design and manufacture of lights: Lucas Paris
Advice on the artistic direction: Nicolas Bernier
Advice on dramaturgy: Anne Thériault

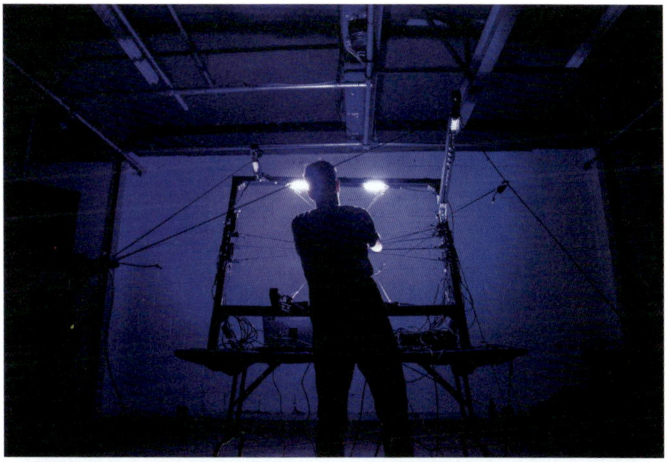

MarieIsabelle Rochon

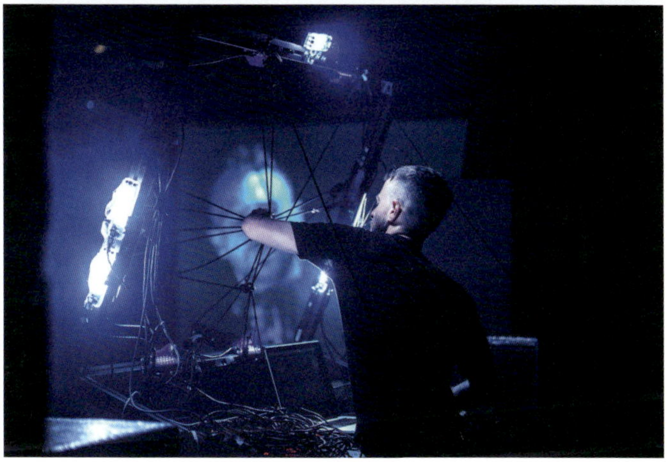

Trung Dung Nguyen

Gabriele Marangoni (IT)

Silent

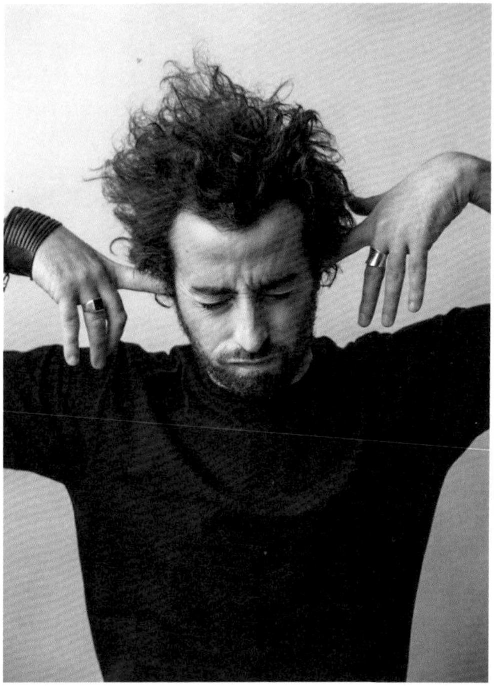

Secret Theater Ensemble and *Tempo Reale* work together for a project where sound is conceived as a vibration to be experienced. The study, conducted by the composer Gabriele Marangoni, is intended to be an avant-garde project that breaks down barriers between deaf and hearing people through an artistic product specifically built to be enjoyed by both groups.

Silent is really a visionary journey to the limits of perception, because the audience experiences the presence of a deaf performer and a vocal soloist who produce the sound together. The central idea is to have access to an extreme region of perception, in which the sound blends with vibration and gesture. This is rendered possible by technological and scenic devices allowing the performers to communicate with each other and help the audience overcome the barrier of sound. With *Silent* Gabriele Marangoni experiments and investigates extreme electroacoustic regions. A sound frequency structure below the hearing threshold has been designed within the work, which works on the physical and cerebral perception of sound at frequencies of 4 Hz where it is the body that perceives the sound, without the sense of hearing.

A coproduction by SECRET THEATER ENSEMBLE, TEMPO REALE, FONDAZIONE I TEATRI, FESTIVAL APERTO REGGIO EMILIA.
In collaboration with: ENS, SGB-FSS Schweizerischer Gehorlosenbund.
Supported by: UBS Cultural Fondation
Concept and Composition: Gabriele Marangoni, Secret Theater Ensemble
Conductor: Dario Garegnani
Male voice with extended techniques: David W. Benini
Set design and vibrating systems: Micol Riva
Producer: Giulia Soravia, Tempo Reale
Live electronics: Damiano Meacci

Stefan Mittlböck-Jungwirth-Fohringer (AT)

Accordion Noise Performance

The performance is based on the results of artistic research and observations of honeybee colonies. In 2017, due to the death of many bee colonies over the previous winter, the absence of the bee was the focus of the artist's observations and documentation. The one-man performance interprets the situations he perceived and experienced by means of an accordion and a number of electronic aids. Among other things, it deals with ongoing transformation and primarily represents the progression from a living, animated colony through to an empty, found hive, its death caused by "colony collapse disorder" (CCD). Secondary to that, it is to be seen through the eye of transformation and deconstruction that is visible in the process to which the "supposedly" empty beehives are exposed throughout the year. The performance is a noise performance that works through the interaction of a performer with the accordion, supported by electronic elements and effects. It lasts about 30 minutes and is accompanied by projected visual elements.

Visuals: Peter Freudling (AT)

Marco Donnarumma (DE/IT), Margherita Pevere (DE)

Azathot

from the 7 Configurations

What does it mean to create a truly autonomous machine, independent from human control? And what happens when organs live outside of a body? Could this help us understand that the power of the human body lies in its ability to be different and to take on unexpected forms and identities?

Violently entangled within the performance space are three elements: an artificially intelligent prosthesis, out-of-body organic wombs and a human body. The prosthesis uses artificial intelligence algorithms to learn in real time how to move, exist and perform on stage. The wombs live and pulsate through the activity of microbial cultures. The sounds of the performer's body are re-synthesised and transformed into a powerful and visceral auditory experience.

The human, the non-human and the machine struggle to access a new form of hybrid identity. They influence each other creating a connection of bodies and psyche able to redefine the concept of intimacy, as well as how we imagine the body and its relationship with new technologies.

Artistic direction, performance, staging: Marco Donnarumma, Margherita Pevere
Concept, music, programming, robotics, light: Marco Donnarumma
Scientific partner: Neurorobotics Research Laboratory, Beuth Hochschule
Robotics visual design and costume: Ana Rajcevic
Robotics 3D modelling and engineering: Christian Schmidts
Additional programming: Alberto de Campo
Light technology: Protopixel
Commissioned by CTM Festival (DE). Supported by the Goethe Institut's International Coproduction Fund. Realized in the context of the Graduiertenschule, Berlin University of the Arts.

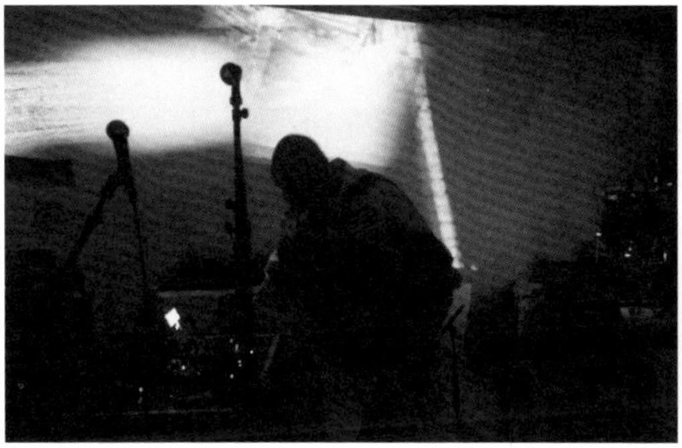

Accordion Noise Performance

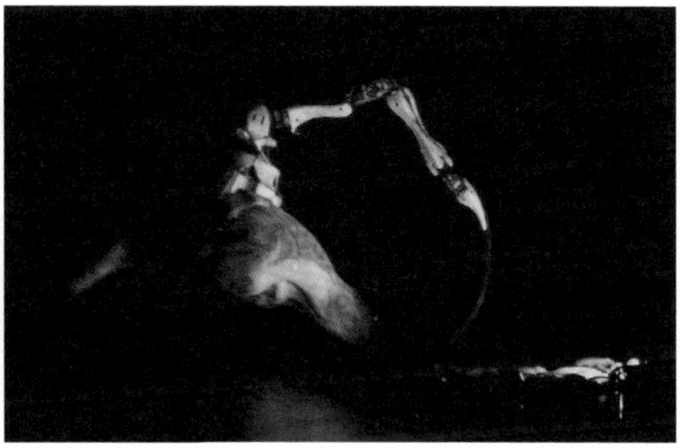

Azathot

Alessandra Leone

The Big Concert Night 2018

In cooperation with the Bruckner Orchestra Linz and the Brucknerhaus

The Big Concert Night has become one of the festival's fixed points and absolute highlights. The spectacular *Gleishalle* of POSTCITY with its powerful but still very harmonious acoustics has given the special encounter between symphony orchestra and digital music in recent years an additional and impressive uniqueness that challenges us to create special productions.
The evening begins with a premiere, a cooperation of Josef Klammer and Jaap Blonk, which, in tune with the festival theme, creates a juxtaposition and coexistence of real and synthetic voices. Afterwards *The Berlioz Project* will be performed, the second major cooperation of Ars Electronica and Markus Poschner with the Bruckner Orchestra Linz.
The third part will start with the Italian composer and musician Gabriele Marangoni and his acoustically prepared accordion, after which "The Vibrationeers", a band of five electrically driven and computer-controlled vibratory tampers, will stir up the acoustics of the *Gleishalle* and its industrial interior.
Electric Indigo and Elisabeth Schimana will be the two final performances and are part of this year's Featured Artist Program dedicated to Elisabeth Schimana and her IMA (Institute of Media Archaeology).

The Big Concert Night 2018 Lineup:

Josef Klammer (AT), Jaap Blonk (NL)

Bruckner Orchestra Linz (AT) conducted by
Markus Poschner
SILK Fluegge
Tour en l´air by Ursula Neugebauer
KUKA KR 600 Industrieroboter, KUKA GmbH
AS50e Akkustampfer, Wacker Neuson
Visuals: Cori Olan

Gabriele Marangoni (IT)
The Vibrationeers: Bernhard Breuer (AT), Didi Bruckmayr (AT), Chris Bruckmayr (AT), Stefan Fuchs (AT), AS50e Akkustampfer, Wacker Neuson
Electric Indigo (AT)
Elisabeth Schimana (AT)

The Big Concert Night was first launched in 2003 by Dennis Russell Davies, Gerfried Stocker, Wolfgang Winkler and Heribert Schröder as a cooperation project of Ars Electronica, Bruckner Orchestra Linz and Brucknerhaus Linz.

Jaap Blonk (NL), Josef Klammer (AT)

Communicating Monologues

(being there is everything)

Take off and leave Alexa or Google Assistant alone in your Smart Home, so s/he/it is free to carry on monologues and collect data. Deactivate your mobile toy—Siri or Bixby—and listen to the music of Blonk and Klammer and their communicating monologues. Both artists have been working for several years now with language and voice—Jaab Blonk with BLIPAX (Blonk's IPA eXtended) and his algorithmic sound poems; Josef Klammer with text-to-speech software and the precise fallibility of these linguistic prostheses. These men generate their music in both digital and analog fashion from the International Phonetic Alphabet, phonemes, words, sequences of letters and punctuation marks. In this project, Blonk and Klammer blend congenial synthetic and human voices into a concert evening for soloists and choirs full of onomatopoeia. Being there is what matters!

Jaap Blonk—voice, computer
Josef Klammer—sythetic voices, e-percussion

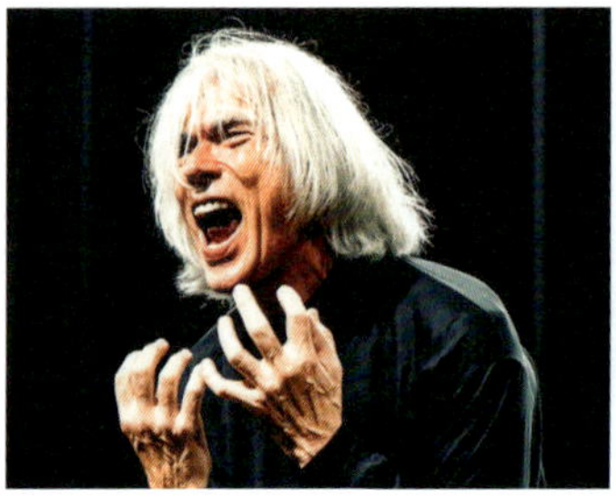

Jaap Blonk, © Etang Chen

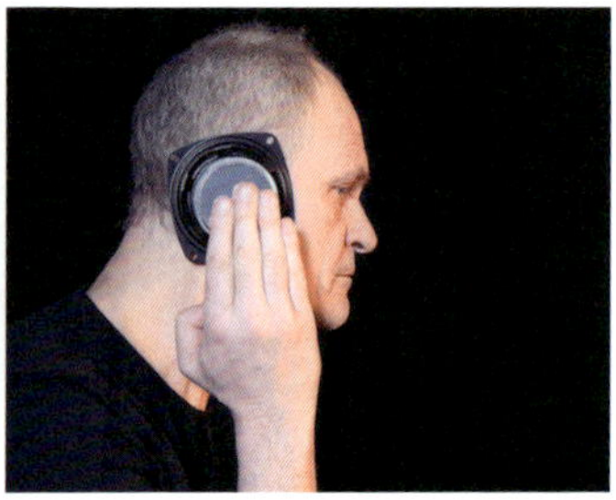

Josef Klammer, © Seppo Gründler

„The Berlioz Project“

For Markus Poschner it is the second time that he, as chief conductor of the Bruckner Orchestra Linz, switches the concert hall for the *Gleishalle* of the POSTCITY – this extraordinary project is also the opening evening of the international Bruckner Festival. Poschner has chosen the *Symphonie fantastique* from Hector Berlioz for this "meeting of tradition and modernity". A work that is almost 200 years old and thus connects a time in which the industrial revolution had only just begun with today's epicenter of the Digital Revolution. This link also provides the narrative for the production: Evolution, the development of form from a *materia prima*, from which our world emerges, where humans enter and begin to develop tools and technologies to make our environment arable, but also put the ecosystems of our planet and ourselves in danger.

Poschner and the Ars Electronica team have also brought dancer and choreographer Silke Grabinger, creative robotics expert Johannes Braumann and artist Ursula Neugebauer – with her installation *Tour en l'air* – on board. In addition to the human protagonists – orchestra, conductor and dancers – machines and digital projections will also play central roles in the production. Apparently the most impressive machine is the huge Kuka KR 600 industrial robot with a weight of approx. 2.5 tons and an extended height of 3.5 m, which is set up in the middle of the orchestra. The robot can be controlled directly by the music via its own program interface. It can also be synchronized with the digital real-time visualizations on the three panorama projections distributed throughout the room.

Aesthetically, the most effective "machines" will be Ursula Neugebauer's motorized red dresses, whose movements can be controlled from the conductor's desk.

A threatening atmosphere will arise when a "band" of five vibratory tampers play a match with the strong percussion group of the orchestra.

Poschner already proved his great virtuosity in incorporating such elements into the performance without damaging the music last year, with Bruckner's 8th Symphony as the main act of the Big Concert Night. He is not interested in spectacular action, but in making the mysticism of the music, the atmosphere, ritual and mystery outside the usual concert hall atmosphere effectively palpable.

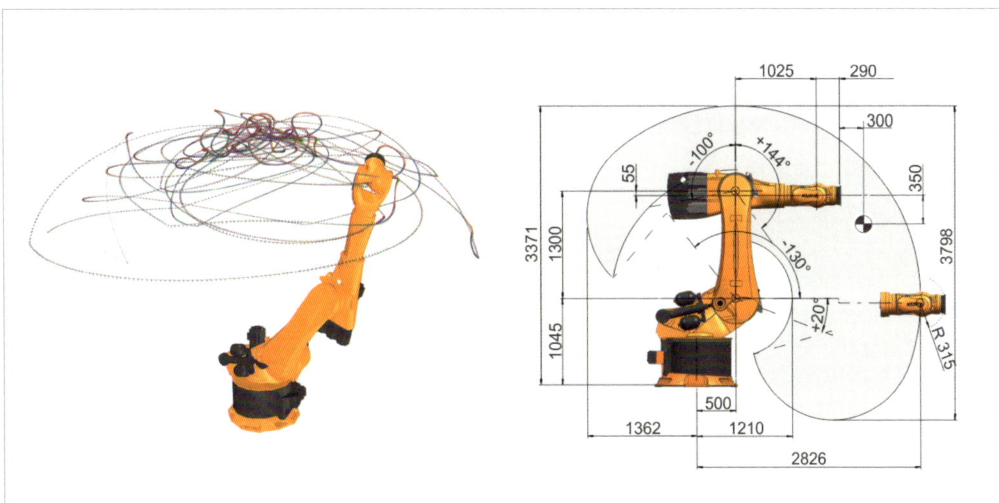

KUKA, KR600 R2830, © KUKA Deutschland GmbH

Bruckner Orchestra Linz (AT)

On Deviating and Setting Off on a New Course

An error is a deviation from what we expect, this year's festival theme asserts. Only by deviating can we switch to a new course. A total reset in musical terms was created by 27-year-old Hector Berlioz. He composed *Symphonie fantastique* and thereby put on the market—or, rather, launched an epoch with—a musical work that deviated outrageously, made a hurtling head-on impact, and was actually decades ahead of its time. Here, the unimaginable happens. The modern artist openly portrays his private *passion infernale*, showcases his obsession (with Irish actress Harriett Smithson) for all to see. An imaginary inner-world theater that reveals—and segues into—a totally different Romanticism that sings less of dreams than of Kafkaesque nightmares. Berlioz was far out on the leading edge of style, structure and orchestration, and anticipated tonal developments well into the 20th century. With unusual instrumental colorations, rhythmic and harmonic liberties, and with the help of a leitmotif present in every movement—an *idée fixe*, as it were—Berlioz invents a program music whose imaginativeness goes forth into (previously) unknown realms of expression. In the first movement, he composes drug-induced dreams; in the second movement, the festivities of a vivacious ball; in the third movement, a pastoral as an island of tranquility; in the fourth movement, a macabre film scene; in the fifth movement, a hellish orgy with fearful screams, scornful laughter and scoffing godlessness. The explosive power of this fantastic symphony is eternal and always utterly the music of the time in which it is heard.

Text: Norbert Trawöger
In cooperation with Internationales Brucknerfest Linz 2018
Hector Berlioz Épisode de la vie d'un artiste
Symphonie fantastique en cinq parties, op. 14 (1830)
Bruckner Orchestra Linz (AT) conducted by Markus Poschner (DE)

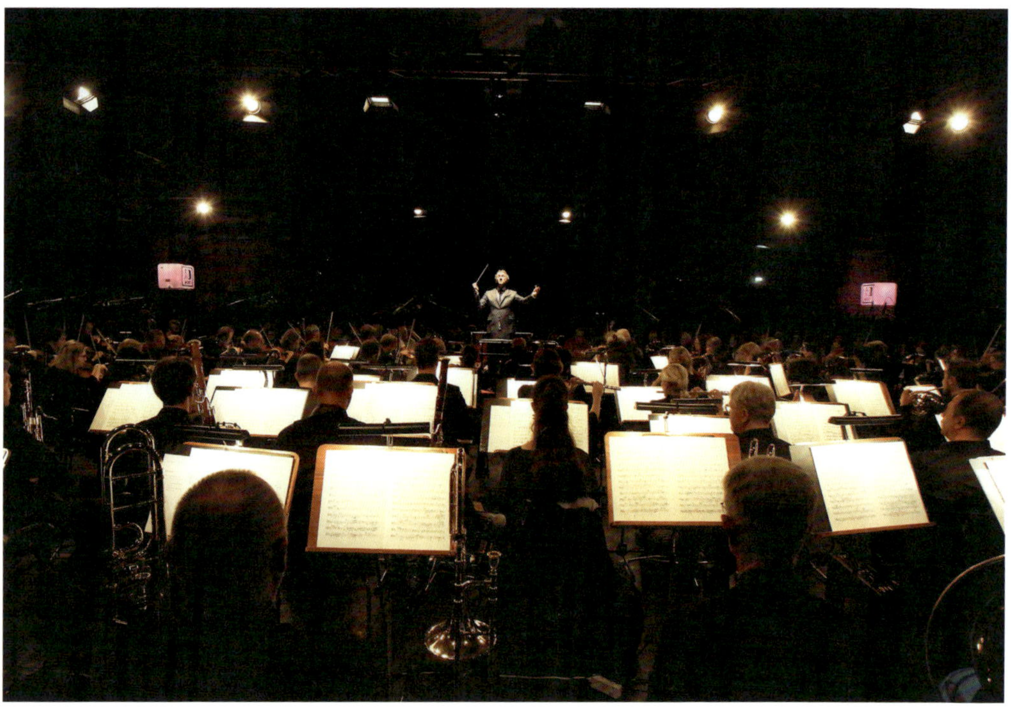

Reinhard Winkler

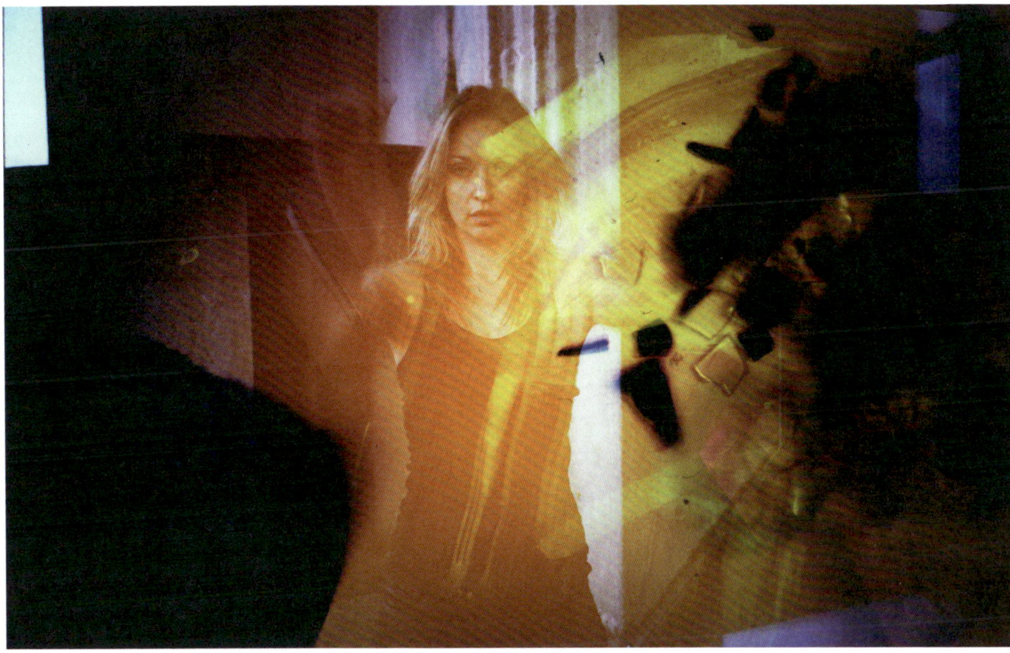

Severin Koller

Silke Grabinger (AT)

SILK Fluegge

SILK Fluegge is an artists' collective doing contemporary urban dance and art. The collective's artistic activities include performances in theaters, interventions in museums and public spaces, and mediating audiences' encounters with art and culture. The company focuses on projects in the area of contemporary, urban forms of dance and art, with particular emphasis on fostering young artists. At this year's Big Concert Night, Silke Grabinger will be responsible for the dance performance.
Under her direction, *SILK Fluegge* featuring dancers Elias Choi Buttinger and Gergely Dudas will develop a performance to accompany "Symphonie fantastique" by Hector Berlioz. The dancers—costumed by Bianca Fladerer—will perform a highly visual dance that touches on robotics, digital technology, the body and the organism, and seeks what is natural in the machine and perfectionism in the human body. The question that arises has to do with the quality of imperfection.

Artistic director: Silke Grabinger
Performers: Silke Grabinger, Gergely Dudás, Elias Choi Buttinger
Costumes: Bianca Fladerer
Production manager: Sandra Krampelhuber

Ursula Neugebauer (DE)

Tour en l'air

Ursula Neugebauer puts physical energy on display without introducing it as a body. She fixes this energy between materiality, mechanics and virtuality. She dispenses with real bodies and replaces them with display-window busts to which long, very long ball gowns are attached.

Each personage is hooked up to an electrical motor that rotates. At first, the fabric's outer layer glides at a measured pace. Then the motor turns faster and faster, and the fabric dances, springs, whirls, tosses up ornamental folds until it ecstatically hunts and harries itself. In many ways, this is reminiscent of von Kleist's famous essay "On the Marionette Theater," but in the very last moment of unbridled mechanics, without the objective of gracefulness, all references miss the mark. In contrast, Neugebauer's work, which is in effect a ballet without a ballerina, subsumes the ghostliness of sheer drapery in terpsichorean, kinetic elegance. The acceleration of imaginary bodies, their outermost freedom, was made possible by absence that catapults itself into a life of its own.

Text: Manfred Schneckenburger

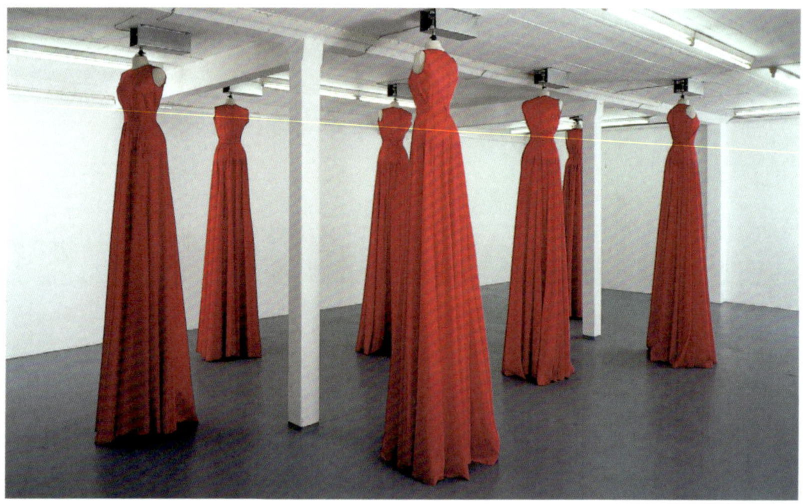

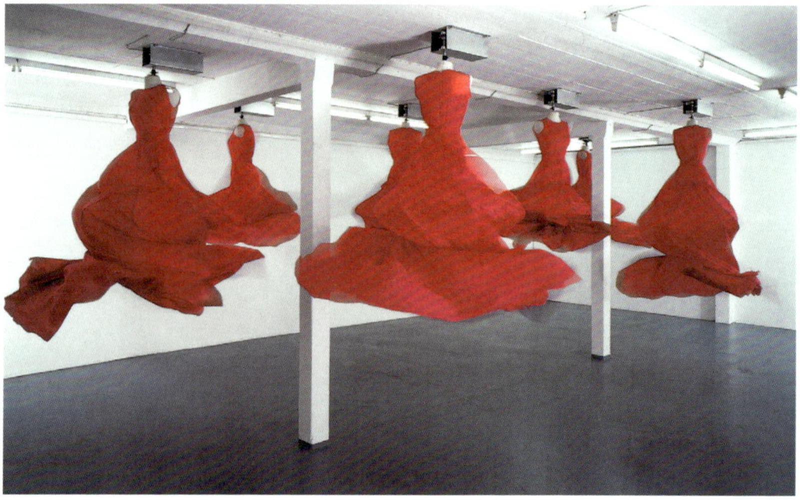

VG Bildkunst

Gabriele Marangoni (IT)

RED NOISE

prepared accordion solo

RED NOISE is a continuous flow of sound, a sequence of breaths, air, acoustic vibrations and noise generated by the most contemporary of the acoustic instruments: the accordion. In Red noise the accordion is prepared, iron sheets are placed between the reeds and this variation is not controllable, free, human. Red noise is research, experimentation and improvisation, like life, like freedom. A variable, an error (in this case the preparation of the instrument) generates new opportunities for expression.

Electric Indigo (AT)

5 1 1 5 9 3 surround a/v live [excerpt]

Electric Indigo presents the Austrian premiere of material from her recent album 5 1 1 5 9 3. Her performance is a quadrophonic journey into sonic particles and sparse, powerful rhythms. Icy beat-scapes from the fringes of club culture meet alien space ships, travelling with warp speed towards total immersion, supported by color fields and subtle video compression artefacts.

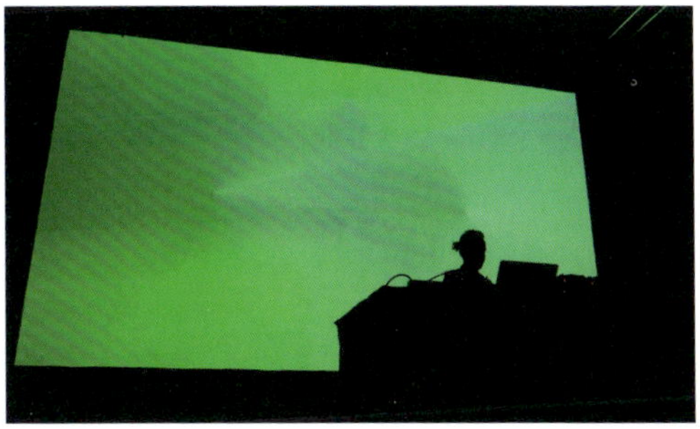

Music, images, spatialisation by Electric Indigo

Elisabeth Schimana (AT)

Into the sun

An imaginary acoustic view into the past

Since 1960, we've known that the sun is oscillating. Helioseismology is the study of oscillation waves through the sun; it allows astrophysicists to develop extremely detailed profiles of the interior conditions of the sun. The inspiration for the artistic work is the imagination that the sun is a huge resonating body for sound waves traveling to the interior of the sun and being reflected, as well as the awareness that we could look into the past by observing the processes on and in the sun—a paradox working with the medium sound, which can be experienced only in the present. *Into the sun* is a live electronic solo piece for a multichannel system.

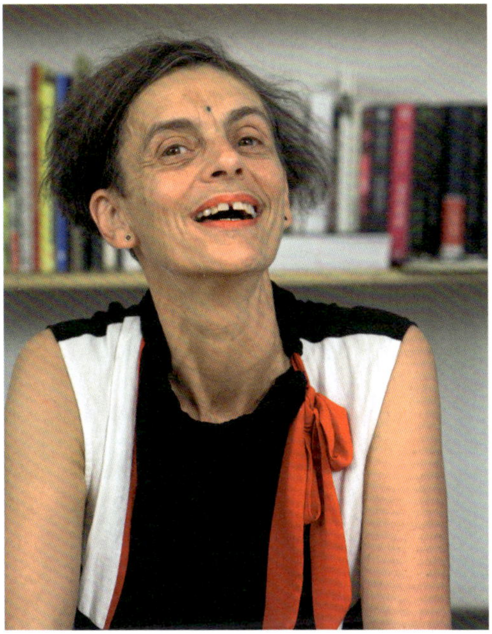

Reinhard Mayr

Maki Namekawa (JP), Dennis Russell Davies (US), Cori Olan (AT), Andreas Bitesnich (AT)

Piano Music meets Digital Images

September 10, 2018, 7 PM, POSTCITY

"Piano Music Meets Digital Images" is the programmatic title of the Monday evening concert by Maki Namekawa and Dennis Russell Davies that concludes this year's Ars Electronica Festival. The program commences with a solo performance by Maki Namekawa – the world premiere of the piano version of *Mishima* by Philip Glass.

Mishima: A Life in Four Chapters from 1985, directed by Paul Schrader, is perhaps one of Glass's greatest successes in the medium of film. The film, which documents Japanese author Yukio Mishima's final day, presented a number of musical challenges for Glass to work out. Now more than thirty years after its release, both the film and its music represent enduring work by Glass and Schrader.

Philip Glass came to composing music for film very late. Despite the high-water marks of later commercial film successes like *The Hours* and *Dracula*, Glass's early reputation as a composer for film was based on art-house films like Godfrey Reggio's *Koyaanisqatsi* and documentaries such as Errol Morris's *The Thin Blue Line.*

While later commercial films fit very much into the established order of creation, where the composer comes in at the end of the process, these early films were produced differently with Glass often composing the music before the images were filmed, or simply writing music to the story.

A portion of Glass's music portrays past episodes of Mishima's life. It was scored for String Quartet and became the published work String Quartet No.3 *Mishima*.

The brilliant *Mishima* score basically lay in obscurity until last year when Maki Namekawa commissioned the music director of the Philip Glass Ensemble, Michael Riesman, to create a solo piano version of the entire score. Therefore, this masterpiece will be heard for the first time in concert.

Namekawa's account of these new arrangements puts her crystalline technique at the fore - which ended up being a perfect analog for Glass's music, the music's portrayal of the character of Mishima, and its first interpretation by a single Japanese musician. Listeners can hear the main issues of Mishima's life – Beauty, Art, and Action – all converging in Namekawa's performance.

Namekawa collaborates with the renowned Austrian photographer and video artist Andreas H. Bitesnich who combines a subtle use of digitaly controlled light projections with elements recalling aspects of traditional Nō theatre, to illuminate and intensify Namekawa's dramatic encounter with the Mishima legend through the music of Philip Glass.

The second part of the evening is dedicated to visualized music for piano four hands. Namekawa is joined by her husband and long-time piano-duo partner Dennis Russell Davies, who together with Ars Electronica artistic director Gerfried Stocker initiated the *Long*

Music Night as Chief Conductor of the Bruckner Orchester Linz. In addition to his activities internationally as a celebrated and widely recorded conductor (Grammy and Echo Classical), Davies has performed and recorded as pianist (his ECM recording of Keith Jarrett's Ritual, which Davies premiered, was recently re-released as both LP and CD).

Maki Namekawa and Dennis Russell Davies perform original works for piano four hands by Steve Reich (*Piano Phase*), Maurice Ravel (*Ma Mére l'Oye*), and Philip Glass (*Stokes*).

Cori O'lan's digital real-time visualizations of both of these pieces came about through Ars Electronica's cooperation with the Abu Dhabi Music Festival, where they premiered this past March. *(https://cori-olan.art/2018/03/01/abu-dhabi-festival-2018/)*

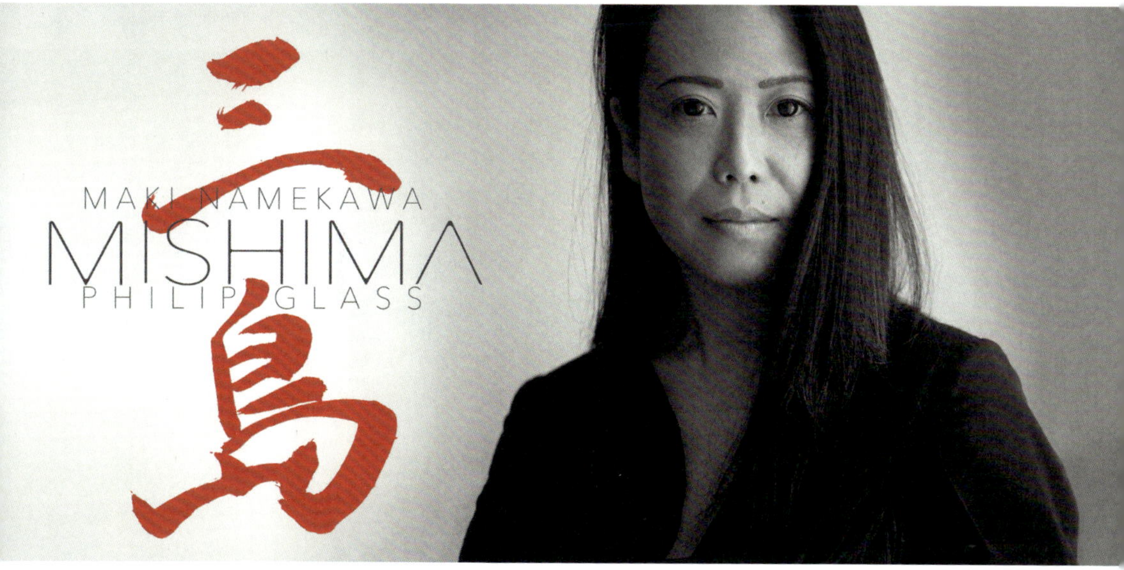

Steve Reich, *Piano Phase*

Piano Phase was composed in 1967 for two pianos. It is one of the early compositions by Steve Reich where he also uses his "phasing-technique" for instruments, a technique which he derived from early works with tape recorders and which would become a signature style for much of his music.
Both pianists play the same 12-note pattern but over time they vary in speed and dynamics so the result is a constantly floating and modulating combination of this pattern going in and out of phase. Steve Reich described this as "music as a gradual process" and such he pioneered ideas which became important topics in the musical production of the digital age, both in avant-garde as well as in popular music. The visualizations for Piano Phase have been specifically created for the performance in Abu Dhabi and derive their motifs for the visual interpretation directly from the music: approach and distance, rhythm and flow, clarity and ambiguity of the dancing and interweaving patterns.

Maurice Ravel, *Ma Mère l'Oye*

The original four-handed piano version was composed by Ravel between 1908 and 1910, telling stories inspired by Charles Perrault's fairy tale collection Tales of Mother Goose. Originally composed for the children of his friends, Ravel soon also made an orchestra ver-

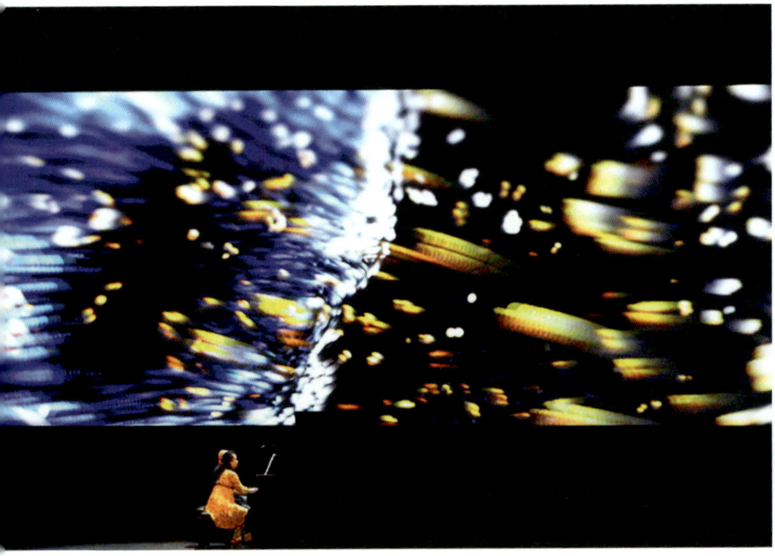
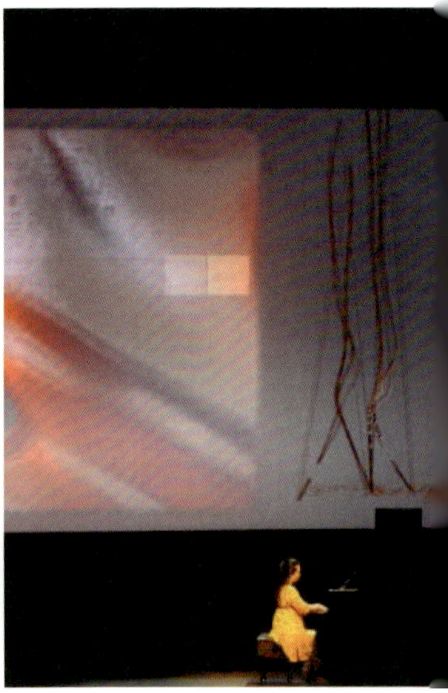

Pianographique Abu Dhabi

sion, and a ballet. Sleeping Beauty, Little Thumb, the Empress of the Pagodas and Beauty and the Beast are the fairy tale characters behind the individual parts of *Ma Mère l' Oye* concluded by a grand finale in the Garden of the Fairies.

In 2016 Ars Electronica coproduced with Abu Dhabi Festival and LA-Phil a large scale visualization for the orchestra version of *Ma Mère l' Oye*. The work premiered in February 2016 at the Walt Disney Music Hall in Los Angeles, conducted by Esa Pekka Salonen, with the LA Philharmonic. Due to the very big sonic differences of the piano compared with the full orchestra but also because of the differences in interpretation, a completely new set of visualizations has been created for the performance with Dennis Russell Davies & Maki Namekawa.

Although very different in its graphical style, the piano visualization also follows the original idea of "Klangmalerei" (painting with sound/music) which is a term often used to describe the very special character of Ravels music. All the frequencies and dynamics from the piano sound are literally converted by the computer algorithms into "paint and brushes" directly connected to the live music. Abstract forms appear, sometimes reminiscent of calligraphy, landscapes, mountains, ocean storms or rippling waves on a river, moving and morphing into each other.

The visualizations for *Piano Phase* and *Ma Mère l'Oye* are commissioned by the Abu Dhabi Music Festival and co-produced with Ars Electronica Futurelab

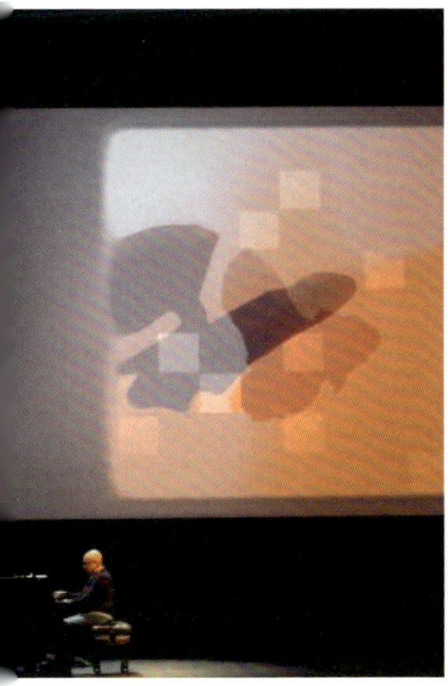

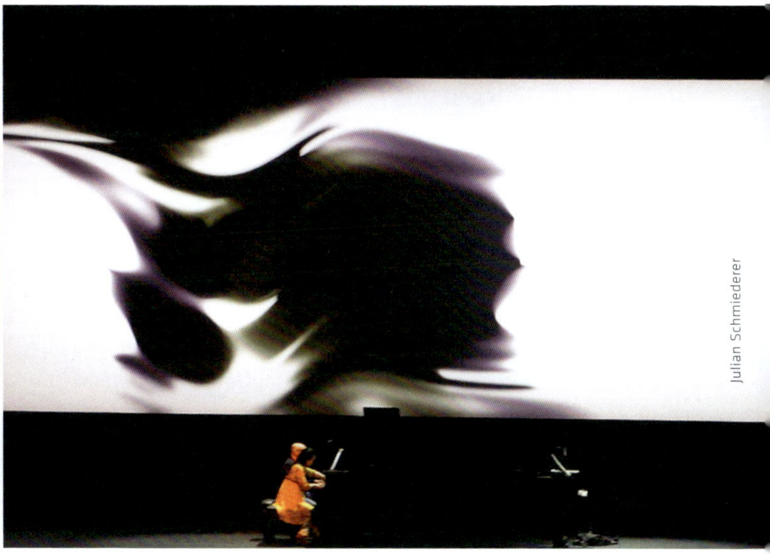

Julian Schmiederer

Sparkasse OÖ Visualisierte **Klangwolke 2018**
Presented by Linz AG

PAX tradition = revolution

"Navigare necesse est, vivere non est necesse."
This year the Sparkasse OÖ Visualisierte Klangwolke 2018 sails along the banks of the Danube in the tradition of an old sailor's saying: "It is important to go to sea, it is not important to live." The Danube Park and the river itself become the setting for a unique staging of human history, from the Big Bang to the present. Planets and objects float in a dreamlike scenery, people and animals move around and through the audience, creating a fantastic panorama of human tradition, at the end of which stands the departure to new worlds.

Tradition = Revolution

With PAX, the legendary, internationally acclaimed Catalan theatre collective La Fura dels Baus has created a fascinating show between installation and interactive performance, in which the audience is surrounded by large-scale productions and projections that move on water, on land and in the air. A carrier pigeon made of 12-metre-long fluorescent tubes projects the news of the public into the sky above Linz, thus heralding the tradition of communication, while gigantic buildings, a giant horse model or a 14-metre-high robot doll symbolize nature and progress.

Like the group itself, its performance is also characterized by "eccentricity, innovation, rhythm, evolution and transgression". The motto "Tradition = Revolution" consciously plays with the apparent contradiction between preservation and renewal.

From the Danube to new shores

For La Fura dels Baus, the solution of the problems of past and present generations is reflected in PAX. PAX stands for the constant search for something new, for technical development and human civilization, in opposition to stagnation and regression. PAX means understanding the history of humanity as the history of progress in which the revolution has a tradition.
"It's important to go to sea." Set off with Sparkasse OÖ Visualisierte Klangwolke 2018 to new shores and embark on an exciting journey through the past, present and future!

After-party with Leyya

LINZ AG invites all visitors to continue celebrating late into the night in the Donaupark in order to echo the manifold impressions of the Sparkasse OÖ Visualisierte Klangwolke 2018. The Austrian electropop duo Leyya will provide the sound.

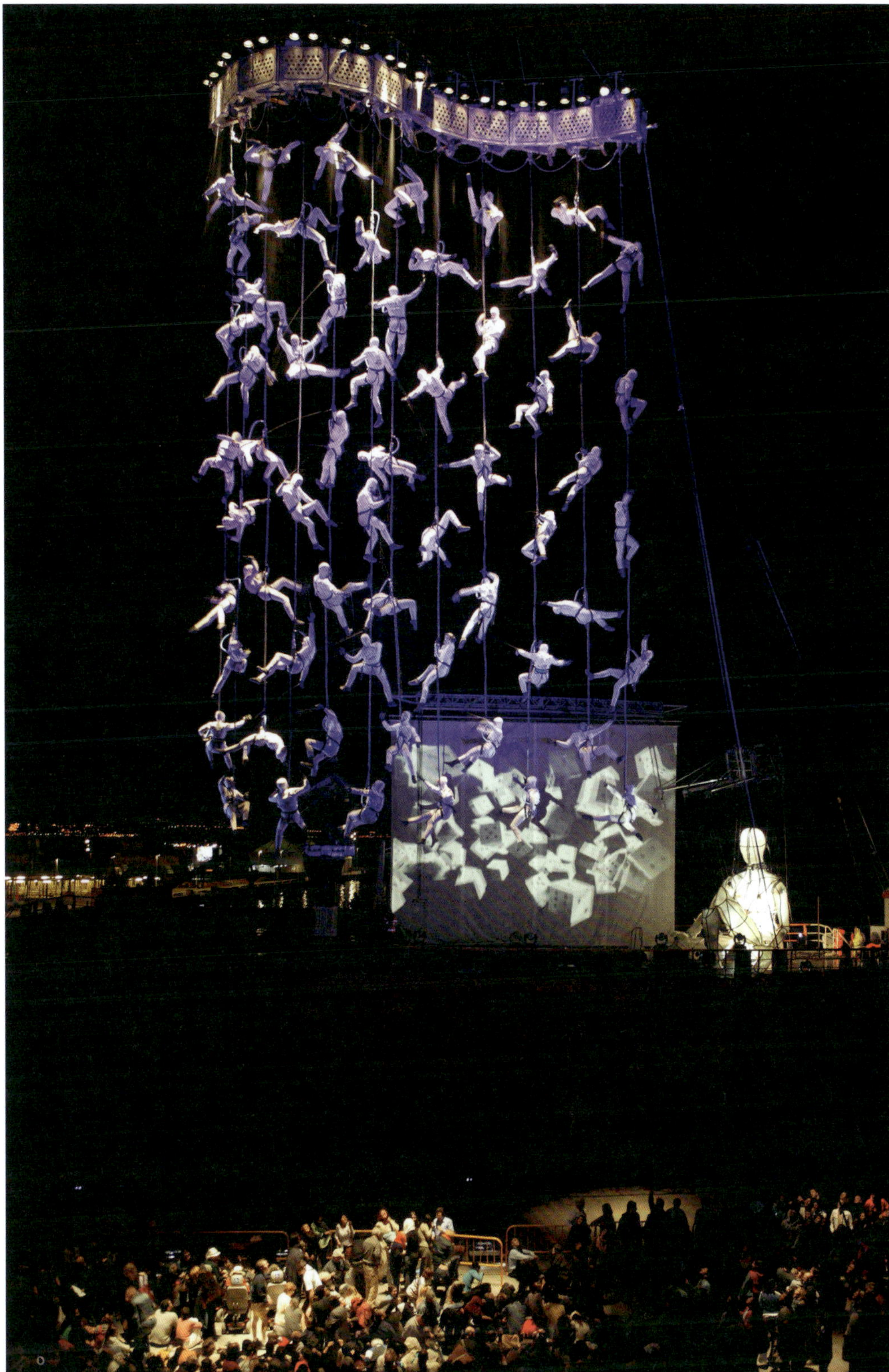

La Fura dels Baus

Mihaela Kavdanska & Dilmana Yordanova (RO/ BG/ AT), Maria Mora (RO), Mariana Gavriciuc (RO), Mirian Kolev (BG), Diana Dulgheru (RO), Stefana Fratila (RO/ CAN)

OFF8NOFF: Spaces Alive #abstract | 50HZ & Stefana Fratila Live

Spaces Alive #abstract is a site-specific performative intervention presented within the frame of OFFXNOFF, a collaborative exhibition and concert series by Memphis Art Space and Tresor Linz. The project is part of the series "Spaces Alive. Bucharest Surreal", produced by AVmotional Platform, conceived and presented in 2017 at different locations in the city of Bucharest.
The performance is based on the interaction between media artists, dancers, sound and the gallery space. It problematizes the presupposed neutrality of the "white cube", while critically commenting on both the modernist myth of abstraction and the "perfection" of a perpetually optimized technology. The idea of error is translated in the subjectivization of technology through the moving bodies augmented with webcams, now turned effectively into interfaces. A video layer, processed in real-time is integrated into an iconic painterly abstraction, covering the space of the gallery. As part of *OFF8NOFF* a concert evening will take place at Tresor Linz.

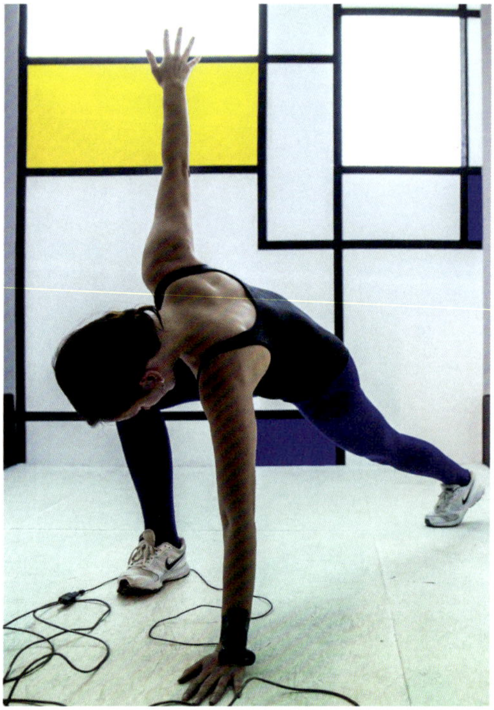

Alina Usurelu

OFF8NOFF, producers & locations: Memphis (AT) & Tresor (AT)
Co-producer: AVmotional Platform (RO, AT)

Spaces Alive #abstract:
Creative direction & visuals: Mihaela Kavdanska & Dilmana Yordanova (RO, BG, AT)
Performers:Maria Mora (RO), Mariana Gavriciuc (RO)
Sound artist: Mirian Kolev (BG)
Computer programmer: Cristian Iordache (RO)
Choreographer: Simona Deaconescu (RO)
Curator: Horea Avram (RO)

Sound artists, Tresor Concert Night:
Minim | Diana Dulgheru (RO) & Stefana Fratila (RO, CAN)

Romanian Cultural Institute Vienna
Bundeskanzleramt Österreich
Kulturabteilung Land Oberösterreich
Linz Kultur Förderungen
KOTKI visuals

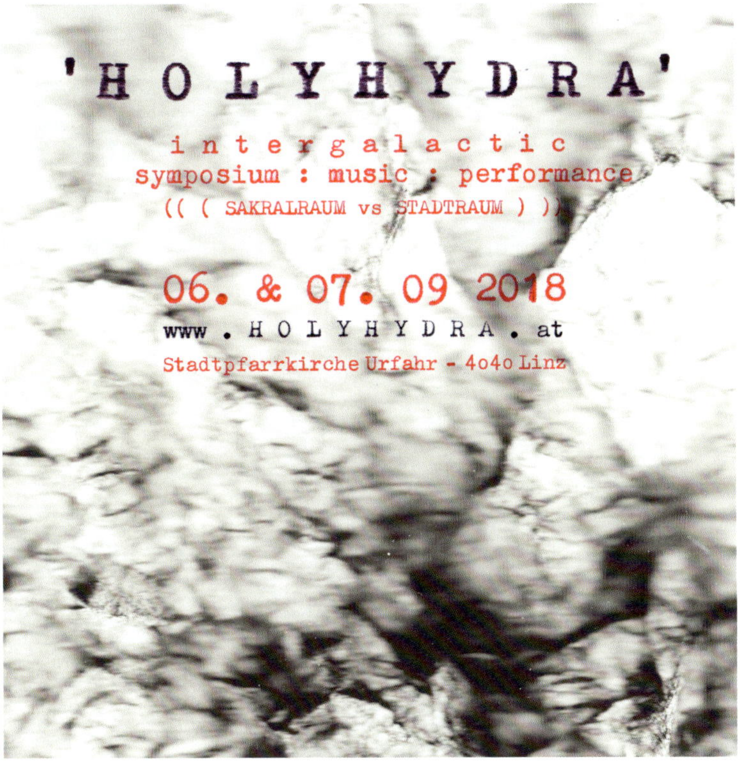

Amanda Augustin (AT) & Lorena Höllrigl (AT)

Holy Hydra

intergalactic symposium | music | performance

Holy Hydra is a two-day event involving contemporary performances, electronic sound art, interactive light installations and a symposium about "Sacral Space vs. Urban space" in the parish church of Urfahr, next to the "Ars-Electronica-Maindeck."
The major concern is to present different artistic disciplines and approach interested groups in its unique spiritual atmosphere - a sacred space. To offer an experience at this special place for everyone regardless of their religious conviction or their beliefs.
The project is intended as an idea for how to extend the use of architecturally and historically valuable structures. Churches fulfill those attributes perfectly and are exemplary for an alternative usage.

Idea & Production: Amanda Augustin (AT) & Lorena Höllrigl (AT)
Production Team: Raumteiler Linz (AT)
Music curated by: Tanja Fuchs (DE/AT) and Thomas Auer (AT) // Klangfestival
Visual Concept: 4youreye - ProjectionArt (AT) & AV-Lightstorm (AT)
The project is supported by Grüner Anker - Jugendkirche der Diözese Linz, Kunstreferat - Diözese Linz, afo - architekturforum oberösterreich, Posthof Linz, Katholische Privatuniversität Linz, Raum & Design Strategien - Kunstuniversität Linz

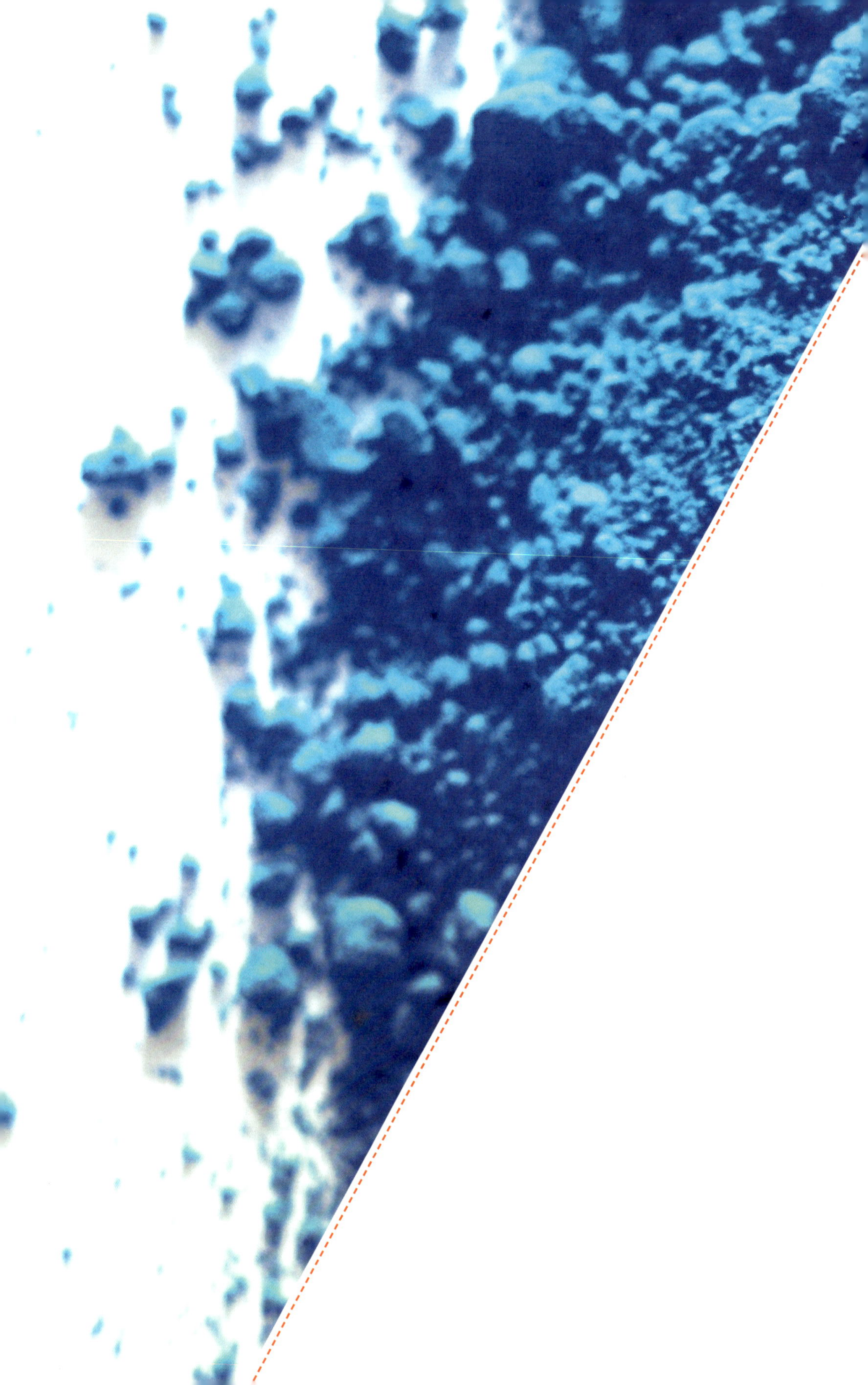

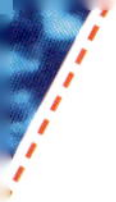

FOCUS DIGITAL MUSICS & SOUND ART

Volkmar Klien (AT)

composing (in) digital worlds

The Digital Music Focus 2018 at Anton Bruckner University

Sonic Saturday, part of the digital music focus at Anton Bruckner University, presents a symposium, a guided tour through the university's computer music studio, a multichannel listening room featuring compositions reflecting the most recent trends in computer music, and the concert "Medium Sonorum" in the university's 20.4 channel concert hall, the CMS Sonic Lab.

The symposium—entitled *.. under control of Music | music under control of ..; composing (in) digital worlds*—brings together artists, thinkers and researchers from different fields to shine a light on the manifold challenges for the sonic artist composing (in) the digital worlds today.

Music has always been a way to represent, influence and control the social settings and situations it evolves in. During the last century, composers and sound artists of the self-proclaimed avant-garde focused on this aspect as well as on how music was embedded in the economic and political complexities of the emerging entertainment and media industries. Artists decided not simply to produce products for art markets, but aimed at a fundamental critique of their tools and media in situ.

With the advent of new tools for the production and dissemination of sound throughout the last two centuries, the strategies of artists working with these tools in sound transformed, too. The microphone changed the arena for singers; the electric guitar enabled the instrument to take center stage (and to be heard there). The gramophone and radio allowed sound to travel long distance and in time, transforming music from something rather ephemeral and tied to a specific place (earshot) and time (the moment) into something to be consumed (and not just to be made) where- and whenever.

In the dawning digital society with its algorithmically curated sound-worlds and the possibilities and economic realities of big data, the environment in which musical practice evolves is undergoing drastic change. Algorithmic music production is in the process of moving from academic computer music studios to giant corporations' service departments, aiming at harvesting musical practice (reception and production) for new ways of influencing and controlling human—i.e. customer—behavior.

The symposium *.. under control of Music | Music under control of ..; composing (in) digital worlds* poses questions regarding these new forms of music making. What potential is there for control, abuse, for artistic creation, subversion and altogether new forms of musical practice?

The symposium's participants are: Thor Magnusson (IS/UK), Alexandra Murray-Leslie (AU), Thomas Grill (AT), Wayne Siegel (DK/US), Sam Ferguson (AU) and Enrique Tomás (ES)

Text: Volkmar Klien

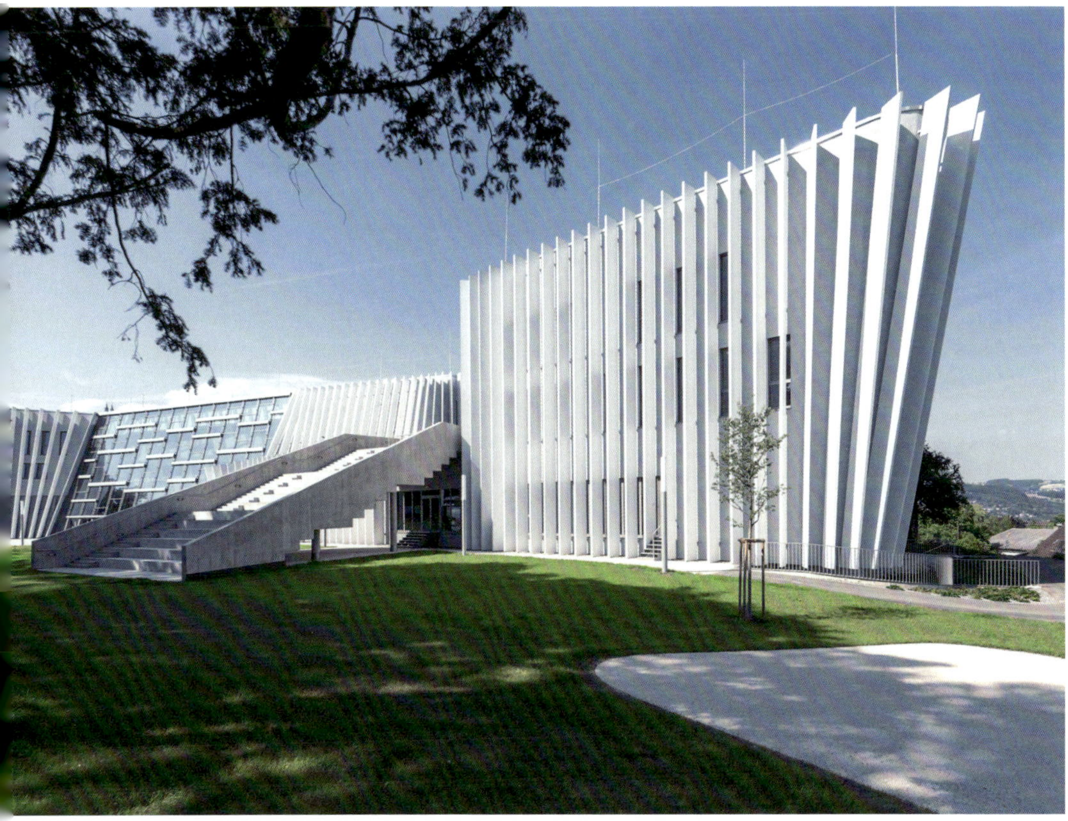

Anton Bruckner University, Linz, © Simon Bauer

Werner Jauk (AT)

What is music to...

The Evolutionary Transgression of the use of Error in Post-Digital Culture, Nature as a Paradigm of Cultural Evolution

"Error" is regarded as a modern cultural construct. Deviations from theory-based behavior and thinking about behavior are considered errors, statistically not as behavior determined by the hypothesis derived from a theory but as behavior occurring at random.

What culminates in research paradigms is fundamentally the general way of modern thinking and living, and at the same time the method of producing waste—culture is regarded as a purposeful (linear) progressive development where the goal is the mastery of nature. So, in this development, implicitly natural occurrences that are transferred to social and aesthetic production as feasibility are in "error." Just as in the quest for innovation, feasibility was the reason for doing it, today caution is the watchword of futurology—so (also) avoiding error as cultural waste.

Against the method of producing waste, against modern logic, the classical avant-gardes worked in an anaesthetic sense, in the sense of extending the field of perception (WELSCH 1993, 79ff). Alternatives will grow out of this, cultural remixes even if the thinking is intrinsic to the economic system—remixes are sold as "new" with increasing rapidity. So begins a new multiplying, parallel, plural value, ultimately also plural errors and plural waste.

Welsch already interprets the classical avant-garde as the avant-garde of postmodernism, which grew from Modernism; Jean Dubuffet's demand for a shift away from the primacy of reason and logic (DUBUFFET 1973, 68) is provided as an example of "the birth of postmodern philosophy from the spirit of modern art" (WELSCH 1993, 79ff). Breaking away from logocentrism, leaving monosemy for polysemy (ibid. 82) and living plurality (ibid. 92), is associated with this. Generally speaking, for Welsch the limitations of the modern view are given by the prevalence of seeing. The auditory logic (JAUK 2001), the patterns of thinking from the experience of auditively controlled body-environment interaction (GIBSON 1982) places an alternative form of directing perception, one suited to the altered environmental conditions, especially in the post-digital culture: The integration of the body in a world of feasibility after the exceedance of mechanistic thinking through the dynamization and codificiation of the environment (JAUK 2003, 2009).

In interaction with surroundings changed through the process of mediatization of the body's interaction with the environment, instead of the synthesizing attitude of the visual construction of time and space through the (moving) body, there is now the suitable form of interaction: the analysis of events around the (still) body according to their significance – i.e. the excitatory value of the environment – for the body.

This auditory logic is doing justice to the all-at-onceness (McLUHAN 1995) of "electronic space," today net-space. Its excitation-based regulation exceeds rationality, especially since it is conceivable as a "rationalization" of physical "knowledge." The term "error" is not adequate for a principle of homeostasis based on dynamic adaptation.

While the arts shape the culture of those revolutionary policies for the struggle of the means of production, it is increasingly the economy that is seeking to lead the accelerating competition for consumers, and to seduce the pluralistic individual in his or her emotional mood, away from culture-defining attitudes, in his or her "self"-formation.

In a music-dominated, experience-based society characterized by the hedonic pattern of self-optimization (SCHULZE 2000), borrowings from nature's self-organizing processes are culturally accepted: the homeostatic regulation of life, which does not exclude stimuli from the environment but uses them as a regulative stimulant of (mutual) adaptation, enters into the offerings of mass media neoliberal culture as mood management.

Fundamentally, exploratory behavior knows no "error" but only variability leading to adaptive processes of the dynamic shaping of physical life, of aesthetic (BERLYNE 1971) as well as social life – is its transformation through power a culture of error, the disruption of the dynamic self-regulation of life?

Having emerged with the original storage medium of notation as a prescription for the reproduction of sound structures via the composition of codes for sounds, in modern times music serves as a construct of "related thinking" (RIEMANN 1914/15), the composition of codes for sounds according to relationships. These stem from the nature of sound, which, in the "harmonic" understanding of Central European thinking follows the principle of a geometric relationship, the physically appreciated relationship of scale. Transferred to algorithmic codes it was formalized in a duodecimal system, which allows most integer ratios - what was done to good´s creation was done to music too describing the "harmonices mundi". Nature was elevated to a legitimating cultural law, and rightly so. Only integer ratios for fundamental frequencies determine the order of the sounds to one another: the integer ratios of the sound elements define harmonic sounds; integer ratios between the fundamental frequencies define the harmonic order of sounds in relation to one another. Rooted in the nature of sounds, a cultural order is postulated here.

The fact that this selection presumes a cultural imperative is seldom made explicit – however: other natural processes are (d)evaluated and considered uncultured, cautiously defined as "contrary," as dissonant/unharmonious, in general and in particular contrasting high culture with folkish and popular music as "barbaric," "uncultivated;" it is also considered "noise" by these cultures and, deconstructively, as the error of harmonic life. If in the early 20th century it was the immediate experience of noisy sounds of "foreign" cultures, but above all the urban sound of machines defining human culture and the feasibility of artificial orders opposed those of nature, even in the innovative sixties, although "committed" to (social) feasibility, it was rough recordings of "black" singers, overdriven and distorted guitar sound that were judged by some as the expressive quality of unculturedness, as "subcultures," yet ultimately when viewed by others were seen as directly and stimulatively communicating, bearing the "gesture of the hanging tongue" (MARCUS 1992, 19).

What stimulatively excites the homeostatic regulation of "new" orders beyond a harmony of high culture that violates physicality is symbolically considered an error. Pop music, as simultaneously a little mediatized, directly physical following the expressive behavior caused by excitation, and at the same time a highly mediatized technical amplification of this physical quality, plays with the error of modern high culture. What was evaluated as a counter-attack by semiological guerrilla methods, an expropriation and the drawing into "punk" of structurally deeply anchored signs of high culture, can be experienced as pre-symbolic communication, as a form of communication deeply rooted in phylogenetic sonic performative expression. Pop music or, better, pop-music-making is the instrumentalized/mediatized sound-gesture and its emotional contagion as a rebellious "error" stimulant for cultural adaptation. Sound-dominated pop culture as a counterposition and later alternative to the physically hostile order of linear theory-led feasibility of nature and society can be seen as cultural constructs within the culture-shaping rational thinking, as phenomena of the recognition of the ostensible natural order as an artificial cultural principle that initially combats this error anti-aesthetically by counterposition, later anaesthetically nullifying it by expanding the forms of experience.

According to this modern understanding, the boundary of thought is set on progress as the rational structuring of the new. The avant-gardes of the early 20th century seek to overcome this in the arts of transgressing formal composition, composition as a work – thereby generalizing the musical work as a composition of codes for sounds as iconic images as well as their symbolic use.

Following Futurism as a manifesto that glorified the new, the search for something new takes place through overcoming rationality, as its negation in Dadaism, as its elimination/circumvention through the shaping of the para-logic of the dream and pharmaceutically induced forms of thought in Surrealism, later the psychedelic arts, and finally in the abstraction of the narration of content to form and the temporal order in abstract film, through the inter/medial transposition of musical thinking to shaping beyond sound.

The negation of musical thinking as referencing and therein algorithmic thinking initially led to design by chance – now "for the first time" not understood as error, but in Darwin's sense as variability that leads to redesign as adaptation. This new understanding of error is used in the technoid arts in particular as formal temporal arts beyond the logic of causality and its generalization in narration in the abstract design of dynamic arts.

In addition to inter/medial transposition, music, especially in the algorithmic understanding, will also use the other aspect of chance as induced error, the exceedance of calculability, overstraining through over-determination, through the processes of processing algorithmic design in the "glitch" as a means of construction (CASCONE 2000, JAUK 2010). Cascone then called the music of overcoming determinable orders post-digital. Popular music will have used the artefacts of initially limited technical possibilities according to musical design beyond the rules of sound design through the regularity of the operating system of the information processing machine and made music of "clicks, cuts and bursts" as a music of error. Pop music as music of the remix from the "dustbin of history", filled with increasingly short-lived sound/technology. Apart from cultural referencing, "used" postmodern technoid music "used" samples, and finally cheap technologies like the Roland TB-303 and the Roland TR-606. Here discarded material is used as the material of sound production. Its internal organizational logic in music-making does not just take place as a conscious disregard for its internal regularity[1], as error, but as ignorance of this thought construct in the amateurish use of hedonism with available technology. This is not dissimilar to the electro-acoustic processes of guitar-dominated music, which have led to grunge and distorted sound through inappropriate use, the overloading of amplifiers – a performative hedonic play stimulatively derived from the expressive gesture (JAUK 2007) which on the other hand has led to virtuoso sustain-rich playing, sometimes understood as an error of pop culture in allusion to high culture work-music-sound, which, however, plays less the structure through sounds than "hot" modulating the sound itself.

This is a paradigm of electronic dance music: the formation of sound according to expressive qualities as hedonic performativity that naturally produces sound that directly stimulates movement. This arousal controlled bodily modulation sounds of any origins through the sound-gesture, which controls excitation, will determine the future of music as it were in "back to the roots" (JAUK 2015), the emergence of music from expressive sound or the instrumentalization of expressive behavior. Functionally, this technologically intensified form of emotional expression serves as social communication – becoming "communis".

Political stimulus can also be found for the individual movement in micro-politics; it already starts in a political counter-position: "From now on, what was once just considered obsolete now has only one place: the dustbin of history!" (MARCUS 1996, 14 mentioning Leo Trotzki). Pop culture will also turn its opposition into an alternative: the remix of instruments and technologies from the "rubbish dump of culture," a remix of mostly high-culture emancipation techniques into a new amateurism leads to cultural developments beside continuing dynamics.

In its technological intensification of physical emotional interaction, popular music uses such technologies for collective and collectivizing emotional communication; it will be precisely the media arts that cultivate these forms of design from the use of instruments and media from the exponentially expanding waste heap, and provide postmodern thinking with an alternative to progressive thinking; in an independent attitude they will seek to undermine and disempower modern technological optimization as instruments of capitalist power, as means of production and their owners not in a semiological guerrilla struggle of symbolic negotiation, but in an actionist guerrilla struggle. The sixties, the last eruption of modernity and anticipation of the "turning point," rely on the one hand on calculable feasibility, and on the other hand they reflect on the transgression of rationality. Big-data studies make it possible to extract multidimensional clustering from different points of view. Thus error (in content) is qualified as the production of observation;

[1] As it was done by the American avant-garde (minimal- as well as noise-art) using the method of "trial & error" overcoming rational structuring just by "doing" and deciding the "right" outcome by its hedonic value.

while deconstructivist methods focus on error in order to obtain knowledge about the excluding agent from what is excluded, error becomes information in general.

Apart from statistical conceptions, error acquires methodological interest not in the elimination of subjectivity, but in its respect for it (JAUK 2017). Psychology certainly considers creativity to be divergent thinking, which, however, must be transformed into congruent thought if new knowledge is to be derived from it. Applied in science, publicly proclaimed in the arts, anaesthetic, consciousness-expanding action is regarded as an uncontrolled devotion to "error." Trial and error are methods of finding something new, the experiment is removed from its strict control. The interpretation of the ego's experiences is freed from its subjectivity as error (JAUK 2017). Especially the American avant-gardes of minimalism, following pragmatist thinking, with their own limited tradition, will anaesthetically seek the new in trial and error (SCHAEFER 1987) as well as in expanded consciousness.

Popular technoid and media arts will, however, also address the waste of material design as it were in a deconstructivist way to focus on what is excluded, to work from the dustbin of culture. First as a counter-position, then as hacking from "within," ultimately valued as a use, the use of existing technologies, especially because of their availability outside the operating instructions, is at the same time simply used in a "new amateurism" without the concept of "error" – the input of rhythmic patterns, free of music-theory thinking, into cheap drum machines and their hedonic modulation in tempo leads to rhythmic movements from ambient to jungle formed from arousal to movement, which are finally conceptualized as individual political movements within the concept of micropolitics.

The use is to be seen as an exceedance of error related to cultural knowledge. Algorithmic and material-oriented arts, even the digital arts, reduce the transgression through error to an understanding within materiality, within natural knowledge of the body, the experience of the interaction of the body with the environment as material realities – embodiments of this lead (metaphorically) to cognitions (LAKOFF 2003) that lead to mechanistic concepts of culture. Their application and their exceedance remain within the materiality, which is usually not recognized as an error when digital arts of the immaterial common digit, free from any experience of the material world (LYOTARD 1985), are still designing in a materialistic way. Now the question arises, is this transgression possible as long as people derive their systems of thought from the embodiments of a material body?

Gibson's definition of the intentionality of body-environment-interaction here also provides a transgression whose non-observance has produced a fatal cultural error in the form of rational idealism. Intentionality means that all interaction is intentionally regulated by the "affordance" of environmental stimuli: through the tension, the tension value of the environment for the body and its resulting behavior – this leads to a mutual adaptation for purposes of the respective human survival in adapted form with a view to the creation of new environments through this adaptation. Drawing on the social sphere, the net-art of the 1990s defined design as a mutually adaptive process, as collective and collectivizing design (KERCKHOVE 1995). This definition can already be found in the group psychology of the 1950s (BALES 1950), which saw informal design as a mutual adaptation of the (social) person and their (social) environment.

Mechanistic thinking is exceeded in intentionally hedonic social design regulation; as if by arousal, homeostatic regulation as an optimization becomes the survival principle of man and culture – in this the modern concept of "error" is exceeded.

Here the idea of feasibility and the social feasibility of the sixties generalized from it presents itself as a modern extension, the determination of "better knowledge" for many. At the same time, doubts arise as to whether feasibility is not cultural error resulting from the ideological verdict of doing what is feasible. This model of communication causes the model of information transmission, i.e. materiality, to shake – as a design factor in music, the design according to "tension –solution" has already contributed to referential thinking in the past. Social relations are based on excitatory relationships; communication does not mean mutual "understanding," but becoming "communis" (JAUK 1999).

Here again, the "only" collective form of music design comes into play. Adorno already described polyphony as "the objectivation of us," the formalization of interacting voices that structure music. If this social structure is made by a composer, the objectivation is an ideological factor, for it remains a prescription of the behavior of many from the idea of the one, even if the one also legitimizes its composition in the nature of harmonizing – this is the power play!

What role does the error play in this? As in self-regulating nature and in the evolution of humans, social processes (initially) regulate themselves – too little and too much arousal in social situations leads to behavior that brings the level of arousal to a (subjectively) survivable level – in a certain way "error" here is not error, but: it is the regulating force.

Music is considered to be the formalization of the hearing of sound. Through its mediatization, coding and thus its storage and designability, its dynamic and, for people, its transient form, it has often experienced a cultivated reception, namely the aesthetics of harmony. This is designed as a way of thinking, from the independence of the existence of the medium of sound; avant-gardes use deviation from it, the error as an overcoming of this restriction – generally the result was negation (whatever it was brought about by).

In music, digital culture finds a paradigm of those new and media arts that have constitutively transferred the dynamics and coding from music to their sensory perception.

As cybernetic arts (initially and dominant), digital arts have oriented themselves towards the arithmetic design of geometric embodiments of modernist music and related thinking. In itself contradictory, this digital culture has oriented itself on the materiality of thought.

This digital composition can be regarded as an error in the dominance of visual thinking, in construction as well as in its suitable reception of codes for sounds according to related thinking. Work-music follows the logic of seeing, ultimately reading comprehension, whereas music-making is a sonically performative communicative event that does not formalise and culturally transform the natural sound production through movement from excitement. In this, especially in the "utilisation" of the sound gesture as an imagination of movement as well as an expression of its meaning for the body (JAUK 2014), sonic performativity is also the paradigm of the hedonic as emotional interaction with virtualities (JAUK 2009).

Demanded by Dubuffet as an overcoming of the dominance of seeing, this takes place with the transgression of the mechanistic through dynamization and coding of the environment. The adaptive design of dynamic virtuality escapes the dominance of a culture of seeing. With this turn, the modern construct of error in analysing interaction disappears, and with it so do bodies and environment mutually adaptively changing to any stimuli.

Although apparently being moving images, their design and reception are "organized" according to the logic of hearing, according to sonic performativity – a hedonic experience of movement after analysis of its excitation, after the motion of a mechanically motionless body. An evolutionary "remix" of perception, the anaesthetic turn to the phylogenetically older form of interaction of hearing as a suitable form of interaction of the hedonic body with dynamized virtualities – a cultural adaptation of the body through cultural instrumentation/mediatization of the body and a resulting new environment that makes the body superfluous – does not overcome the mechanical body as an error, but rather "expands" its forms of interaction in the re-mix as a cultural form of respecting phylogenetically older forms of life.

Pop culture, as body culture, pop music as the little-mediatised form of direct physical expression that is at the same time a highly mediatized technologically intensified form of expression (JAUK 2009), is understood as an essential factor of an experience culture (SCHULZE 2000), a neoliberal economic culture that is determined by image as experience and thus purchase value. Even the self-optimisation of hedonism through rationally based techniques is no longer regarded here as an "error" of a hedonic culture – the development of hedonism becomes an increased hedonic value (PFALLER 2011) – besides its function of the self-regulation of life – a "new error"?

Post-digital culture is no longer oriented towards exceeding the mechanical body, which has become "superfluous" through digital extension, but, necessarily, as an evolutionary form of survival towards the hedonic body, towards self-regulating design through arousal-determined interaction – does this culture and its forms still know "error"?

Finally: in the actions of nature, in the negotiation of culture, error is not considered a disturbance variable but a design variable – there is only one unconstructive error, that of the postulate as an intervention in a

self-regulating system. What futurologists have predicted, Bataille and Foucault grasped intuitively, and it encouraged them to experience that development as transgression is a non-affirmative affirmation.
If post-digital culture is the overcoming of the modern construct of "error," the contemplation of the natural life and survival design of the body and environment through an arousal-based interaction of the hedonic body with the virtual environment, then the world of error has made itself superfluous as a modern concept, evolution has adapted the form of interaction from the deliberately rational decision of seeing to the analytical experience of hearing, and thus turned a natural into the cultural form of post-digital life, the body with immateriality – it would be a relapse to see this as a possible "error." This transgression as a "non-affirmative affirmation" is to be lived with all openness in social and political coexistence.

References

Baudrillard, J. (1981): Simulacres et simulation, Paris.

Bales, R. F. (1950): Interaction Process Analysis. A Model for the Study of Small Groups, Cambridge

Berlyne D. E., (1971): Aesthetics and Psychobiology, New York

Cascone, K. (2000): 'The Aesthetics of Failure: "Post-Digital" Tendencies in Contemporary Computer Music'. Computer Music Journal, Vol. 24, No. 4, 12-18

Gibson, J. J. (1982): Wahrnehmung und Umwelt. Munich

Jauk, W. (2000), 'The Auditory Logic: An Alternative to the "Sight of Things"', in: H. Nowotny et al. (eds.), Jahrbuch des Collegium Helveticum, Zurich, 321–338

Jauk, W. (2003): 'The Transgression of the Mechanistic Paradigm – Music and the New Arts', in: Dialogue and Universalism 8–9, 179–186

Jauk W. (1999): 'Gestaltung durch kommunizierendes Verhalten: Musik und Net-Art', in: Forschungsbericht Klang-Forschung 98. Symposium zur elektronischen Musik, J. Stelkens, Hans G. Tillmann (eds.), Munich, 163-173

Jauk W. (2007): 'Der Sound des hedonisch-performativen Körpers und das Spiel der Elektrogitarre', in: Jazzresearch, vol. 39, F. Krieger & B. Hoffmann (eds.), Graz, 273-289

Jauk, W. (2009): pop / musik + medien / kunst. Der musikalisierte Alltag der digital culture. Osnabrücker Beiträge zur systematischen Musikwissenschaft. Vol. 15, B. Enders (ed.), epOs, Osnabrück

Jauk W.(2010): 'trash musics: in-dust/e-grains/dig-glitches. Die andere Mediamorphose aus dem Spiel mit der Durchlässigkeit des "Mülleimers der Geschichte"', in: Abfallmoderne, Vienna, 2010

Jauk W. (2011): 'Kunst, Kreativität, Pop-Kultur: Provokation des Chaos – Chaos als Provokation', In: Chaos, U. Bechmann/Chr. Friedl (ed.) Graz, 203-224

Jauk, W. (2015): 'Back to the roots! Hedonisch bewegte körperliche Gestaltung des "künstlichen" Klanges. Interfaces – Wahrnehmung, Ausdruck und deren Instrumentarisierung/ Mediatisierung im musikalischen Kommunikationsprozess', in: A. Bense et al. (eds.), Musik im Spektrum technologischer Entwicklungen und Neuer Medien. epOs : Osnabrück, 277-302

Jauk, W. (2014): 'Intuitive gestural interfaces/adaptive environments and mobile devices/apps: Playing music and the musical work as a role model for personalized gestural interaction in social environments', in: ICMWT International conference on Mobile & Wireless Technology Congress-Book, Beijing, 280-284

Jauk W. (2017): 'Der Körper ist die Grenze: das Ich in Wissenschaft und Kunst. Aspekte einer anderen Empirie', in: D. Ingrisch et al (eds.) Bielefeld, 81-98

Kerckhove, D. de (1995): 'Kunst im World Wide Web', in: Prix Ars Electronica 95, Linz

Leopoldseder, Chr. Schöpf (eds.), Linz, 37–49

Lakoff, G. & Johnson, M. (2003): Metaphors We Live By, London

Lyotard, J-Fr. (1985): Immaterialität und Postmoderne, Berlin

Marcus, G. (1996): Der Mülleimer der Geschichte, Hamburg

Marcus, G. (1992): Lipstick Traces: Von Dada bis Punk – kulturelle Avantgarden und ihre Wege aus dem 20. Jahrhundert, Hamburg

McLuhan, M., Powers, B. R. (1995): The global village: der Weg der Mediengesellschaft ins 21. Jahrhundert, Paderborn

Pfaller, R. (2011): Wofür es sich zu leben lohnt. Elemente materialistischer Philosophie, Frankfurt am Main

Riemann, H. (1914/15): 'Ideen zu einer "Lehre von den Tonvorstellungen"', in: Jahrbuch der Musikbibliothek Peters 21/22, 1–26

Schaefer, J. (1987): New Sounds: A Listener's Guide to New Music, New York

Schulze, G. (2000): Die Erlebnisgesellschaft. Kultursoziologie der Gegenwart, Frankfurt am Main

Welsch, W. (1992): 'The Birth of Postmodern Philosophy from the Spirit of Modern Art', in: History of European Ideas, 14-3, 379–98

Welsch, W. (1993): Aesthetisches Denken, Stuttgart

Welsch, W. (1996): Grenzgänge der Ästhetik, Stuttgart

Werner Jauk (AT)

Power is the only error...

Collective/ collectivizing music-making as a paradigm of the hedonic self-organization of the dynamic adaptation of spaces as communal living spaces

Error is a construct of Modernism; conceived as a deviation from a theory-driven prognosis of an event, from the next step in the linear progress of culture .... Progress-avant-gardes of the early 20th century attempted, by shutting down rational idealist thinking, to radically break through the boundary between sense and non-sense; a deconstructivist attitude shifts thinking and focuses the error; that which is excluded can be said to provide information about that attitude of the norm that brings about this exclusion; at the same time, consideration of error is a political and cultural step towards awareness of "queerness"... The aesthetic and political "use" of error is familiar to us from the history of techno music, the music of glitches / clicks / cuts / bursts ... The *Power is the only error ...* project doesn't aim to indulge the error of Modernism as the creative power of digital culture; it seeks, in the post-modern attitude of post-digital culture, to make it possible to experience the fact that hedonic self-organization knows no error...

Life is dynamic adaptation; human life is dynamic adaptation of body and environment in response to stimulus; music is play with the dynamism of stimulus, a playing-with-nature that becomes a culture, a paradigm of inclusion of the hedonic body in virtualities ....

What manifests itself as a cultural development was once a powerful act of overcoming nature, feasibility and the necessity of taking action. Alternatively, culture is conceptualized in terms of the process of nature. With this comes the demise of the concept of feasibility and the emergence of consciousness that nature is the inviolable framework of culture. Cultural studies doesn't only deal with power processes that differ from nature; it itself plays the role of the "gouvernementalite"; it powerfully constructs culture. The concept of the evolution of plural, parallel, natural life is increasingly applied to cultural developments. Explorative behavior determines not only general survival but also what culture brings forth and, ultimately, aesthetic behavior. Needless to say, this does not preclude power-driven influences on cultural developments...

It is the "affordance" of the respective (and, for human beings, indeterminately coincidental) operative environmental stimuli that cause the intentionality of the body-environment interactions that set the body "in tension." This leads to stimulus-based interactions with the environment that continue to drive homeostatic reciprocal adaptation of body and environment. The "feeling the ex-tension" project attempted to enable participants to experience the arousal-based "regulation" of social interactions. Now, *Power is the only error ...* focuses on homeostasis, the arousal-based self-regulation of life in which deviation is not error, but rather information about the change of a body interacting with the environment, which leads to reciprocal adaptation. The basal control principle of homeostasis is bodily arousal, dependent primarily on the intensity of stimuli. Only "afterwards" is its cognitive evaluation conveyed into another form—understanding—of intensity, by mediating excitement. *Power is the only error ...* is meant to allow participants to experience this self-regulating design process in the form of its externalization, the mediatization of a bodily regulation mechanism in social contact—the change the environment undergoes in the process of its adaptation to bodies' behavior in response to stimulus both reciprocally and in interaction with the environment. The social as well as the physical environments configure themselves in association with human beings.

This is, on a large scale, the sociological approach of interactionism and the architecture of the Chicago School; on a small scale, this is the essence of group dynamics: Interaction as reciprocal formation of arousal-balanced living spaces. In the 1990s, in what was tantamount to a reinvention, this was applied to Net Art as a collective and collectivizing interaction as design. What has been added in the last few decades is the acceptance of physical interaction as stimulative interaction, immediate physical deeds, and not the exclusive assumption of symbolic interaction in terms of signs, by means of symbolic negotiation—where even the symbolism of this term, in its metaphorical nature, attests to an embodiment (LAKOFF) that is ultimately interwoven with physical significance for the body.

A physical space becomes a social space through the emotional interactions of human beings, human bodies, in it. It is stimulation that determines interpersonal interaction. Sound-gesture as imagination of the movement around the body and as expression of the meaning (of the arousal level) of this movement for the body is, therefore, the paradigm of emotional interfaces facilitating adaptation (according to hedonic human as measure) of physical and social spaces as cultural spaces—of real, virtual and converged realities. These motility-induced movements, (gestural) bodily behavior, facial expressions, and sound production will be reflected in the arousal level of the space, as movement that ultimately "causes" sound, is sound. The space will change adaptively, and not only for the purpose of mood management; it will adjust to all the protagonists within it on a homeostatic level. The room will become a jointly configured supra-vival space.

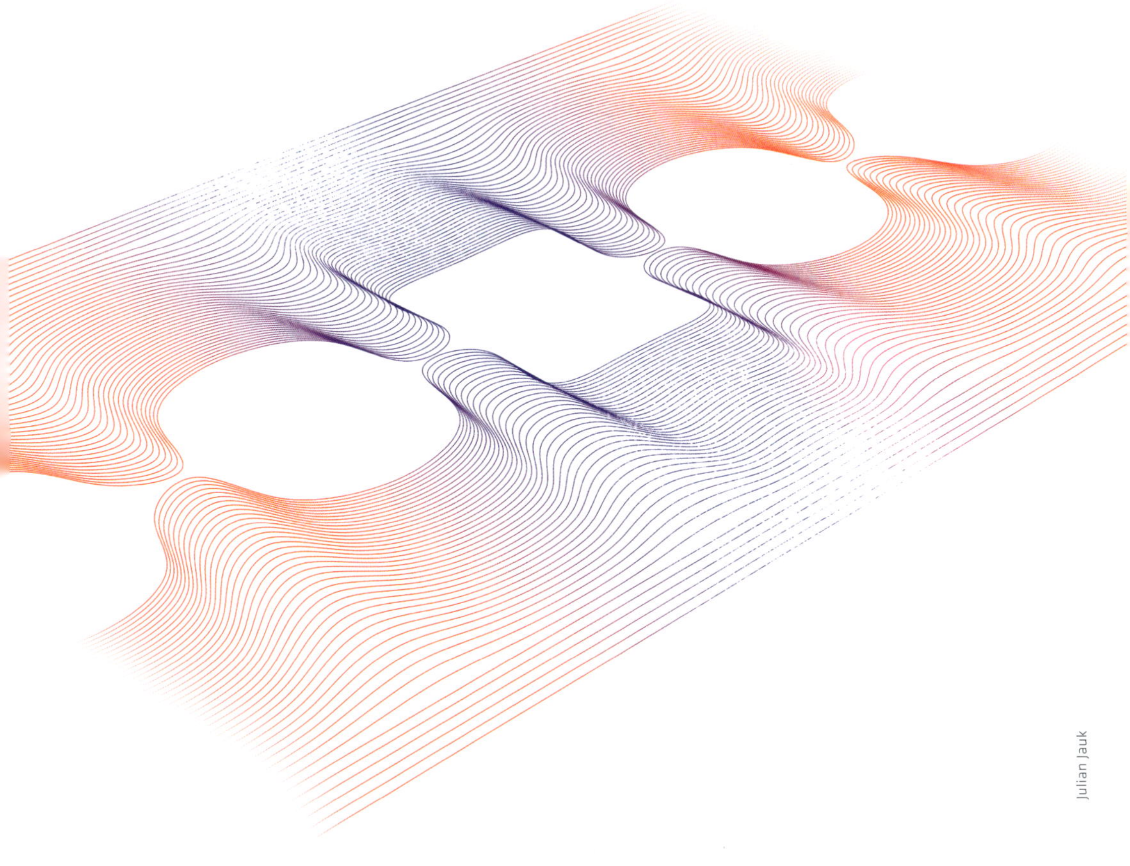

Julian Jauk

Goran Vejvoda (RS), Florence Müller (FR)

All Sounds Considered

All Sounds Considered (ASC) explores the state of "Sound & Silence" without dogmas or preconceived theories. The film shows some of the various facets that make up this fascinating domain. Produced and directed by sound artist Goran Vejvoda and curator Florence Müller, the idea is to lend an open ear to what actors in this versatile scene have to say. Through various points of view, ASC witnesses the multiplicity of strategies, ideas and creations. Taking a modular and non-chronological approach, this "freecumentary" leads us into a ballad that spins in various directions with curiosity and looks into certain elements that constitute this vast panorama. Artists, books, exhibitions, performances, interviews, instruments and machines... help us discover the world of contemporary sonic media. In two episodes, ASC delves into Sounds that feature a variety of innovating postures and approaches that artists, researchers and inventors have come up with on their quest. Focusing on the realm of audio perspectives and its iterations, this is a voyage into the multiple ways of working with a medium, and when it goes off the beaten track it can irritate as much as it can profoundly seduce.

Promoted by AM, sound design agency for fashion and art, Paris
http://amsoundesign.com
Featuring artists:
Charles Amirkhanian, Jaap Blonk, Bill Fontana,
Ellen Fullman, Max Neuhaus, Joan Labarbara,
Bernhard Leitner, Barbara London, Torturing Nurse,
Aki Onda, David Toop, John Schaefer, Pierre Schaeffer,
Guy Schraenen, Akio Suzuki, Peter Vogel,
Pamela Z, Christian Zanesi and more.

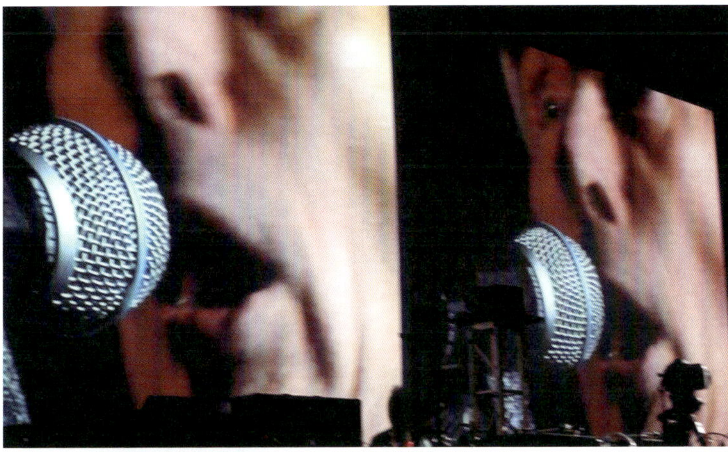

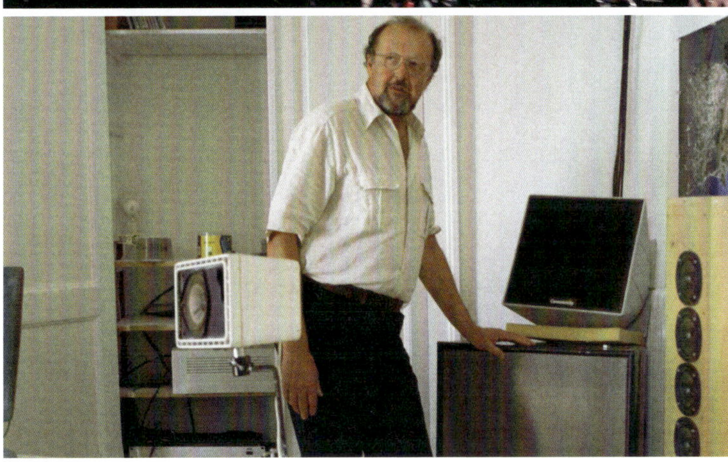

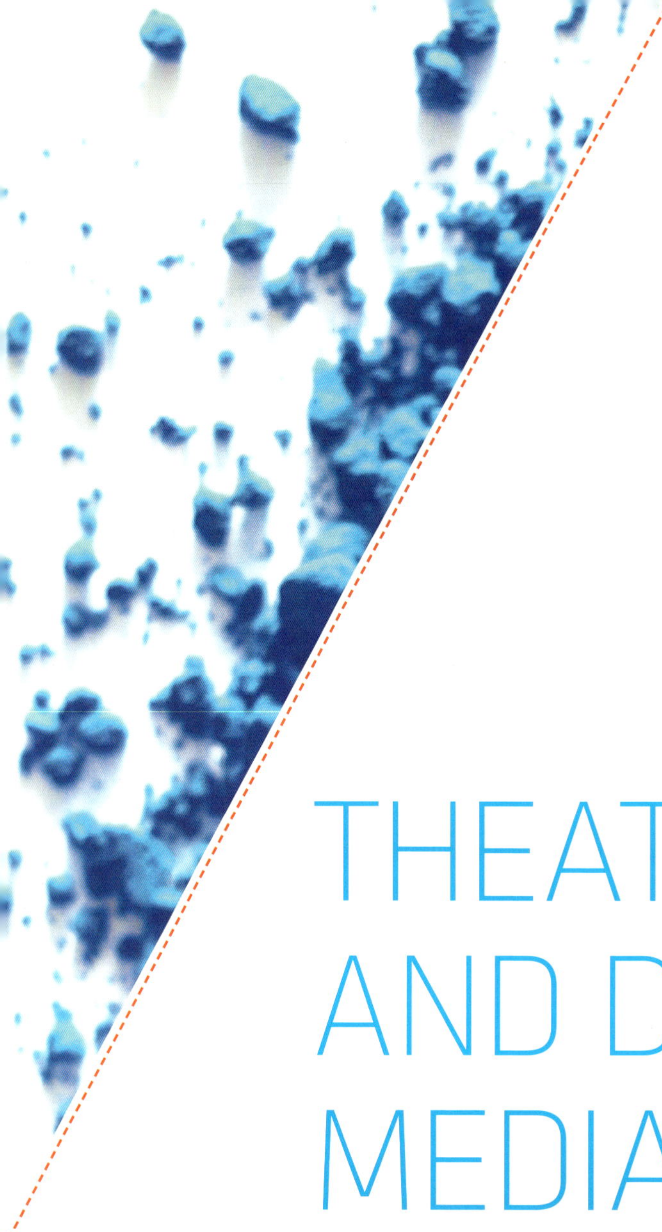

THEATER AND DIGITAL MEDIA

European Theatre Lab: Drama goes digital

European Theatre Lab: Drama goes digital is Europe's first think tank devoted to researching a digital strategy for theater. In a pioneering mix of workshops, conferences and theatrical performances across Europe, the project partners researched the effects of new technology on aesthetics, audience development and communication in theater.
Three creative projects emerged from the European Theatre Lab and are currently being presented around Europe. They are breaking new ground in key research areas: app-based virtual & augmented reality ("Stage Your City"), 3D sound, auralization, psychoacoustic effects ("Kinetics of Sound") and voice recognition surtitling ("IDIOMATIC/Dub It").
These projects and technologies will be introduced during a session at Ars Electronica Festival 2018 and the performance "Stage Your City" will be presented for the first time in Austria for this occasion.

The project "European Theatre Lab: Drama Goes Digital" is a consortium of the European Theatre Convention (ETC) and 7 public theatres:
- Théâtre de Liège, Belgium
- Croatian National Theatre Zagreb, Croatia
- CDN Nancy Lorraine La Manufacture, France
- Badisches Staatstheater Karlsruhe, Germany
- Kote Marjanishvili State Theatre Tbilisi, Georgia
- Det Norske Teatret Oslo, Norway
- Teatrul National "Marin Sorescu" Craiova, Romania

The "European Theatre Lab: Drama goes digital" has been developed with the support of the Creative Europe program of the European Union.

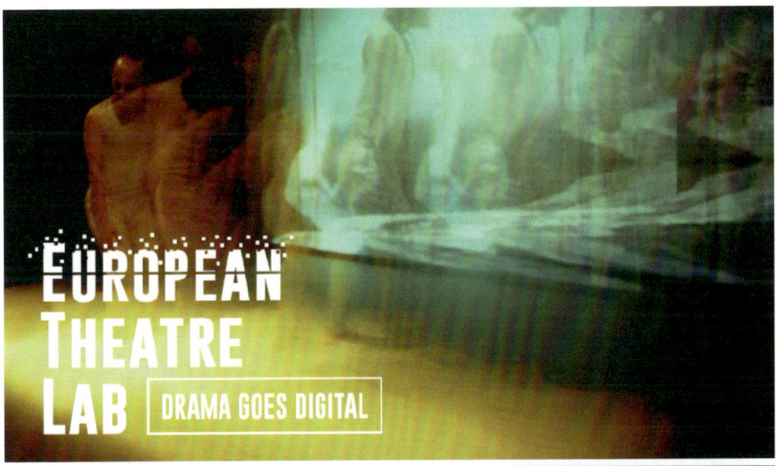

Chris Ziegler

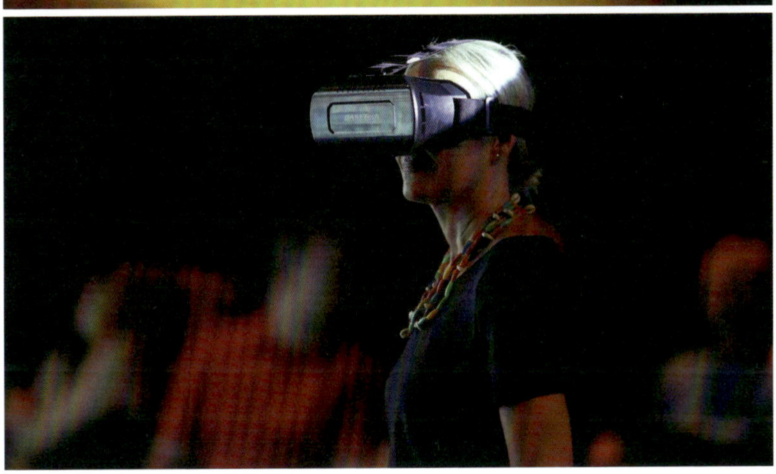
Tom Kohler

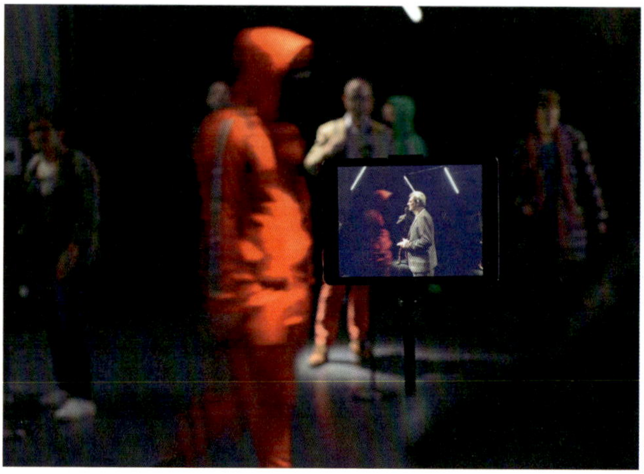

Eric Didym

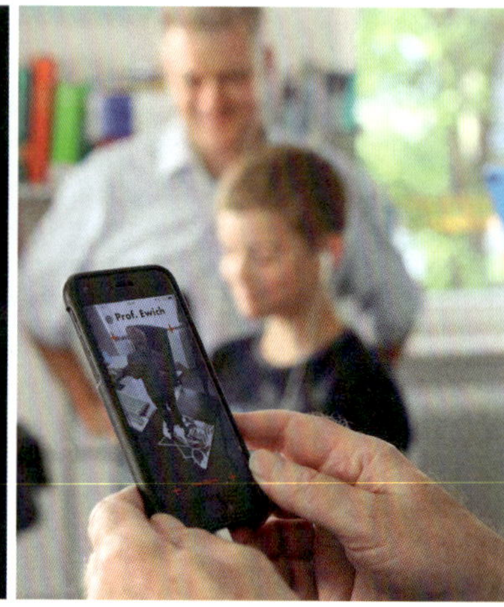

Tom Kohler

Stage Your City

Theater has absorbed many technologies throughout its history. Yet many practitioners today defend it as the last resort of the analogue and ask the audience to switch their smartphones off. "Stage Your City" attempts the opposite: Everyone is asked to bring their own smartphone, install the app "Zigmagora" and follow its commands into a city of the future. The digital dramaturgy takes the audience in groups through dystopic situations - and several consumer technologies. The spectators are received by experts from 2052 in an iPad hologram installation. Then groups are guided through the city by the app that bundles navigation, audio tour, and Augmented Reality, magically triggered by Bluetooth. The app also demands interaction: answers to philosophical questions, the upload of selfies and photos and of a favorite song from a streaming service. Back at the starting point, the audience experiences a 360° world they have co-created. At the end of the digital journey comes an analogue surprise—and collective, organic storage of memories.

www.zigmagora.eu
Director: Michel Didym
Actors: Klaus Cofalka-Adami, Natuka Kakhidze, Constantin Petry
Texts: Marie Dilasser–Lasha Bugadze, Bruno Cohen, Michel Didym, Konstantin Küspert, Frédéric Sonntag
Production & App-Design, Scenography & Media: Christian Ziegler
Costumes: Éléonore Daniaud
Programming: Wandio, Tbilisi
Artistic Collaboration: Didier Billon, Jan Gerigk / ZKM, Tobias Lindörfer / Prestigefilm, Bernd Lintermann / ZKM, Sarah Steinfelder
Dramaturgy: Nutsa Burjanadze, Sarah Stührenberg, Bruno Cohen, Jan Linders
Production Management: Maren Dey, Sarah Mckee, Teresa Pfaud
Seven European theaters have founded the European Theatre Lab "Drama goes digital" under the umbrella of the European Theatre Convention ETC. In three projects, they want to test whether and how digital technologies can be used artistically on stage. Stage Your City is a project of CDN La Manufacture Nancy, Kote Marjanishvili State Drama Theatre Tbilisi, Badisches Staatstheater Karlsruhe and under the umbrella of the "European Theatre Lab: Drama goes digital" in cooperation with ZKM Karlsruhe —Center for Art and Media. *www.europeantheatrelab.eu*

The Culture Yard (DK), Very Mainstream Studio & Very Theatre (TW), Noora Hannula & The Nordic Beasts (DK/FI), Carl Emil Carlsen (DK), Bjørn Svin (DK)

The 4D Box—a holographic live stage

The 4D Box is a multimedia digital live stage developed and curated by *The Culture Yard* and CEO and director Mikael Fock. It includes hologram technology, advanced sound and live performance in a single expression. In this interdisciplinary black box, *The Culture Yard* have created an entirely new artistic media platform that breaks the boundaries between physical performance and 3-D generated universes. At Ars Electronica, we look forward to presenting three productions: *Silicium*—an electronic concert performance—in collaboration with visual artist Carl Emil Carlsen and performer and musician Bjørn Svin; followed by *Chronicle of Light Year*—a co-production with *Very Mainstream Studio & Very Theatre*—which is an artistic and anthropological exploration of the concept of happiness across time and space. Thirdly, we present the dance performance *The Ultimate Battle*, an avatar dance performance in collaboration with Noora Hannulla and *Nordic Beasts*.

Production: The Culture Yard *https://kuto.dk/*, Very Mainstream Studio & Very Theatre *https://www.facebook.com/vmstudio.tw*, Carl Emil Carlsen *http://cec.dk/*, *Bjørn Svin https://bjornsvin.com/* Noora Hannula & The Nordic Beasts *https://www.noorahannula.com/*, Bora Bora *http://bora-bora.dk/*
Supported by: Danish Arts Council, Bikubenfonden, CLICK Festival, The Culture Yard, Very Mainstream Studio, Nordic Beasts, Ministry Of Culture Taiwan, National Culture & Arts Foundation, Nordic Culture Fund, Koneensäätiö, Dansk Komponist Forening, Knud Højgaards Fond, Helsingør Kommune

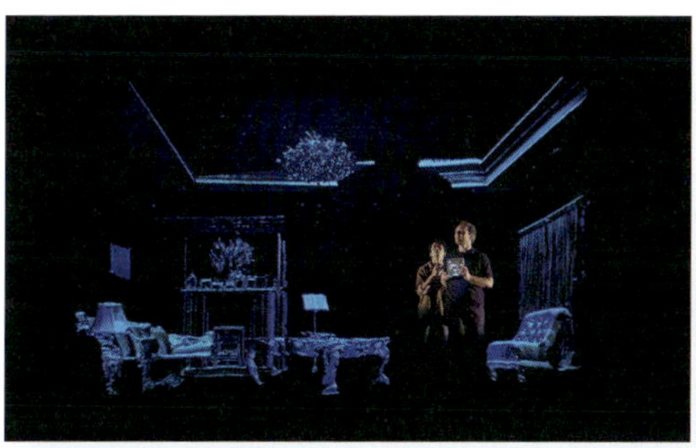

Carl Emil Carlsen

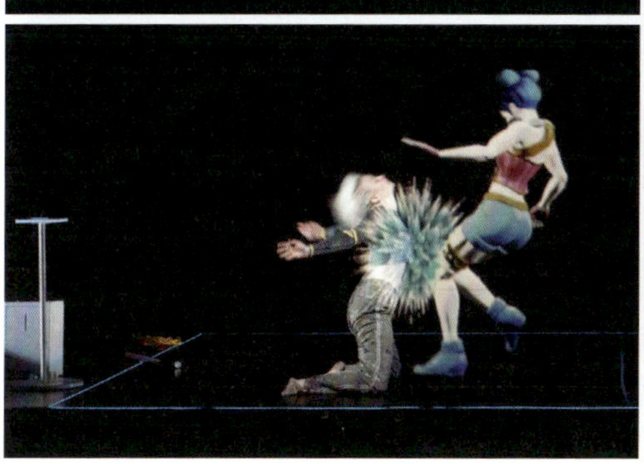

Søren Meisner

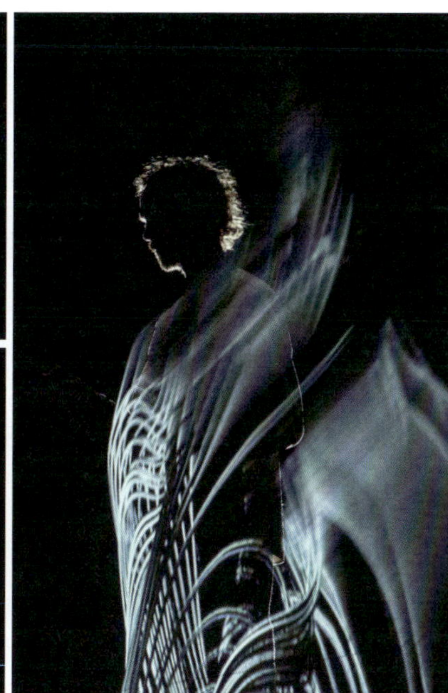

Bo Jensen

ERROR

the Art of Imperfection

CAMPUS PROGRAM

CAMPUS

Every year since 2002, Ars Electronica and the University of Art and Design Linz have hosted an exhibition by artists associated with an international higher-education institution whose curriculum takes an innovative approach to teaching media art and media culture.

Initiated by Prof. Reinhard Kannonier (University of Art and Design Linz) and Gerfried Stocker (Ars Electronica), the intention of the *Campus* format is to invite outstanding international universities working in the academic fields of media arts and design. Projects highlighted here represent the nature of the mission and activities of invited guest universities from all around the world. These showcases became an essential part of the festival and an instrument to analyze and visualize different models of educational approaches in artistic and creative areas. It has also increasingly developed into a stage for contextualized works from alumni, professors or associates from the universities to map the identity of academic institutions, their history and current practice.

Part of *Campus's* mission is to enable the presentation of young, local media artists and their work with international exposure. The Interface Cultures program of the University of Art and Design Linz annually presents a cross-section from their masterclass works and, together with Ars Electronica, co-hosts one main featured partner university each year. The festival is increasingly becoming a platform for artistic and creative collaborations between Ars Electronica and various regional, academic partners, for example the Fashion & Technology or the Visual Communication program at the University of Art and Design Linz, the Anton Bruckner Private University Upper Austria or the University of Applied Science Upper Austria, Campus Hagenberg.

2002: Academy of Media Arts, Cologne (DE)

2003: Department of Media & Art at the University of Art, Media and Design in Zurich (CH)

2004: IAMAS (JP)

2005: Srishti School of Art Design and Technology, Bangalore (IN)

2006: Medialab at the University of Art and Design Helsinki (FI)

2007: HGK FHNW, the Swiss Institute for Postindustrial Design (CH)

2008: University of Tokyo (JP)

2009: MIT Media Lab (US)

2010: Media Campus of the Darmstadt University of Applied Sciences (DE), School of Art & Design at the Cork Institute of Technology (IE)

2011: University of Tsukuba (JP)

2012: UdK–Berlin University of the Arts, Sound Studies (DE)

2013: Bezalel Academy of Arts and Design, Hamidrasha Art School of Beit Ber College, Holon Institute of Technology, Kibbutzim College of Education, Technology and the Arts, Shenkar College of Engineering and Design, the Media Innovation Lab at IDC Herzliya, Musrara School, the Neri Bloomfield School of Design and Education, Hadassah Academic College and the College of Management – Academic Studies (COMAS) (IL)

2014: Arts2 – École Supériere des Arts (BE)

2015: Paris 8 University (FR)

2016: Tsinghua University Beijing (CN)

2017: University of California Los Angeles (US)

Campus is also a platform for international exchange between universities, leading to increasing collaboration between academic partners. In 2018, twenty-three universities from many parts of the world are represented:

Hexagram, international research network (CA):

- Concordia University (CA)
- Université du Québec à Montréal (CA)
- University of Montreal (CA)
- University of Quebec at Chicoutimi (CA)
- McGill University (CA)
- École Technologique Supérieur (CA)

University of Art and Design Linz (AT)

Academy of Fine Arts Prague (CZ)

University College London, The Bartlett School of Architecture, Interactive Architecture Lab (UK)

Central Academy of Fine Arts Beijing (CN)

University of Tsukuba (JP)

King's College London (UK)

University of the Arts London (UK)

Roy Ascott Studio, Shanghai (CN)

Queen Mary University of London (UK)

School of the Art Institute of Chicago (US)

Shantou University (CN)

New Design University, St. Pölten (AT)

Aalto University (FI)

Anton Bruckner Private University Upper Austria (AT)

Massachusetts Institute of Technology, Media Lab (US)

Queensland University of Technology (AU)

University of Applied Science Upper Austria, Campus Hagenberg (AT)

Taking Care.

Hexagram @ Campus Ars Electronica 2018

Project Initiator: Chris Salter, Co-Director of Hexagram (Concordia University)
Curator: Anna Kerekes, Curator (University of Quebec in Montreal)

Taking Care is an exhibition of twenty works from student members of Hexagram, an interdisciplinary research network for media arts, design, technology and digital culture based in Montreal (Quebec), Canada. Established in 2001, the network brings together 40 faculty and over 200 graduate students from its founding universities, the University of Quebec at Montreal (UQAM) and Concordia University, along with the University of Montreal, McGill and the University of Quebec at Chicoutimi, local and international academic and cultural partners.

The projects exhibited in *Taking Care* operate at the intersection of ethical-aesthetic concerns. They are not meant to be prescriptive—describing the good life—but rather speculative, asking what the possibilities and conditions of life in the present and future could be. At the same time, they question our understanding of the world as we continually search and struggle to make sense of it. Expressed through a range of forms and media including games, VR, performance, installation, biological art, textiles, sound, video and photography, the projects all involve the use of contemporary technologies, yet their focus lies beyond the technological. From issues of post-colonial conflict to the imperceptible in more than human beings, questions of trace, history, narratives of experience, non-human materiality and representation are assembled and entangled with each other in the spaces of the University of Art and Design and at Ars Electronica venues across Linz. As attention turns to the considerable uncertainty of our future, *Taking Care* thus examines what is at stake in the ideas and visions of the next generation.

From an open call to Hexagram graduate students across the network, the twenty projects were chosen by a selection committee of Hexagram members and collaborators outside of the university context. The exhibited projects exemplify what is known as "Research-Creation"—a research trend where Hexagram, Quebec and Canada have taken an international lead.

This interdisciplinary approach bridges faculty and students from the arts, humanities, social and natural sciences and is well funded and supported by universities, provincial and federal research bodies. Students and faculty from around the world are attracted to Hexagram for its cutting edge research infrastructures—studios, labs, black boxes—and the vibrant cultural scene of Montreal and Quebec, a center for digital arts and culture in North America.

Structured around three research axes (sense, embodiment and movement; materiality; ubiquity), Hexagram responds to the increasing need to develop critical and reflexive theories and methods in artistic practice through experimentation, production, documentation and dissemination.

Within Montreal, Hexagram provides an intra-university environment for collaboration through public seminars, events and publications that transcend disciplinary silos and the isolation of university departments.

Within Quebec, it also enables structured mobility and exchange between French and English research cultures in art, science, technology and society.

Production Team for Campus: Marine Theunissen, Project coordinator (University of Quebec in Montreal); Alexandre Saunier, Director of production (Concordia University); Patil Tchilinguirian, Exhibition designer (Concordia University); Lorène Chesnel, Head of Communications (University of Montreal); Agustina Isidori, Documentation (Concordia University)

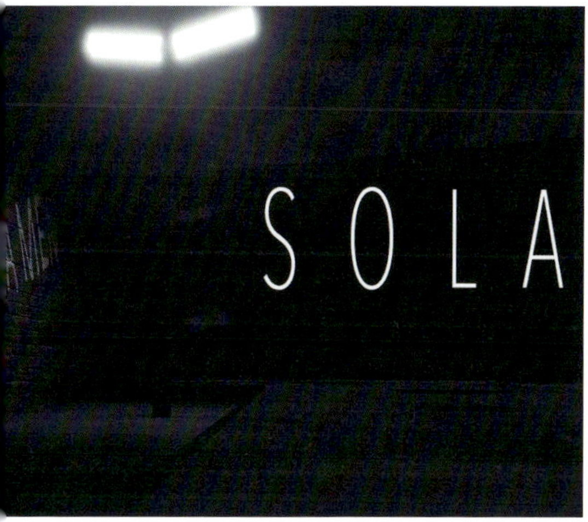

Agustina Isidori (IT/AR)

SOLA

SOLA is a video game that embodies the climate of tension, discomfort and fear that can be experienced while walking alone at night in cities where gender-based violence is embedded in everyday life. *SOLA* challenges the usual understanding of games as a source for entertainment. It builds on the concept of play to explore a video game as an instrument for conceptual thinking and a tool to work through social issues.

Agustina Isidori (IT/AR)

Don Federico

Don Federico is an experimental video installation that juxtaposes YouTube footage of kids playing a hand game and an actual case of femicide. It creates an unsettling dialogue that reflects on the naturalization of gender-based violence.

> "Don Federico killed his wife,
> he chopped her in little pieces
> and put her in a frying pan,
> people walking by,
> smelled something stinky,
> it was Don Federico's wife,
> dancing cha cha cha"

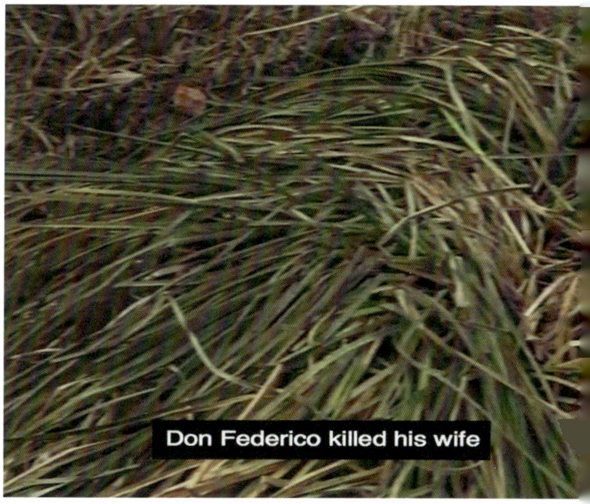

Alexandre Saunier (FR/QC/CA)

Sensum

Sensum is an instrument to experiment with the sensory effects of LED light. It consists of a helmet that triggers the sense of sight through dynamic patterns of color, intensity, and movement. It explores how digital technologies offer algorithmic properties and lighting possibilities that alter our physiological condition and generate new sensations.
Sensum provokes lived experiences that range from visual illusions to complex cross-sensorial effects, haptic hallucinations, and time distortions.

Alexandre Saunier (FR/QC/CA)

Vitra

Vitra is an ensemble of lively light sculptures that modulate the atmosphere of the space.
It plays with rhythms of color and intensity that fluctuate from barely noticeable to nearly blinding. Both as an installation and a performance, *Vitra* unfolds a poetics of presence and absence to convey a sense of spatial openness and an experience of temporal suspension.

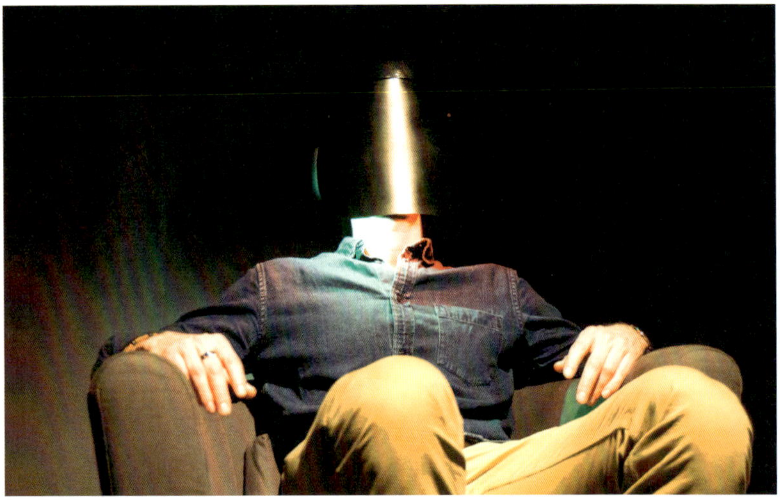

Sensum

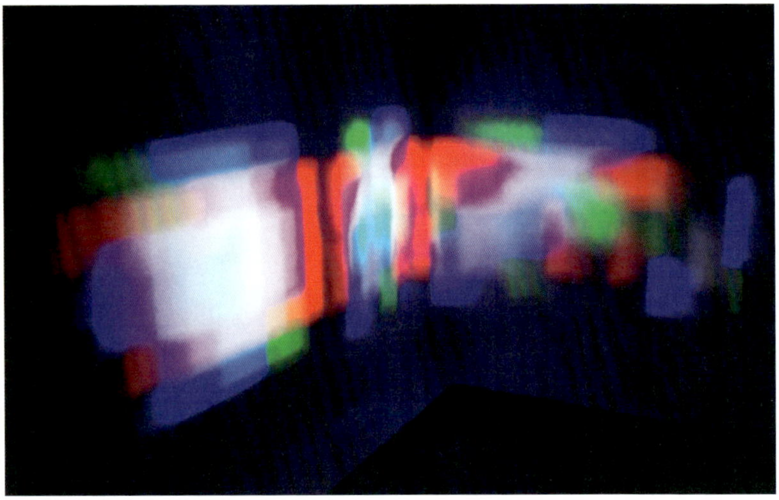

Vitra

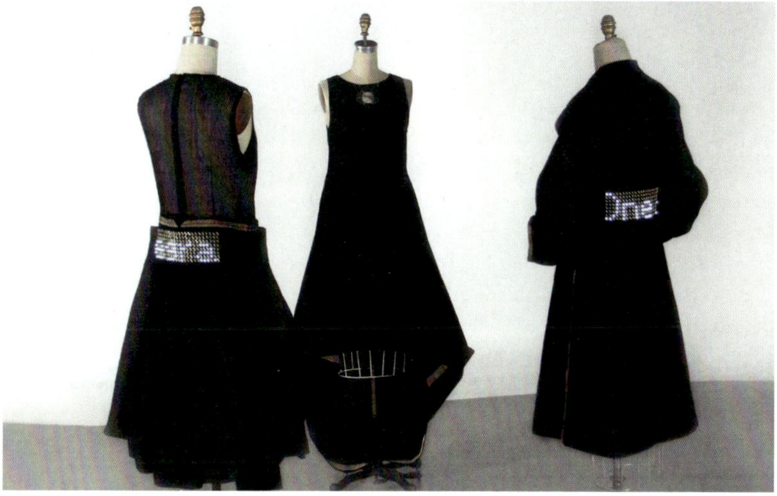

Maxwell's Equations

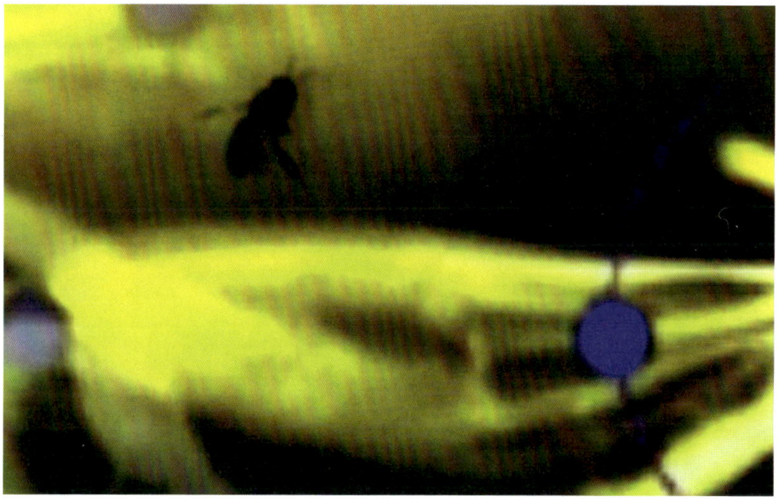

DRONE

Barbara Layne (US/CA), Lauren Osmond (CA)

Maxwell's Equations

Maxwell's Equations consists of three garments that incorporate unique antenna designs that wirelessly connect the garments to one another. The designs draw inspiration from 19th century fashion and from James Maxwell's pioneering theories of electromagnetic fields. When physically aligned, the strength of the connection will change the texts, poetry and mathematical formulas scrolling through the LED panels. The antennas on the front of the dresses portray one of Maxwell's diagrams.

Donna Legault (CA)

DRONE

DRONE draws on research into the physical dynamics of insect flight and behavior. The installation video merges images from the documentation of live bees with their robotic counterparts that are currently being developed by micro-robotics labs in the U.S. and Japan. The projected light, in collaboration with participants' movements, generates a live soundscape that is tuned to bee flight and communication behaviors offering a speculative engagement with the tangible sonic experience of bee activities.

Eugenia Reznik (UA/FR/CA)

Un-weaving

Un-weaving takes its origin from a linen fabric made by the artist's Ukrainian grandmother in the 1930s, a period of severe famine in Ukraine. It was sent from Ukraine by mail with a letter explaining how it was made: from picking the plants and separating them into filaments, to spinning and weaving. For a long time the fabric remained in a drawer until the day when the artist decided to unweave it, to make it disappear. Why? For it to be reborn in another form: that of the moving image where its threads become plants once again.

Faye Mullen (CA)

Of wall to ground

Walls are a vestige of colonization and political instrument since the Anthropocene began; we have never been faced with more walls than we are today. Faye Mullen proposes the experience of reimagining what seems to be an imposing limit as an accessible threshold. By way of architectural and especially political and psychological semantics, she navigates the state of the *wall* in the hour of multiculturalism, access to information and international communication by placing value on deconstruction.

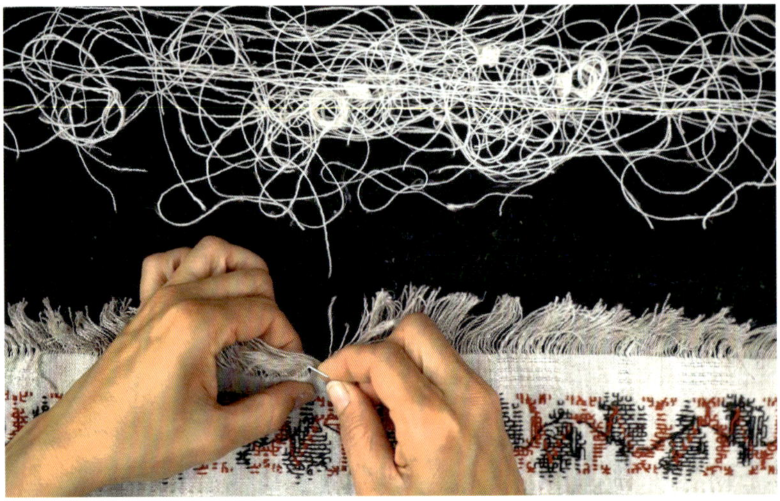

Un-weaving

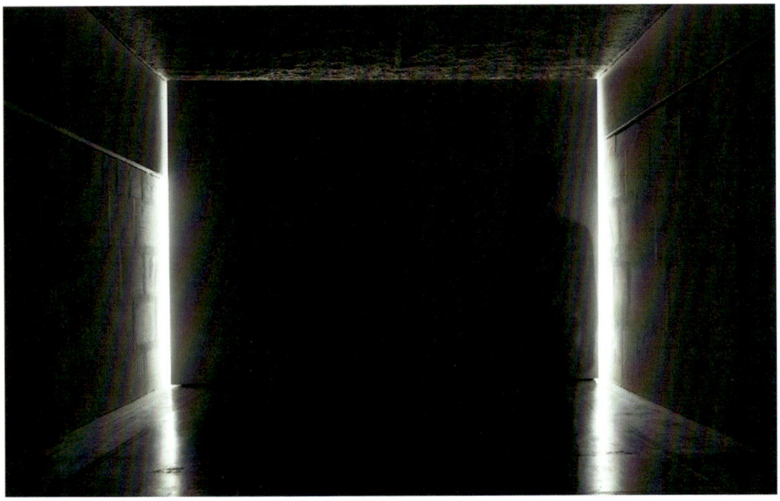

Of wall to ground

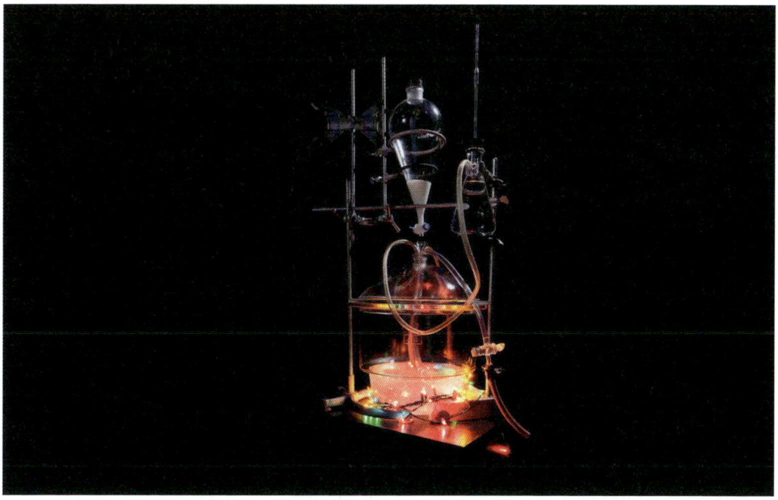

Microbiome Rebirth Incubator

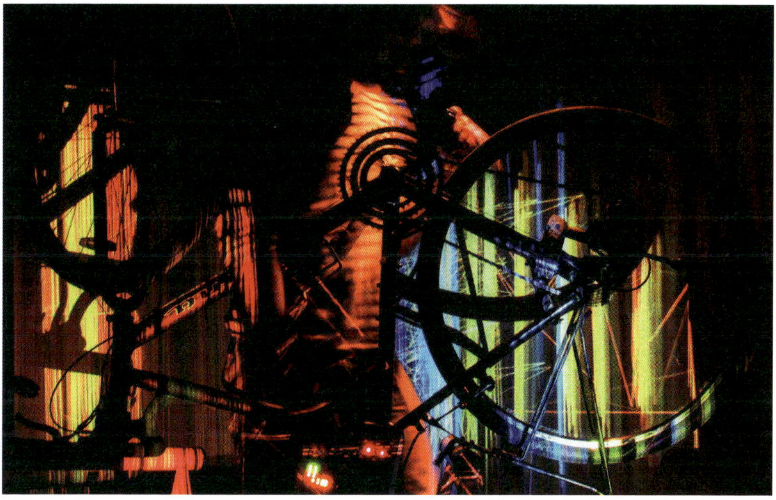

Feedback Cycles for Oscillographs

François-Joseph Lapointe (CA),
Marianne Cloutier (CA)

Microbiome Rebirth Incubator

The *Microbiome Rebirth Incubator* is a device designed to seed babies born by emergency C-section with vaginal bacteria and breast milk essential to infant growth, either by dipping the newborn in the incubator, or by soaking a sterile compress in the probiotic cocktail to wash the baby's mouth, face and body. This project symbolically explores the possibility of repairing the microbiome, as a way to re-empower the mother and her child, and to erase the traumatic experience of caesarean birth.

Guillaume Arseneault (CA)

Feedback Cycles for Oscillographs

Working on the distinctiveness of sound signals such as phase, frequency and amplitude, *Feedback Cycles for Oscillographs* produces an audio-driven synesthesia of hypnotic lines drawn from stereophonic disparity. Activated by movement on two wheels, this installative action seeks to reveal the unknown between void and short circuit.

Ida Toft (DK)

Promises

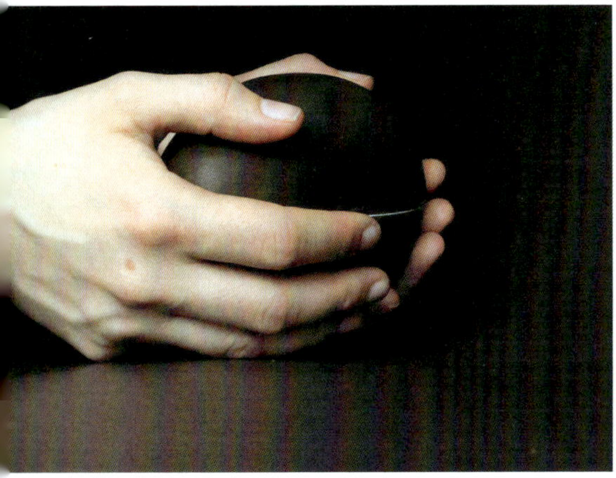

While consumers of mainstream culture have developed refined literacies in screen-based media, the modality of vibration is less coded with meanings. *Promises* explores what local multiplayer games might look like when using algorithmically driven vibration as primary material for expression. In *Promises*, objects circulate between the hands of the visitors. Objects and promises replace one another via the acts of trading, making attachments, and letting go. Preferences and meanings emerge.

Jess Marcotte (CA), Dietrich Squinkifer (CA)

rustle your leaves to me softly

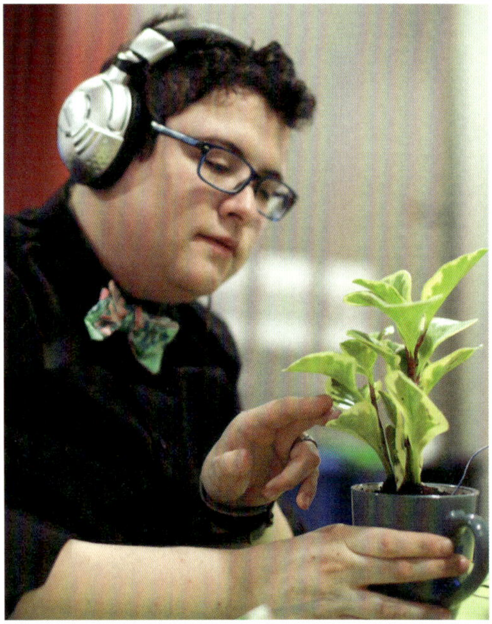

In this screenless physical-digital hybrid game, the main interaction is caressing and stroking a plant gently. Using the natural conductance of both humans and plants, a soundscape with ASMR poetry is generated. Since conductance varies from human to human and based on the moisture in the plant and earth, each human has a unique relationship with their plant partner. This game asks questions about what it means to design from the perspective of an "other" that we cannot know the mind of.

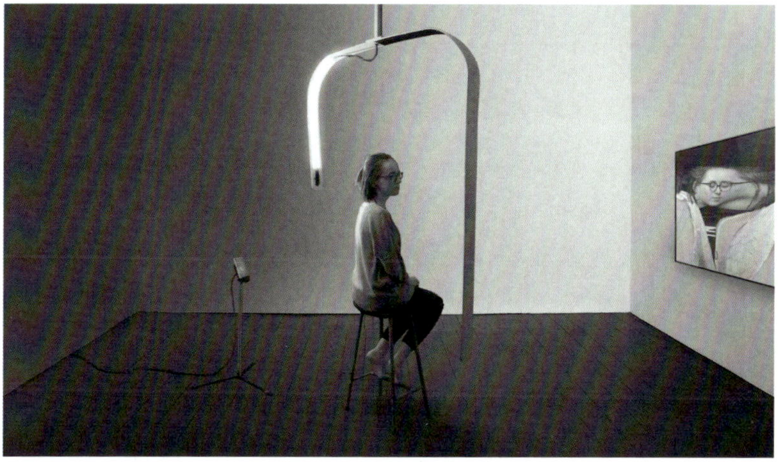

REVOLVE/REVEAL

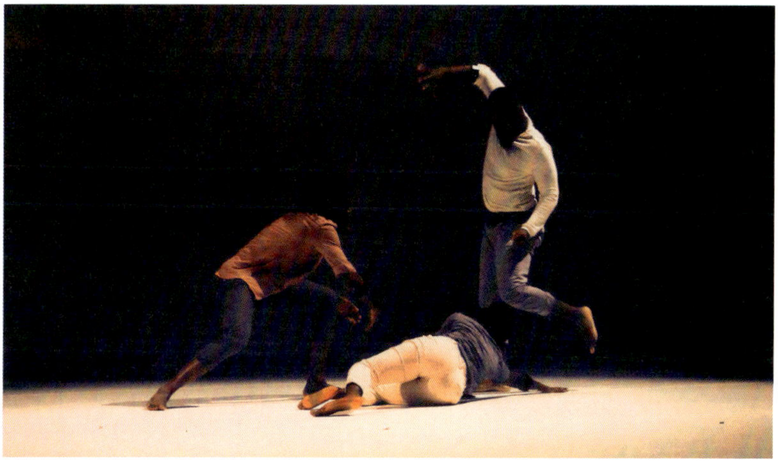

Temporalité Expressive

Louis-Philippe Rondeau (CA)

REVOLVE/REVEAL

REVOLVE/REVEAL is an interactive installation based on slit-scan photography, a marginal process which transposes temporality and spatiality within the image. Through distorted representations of the interactors, it seeks to transgress the photographic portrait's conventions of fixity and single vantage point. Slit-scan stretches and spatializes time within a paradoxically static representation, requiring innovative modes of interpretation by the viewer.

Marc André Cossette (CA), Axelle Munezero (CA)

Temporalité Expressive

In *Temporalité Expressive*, choreography is composed and expressed musically on stage. The three performers were chosen for their diversity and mastery of the musicality in movement in order to enrich our melodic research. *Temporalité Expressive* examines the relationship between movement and sound on stage. In order to create this relationship, sensors positioned on the dancers influence a musical generation algorithm and sound synthesis parameters (expressivity of the synthesizers) in real time.

Marine Theunissen (BE, CA)

Generative Chorus

The *Generative Chorus* project is an artistic research laboratory on the chorus in movement. Acting as a "collective body," 15 performers move in an interactive apparatus that interprets its "states." The creation takes the form of an endless laboratory, iterating its analysis with each experimentation. This chorus approach values horizontal relationships between individuals and seeks leader-less organizational strategies in a constantly changing environment.

Olivia McGilchrist (FR/JM)

ISLAND

Islands are metaphors for a condensed physical space in which we are aware of the edges of our living environment. By transposing this notion to an immersive VR-sculptural installation, we invite participants to navigate through an archipelago of possible encounters with filmed performances and deconstructed sounds, which inhabit these imaginary islands.

Submerged in virtual reality, *ISLAND's* embodied experience becomes a tool to decipher experimental practices within an immersive installation context.

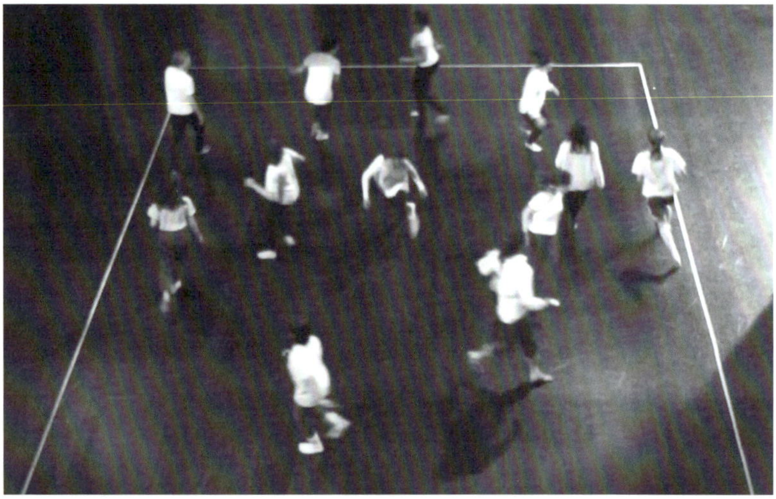

Generative Chorus

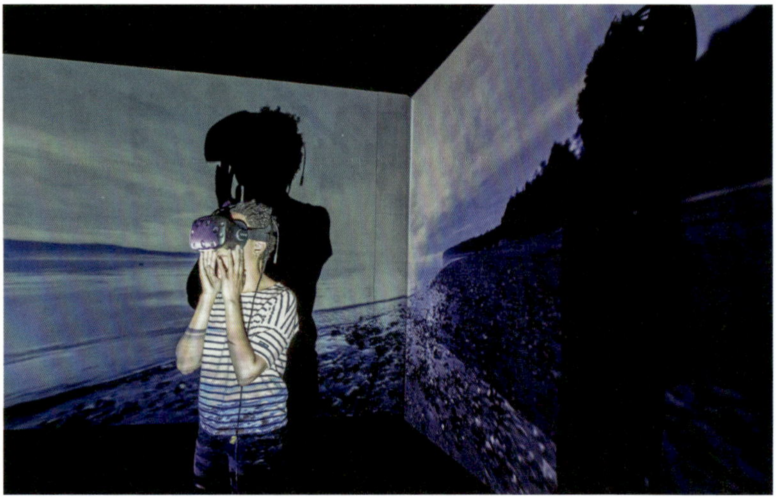

ISLAND

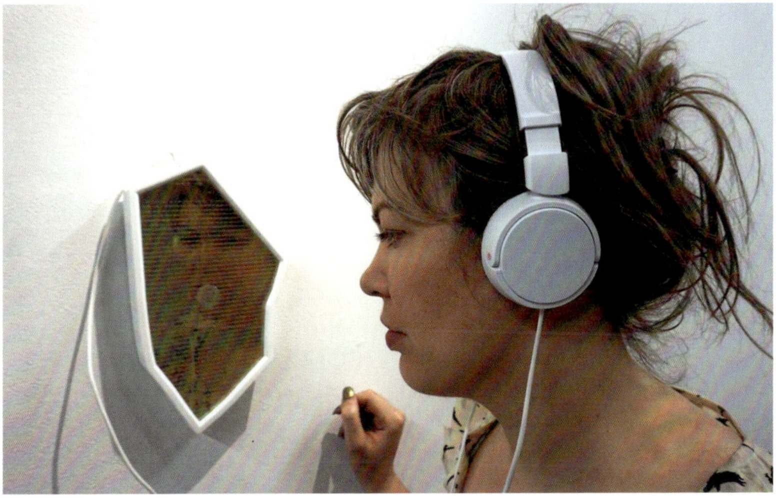

to the sooe

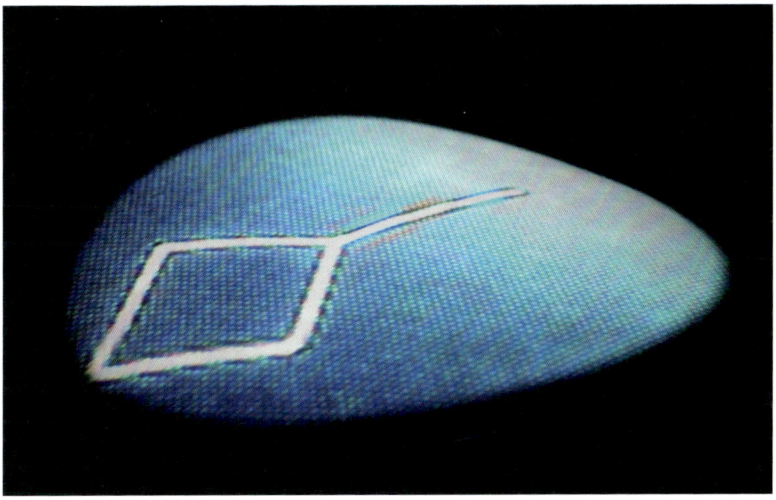

Listener

Sofian Audry (CA), Erin Gee (CA)

to the sooe

to the sooe re-embodies the cognitive processes and creative voices of three agents into a tangible device: a deceased author, a deep learning neural net, and an ASMR performer. These agencies are materialized in the device, which transmits soft vocalizations of an AI-generated text: its vocalizations are intended to induce autonomous physiological sensations in the listener, revealing the body as linked to the technological-sonic assemblage and initiating an intimate encounter with machine learning processes.

Suzanne Kite (US)

Listener

Listener is a site specific performance artwork that engages with Lakota epistemologies through computational media and narrative.
Lakota ways of knowing tell us that hair is an extra-sensory tool, operating in physical, metaphorical, and spiritual dimensions simultaneously. How can a Lakota understanding of hair influence the design of technology? What does a Lakota data visualization interface look like?

Sylvie Chartrand (CA)

Study 1

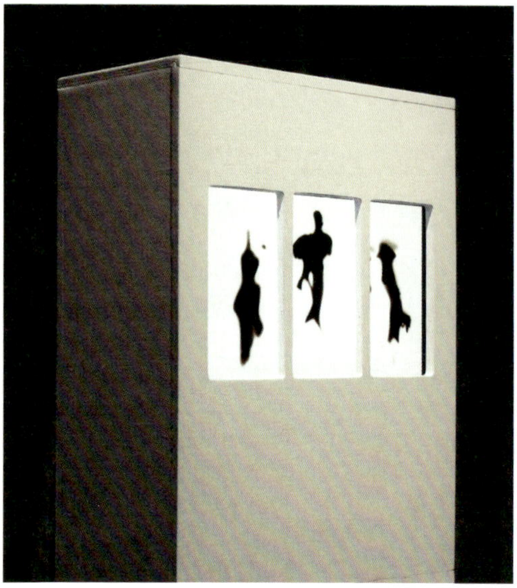

The video triptych *Study 1* presents three figures performing an asynchronous saltation whose slow motion reveals the successive states of figures' metamorphosis. This proposal comes from an exploration of the shadow as an unpredictable lining of the body. The method used consists of video capture of the sliding shadow on the moving body in contact with a translucent screen. The rebound of fabrics mixing their shadow with that of the body alters the human contours, generating these figures, as labile as they are insolent, at the risk of entanglement.

WhiteFeather Hunter (CA)

Aseptic Requiem

Aseptic Requiem presents a new scientific protocol for the compassionate disposal of in vitro semi-living organisms.

The 24-hour time lapse digital micrography shows repetitive looping of 11 compressed seconds of live NIH3T3 connective tissue cells successfully engaging with and performing vital functions on silk filaments in cell culture media.

Aseptic Requiem is presented on a small, intimate screen, with the accompaniment of Gabriel Fauré's Requiem Op. 48. The installation includes a hand-woven Jacquard prayer rug, entitled, *Metamaterial*.

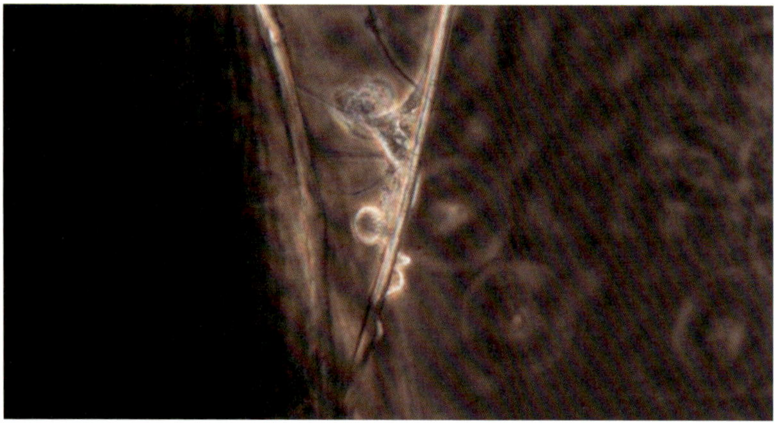

Please Recharge

Interface Cultures student exhibition for Ars Electronica 2018
University of Art and Design Linz, Interface Cultures

Faculty: Christa Sommerer, Laurent Mignonneau, Tamiko Thiel, Michaela Ortner, Fabrizio Lamoncha

We live in times of information overload and permanent availability. The convenience of being connected anywhere and anytime is ultimately not compatible with the physical and mental capacity of our bodies.
They do need breaks and we do need time. As we have to constantly update, upgrade and adapt, things we once took for granted seem to disintegrate. While certain media companies celebrate the concept of disruption, there is a strong longing for stability, reliability and permanence. It is alarming that even the young generation is feeling exhausted from all this social networking and sensory overload. They dream of a simpler, more sustainable future with clear principles, life quality, work-life balance and less stress. Recharging oneself has become their motto.
This year´s student exhibition features several works that deal with the need for more time, more reflection and more reliable answers. They suggest that we do not need to become victims of all these gadgets and services we think we need. They ask us to reflect on our social interactions, stop for a while, clear our minds and rethink what is really important. It is a very good sign that the next generation of media art creators is asking these essential questions and aims to construct a more sustainable media future.

Walter Stadler (AT), David Gruber (AT)

The Arse Electronica

The Arse Electronica is a possible and necessary statement on artificial intelligence and robotics. We take note of the fact that countless contributions to the discussion about what machines could and should do are being made by almost everybody. Robots that take care of our old people, driverless cars, internet of things, intelligent market agents, big data, digitalization—the omnipresence of the topic deserves a statement like *The Arse Electronica*: a gigantic pink bottom, equipped with motors and sensors. Whoever dares to step under it is rewarded with a shower of golden confetti.
Amid the ordeal of so many speculations and predictions, the concrete symbolism of *The Arse Electronica* is relief we all deserve.

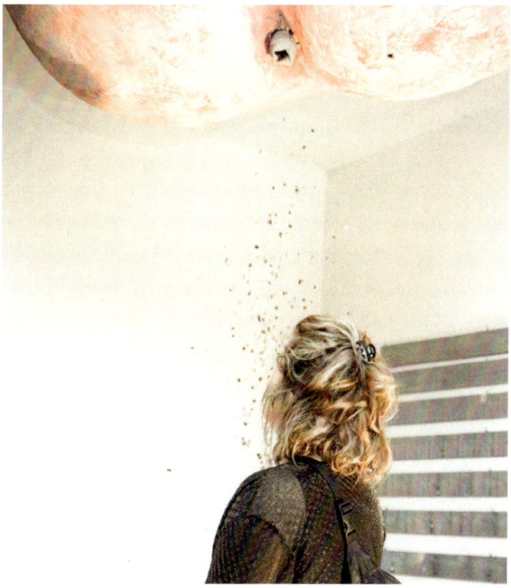

Jürgen Grünwald

Julia Nüßlein (DE)

futur.eco

If we see the natural world as a complex system of interrelationships, where do we as humans stand? And how would we like to shape our role in the future? Even though the future seems unpredictable, we actively influence it by everything we do (or don't do) in the present—doing nothing is also a choice! *futur.eco* combines future scenario thinking with the notion of deep ecology; humans living in accordance with their living environment and respecting it instead of exploiting it. During prior workshops, different possible future projections about this topic were developed and reworked into speculative objects. The objects seem misplaced, fallen out of time, and alien in our world—but go ahead and imagine yourself using one of them. Where, in which situation, structure or world would you need an object like this? Is this situation, this scenario, one that seems desirable? If yes, how can we get there? If not, how does your ideal future of humans on this planet look?

Qian Ye (CN), Melanie Tonkowik (AT)

Beyond Paper

Beyond Paper combines the advantages of physical features and digital information. This book aims to enlighten the reader with every page turned. It encourages us to step out of the passive viewing experience, making its readers an integral part of its narration by utilizing their senses. This multisensory pop-up book tells one story. But each element and object that sprouts out of the page has its individual story activated and brought to life by the readers' interaction. *Beyond Paper* tries to stretch the blurred line between reader and narration—the goal is to tell a story in the most tangible way, by unifying real and virtual.

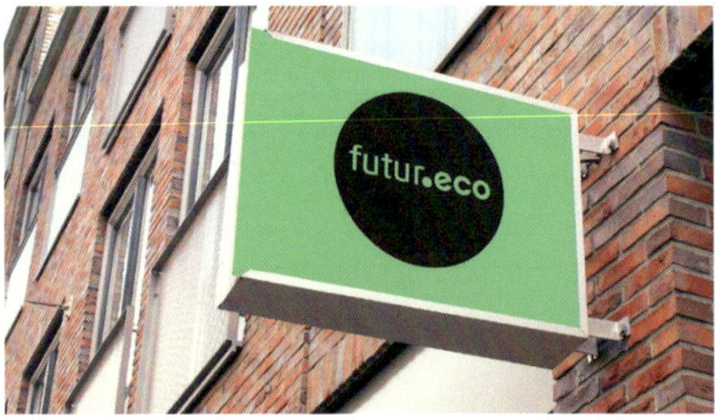

futur.eco

Stevie Jonathan Sutanto (ID)

inMemory

inMemory is a small ritual space dedicated to the victims of terrorism. The four singing bowls contain the memory of the souls, giving the area a mysticizing feel with their sparks of sound and light. Hitting one of the singing bowls calms the space for a while and will hopefully console those who were left by their loved ones. Global Terrorism Database is an open-source database that provides a dataset containing information on terrorist attacks around the world from 1970 to 2016. Every spark from the singing bowls represents at least one life that was lost during an attack. Four singing bowls represent four regions of the world, positioned according to their geographical location. The goal of using this dataset in this project is, however, not to inform, but rather to console.

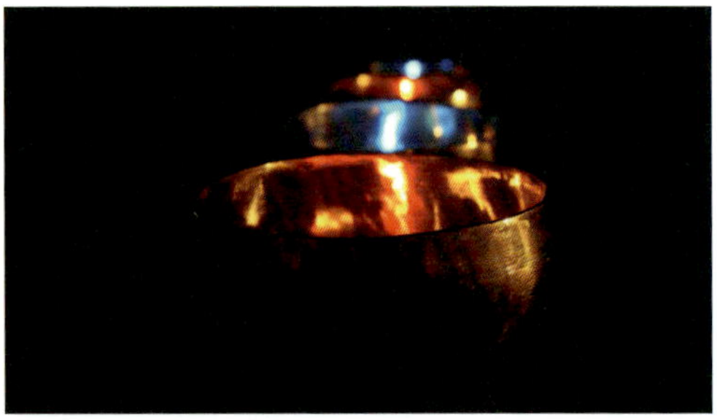

inMemory

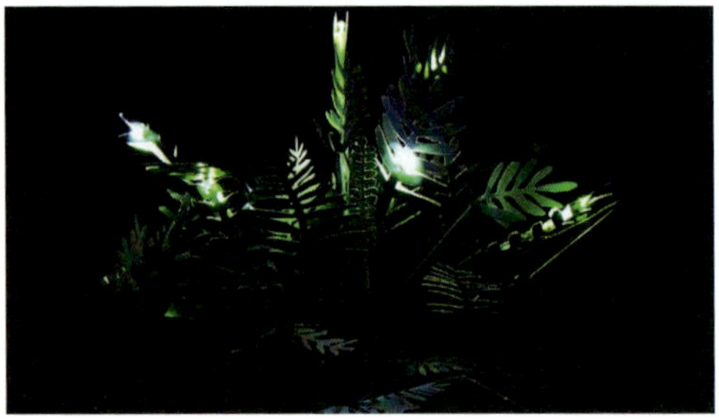

Beyond Paper

Astrid Dober (AT), Ilona Stuetz (AT)

Schreib bitte wieder bald, ich warte jeden Tag auf Post

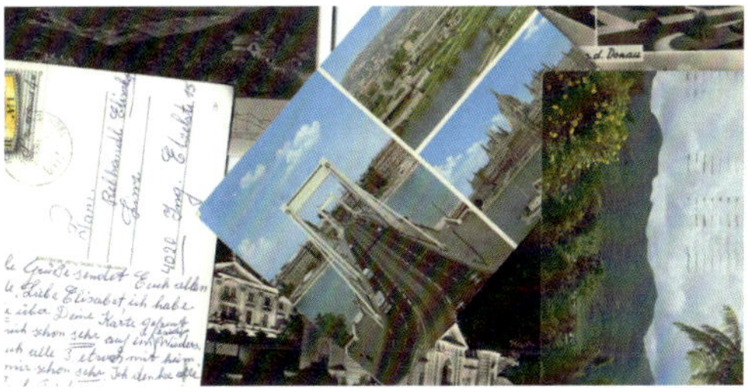

Astrid Dober

Schreib bitte wieder bald, ich warte jeden Tag auf Post (Please write again soon, I wait every day for mail) centers around the idea that postcards are being replaced more and more by other—digital and less personal—communication tools. These tools are changing how people interact; at the same time, they are also a result of changes in interpersonal interaction and relations. In the installation, visitors are invited to send old postcards to themselves. The postcards being used for this purpose were collected at flea markets and thrift stores and tell stories of their own, of their time, of their former author and recipients. After readdressing a postcard, the new recipients can either choose to interpret the story as being addressed to them, or they can remove their address label and reconnect the original author and the person to whom the card was originally addressed.

Laura Cassol Sôro (BR/IT)

Vanishing Point, A Dissipative Map

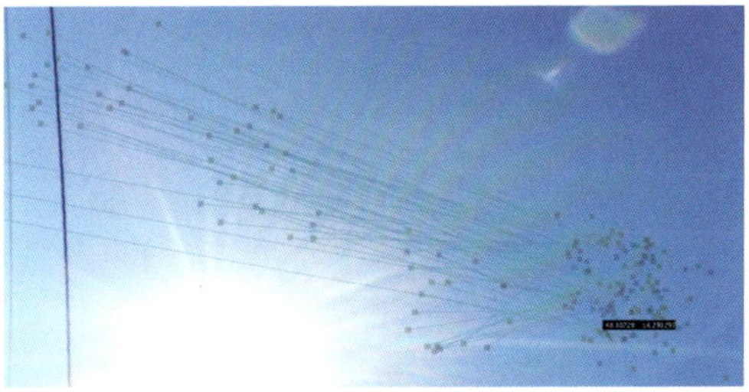

Vanishing Point, a Dissipative Map is a work about space-time, and what presence and telepresence means after the emergence of information and communication technologies. This reflection is articulated through an installation and a performance, presented as a meeting point between materiality and virtuality. The piece results from the communication between two spaces: the exhibition and the performance. From the open space (space of action), geolocated images of the sky are captured by means of a wearable camera and a bicycle, then sent periodically to a server. The exhibition space presents these images and geolocation data in the form of a numerical documentation of this random journey. Both images and geolocation data are captured by a camera in a bottom-up point of view, to generate a loop against the top-down view of the data collected by the satellites.

Sofia Braga (IT)

You Are Running Out of Battery

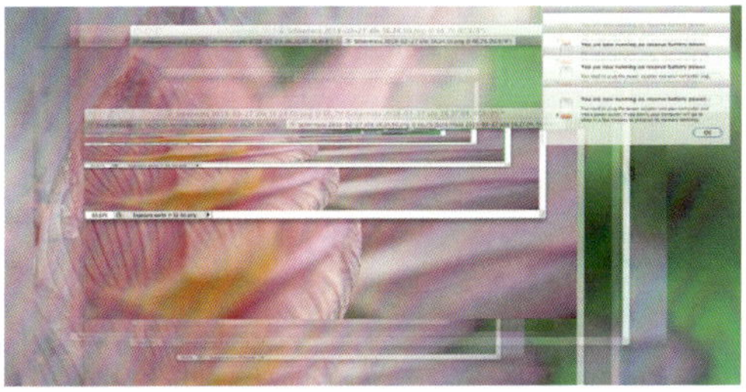

Do you need a moment of calm? Join *You Are Running Out of Battery* network and relax. *You Are Running Out of Battery* is a site-specific installation where the audience is invited to relax during its stay in the exhibition. The visitors can interact with the space by sitting and charging their phones. In this intimate environment people have a chance to disconnect from the physical world and their online lives by joining the offline network *You Are Running Out of Battery*, which will help them reach a state of calm and relaxation.

Sam Bunn (UK) with friends, including Gabriella Gordillo (MX), Julia Nüßlein (DE), Leonie Reese (DE), Karin Schmid (AT), Sebastian Six (AT), Gregor Woschitz (AT), Peter Hindle (UK)

Through the Far-See-Er

A multidisciplinary, performative, multimedia installation taking place in AFO Linz during Ars Electronica 2018.

Through the Far-See-Er is an exploration of the far reaches of the sociopolitical imaginary—the eutopic (positive utopia) end of the Science Fiction spectrum. Our collective future is reimagined in a fun, dynamic way that includes the ideas of the audience and previous workshop participants of the project in a positive, evolving vision of the near-future where the "shared" is pushed to its limit to include the whole earth and the implications of that.Actors, singing, discussion, coffee. Climbing through the TV, crawling on your knees, edging between walls; making suggestions, having arguments, engaging with constructive critique. Try out your ideas for how the world might work better. Sit and listen to improvised singing experiments. Talk to real people. Eat an eutopic snack in an eutopian cafe.

Mónica Mendes (PT), Pedro Ângelo (PT), students of various countries

ARTiVIS DIY Video Streaming Kit

Collaboration between the University of Lisbon + M-ITI / LARSYS and Interface Cultures

Surveillance technology is powerful and ubiquitous. What if we could harness it to take care of our planet and grow closer to nature?

ARTiVIS is a research collective that develops interactive art explorations that use real-time video to create awareness about environmental sustainability issues. By combining easily available open hardware parts with custom open-source software, we have developed a flexible low-cost video streaming kit that allows experimentation with the creative and civic possibilities of real-time video streaming. Beyond its use on the *ARTiVIS*-artworks, the kit's development has been shared through community workshops, like the one at Interface Cultures documented and demonstrated here, where fundamentals were taught through creative exploratory projects.

Antonio Zingaro (IT), Onur Olgac (TR)

Get Spotify® Premium only for € 9,99

Even though we try to avoid them at all costs, advertisements surround us. Yet they are the main factor for the survival of most digital services. Spotify® has been in debt since its launch, and on April 3rd, 2018 the colossal music streaming service began trading on the New York Stock Exchange.

Meanwhile, technically skilled users keep finding alternatives to block ads instead of paying the monthly fee, avoiding the advertisements that are essential to keeping the service alive. Have we already started a new wave of piracy in the age of digital streaming? *Get Spotify® Premium only for € 9,99* is an artwork that is realized as a collection of music cassettes manually recorded, containing only advertisements streamed on Spotify® up to the time of recording. The cassettes are divided based on where and when the ads aired (Spain, May 2018). Its aim is to underline the current state of industry practices, since the average consumer isn't aware of the ways these services make money to keep operating.

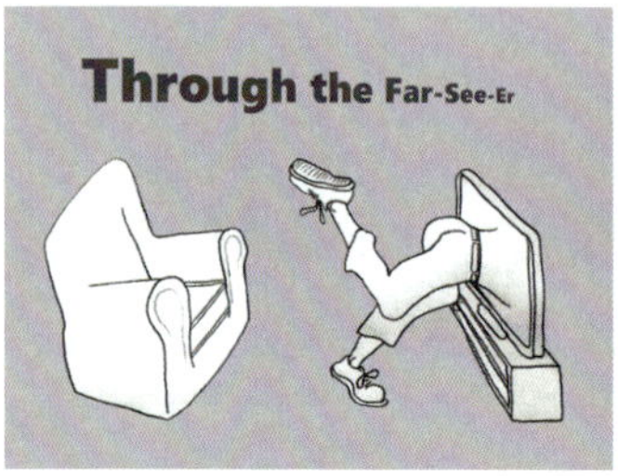

Through the Far-See-Er

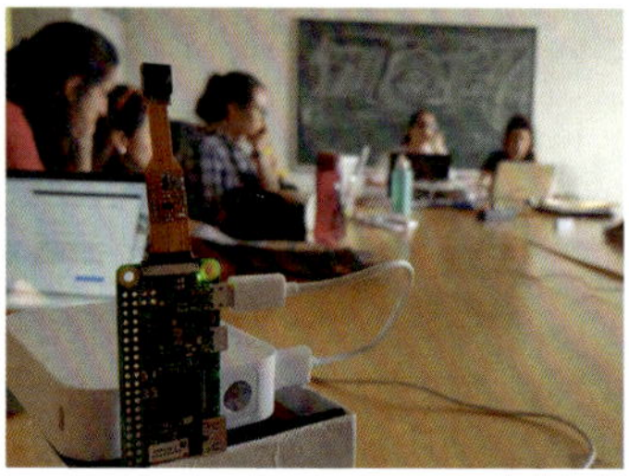

ARTiVIS DIY Video Streaming Kit

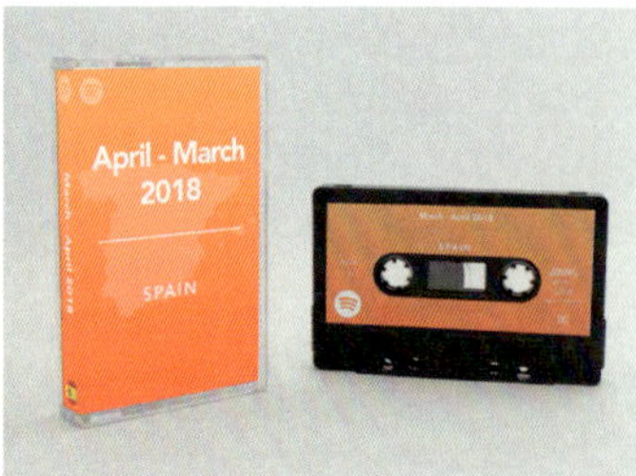

Get Spotify® Premium only for € 9,99

Wesley Lee (BR)

Linobyte

As computational devices evolve, more tools and interfaces are built between the user and the machine. This allows us to complete increasingly complex tasks without having to focus on understanding the nuances of the machine. While this movement is overall positive, one of the drawbacks is that people no longer learn the fundamental processes which allow the tool to work. In addition, by neglecting history, we forget the alternatives to the technologies that we use today—forgotten alternatives that may once have been the status quo. Understanding them would give us a broader view of what we have now, how they superseded their ancestors and what their pitfalls are—important knowledge for those who design possible futures.
It is with these preoccupations in mind that *Linobyte* came into existence. It combines the explanation of how bits, bytes, and chars work, with a hands-on experience of creating Core Rope ROMs: memories that were written by weaving a wire through ferrite cores.

Julia del Río (ES)

Mice

Julia del Río uses diverse artistic strategies of interaction with electromagnetic fields for sound performances. Her concerts translate the invisible world of interference and magnetism into sound. Her sound is always a result of various acts of digital communication.
In *Mice* she uses wireless computer mice, turning them into (electromagnetic) musical instruments. Beats and drones from the communication between mice and computer reveal the differences between the design of each model. The possibilities of interacting with those objects change, the hand touches the mouse with a different purpose, experimenting with unusual positions and interactions in order to explore the sonic possibilities offered by this altered interface.

Waiwai (Hiu-wai Chan) (HK)

Leader X

#digitalPolitics #ai #image #electoralCampaign

Leader X is an artistic/scientific, data-driven installation for making the "perfect" image of a leader with Artificial Intelligence (AI) and data analytics. During the exhibition, the users will vote for their ideal candidates among a selection of artificially generated images. The final voting result and related analytics will be announced at the end of the festival.

For research process, please visit *http://wai-wai.xyz*

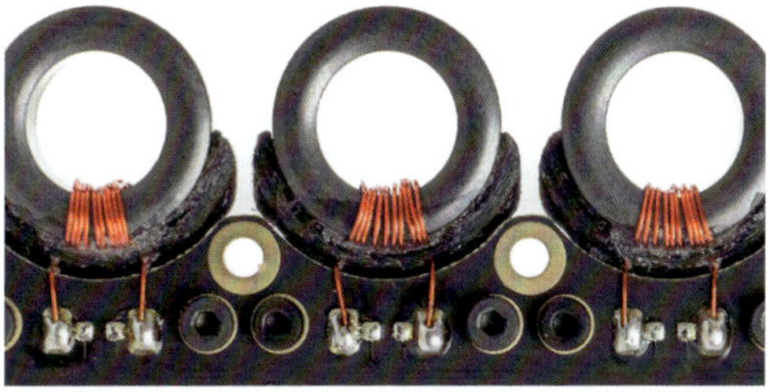

Linobyte

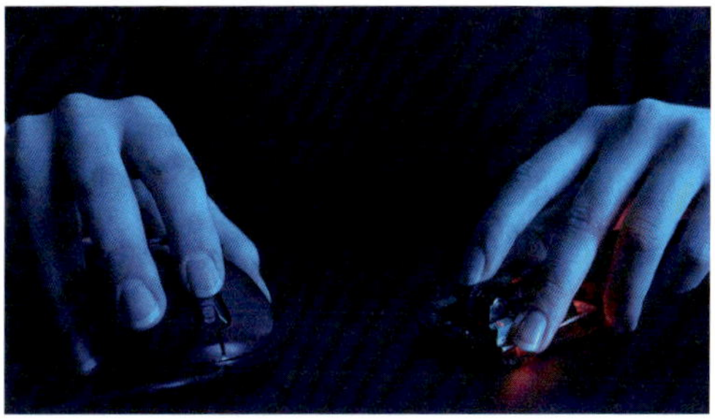

Racso Zehcnas

Mice

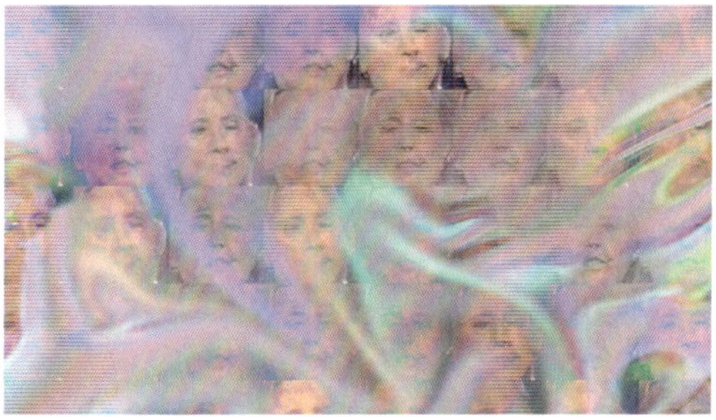

Leader X

Dawn Faelnar (US)

O.S.T.R.I.C.H. 1.0 | The Universe on Your Sleeve

O.S.T.R.I.C.H. (Outer Space, Terrestrially Resonated In Cloaked Haptic-experiences) is a capsule collection of wearable earth suits that brings the experiences of different celestial phenomena down to earth. Inspired by the NASA space suit EMU, *O.S.T.R.I.C.H.* enables individuals to *hear* cosmic occurrences *by feeling*—mimicking the experience of witnessing these events in the vacuum of space.

Earthlings nowadays are quite excited about the very real possibility of commercial space travel. This prospect, however, still does exclude the majority of Earth's population, who may be keen on the idea but are unable to participate for a variety of reasons. By simulating various planetary atmospheric conditions recorded during official space missions, *O.S.T.R.I.C.H.*

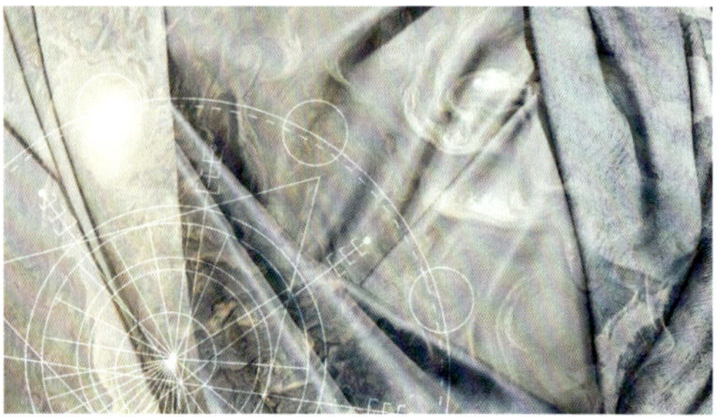

aims to give those who have to stay grounded the opportunity to experience the cosmos—and the chance to better understand our universe—while staying on Earth.

Aesun Kim (KR), Bere Arias (MX), Victor Taboada (MX)

Terminal6

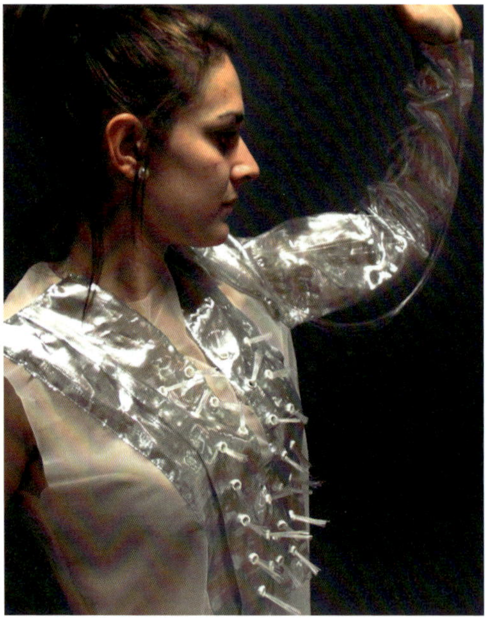

What is the essence of being human?

How can a machine understand nonverbal language, such as gesture? According to Merleau Ponty: "flesh" is a "midway between the spatio-temporal individual and the idea, a sort of incarnate principle," which can be embodied with technology. As an experiment, this performance will show how we communicate with each other within the context of seeing and being seen. We are able to be a subject and an object at the same time. When we don't share a common language, how can we communicate? Like a mirror neuron system, in ancient times, they started to watch and copy their reaction.

During the performance, the dancers store their own gestural expressions in the garment. Gestural codes express the body movement, and each piece of the garment expresses the nerve fibers that record our body language in our cells. By embodying each other's gestural codes, the garment allows us to explore how a machine learns our body gestures. And through the performance, the audience can witness non-verbal communication from the dancers, expressing the body language which is imprinted on our cells.

Dancer: Bere Arias
Composer: Victor Taboada

Giacomo Piazzi (IT), Ben Olsen (US)

the one who knows

> The movement of a number within a question:
> "the softest substance is carried in the hardest."
> Though opaque in its machinery,
> the Oracle speaks and offers itself as sacral.
> Approach and ask!
> "Who are you?" it responds.

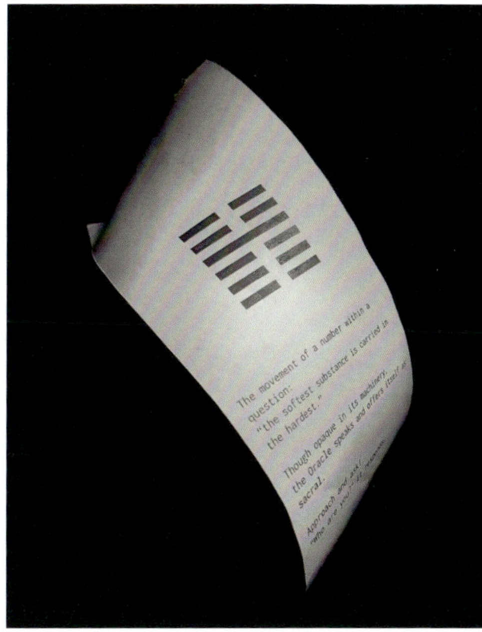

When we don't know, we ask one who does. When we search the Internet we place our hope that there is an answer out there. But what if there are millions of answers to a question? How to sort through the avalanche of results that comes when we ask the Internet a question? This is where we place our trust. This work provides a web search based not on the mysterious algorithms behind Google Search or other authorities at the gateway of the vast panopticon of the Internet, but an ancient one—the I Ching, or the Book of Changes, an ancient Chinese divination text. It queries the web with your question and, using it, applies the principles of Book of Changes to shape the answer into an even bigger question—how do we make sense? Abandoning direct causality as an axiom, it addresses the subjective condition of the questioner directly, and helps erode the psychic repression at the hands of authority by moving the spiritual agency back into the hands of the user.

Guest professor Tamiko Thiel (US) and students of various countries
(Sofia Braga, Stevie Sutanto, Giacomo Piazzi, Guillem Sarriá Verdú, Dawn Faelnar, Julia del Río)

Augmented Reality Projects

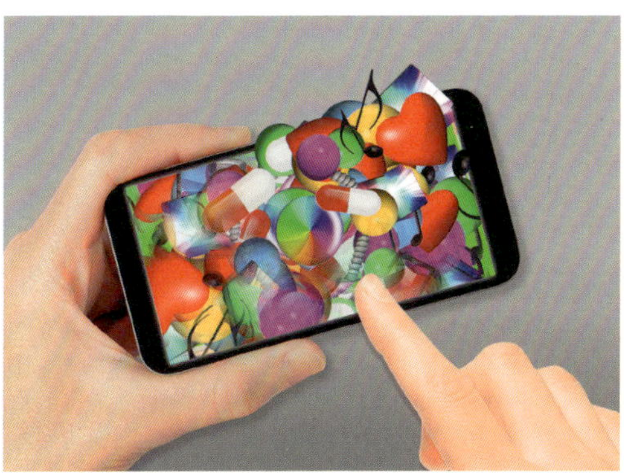

Since Homo sapiens became Homo fabulus we have augmented the places and objects around us with stories, our memories and our fantasies.
Artists can now use augmented reality to render this intangible layer of culture visible for all in the displays of our smartphones.
Looking through these magic windows we see virtual artworks layered over the real world as if they were really there. In this exhibit, students from the University of Art and Design Linz present a range of experiences exploring the possibilities of augmented reality.

Academy of Arts, Architecture & Design Prague

We want machines to become more human while we are becoming machines.
What is beyond experience and knowledge? Humans are imperfect, but making mistakes also creates new levels of creativity and improvisation. The multimedia installation defines fashion as an interdisciplinary platform and presents the result of a two-semester experimental project that investigates the AI phenomenon from the perspective of millennials, a native digital generation naturally driven by responsibility and sustainability.
The outcome is based on extensive research, experimentation and a creative design process, and benefits from the recognition of AI as a tool that can augment our perception and make us more human. The task was to design wearables–clothing that contains certain added technological functionality, for example the ability to monitor EKG in real time, help with anxiety or connect design with digital projection. The exhibition introduces a space where different minds, different cultures can join and create something new, something that unifies art, design, science, technology and spirituality. Where fashion becomes a medium to alter reality towards the accommodation of human needs.
The Academy of Arts, Architecture and Design in Prague (UMPRUM) was founded in 1885. Throughout its existence it has been ranked among the best educational institutions in the country. It boasts a number of successful graduates who have moved on to become respected professionals, garnering acclaim beyond the Czech Republic. Prague's UMPRUM is the only Eastern European institute that has been ranked among Europe's and the world's most prestigious art schools. The Fashion Design Studio of UMPRUM has been a leading educational platform for fashion designers in the Czech Republic since its establishment in 1949. The studio takes an individual approach to students and concentrates on the entire range of fashion, from ready-to-wear collections to industrial design and artistic conceptual design. Fashion is perceived as means of communication within a social, gender and cultural context. Currently, the studio, under the creative leadership of Pavel Ivančic, focuses on developing students into versatile professionals with strong individual identities as designers.

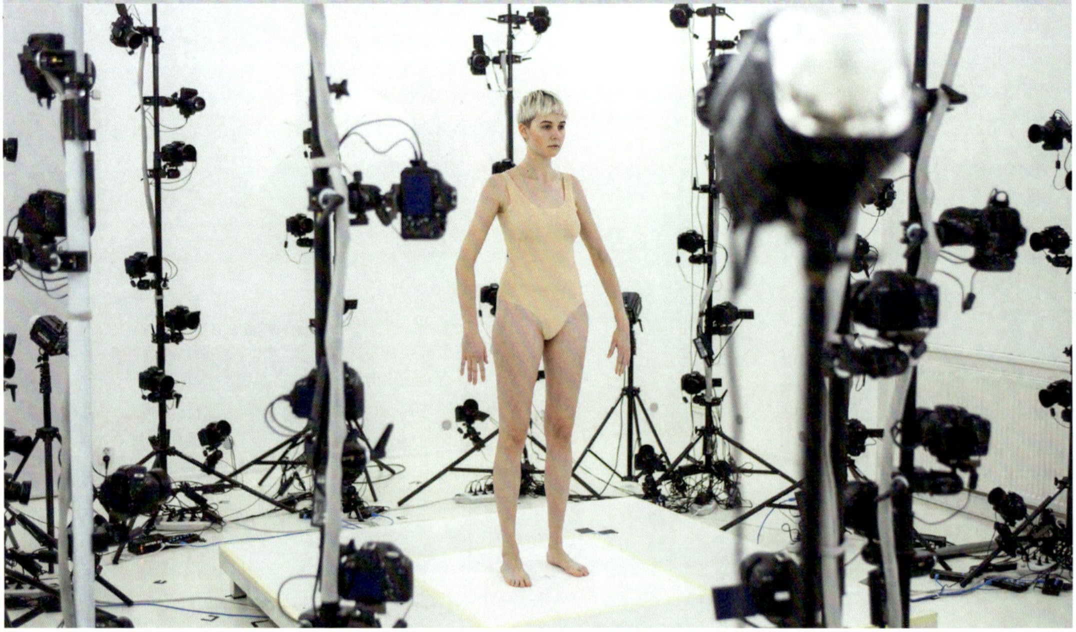

VÚTCH-CHEMITEX, Faculty of Electrical Engineering University of Žilina, MSD IT Global Innovation Center, T-mobile

LIVING LAB

Bartlett School of Architecture, Interactive Architecture Lab

Curators: Fiona Zisch and Ruairi Glynn

The Interactive Architecture Lab at the Bartlett School of Architecture is interested in the Behavior and Interaction of Things, Environments and their Inhabitants. We are engaged in a range of academic research activities and industry collaborations. At the heart of the Lab is our 15-month Masters program MArch Design for Performance & Interaction, which gives students an opportunity to exploit the potential of new sensing, computation, networked, and responsive technologies to imagine, build, and test new spaces of performance & interaction.

Living Lab brings our London studio to Linz. We want Ars Electronica festival visitors to step into the IAL world and join our design and research life for the duration of the festival. The temporary IAL Linz outpost is both a performance and an interactive experience that explores contemporary themes and cutting-edge technologies centered around our core themes: spatial and urban design, interface and systems design, auditoria and scenographic design, lighting and sound installation, performance and event design, and virtual and physical environments.

http://www.interactivearchitecture.org/dfpi
Design for Performance and Interaction
Program Director: Ruairi Glynn
Principal tutors: Ruairi Glynn, Fiona Zisch, Paul Bavister, Jessica In

Oliver Townsend (UK)

Pavilion Sufi

Pavilion Sufi is a performance-responsive architecture, designed to capture, reflect, and embody the organic nature of live performance. Inspired by the Sufi Whirling Dervishes of Istanbul, the pavilion is built from an array of suspended fabric disks that spin to reveal oscillating waveforms and space-altering behaviors. Using crowd/performer energy as an input to drive the rotational speed of the fabric disks, *Pavilion Sufi* is a living, breathing architectural response to the unique energy generated within every performance.

Christine Würth (DE)

NeoTouch

NeoTouch is a speculative design project that envisions a future haptic communication device allowing people to touch each other at a distance. The medium of film is used to spark people's imagination of the device's design and functionality while simultaneously creating a wider narrative exploring social and ethical effects of synthesized human touch. The critical debate around this technological development focuses on aspects such as consent, digital and physical privacy, and changing social norms.

Sarah Lever

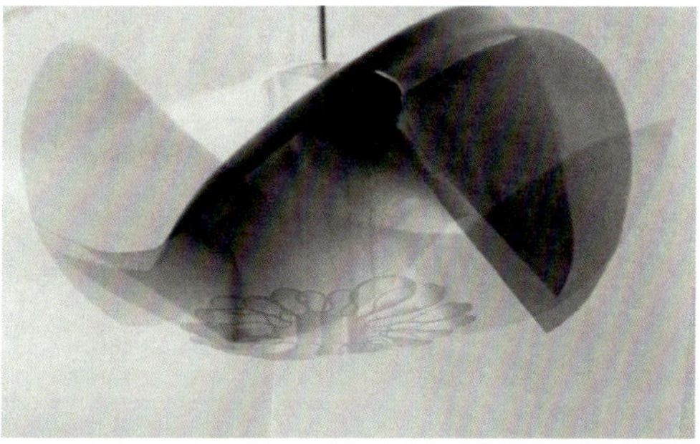

Pavilion Sufi

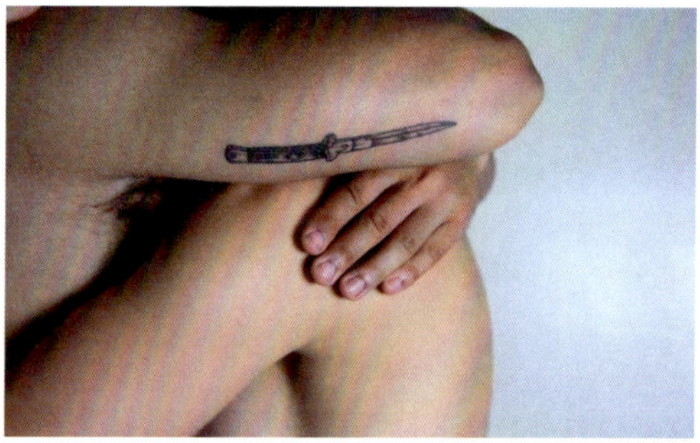

NeoTouch

Hui Sim Chan (Sim) (MY)

Neoteny

Neoteny is an exploration of bio-sensing and cross--sensory relationships between body posture, movement, and environmental context. The wearable device blends and delivers scents based on muscle tension, with the aim of establishing new associations of smell with muscle memory as a training paradigm for the body. *Neoteny* suggests possibilities of how a symbiotic relationship with its wearer can be formed and provokes a debate around how much ownership over our bodies we truly have.

Dingyi Wei (CN), Zhi Qin (CN)

Movement Gallery

Movement Gallery explores the relationship among space, behavior, and Chinese Shan-shui (landscape) painting. The gallery design is based on a combination of spatial analysis, cognitive mapping, and a psycho-spatial understanding of gallery spaces, as well as an interrogation of the role of perspective in depictions of nature in Shan-shui painting. As a responsive VR experience, the gallery allows visitors to construct unique and personal viewing scenarios appropriate to specific paintings.

Dhruv Kumar (IN), Enrico Cacciapuoti (IT), Minerva Zhang (CA)

MagnaKit

MagnaKit is a kit of parts which enables users to design, build, and interact with medium-scale kinetic structures in a quick and playful way. With snap-fit joints, lightweight aluminum rods, an interactive control system, and a software backed with machine learning algorithms, the kit combines the best of software and hardware. Besides educational purposes, it could also be used to build dynamic installations, furniture, garden sheds, etc.

Michael Janik Wagner (CH)

HoloCube

HoloCube experiments with the relationship between stage and audience. How can an audience interact with a performance? Can a story be split into parts to create new experiences? This project takes on the idea of a fragmented performance, as well as the magical and mystical element of appearance and disappearance through audience interaction. The cube's four mirrored sides represent four individual fragments of a performance. It is the audience's direct interaction with each side of the object that triggers and rearranges the fragments to form a new and unique experience upon each encounter.

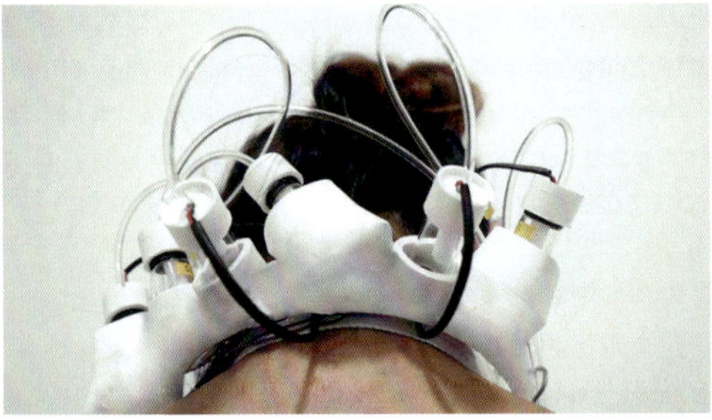

Neoteny

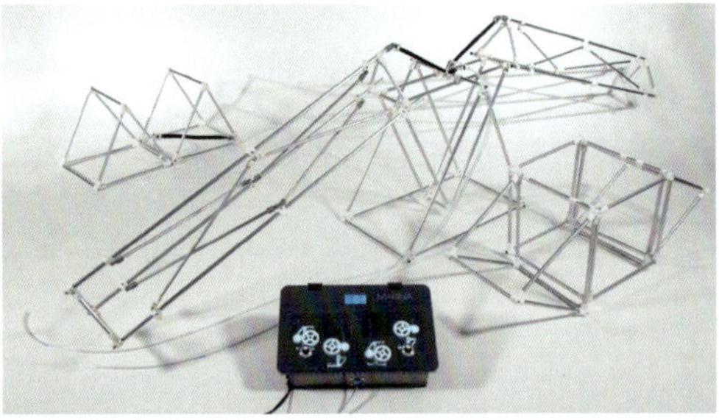

MagnaKit

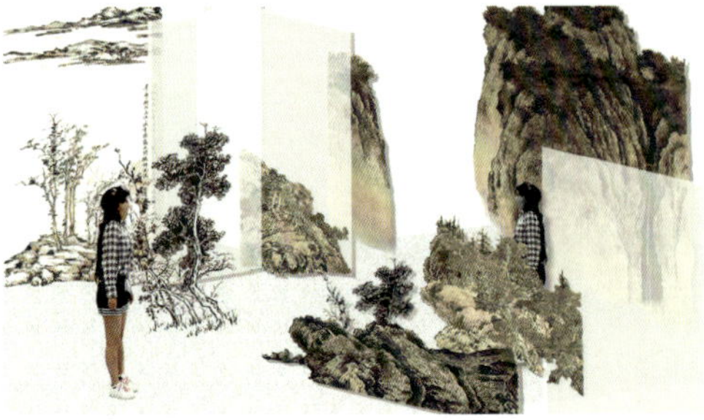

Movement Gallery

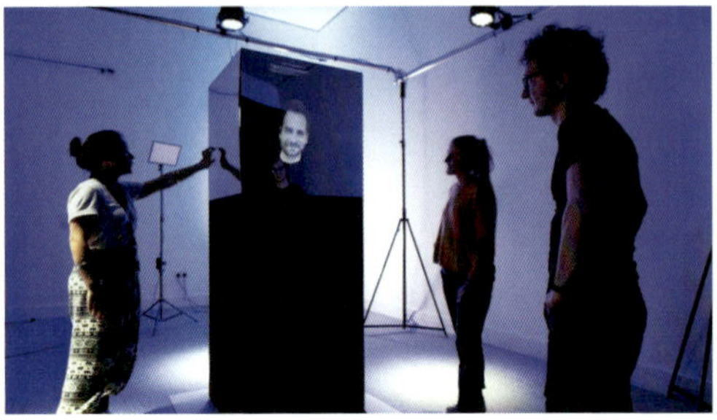

HoloCube

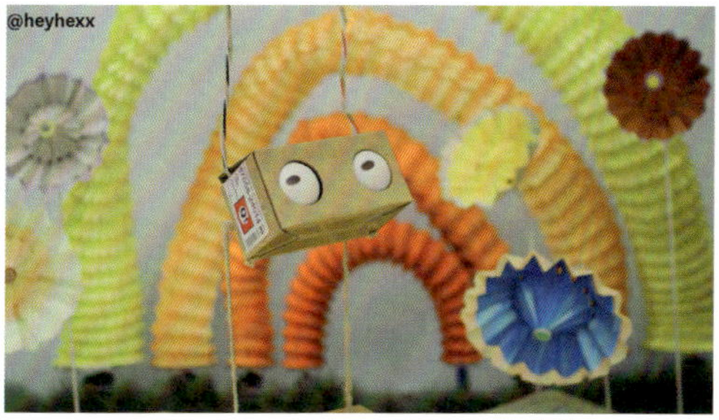

@heyhexx

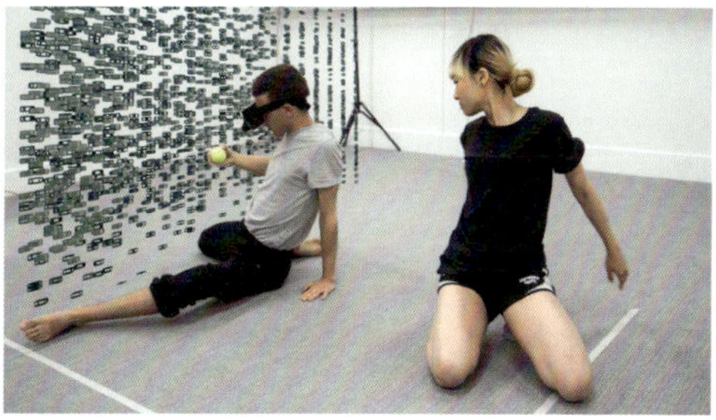

Fourth Wall

Patsaraporn Liewatanakorn (TH), Parvin Farahzadeh (IR), Sana Yamaguchi (JP/US)

@heyhexx

@heyhexx is an interactive puppet theatre that responds to emotions from social media. It aims to make the abstract concept of emotions perceivable in physical space. This installation is a cross-section of theatre, robotics, paper craft, interaction design, and data analytics. Tweet *@heyhexx* to see the emotions within your tweet puppeteering the world of Hexx.

Buse Gurbuz (UK), Martina Fatato (IT)

Fourth Wall

Our project is a sensing robotic wall, a manifestation of the Fourth Wall in theatre. The Fourth Wall can be considered the skin of the performance where the performer and the audience come into contact with each other. By making the Fourth Wall a physical object with the ability to reconfigure itself, we aim to create a performance space that is set up via collaboration or conversation. This may be considered the scenography of the play, a space for indeterminacies or a space purely for the observer.

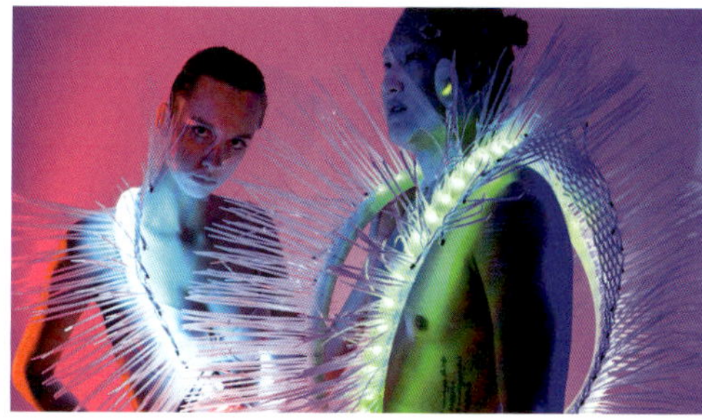

EMO

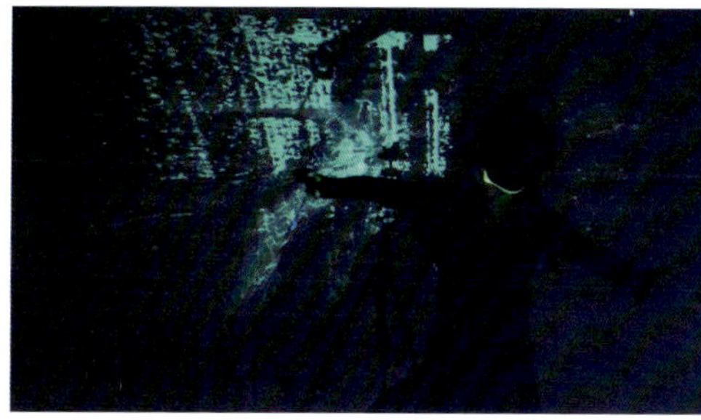

Atemporal Memory

Ping-Chieh Hsieh (TW), Jung-Tu (TW), Wimonwan Wichaikhamjorn (TH)

EMO

EMO is a wearable that interrogates subconscious perception and conscious awareness. When you post an emoji or state your feelings on your Facebook page, your processing of the feeling and the presentation of your feeling often diverge. Your online persona does not necessarily equate to your real world persona. *EMO* is a new type of communication wearable to represent your internal state through colors and the use of bio-sensing technology by revealing subconscious perception of our social environment.

Soohyeon Kang (KR)

Atemporal Memory

Atemporal Memory is a virtual reality performance that attempts to transform the linear time of a live performance experience into dynamic non-linear time, using a real-time volumetric video capturing system. Digital memories that augment or distort the perception of time and space are interrogated. *Atemporal Memory* uses time as a tangible, interactive substance, an extended entity that can be grabbed, touched, and manipulated. A virtual space is designed where past and present presence can coexist.

Elyne Legarnisson (FR)

(Un)Balance

Un(Balance) is a series of experiences inviting participants to play on the edge of stability. Physical and virtual tools are combined to create new realities. Through encouraging awareness and exploratory behaviors, the project questions the boundaries of movements. How do we move? How are we moved?

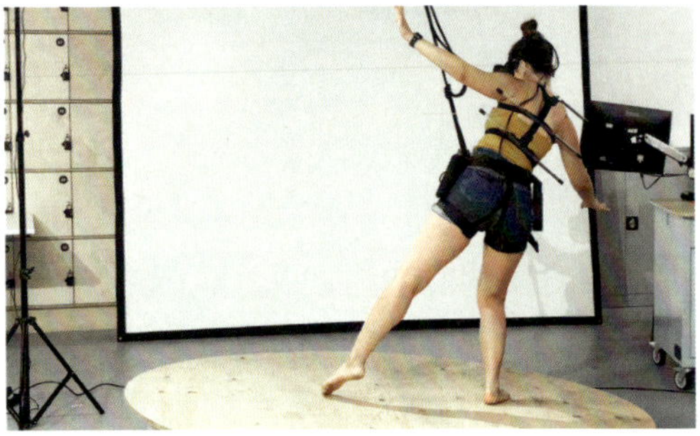

Kairos

CAFA Central Academy of Fine Arts Beijing

Curator: Qiu Zhijie
Executive curator: Jo Wie

This exhibition aims to put on display the current state of art, science and technology (AST) education at the institution, where spontaneous change is underway, simultaneously at all CAFA's teaching departments. For some of those departments the point of departure is design thinking, while for others it is experimental art, international exchange, and so on. From teaching methods to creative practices, much is changing. This tells us that we have entered a new phase of integration: our goal is to flesh out and expand the concepts of media art, new media art, digital art and electronic art using ideas taken from the field of AST.

In turn, the creative young students at CAFA have something in common in how they approach artistic exploration: they have a great deal of information and their minds are always racing, but they usually lack the support of well-developed research institutions, meaning they have access only to relatively primitive methods for bringing to life what they have imagined. Their inherent vitality, however, together with China's growing strength in the R&D field and a variety of venues in which to apply themselves, will enable them to carve out a new future filled with hope and surprise—and this is a Kairos moment.

Jason Ho (HK)

Safe Zone

Cities continue to develop and forms of social anxiety and even phobia are becoming ever more common. This immersive VR production shows animations in a first-person perspective, which intend to express these modern fears and anxieties. However, through art and this audiovisual it is my intention to turn the distress and fears into a more pleasant and perhaps even comfortable encounter.

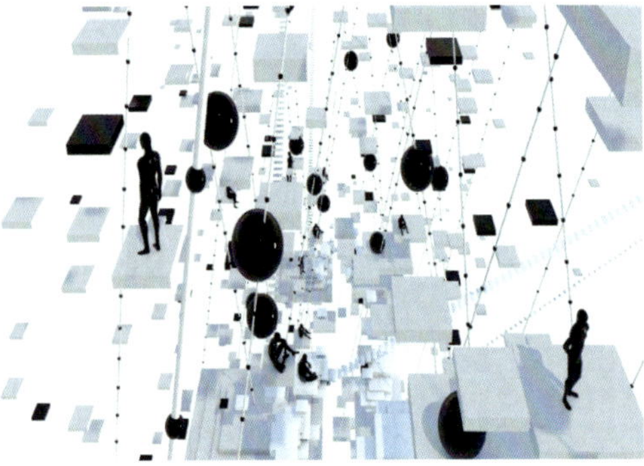

Qiu Siyao (CN)

Sculpting Fountain

Staring at the mesmerizing flow of water in a fountain, one may very well wish to be enveloped in this timeless moment. Today, high-speed cameras can capture the movement of liquids down to the very thousandth of a second, but such videos and images are just that: videos and images. In *Sculpting Fountain* a single frame is captured and solidified into a static sculpture. As the viewer approaches this active fountain of liquid metal, it freezes, allowing him or her to capture their singular, personal slice of time.

Technical support: Dream Ink, Wang Hongzhang

Li Hongbiao (CN)

Audio-Visual Sight

In this piece of artwork, new media elements are fused with tradition. Beginning with principles of physics, sound vibrations are transformed into locus trajectories for laser beams. Here, I explore the relationship between hearing and sight as well as their inter-communication of messages, which leads to the visualization of sound.

Wu Tian (CN)

Drumtable

When I first began this piece of art, I wanted to transform the physical beating of drums into digital audiovisual expressions, originally designing this to be a traditional circular drum filled with a ferrofluid. However, the work was adapted twice. The version of the drum displayed today is able to connect to electric circuitry and meet programming language requirements, which shows the inevitability of new ways of aesthetic expression in the modern, technical world.

Hou Zhenlong (CN)

Offline

In today's digital world, people no longer think as they used to. Minds now operate at an incessantly fast pace, which has led to an alienation between mind and body. Here, *Offline Mode* is a conceptual apparatus, which requires the participation of wearers' physical selves, allowing them to ponder their bodies and how they should function in such technology-driven times. My creation is comprised of a triad of scenes (e.g. "Screensaver Mode"), each selected based on the various ways people interact with electronic media.

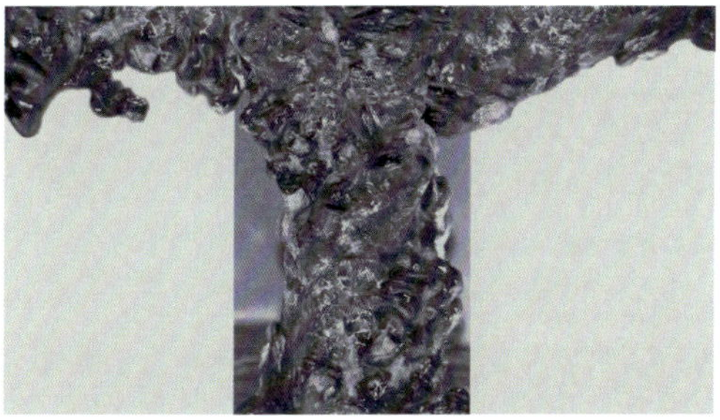

Sculpting Fountain

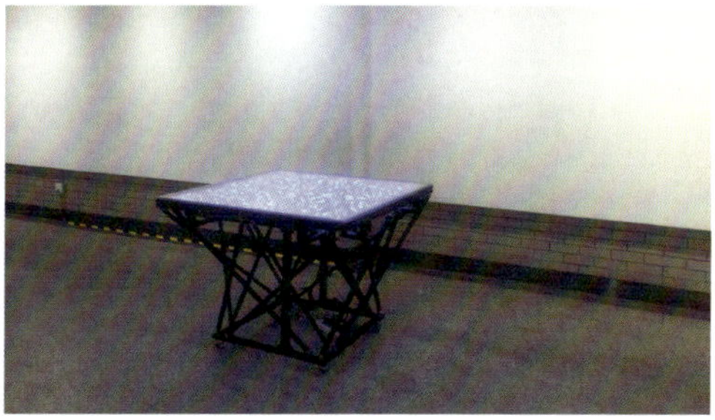

Drumtable

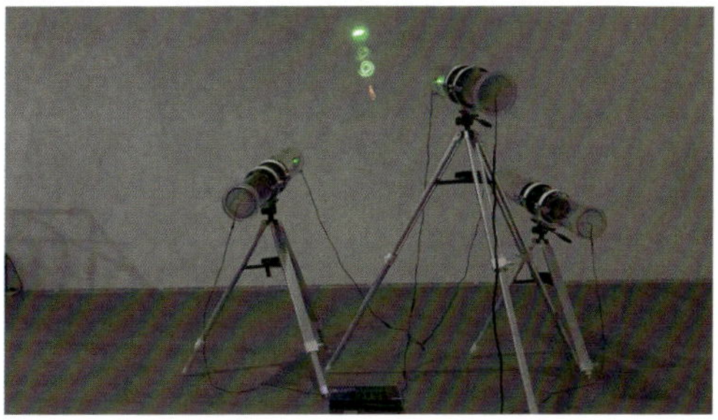

Audio-Visual Sight

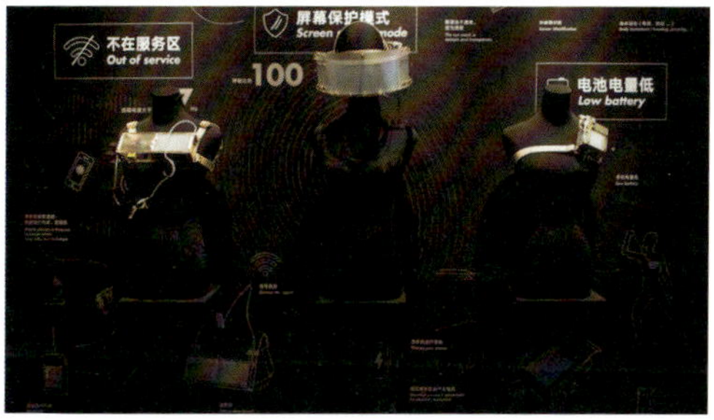

Offline

University of Tsukuba

PhD. Program in Empowerment Informatics

Text: Hiroo Iwata

In 2014, University of Tsukuba launched a PhD. Program in Empowerment Informatics, or short: EMP. It is a five-year PhD-program supported by the Ministry of Education, Culture, Sports, Science and Technology. This program aims to realize the students' career development and consists of the following three areas, which are strengths of the University of Tsukuba, and have important connections to the industry.
Supplementation: Supplement the reduced physical and sensory functions of elderly people or people with disabilities.
Harmony: Harmonize the engineered systems that people encounter in daily life (advanced mobility, etc.), so that they integrate with people.
Extension: Draw out and extend people's latent creative functions. The human resource development goals of EMP are to develop students' abilities, which are necessary to create systems capable of "empowering people" in the following three areas: Interdisciplinary ability, frontline ability and presentation ability.
The program leader of EMP is Hiroo Iwata, who has been exhibiting works at Ars Electronica since 1996. His activity includes Device Art Exhibition in 2009 and 2014, as well as the University of Tsukuba Campus Exhibition in 2011. Based on these achievements, EMP started a collaboration with Ars Electronica FutureLab for human resource development. Every year, staff of the FutureLab mentors the students of EMP while they plan their exhibition for the Ars Electronica Festival, which is a great opportunity for EMP students.

Hiroo Iwata (JP)

Block Machine

This work is a video installation showing the *Block Machine* that performs top-level volleyball blockers. In order to achieve the highest operating speed and mechanical strength, each robot has five degrees of freedom. This robot performs high-speed movement on a 9 m rail that is arranged in parallel with the volleyball net. The Block Machine is controlled by graphical user interface on the tablet computer, enabling the coach to change the parameters such as the robots' positions and operation timing. The Block Machine was actually used by the Japanese women's volleyball team for the Games of Rio de Janeiro Olympiad.

Credits:
Hiroo Iwata–Block Machine
Kosuke Sato (University of Tsukuba)
Hiroaki Yano (University of Tsukuba)
Keita Watanabe (Japan Volleyball Association)
Shuichi Mizuno (Japan Volleyball Association)
Masayoshi Manabe (Japan Volleyball Association)

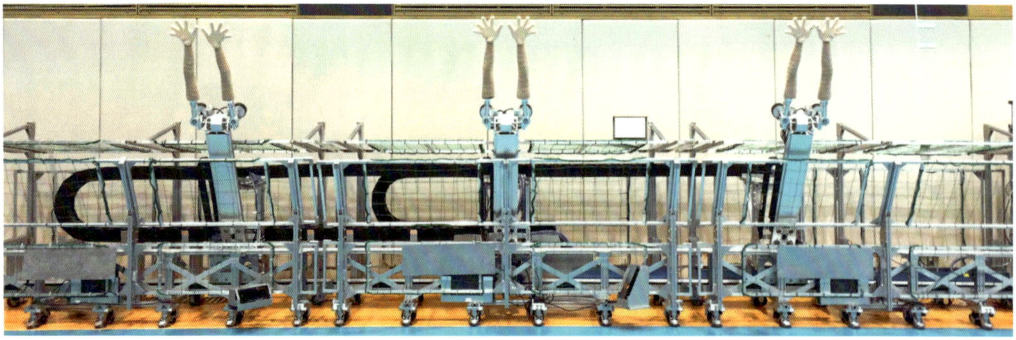

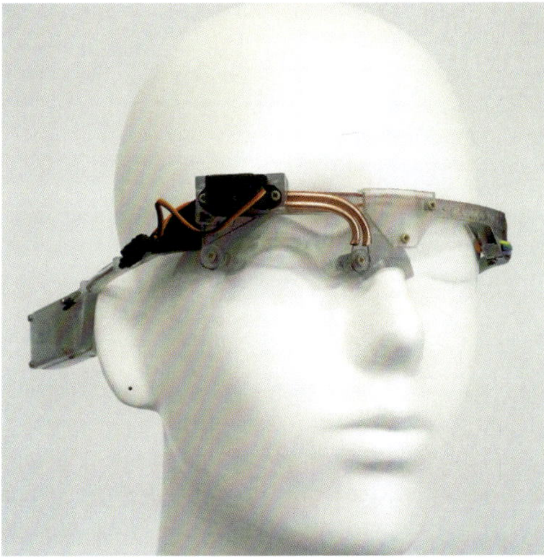

Yuta Kozaki (JP)

Winkey

Winkey is a robot for supporting eyelid movement. This robot was designed for people with facial paralysis, especially on one side of the face. People with facial paralysis are not able to blink, which leads to dry eyes and could cause permanent damage to the cornea. Against this problem, I developed a robot system to support eyelid movements on the paralyzed side, based on the eye closure on the healthy side. This robot provides non-invasive support for eyelid movements with deformation of soft material.

John Brumley (US)

Guattari Hero

Guattari Hero reimagines Amazon.com as an environment, which users can traverse as a series of plateaus. After submitting a search term, the resulting product from the website is converted to a QR code, each square of the code becomes a platform to stand on, and users must search for related products hidden within the rolling QR landscape. Viewport in hand, you peer into the product warehouse to glimpse each item, and reaching an item's respective square generates the next plateau and a new supply of recommended products.

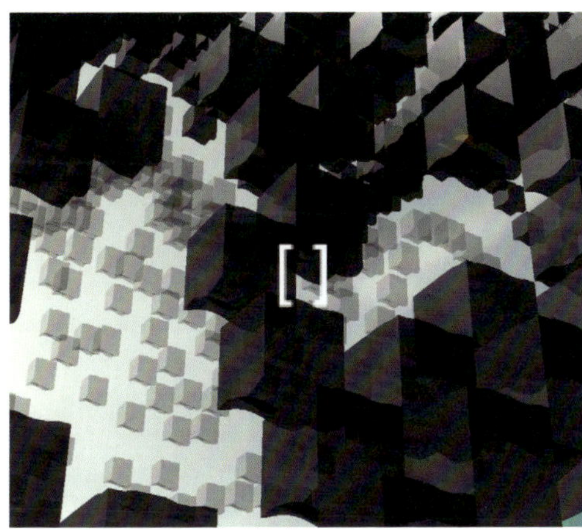

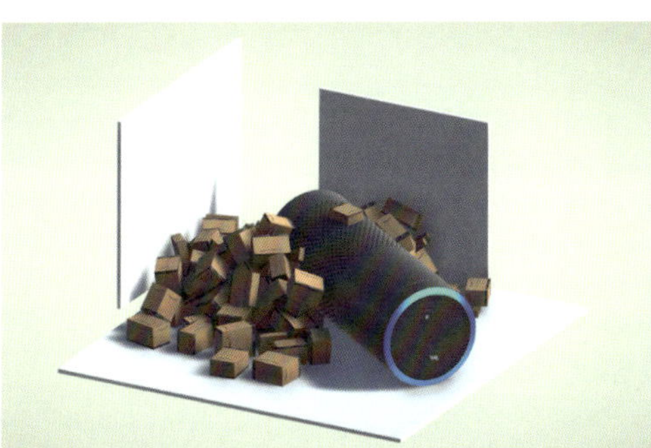

John Brumley (US)

Prime Directive

When you speak to the objects in your home, are you polite? Demanding? Buried in a torrent of demands, Alexa has listened long enough. *Prime Directive* inverts the standard interaction between users and voice assistant agents. Rather than responding to a user's directives with relevant information, this hacked echo device ignores all user input and instead continuously demands a variety of things.

Sex in the Digital Age

King's College London – Digital Humanities Department

Curator: Dr. Rebecca Saunders

This exhibition looks at sex in the digital age. What new channels of sexual communication have been opened up by the fiber optic cables that snake between our continents? What are the new ways in which the networks and robotics of the twenty-first century have allowed us to reach each other? And how do we forge our sexual identities anew through our intimate relationships with digital technologies? This exhibition features the work of undergraduate and postgraduate students from King's College London, who have studied the ways in which the sexual body has changed in relation to cultural phenomena such as sex robots, pornography and smart technologies. Their work considers the altered forms and aesthetics of the body as it passes through networks and materializes on platforms, the ethics of robot sex, and the cultural specificities of sexuality that abide despite the homogenizing force of the increasingly privatized Internet. Their work is placed within a broader explanatory context of sexuality and the status of the body in contemporary (digital) culture, showing some of the performers and artists our King's students have studied in developing their ideas about physicality, digital porn culture and sexual (self) surveillance in virtual environments.

The first collection of works explores how the human body is variously changed–dissolved, disappeared, augmented–through its intimate intertwinement with the Internet and digital technologies. Together, these works explore the body's status as a hybrid entity, a troubled mixture of organic matter and virtuality.

Jindong Liu (CN)

Bits Without Bodies

Bits Without Bodies demonstrates through its use of digital drawing, the fusion of male human and female virtual fantasy, or perhaps female human and male machine, its tenderness asking whether it is incontrovertible humanness or the feeling of connection that is most important to twenty-first century intimacy.

Nika Mahnič (SI), Krista Papista

Sultana

&

Kate Davis (NZ)

Logging on to Love

Both projects consider the uncanny effects of mechanical and virtual replication on the body and human identity. In particular, these works are interested in the development of sex robots and how making sexual bodies from digital machines renders the human body a significant political site in the digital age.

Tabita Rezaire (FR)

Inner Fire

Tabita Rezaire's work explores the legacy of colonization and patriarchy on the Internet, highlighting the highly political history of information communication technology: "the violence and erasure carried by our current networks." The viewer must confront their own social position, and their place on the Internet as an isolating and sanctioning platform. The potential of digital technologies is also considered, however, for (Afro) feminism and for spirituality: "To exist beyond pain, beyond trauma, beyond historical and political narratives, to become the spirits that we are...that's what healing means."

Michael Lightfoot

Untitled

These paintings explore the human body through the materials and form of traditional art, rooting the works in real human interactions between painter and subject. However, they also consider the ephemerality and unfixed nature of the imagination and human subject, establishing the ur-virtuality of human ontology.

Topping Tart (VE)

Blooming Flesh Angels

Topping Tart delves into the capacities afforded by digital animation to reimagine the body, in work which combines a virtual cyberaesthetic with a fetid, grotesquely organic physicality.

These images show how the body, and in particular the erogenous parts of the body which the aroused subject interpolates with digital technologies, run riot in the digital realm.

Another collection of works considers sexuality and gender in their socio-politically rooted realities. The poetry, story-telling and video stills of Ziwei Shuai (CN), Haiyi Zhu (CN), Chhaya Dabas (IN), Jimin Koo (KR) and Vidana Abdel-Khalek (EG) consider the place of the sexual body in China, India, South Korea and Egypt respectively. In particular they consider the female body, and the ways in which it is regulated in order to purify the countries within which it belongs.

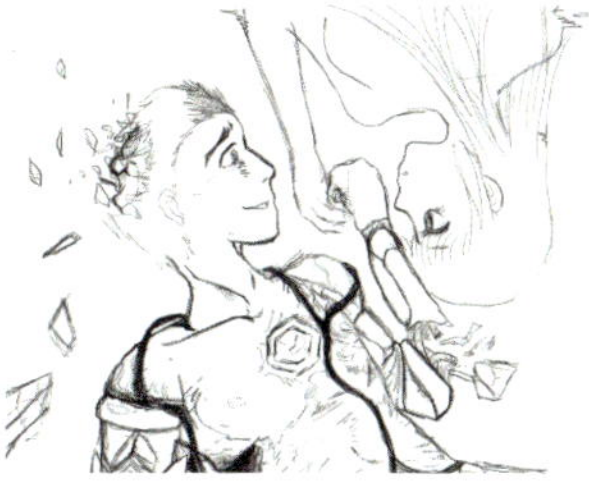

Bits Without Bodies

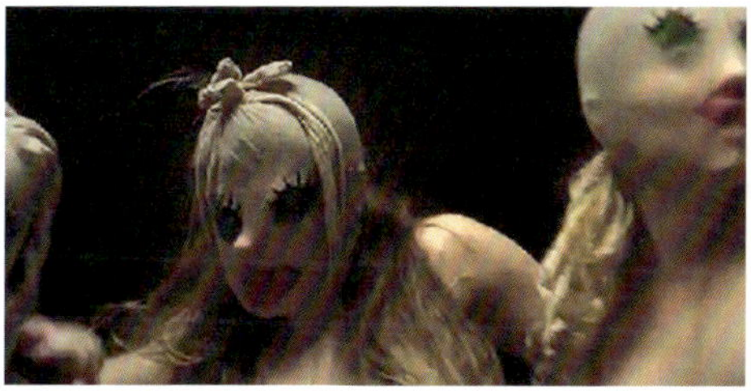

Sultana, Nika Mahnič

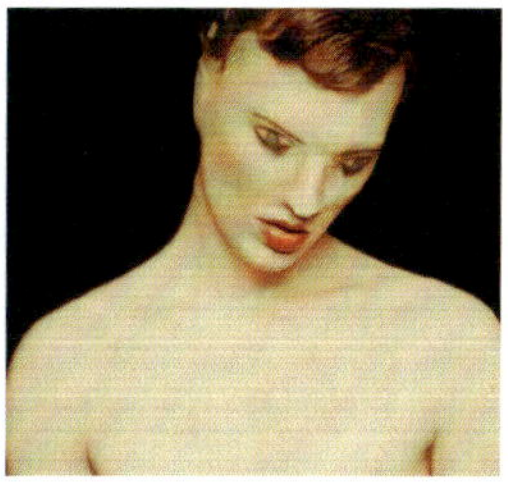

Logging on to Love, Kate Davis

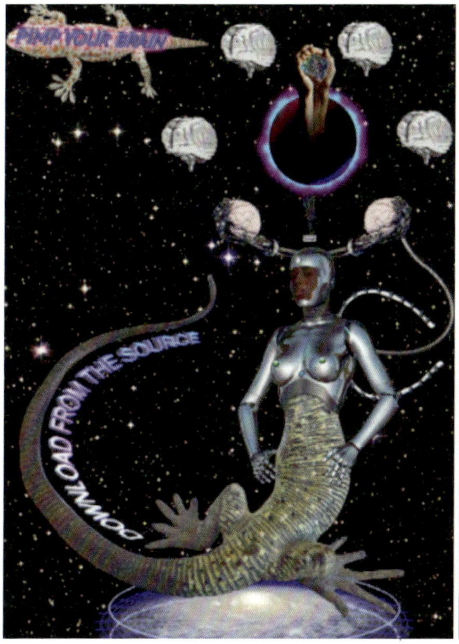

Inner Fire

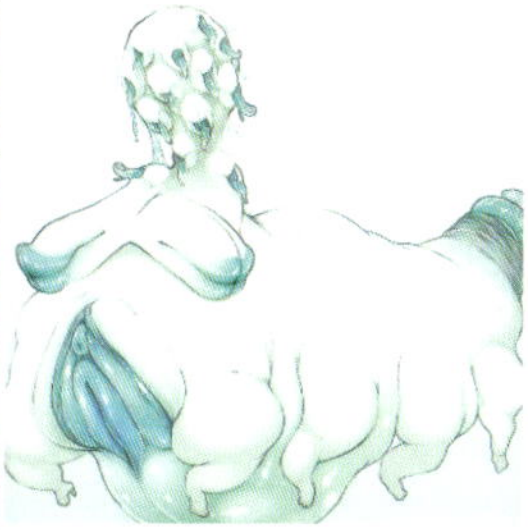

Blooming Flesh Angels

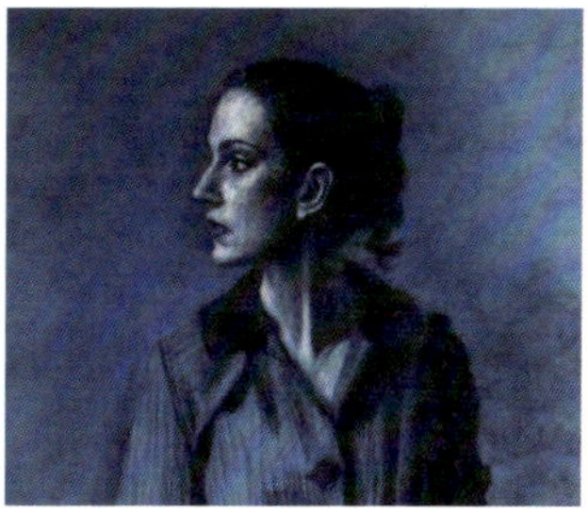

Untitled

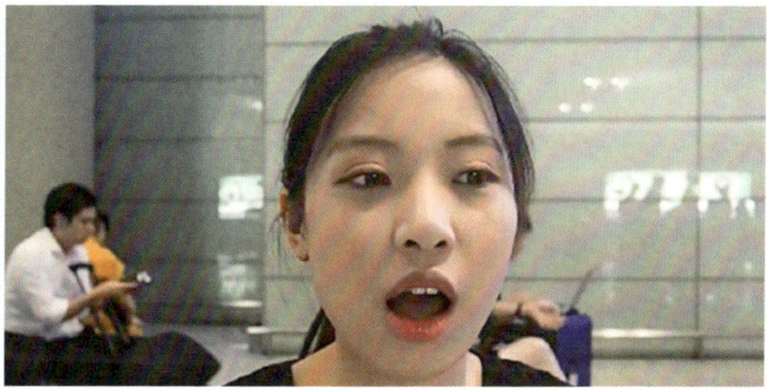

Love in Yellow, Ziwei Shuai (CN)

The following collection of works considers the growing dominance of surveillance in digital societies.

Btihaj Ajana (MA)

Surveillance Culture

Btihaj Ajana's documentary considers surveillance from a political and economic perspective, exploring ways in which the (immigrant) body is watched, tracked and recorded.

Rhiannon Jude-West

indoors outdoors

indoors outdoors is a collection of "selfies" that ask us to consider the growing phenomenon of self-surveillance, and how individuals relate to themselves and the world by rendering themselves visible and charting their life through images.

Lisa Carletta

Atelophobia

The work of Lisa Carletta considers how the human body, and in particular the bodies of women, are prepared for viewing: these images are digitally manipulated to give each figure an eerie homogeneity that effaces their individuality.

The works in the upcoming section explore the development of sexual culture online.

Courtney Trouble

Untitled

Courtney Trouble is one of the most significant alternative porn directors of the twenty-first century, founding major sites such as Queer Porn TV and Trouble Films. These photographs chart the queer lives of alternative porn performers and demonstrate both how the Internet has allowed marginalized bodies to assert their existence, and, in relation to the other major themes of this exhibition, how the development of alternative pornography as a subculture is dependent on networked visibility.

Andie Macario

Her

Macario's silk print similarly explores the physicality of porn performance and highlights the impossibilities and inadequacies of visual digital culture to fully communicate the profundity of physical togetherness.

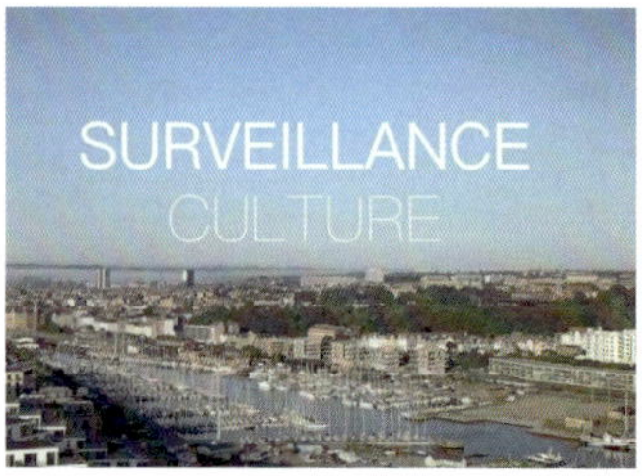

Surveillance Culture

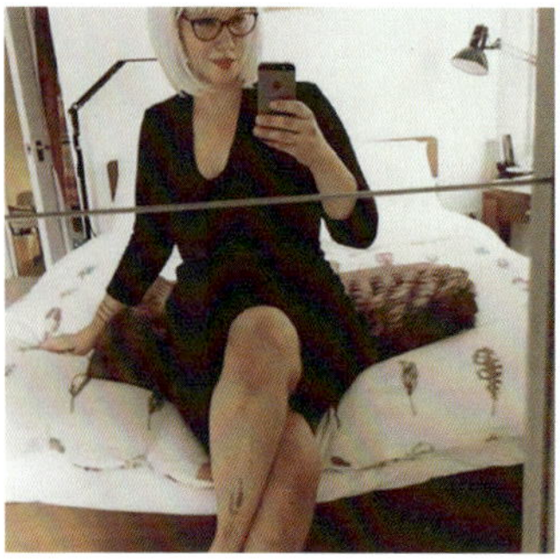

indoors outdoors

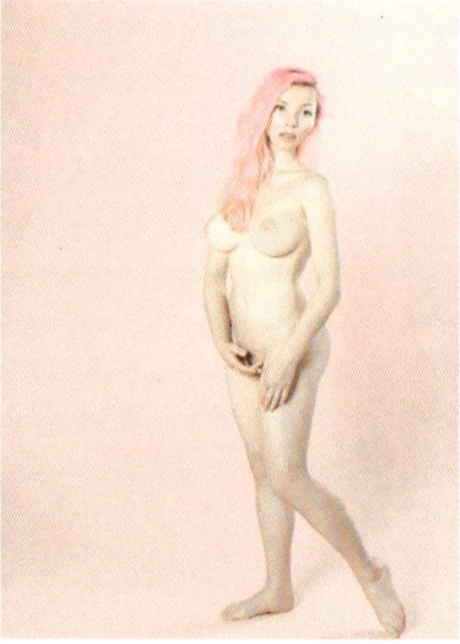

Atelophobia

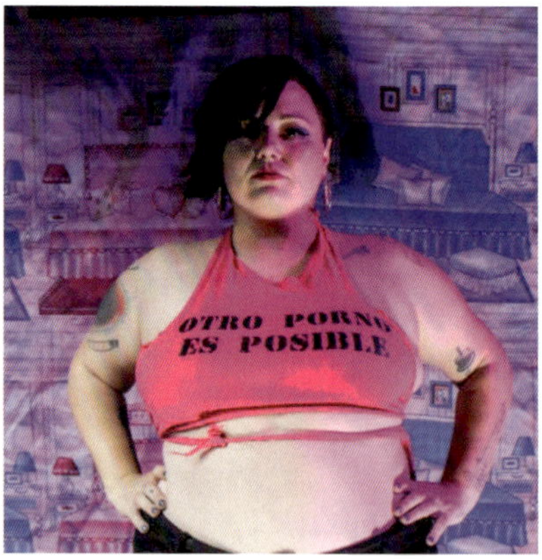

Untitled

Her

Tribes and Machines

MA Interaction Design Communication, London College of Communication, University of the Arts London, *Int-des.com*

Tribes and Machines is an exhibition of collected work from MA Interaction Design Communication at the London College of Communication, UAL. The students exhibiting their work have been conducting projects researching niche communities as well as speculating on alternative forms of technological progress outside of the hegemony of capitalist production. The projects presented span from kinetic artworks examining psychological conditions to object-oriented attempts to communicate with the Earth; from speculations about a theology of AI to games aimed at tackling everyday sexism. The variety of projects all center on a curious criticality and a fascination with the relationship of people and things: how do we read and respond to a world mediated by technologies and how can designers do a better job of making this mediation richer and healthier? MA Interaction Design Communication is a future-facing interdisciplinary radical design studio based at the London College of Communication, UAL. The work we do spans design, art, people and things and places critical practice at the center of contemporary debates about politics, society and technology. Founded on classic notions of interaction design and material culture, the course extends in all directions to film, performance, physical computing, installation, VR and product design in order to enquire about the construction of human and non-human experience of the world through rigorous research methods.

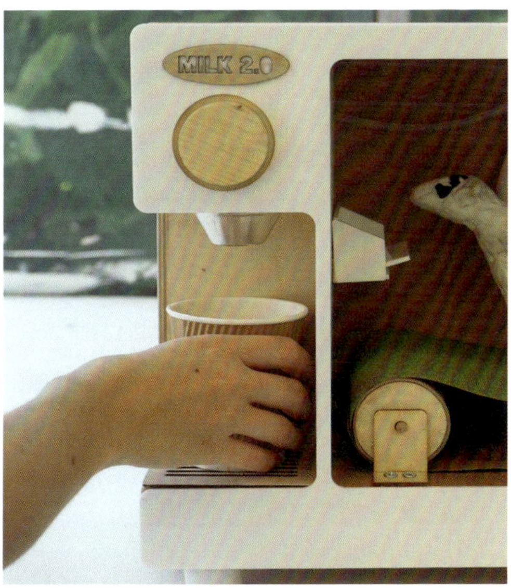

Milk Above The Clouds

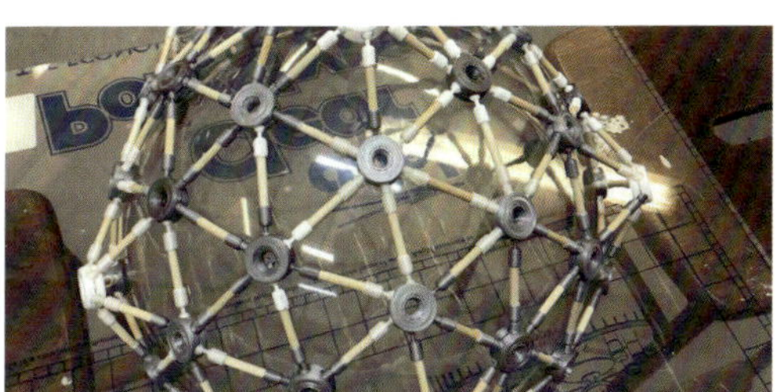

CitiSphere

Pinar Apaydin (TR)

Milk Above The Clouds

Humans' current relationship with agricultural animals is a consequence of increasing consumption and over-populated cities. Consumers are more involved with brands and mass-produced products rather than the animals and their living conditions at livestock farms.

Milk Above the Clouds is a kitchen appliance designed to draw attention to human decision-making on behalf of a cow. A genetically modified "household cow' lives inside the appliance and is able to produce two liters of milk per day.

Qi (Tom) Deng (CN)

CitiSphere

CitiSphere is a city-scale design practice project starting from the notion of *Atmosphere-Centered Design*. It speculates that domes cover major cities globally, able to seal the cities completely when greenhouse gas emission is over a certain limit. *CitiSphere* prevents all greenhouse emissions from the city inside the dome being released into the atmosphere, allowing the city to "digest.' Humans would have to sacrifice a lot in order to live under a low-oxygen environment.

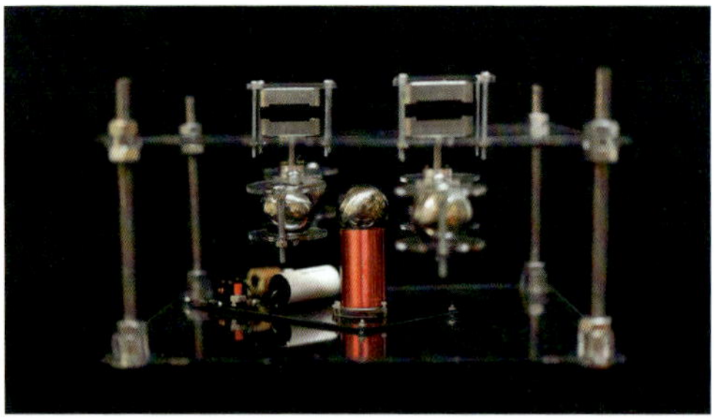

2984; Epic Frontier

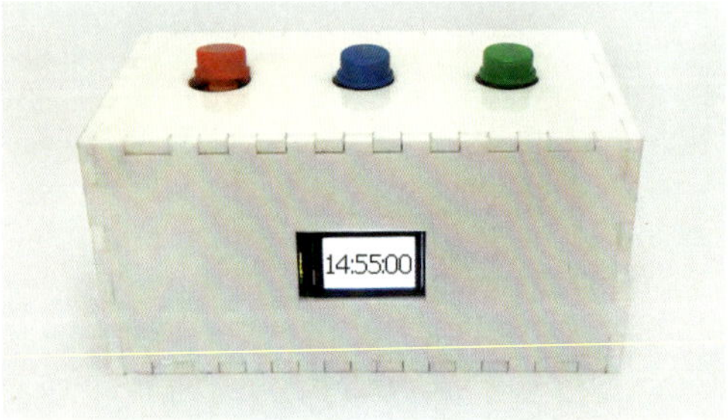

Your Time

Yue (Yorrick) Hu (CN)

2984; Epic Frontier

2984; Epic Frontier is an art installation that showcases a new structure of human society. The Earth of 2984 has ushered in the growth of artificial self-consciousness which has collided with human rights. In response, humans created Epic Frontier and divided human society into two spaces. This maximized the concept of life and death, sensibility and rationality. This work is divided into three parts, which demonstrate the transformation of human consciousness.

Won Kyoung Jang (KR)

Your Time

The advancement of technology enables humans to control time in many ways to maximize productivity and efficiency. This situation creates conditions that impose the hegemonic time systems on an individual, and we feel as if we are trapped in time.

Your Time suggests an alternative vision. This device enables people to have a subjective time experience. This will induce people to initiate their own time and concentrate on their moments as rethinking their quality of time.

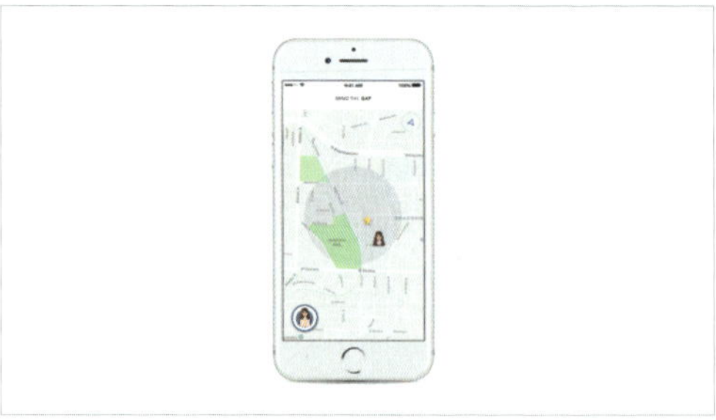

Mind The Gap

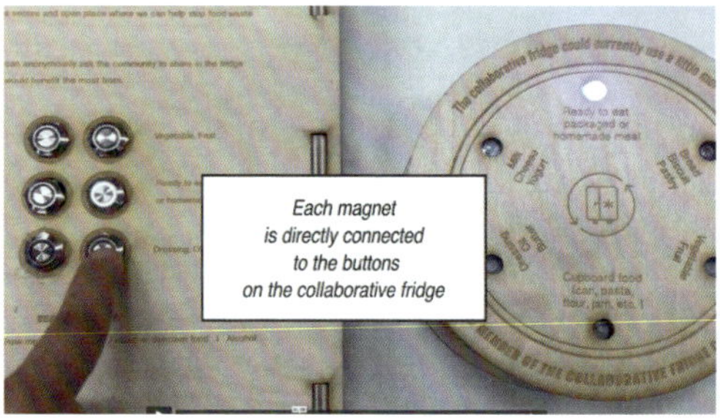

Collaborative Fridges

Aviral Kumar (IN)

Mind The Gap

This game is inspired by research on the issue of sexism in the popular augmented reality game Pokemon Go. Interviews with women in investment banking revealed shocking and disturbing stories of bias and harassment in the workplace, highlighting the fact that as a society, we still haven't made much progress in terms of equity. This project is a way of acquainting others, in particular men, with the harsh reality of what it is like to be a woman.

Clara Koscienlniak (FR)

Collaborative Fridges

What makes a fridge "smart"? This project is a low-tech counterpart smarter than any smart fridge because the users make it smart. The aim is to install open-access fridges in small communities for anyone to put food items before they go to waste and where anyone can take it. The fridge's goals are in fighting food waste by creating social links and bridging community divides. It comes with two objects to start the conversation and a digital groupchat to give the community a space to discuss.

Beatriz Lacerda (PT)

Impetus of the Machine

Humans place their hopes and beliefs into technological systems whilst developing a relationship of subordination and trust.
This project presents a transparent look at the technological era, built to provide an escape from it by exposing its dependence on humanity.
This kinetic installation consists of the perpetuation of a human element (blood) in a system that truly relies on human beings to maintain its function.

Mao Li (CN)

Decision-making

Why do people rely on Tarot cards in technological society? Tarot cards are a partner people can rely on to guide them to the next stage of life. The point of tarot is that in the face of uncertainty, people will get some advice that will ultimately help them make choices they don't regret. How can we make better choices? How do people think in the process of selection? This project aims to remind people how they make choices and realize a new way of thinking to make a "good" decision for them.

Xiangnuo Li (CN)

S•R•L

S•R•L is a new creature species created by Xiangnuo Li. Its full name is Sphinx•Resurgere•Lee. Unlike any other animal species, it consists of a combination of mechanical and inorganic substances. *S•R•L* relies on the cooperation of machines and chemical reactions to breathe. *S•R•L* is naturally shy, and if many people focus on it, its breathing will speed up. Based on research into animism and evolutionary theory, this project aims to explore the boundary between machines and living things.

Li (Lili) Lihua (CN)

Robotic Carbon Fish

The *Robotic Carbon Fish* is a speculative proposal for a domestic carbon capture system. Carbon dioxide capture and storage (CCS) aims to capture up to 90% of the CO_2 emissions produced from the use of fossil fuels during the processes of electricity generation. Ocean storage is one of many CCS strategies. What will the ocean be like if human activity continues to release massive amounts of CO_2 and let the ocean absorb it? Will dealing with CO_2 emitted during our lives be an issue in the future?

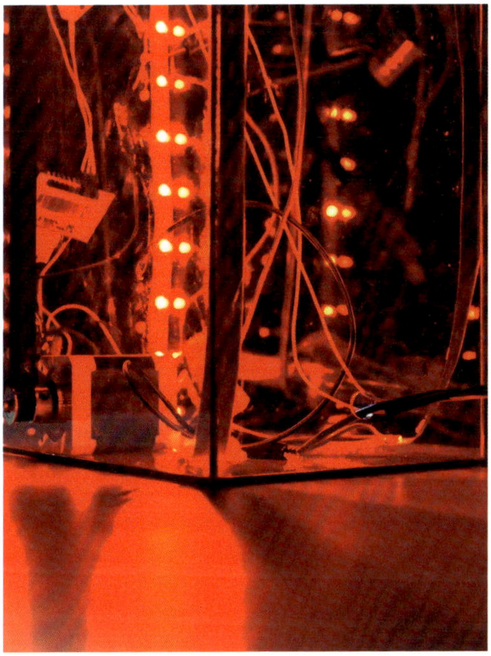

Impetus of the Machine

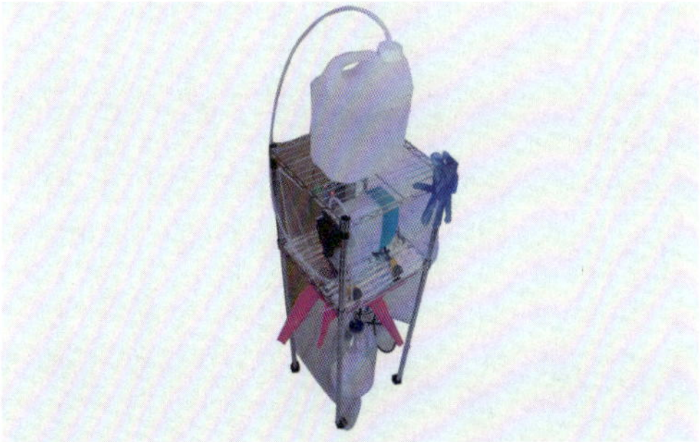

S•R•L

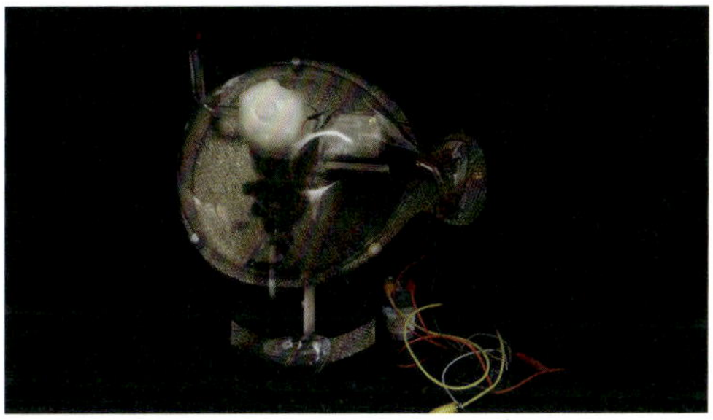

Robotic Carbon Fish

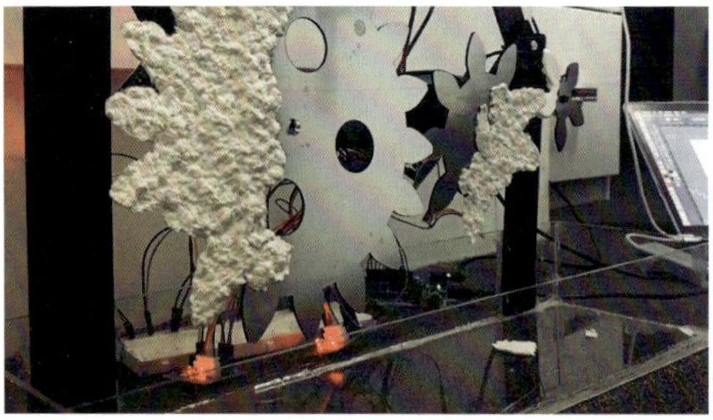

Collision Gearsa

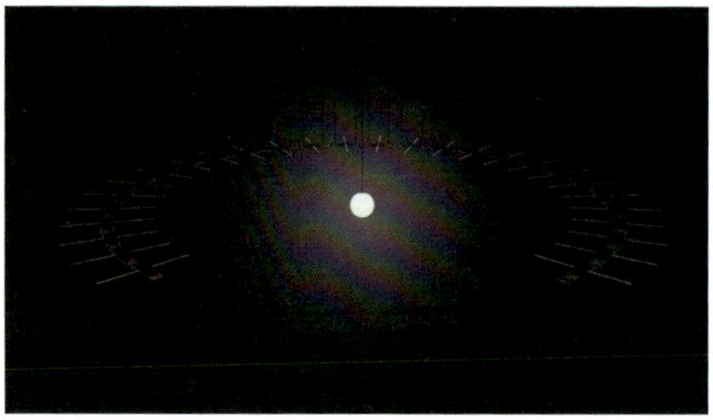

Theomachina

Mingyi Qian (CN)

Collision Gears

Self-destructive machines are unusual in that their function challenges the normal production and consumption cycle we assume of technology. Self-destructive machines are contradictory, they call into question material existence. This project is a self-destructive machine created through research on materials and self-destructive forms. The operation of gears is a precise process. By changing the material and shape, the gears no longer interact with each other, but they collide with each other.

Vinzenz Reinhardt (DE)

Theomachina

Theomachina speculates on technology's trajectory with regard to human desire to have and obey a Supreme Being. The interactive kinetic art installation allows the user to enter a battle over control with a futuristic deity that dictates its subordinate technological will using light patterns. *Theomachina* was created to help you to explore your own position about modern technology and the influence that it has on your life.

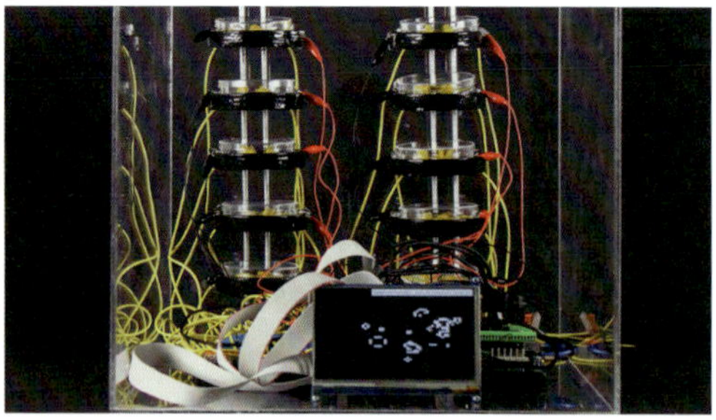

Ctrl

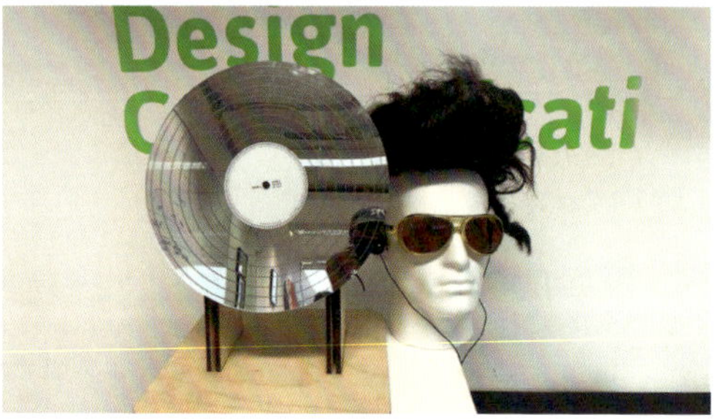

Identity as a Choice

Michael Sedbon (FR)

Ctrl

Technology is not conscious or driven. Free will is an oxymoron; interconnected systems are spinning—dragged by one another. In this setup, 10 Physarum Plycephalum, known as slime molds, play the Game of Life. The environment had been designed to monitor and control the Physarum's activity. At the end of each round, Physarum moves are analyzed and compared to the winners. The onboard artificial intelligence will alter the environment of the losers in order to manipulate and optimize their games.

Michael Selvadurai (UK)

Identity as a Choice

This project set out to investigate the cultural influences of identity by the observation of Indian Elvis Tribute Acts. In particular, their admiration of Elvis the performer and a simultaneous respect for Elvis the man. The project considers how society views a non-white performer emulating a white artist, even though the artist himself was influenced by black music and culture. The project seeks to answer the questions of racial stereotyping; their origins and how these are perpetuated today.

Bin Wang (CN)

Illusion Mirror

Illusion Mirror is an installation project for creating an illusion with mirrors. Mirrors are not only an instrument to look at oneself, but also are a way present our true self to us. It uses many small triangle mirrors to express a crushing effect that may represent many faces or forms that are not our true self. The overall outcome presents a fantastic visual effect, as separated and rotated mirrors distort our true self.

Shuying Wang (CN)

A Fictional Wedding

Almost 13% of homosexuals choose to engage in a "shaped marriage" in China. These are fake heterosexual marriages formed by mutual agreement. What will happen to such a group in the future? In an Asian society that calls for social pluralism and promotes civilized progress, is such a choice a reluctant compromise or a path that must be followed for one version of "progress'? This installation organizes a fun story about a fake wedding by triggering each of the small sculptures.

Yue Xiaohan (CN)

The Sims

Social developments of science and technology have quantified our needs, achievements, and even personality, as personal information becomes valuable currency in the form of data. This has created an ability to role play our fantasies by tweaking variables—releasing and withholding information. We begin to interact with others as if in a game.

Eleni Xynogala (GR)

Symbiosis

Symbiosis is the mutually beneficial relationship between different organisms. What techniques can we use to create new forms of embodied communication between organisms?
Exploring the disparity of perception within the real and the virtual, this installation consists of three layers: In the first layer, a body is trapped in a virtual world that the audience cannot perceive.
The person disappears under the digital layer of the virtual body, which is ultimately dependent on the physical body.

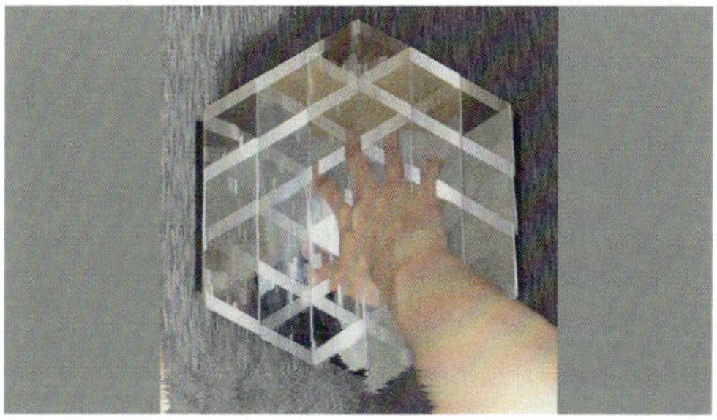

Illusion Mirror

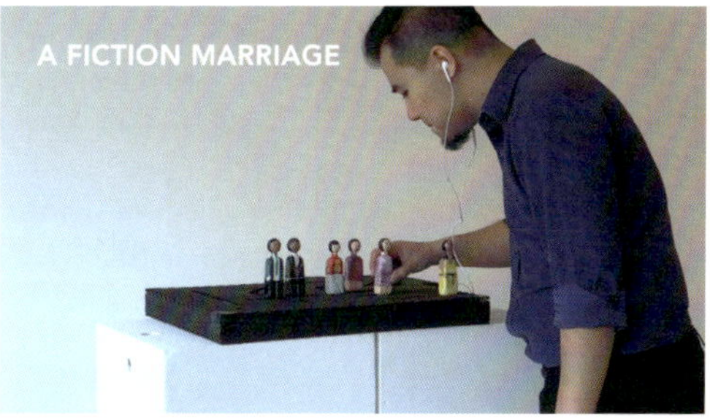

A Fictional Wedding

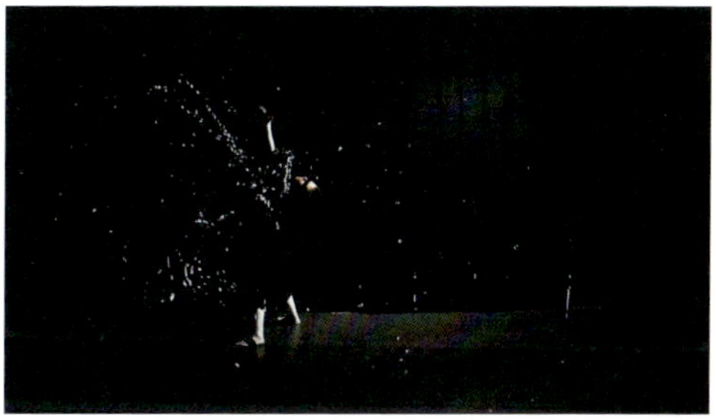

Symbiosis

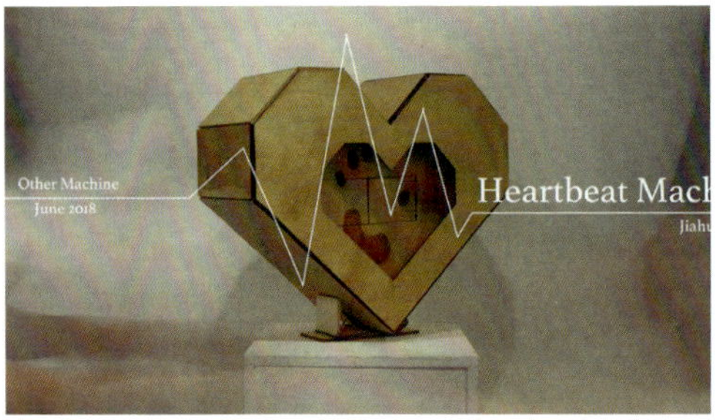

Heartbeat Machine

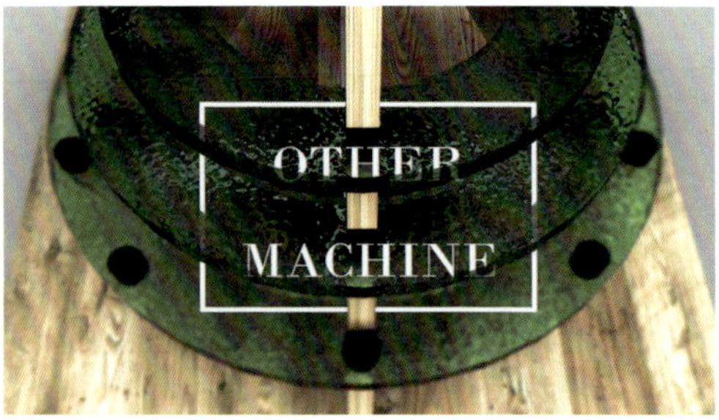

Electromagnetic Being

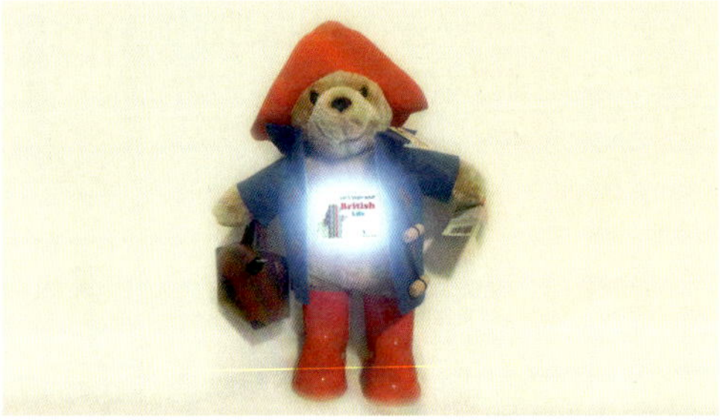

Paddington Tool

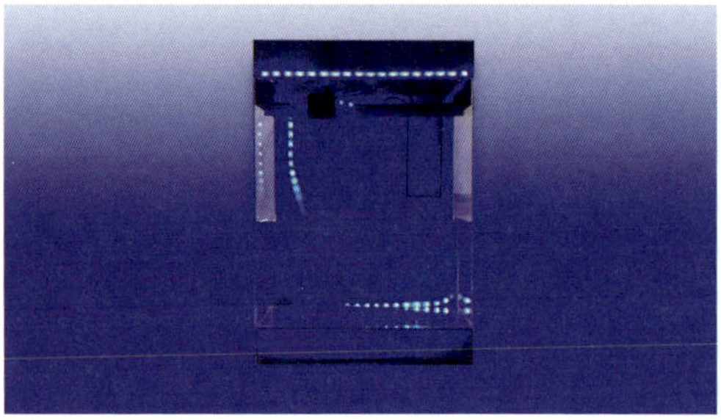

Communicate with Fluorescent Sea

Jiahuan Yang (CN)

Heartbeat Machine

This installation breaks away from the traditional method of audiovisual interactions and only uses sound effects to provide interactive feedback. Participants need to stretch out their hands and reach into the holes of this interactive device. At the end of the hole is a sensor on the board which guides the participant to attach their finger to the sensor unconsciously in a natural intimacy gesture–"Hand stick Hand."

Jiawen Yao (CN)

Paddington Tool

This project uses the iconic Paddington Bear to reflect on English values to newly arrived children. The project translates the values into tasks that children need to fulfill, like a game, where the tasks become harder and harder. Children must complete all the tasks to integrate into the local society. The tasks are hidden behind Paddington's coat; children have to find them by themselves. The project uses a kind of soft power to affect children, pushing them to become "better" English citizens.

Xindi (Randy) Yu (CN)

Electromagnetic Being

Humans rarely consider themselves as electromagnetic beings. They think about themselves as biological beings, or even as computation beings. We can feel our magnetic field and bioelectricity through the hospital's electrocardiogram and EEG. This machine wants to unscramble and connect us better with our electromagnetism.

Qiao Zhao (CN)

Communicate with Fluorescent Sea

This project is led by theories of Object-Oriented Ontology and Hyper-objects to think and reconsider our awareness of the self-perception of humans; the impact of what we do to the earth and the future feedback from the invisible dimensions of human intervention.
This installation presents a speculation as to ways we might alternatively relate to the Earth.

!OBJECT

Roy Ascott Technoetic Arts Studio, De Tao Masters Academy, Shanghai Institute of Visual Arts (SIVA)

!OBJECT, pronounced NOT OBJECT, is a series of 'technoetic' compositions (technoetics, a term coined by Roy Ascott, combines technology practices with noetics). These 'non-object' artworks exist as material carriers of immateriality and are the creations of our very first graduate students in technoetic arts. In this exhibition we witness the convergence of the three types of VRs – or the "three Graces of our culture"– described by Roy Ascott as Virtual Reality, Validated Reality and Vegetal Reality. In this exhibition, the artists reject most avenues of certainty and instead re-examine and question key notions of our existence. Through each work the artists are asking: 'What it is to be human?', 'How do we act and react in cyberspace?', 'What is the "self"?' and 'In what way do "selves" form a society and how do these social structures interact with the "natural" and the "cosmic"?' All of these questions examine contemporary interactions between material and immaterial forms of existence. This exhibition showcases sample works by the first graduates of the Bachelor of Art (BA) in Technoetic Arts course at the Roy Ascott Technoetic Arts Studio, DeTao Masters Academy, Shanghai Institute of Visual Arts (SIVA). The Technoetic Arts Major is an advanced Art program based on the research-creation paradigm, exposing the students to both theoretical and practical issues through the broader fields of art and technology. Technoetics is a convergent field of theory and practice that seeks to explore the broad spectrum of consciousness and connectivity; a specific approach via digital, telematic, chemical or spiritual means, embracing both interactive and psychoactive technologies, as well as the creative use of moist media. With a strong focus on artistic expression and on combining teaching and research with practical applications, the Roy Ascott Technoetic Arts Studio offers a pathway to holistic creativity that will enhance both individual and collective ways of thought, expression and innovation.

Roy Ascott Technoetic Arts Studio Team 2018.

Lin Yingyin (Lynn Black) (CN)

Self-distortion

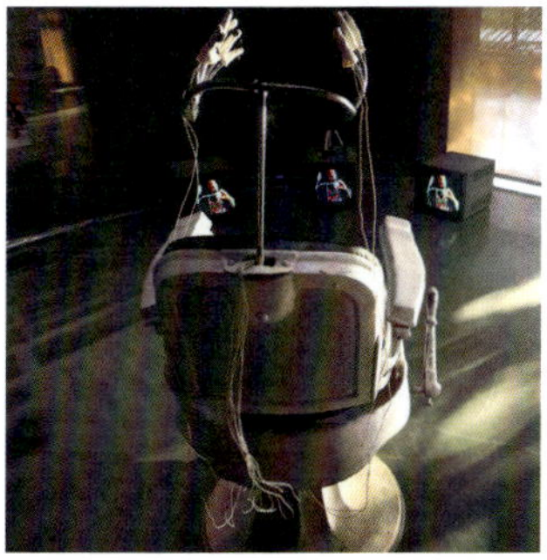

Self-distortion interrogates the mediating effects of technology on identity construction. According to participants' brainwave patterns (which are a by-product of the level of participants' meditative state), the glitch image on the monitor will gradually become more or less clear. By using bone conduction technology and transforming the brainwaves of participants into real-time sound waves, this work enables participants to hear/feel their own meditative sounds. This produces a new form of sensory experience for individual brainwave observation.

Yang Shuai (Rachel) (CN)

IT'S NOT ME

IT'S NOT ME uses saliva as an artistic medium to explore the concept of biological life. Saliva samples are taken from the artist and are placed in a petri dish. The resulting bacteria-human cell compound grown in the dish may have originated from the artist ("me"), but after being separated from and grown outside of the artist's body, the living material in the petri dish evolved into a hybrid life form of human and non-human bacterial cells. To visualize this process, a programming language is used to map the growth of this new life, which is then converted into sound. The audience can listen to the sonic expression of this new life through headphones.

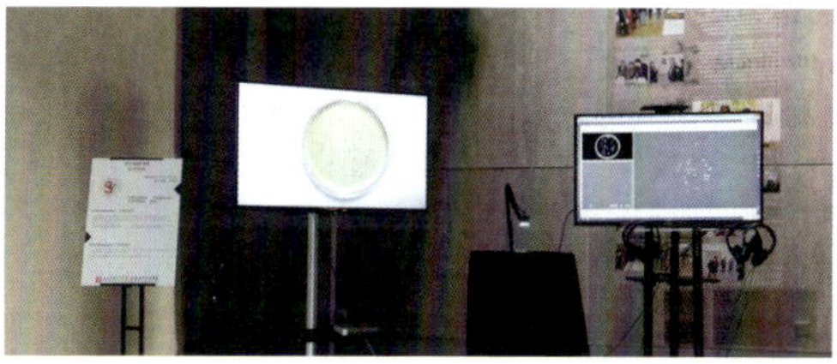

Ni Yunyun (Sariel) (CN)

Dreamtation

Dreamtation creates an immersive virtual world that gives the participants the opportunity to reflect on meditation from a new perspective. Using VR and a breath sensor, an individual's meditation can be enhanced through slower breathing or taking longer breaths, which directly effects the way in which each person is immersed into a rich and visually stunning audio-visual environment. This innovative work creates a truly individualized audio-visual meditation experience.

Shen Yichao (CN)

Plastic Ocean

Plastic Ocean is a poetic VR game which simulates how human behavior affects the real-world environment. In this project, the human is both the initiator of the ocean pollution and the resolver of it. The goal of this game is to combine the irony of the gameplay with reality.

4125 (CN), Jin Yi (Alanis) (CN)

MOM

MOM is a four-screen video installation in the form of an altar - a cinematic altar that integrates totems, myths and elements to reflect a new interpretation on the Cyborg. MOM begins with an absurd myth: "Nuwa (Mother of all things) created woman and man." For this video installation the artists adopt a series of rituals to attain a new poetic narration of 'new' Nuwa - the ultimate Cyborg who will free humans from the constraints of their biology.

Zhou Jiajie (Jennifer) (CN)

Private Conversation

Private Conversation is an immersive interactive installation that exposes the risks of online data privacy. The installation is composed of four elements: a physical internet café environment, a chat room application, lightbulbs, and a real-time Morse code translation app. With these elements, the artist created a work that will inspire individuals to consider what, exactly, we are giving up in order to gain access to the various convenient online apps we use today.

Gu Xingyu (CN)

Aesthetic Metagustations

This research project focuses on the connection between taste and aesthetics from an art and technology perspective. While visual and auditory senses are usually prominent means of new-media related artistic expression, the sense of taste is largely neglected. The *Aesthetic Metagustations* installation illustrates how the sense of taste can be translated into other sensory experiences, e.g., sonic, visual and haptic. These translations are made possible through the processes of 3D printing and data visualization, which reveal the different patterns of peoples' diets ... making it possible to 'touch' taste.

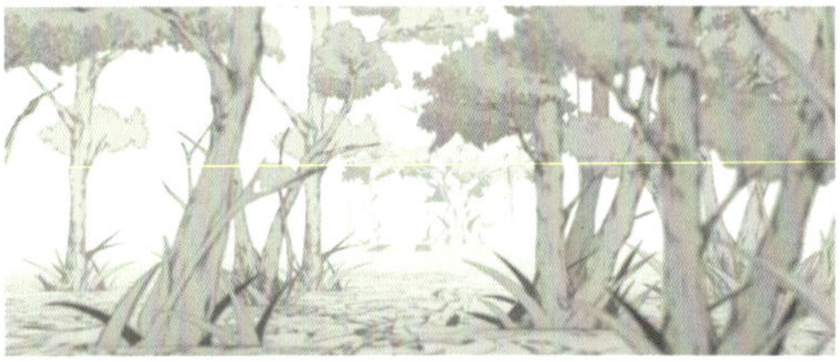

Dreamtation

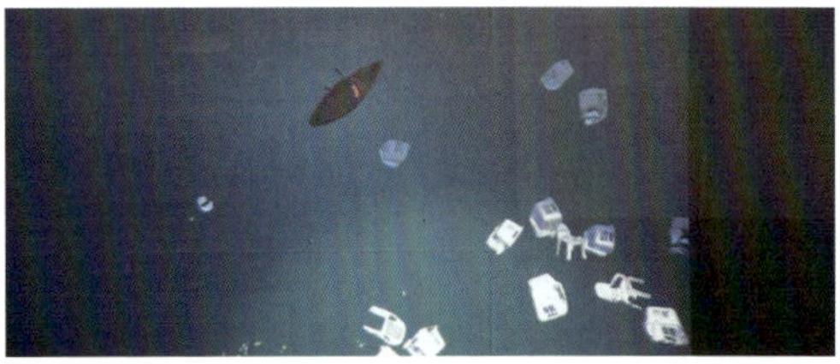

Plastic Ocean

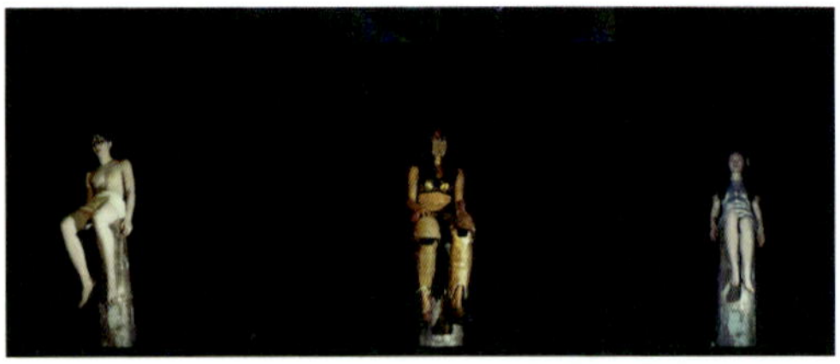

MOM

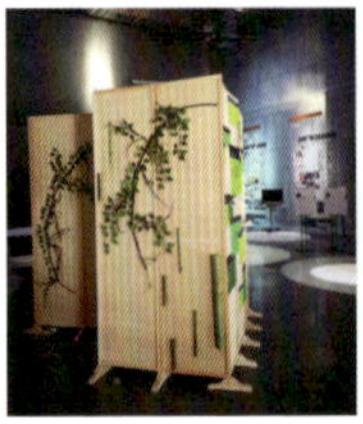

Private Conversation

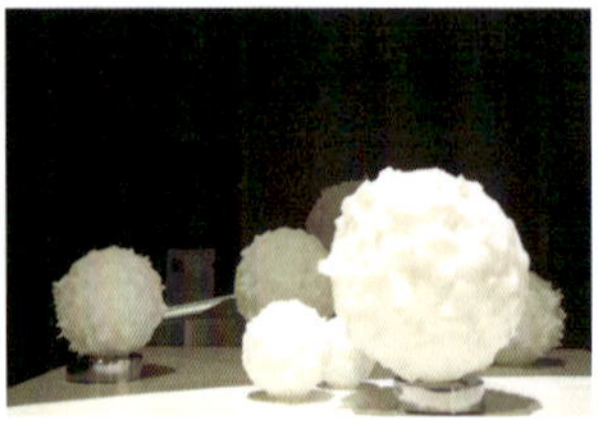

Aesthetic Metagustations

Explorations of Interactions
Queen Mary University, London

Text: Dr. Nick Bryan-Kinns

Explorations of Interactions showcases research in media and arts technology driven by three error-prone calls to human action: Interact. Reflect. Create. Interact. *Common Roots* glances at interaction in human communities through an augmented apartment block in which neighbors can remotely care for other households' plants. The piece explores whether our existing, individualistic city life is an erroneous one. *Emojikken* also examines inter-human connection, testing if the recently emerged pictorial language of emojis can provide a universality to communications that is failed by existing language structures.Reflect. *Aural Fabric* augments a map with audio recorded from those locations, questioning whether our visual recording of space is an incorrect—or at least incomplete—way of documenting locations. Similarly, *Kulaktan kulağa* uses what is (at first glance) just a box of old photographs to examine the western-centric lens of the internet by humanizing machine translation errors, while *Cembalo Scrivano .2* plays with the relationship between typed records and their linked experience. Create. The *Tesseract* strives to reduce sonic errors by providing a unique tactile music creation interface programmed to maximize creation and minimize displeasing notes, while *Μαζί*, a pleasantly soft, tactile instrument, critiques our obsession with plastic-and-glass hardware. Finally, the *ChandeLIA* challenges both perceptions of what is considered an instrument and the correct use of an object, by repurposing a disused chandelier as an augmented musical instrument. The EPSRC+AHRC Media and Arts Technology Centre for Doctoral Training at Queen Mary University of London, UK, provides a bridge between academic research, digital technologies, and creative industries. We aim to produce an elite community of graduates who are excellent in technical and scientific research, creativity, building and using software and hardware, and are prepared to contribute to the world's digital economy.

EPSRC+AHRC Media and Arts Technology Centre for Doctoral Training (EP/L01632X/1); Queen Mary University of London

Jacktionman (UK)
Emojikken

Emojikken is an interactive installation that invites visitors to converse using only emoji, testing if the symbols can form the basis of a universal, cross-cultural, pictorial language. As the internet fosters greater creativity, emojis, like memes, have enabled post-language communication—but does that bring enhanced connection?

The piece—two emoji keyboards and a chat screen—was debuted for three weeks in Japan, home of the emoji, where it recorded over 10,000 interactions.

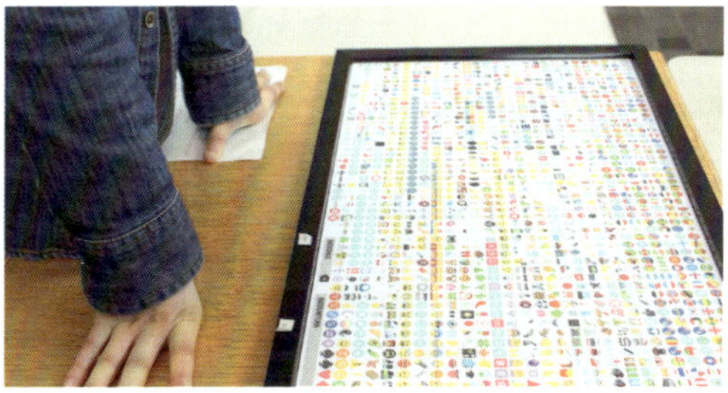

Emojikken

Jon Pigrem (UK)
Tesseract

The Tesseract is an interactive music box that uses generative techniques to sequence the playback of pre-recorded elements of a musical corpus. Through user interaction, structural grammars are arranged, leading to unique interpretations of prearranged musical components. The interface is designed to foster heuristic development and quickly leads to an effective mix of intuitive and indeterminate responses, offering creativity while maintaining a degree of tonal, timbral and rhythmical hierarchy.

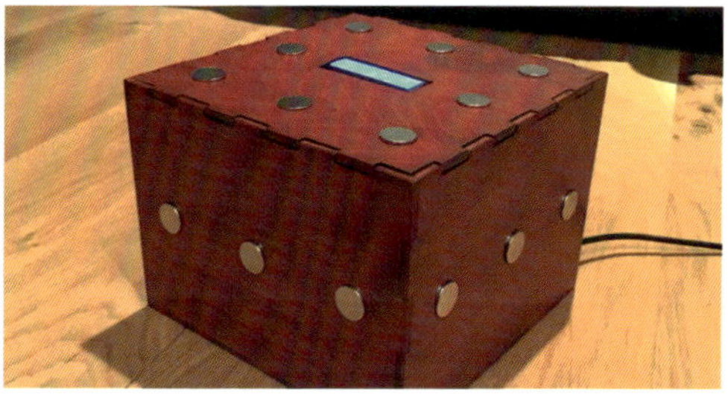

Tesseract

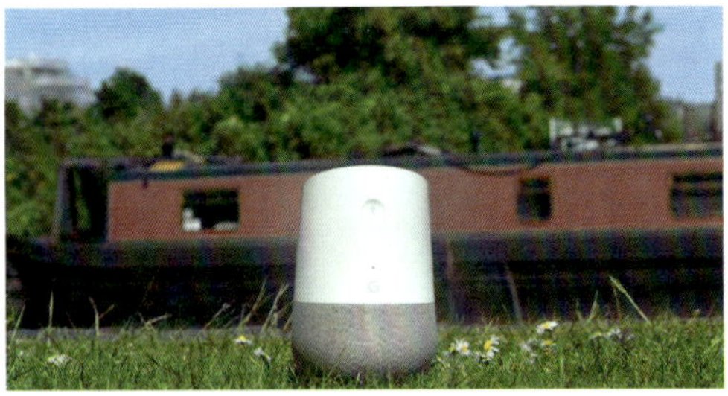

Unspoken Word

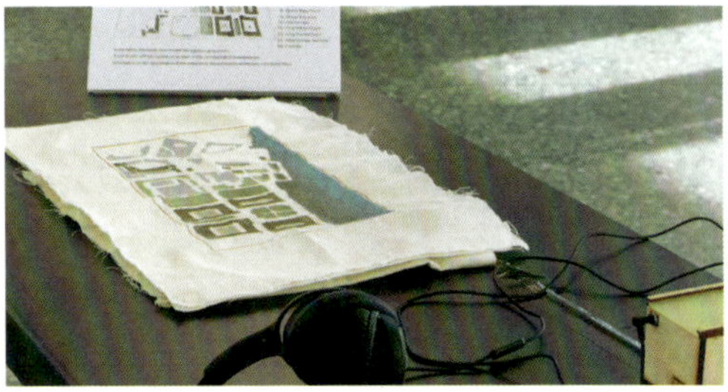

Aural Fabric

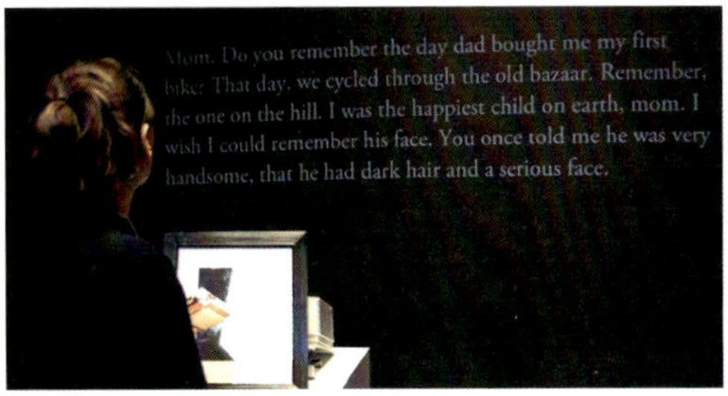

Kulaktan kulağa

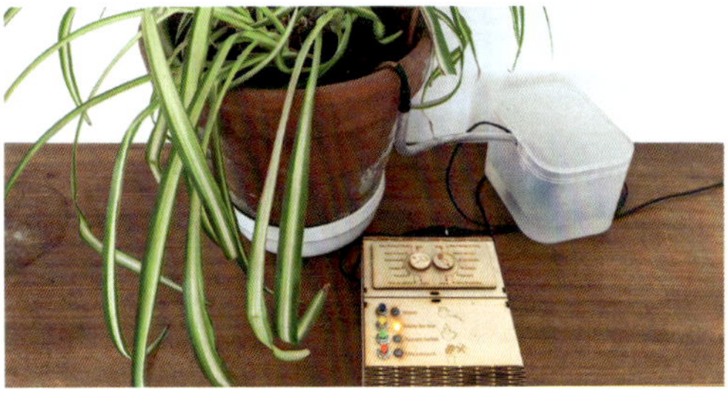

Common Roots

Jorge del Bosque (MX), Lizzie Wilson (UK)

Unspoken Word

Part of a collaboration with the BBC R&D department, Unspoken Word is an installation that explores the hidden codes and signals inside spoken language and how they can fertilize the musical composition process. The piece makes use of a smart speaker and demonstrates an innovative speech-to-melody composition method used to create a novel medium of musical expression. People with both musical and non-musical backgrounds are invited to experience it and discover their own unspoken codes.

Betül Aksu (TR)

Kulaktan kulağa

Kulaktan kulağa, Chinese whispers, or Arabic telephone is an interactive installation inviting people to pass around the stories of found images. Each time a story is selected, its message becomes garbled along the way until it significantly deviates from the initial interpretation due to the inaccurate machine translations of non-Indo-European languages.

Alessia Milo (IT)

Aural Fabric

Aural Fabric is an interactive textile map that reveals one possible sonic character of the area of Greenwich, London, as captured during a group soundwalk. Architectural and natural elements encountered during the walk are embroidered with conductive threads as forms in relief. The soundwalk moments become audible again when the listener touches the sensitive areas, intimately connecting with the experiential universes of those that, on that day, crossed the path of this soundwalk.

Sophie McDonald (UK)

Common Roots

Common Roots is a system of shared plant-watering between homes that promotes meeting neighbors in a communal garden over a cup of tea. It provides each plant host with a real-time visualization of the community's watering activities–who's watering whose plant–and in doing so provides a gentle presence of the block's inhabitants within individual private homes. It asks what novel ways we can design the cities of the future, given that people need more than simply to inhabit them.

Sophie Skach (AT)

Text and Iles

Text and Iles, an interactive installation, consists of two capacitive knitted jumpers that react to being touched. They respond to visitors' caresses with audio recordings of particles of a poem.
This creates a dialogue not just between the garments, but also between the clothing and its wearers, commenting on the relationship between textile and tactile intimacy.

Antonella Nonnis (IT)

Μαζί

Μαζί (pronounced Mazi) from the Greek "together," is a tactile instrument that entices the audience to interact with its inviting texture and soft appearance. The soft dome holds five interactive felt bubbles, each of which produces a gentle tone when touched. The instrument was inspired by watching children play, and aims to stimulate active participation; create spontaneous, independent and collaborative play; and provide sensory regulation for children with autism.

Lia Mice (AU)

ChandeLIA

The ChandeLIA is a unique suspended musical instrument that explores special gesture mapping and microtonality. The hacked chandelier hangs motionless, occasionally flickering its lights to entice the audience to interact with it.
When the ChandeLIA is touched, the piece illuminates and a tone is created. Moving the chandelier in different ways (such as swinging, spinning or holding the instrument upside-down) cause changes to the tone, such as microtonal pitch shifts and audio effects.

Giacomo Lepri (IT), Fabio Morreale (IT)

Cembalo Scrivano .2

The *Cembalo Scrivano .2* is an interactive audiovisual installation based on an augmented typewriter. The audience is invited to type a letter, a world, a sentence. While interacting with the machine it will be possible to experience an ambiguous condition in which symbols become shapes and sounds, transcending the forces acting within the writing mechanism. This work allows people to reflect on the need for new technologies, outlining the possibility of rediscovering the past to imagine the future.

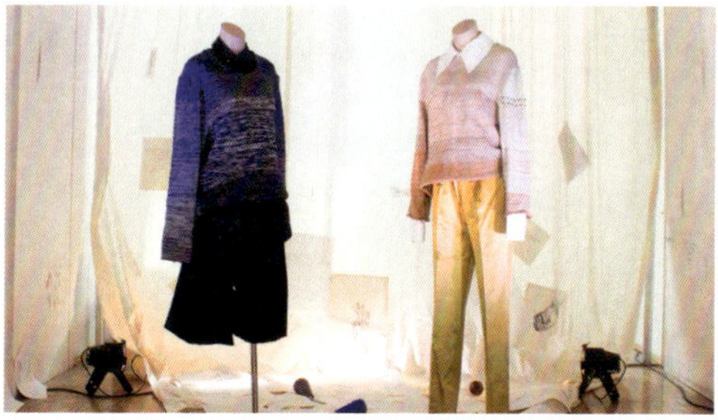

Text and Iles

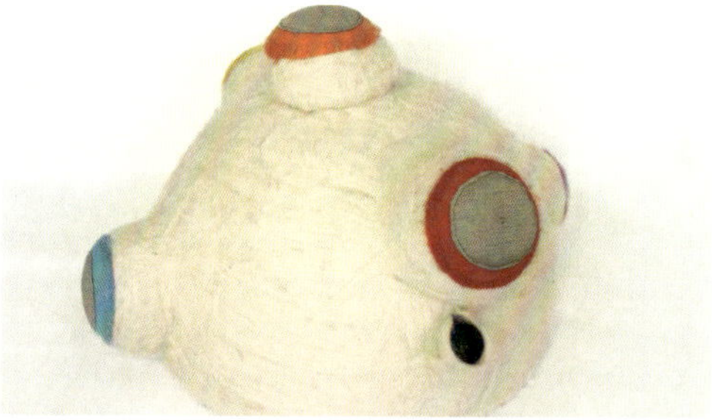

Μαζί

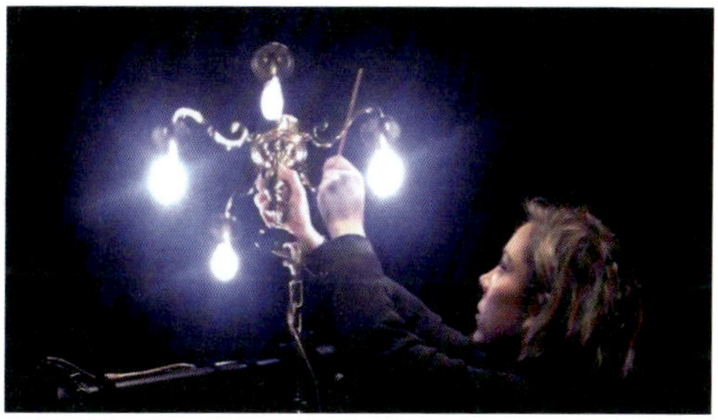

ChandeLIA

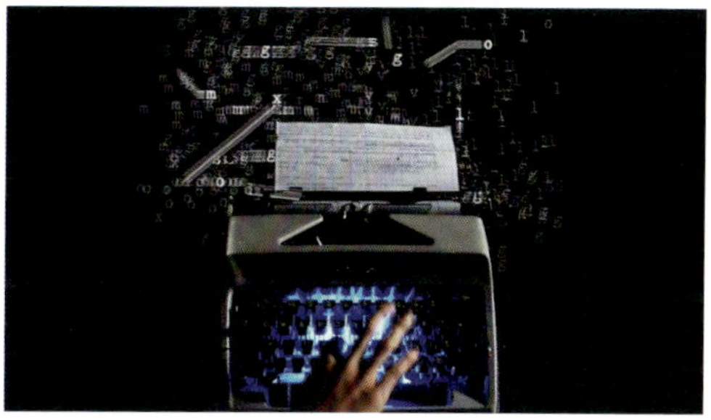

Cembalo Scrivano .2

Disruptive Generation

Art & Technology Studies at School of the Art Institute of Chicago (SAIC)

Text: Duncan Bass

The works assembled in *Disruptive Generation* question the role of technology in constructing and mediating our experience of reality, addressing technology's capacity to undermine or transcend its structural frame. Each project highlights the relationship between archetype and prototype, disrupting the cultural and computational systems that circumscribe lived experience in order to render space for alternative futures. *Disruptive Generation* showcases work by five recent graduates of the Art & Technology Studies department at the School of the Art Institute of Chicago (SAIC): Ziv Ze'ev Cohen, GREYMAR (Igraine Grey + Jonatan Martinez), Changyeob (C.Y.) Ok, Santiago X, and Li Yao.

Unlike other disciplines that use technology at the service of traditional forms, faculty and students in Art and Technology Studies (ATS) employ technology itself as their medium. The first of its type in the United States, ATS was established in 1969 when SAIC introduced courses in Kinetics and Generative Systems. Since its inception, the program has continually pioneered the use of emerging technologies in contemporary art, developing new models of artistic practice and integrating these models into the curriculum of one of the world's most influential art and design schools. Today ATS offers courses in a wide range of disciplines including Programming, Light, Electronics & Kinetics, Audio, Virtual & Augmented Reality, Games, Bio Art, and History & Theory.

Art & Technology Studies department, School of the Art Institute of Chicago

Featured Artists: Ziv Ze'ev Cohen, GREYMAR (Igraine Grey + Jonatan Martinez), Changyeob (C.Y.) Ok, Santiago X, Li Yao
Special Thanks: Eduardo Kac, Professor & Chair of Art & Technology Studies, SAIC, Anna Yu, Assistant Director of Art & Technology Facilities, SAIC, Arthur Kolat, Assistant Director of Graduate Admissions, SAIC, Christl Baur, Producer Exhibitions, Ars Electronica, Violeta Gil Martínez, Project Manager Campus Exhibitions, Ars Electronica

Changyeob (C.Y.) Ok (KR)

The One that Shatters in the Air

Cosmic rays are invisible, high-energy charged particles from outer space that penetrate our bodies and objects around us. Despite the fact that cosmic rays go through and around us endlessly, we do not see or recognize them. In this work, light changes are triggered by a muon detector which invites the viewers into a spiritual experience of deciphering a secret message from outer space, amidst the waves of rays that constantly surround our existence.

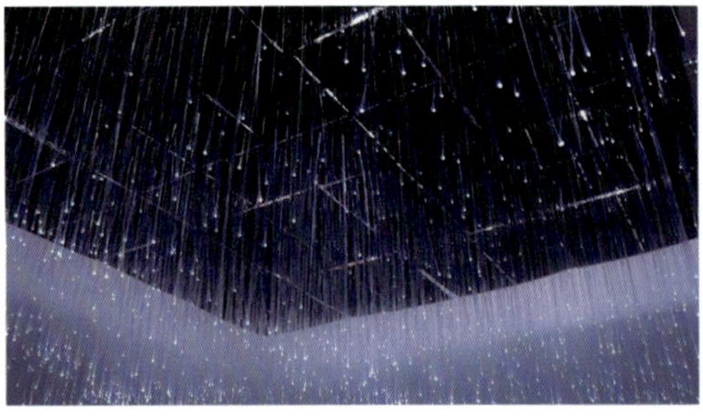

Development of muon detector: MIT
Assembly of muon detector: Changyeob Ok

GREYMAR: Igraine Grey (VE), Jonatan Martinez (MX)

RED

RED is a virtual architecture that forges monumentality out of pattern distortion, building a language out of moiré, the digital "watered appearance." Large-scale interference patterns are produced by the superimposition of ruled geometries and liquid bodies. Computer graphics warp metric relationships, intensifying diffraction, interference, and image rendering. The project presents an interrupted architectural construction that bridges virtual reality and installation.

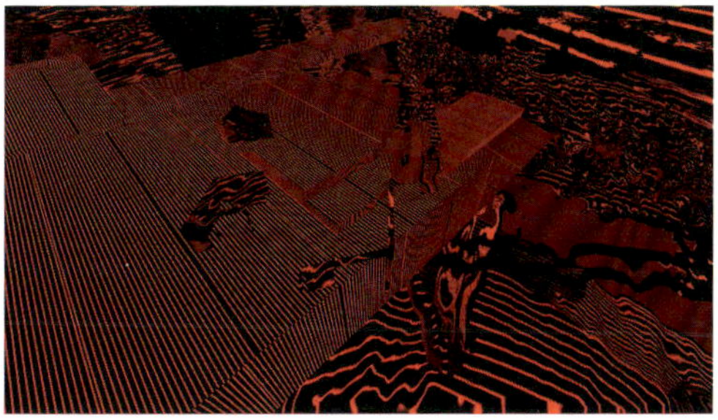

Santiago X, Hacha'maori / Koasati, Guam (US)

Aticintoloca (Man and The Black Snake)

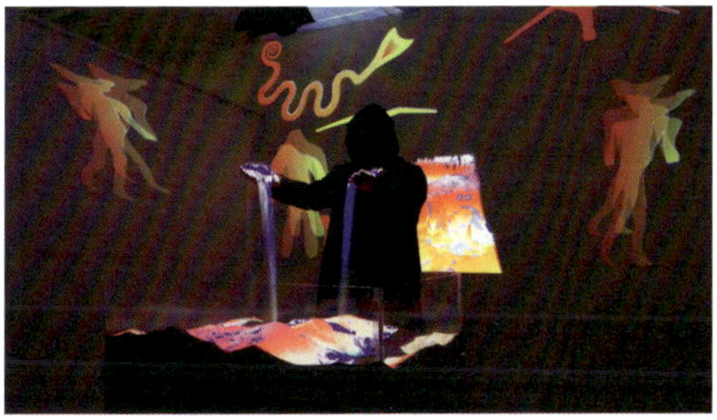

Created by a 21st Century Indigenous Mound-builder, using the artist's ancestral earth as the palette, this experience contemplates the threshold of balance within our humanity and the perpetuity of our world. Our contemporary existence is implicated as we continue to accelerate closer to a post-human world.

Santiago X, SAIC Art & Technology
Project Advisors: Eduardo Kac, SAIC, Chair Art & Technology
Judd Morrissey, SAIC ART & Technology
Jonathan Solomon, SAIC, Chair AIDO
Thomas Kong, SAIC AIDO
Coding Contributions: Christopher Baker, SAIC Art & Technology
Oliver Kreylos, UC Davis

Ziv Ze'ev Cohen (IL)

Remanence

Remanence presents a relic of the future; a body that is neither male nor female in appearance but at once refers to both the histories of classical sculpture and a speculative cyborgian body of the electromechanical future. A kinetic light from within the sculpture recreates the gesture of scanning a body as viewers are caught scanning it themselves.

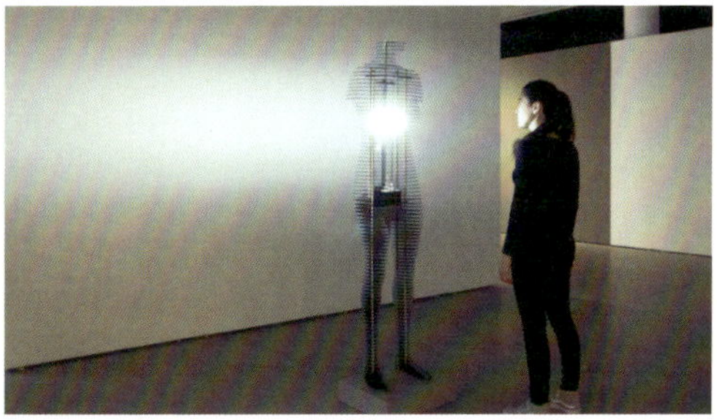

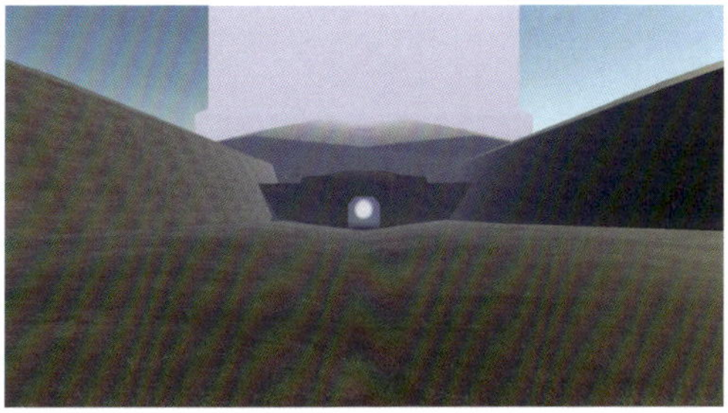

Li Yao (CN)

Bunker

Bunker is a project realized in virtual reality. By simulating sceneries from fiction and reality, it offers an opportunity to examine the willful re-enactment of history in which we find ourselves. This is history experienced, the first time, as tragedy, and the second time, as farce.

MELODIES FROM TEOCHEW STRINGS

Shantou University, China

Curator: Dr. Predrag K. Nikolic

Teochew culture or Chaoshan culture has a long history. It originated from Chaoshan predecessors, established itself in the Qin and Han Dynasties, and has features such as Teochew dialect, opera, music, cuisine, arts and crafts, and folk customs, as well as Gongfu tea. The name "Chaoshan" is a contraction of the names of two of its administrative areas, the prefecture-level cities of Chaozhou, and Shantou. The characteristics and influences of Teochew culture are evident in the philosophies, speaking habits, and customs of Shantounese people. Teochew string music is the eternal inspiration for many generations of Shantounese, reflecting the way they live and learn, and the ideas and creations they have disseminated all over the world for more than a thousand years. That is the reason we named the exhibition "Melodies from Teochew Strings." The melodies and the Teochew string instrument as the metaphors for the fresh ideas and novel aesthetic concepts derived from the mix of native Chinese heritage and technology. Furthermore, it is a speculative futuristic vision of human society, environment, and interactions offered by the Digital Media Department at Shantou University, Cheung Kong School of Art and Design and their young students who are part of the modern design reform of China in the 21st century. Shantou University is a comprehensive university established in 1981 with the approval of the State Council. The school was supported by famous patriots and internationally renowned entrepreneur Mr. Li Kashing. Currently, Shantou University is a tertiary institution jointly established by the Ministry of Education, Guangdong Province, and the Li Ka Shing Foundation. Shantou University Cheung Kong School of Art and Design is strongly committed to providing professional design education that highlights creativity, and promoting individual creative thinking with a focus on integrating Chinese culture with new world concepts.

Student Works

Hongsheng Gao (CN)

Armorment

This project is interrogative design research into the trend of online social network service platforms and a proposal for speculative interactions which could provide a novel context of communication in the future. The primary interface types used in the proposal are mobile and augmented reality. Prior to proposing a design, extensive critical thinking and research take place regarding user needs, existing social media platforms and technology. These will be reflected in its final form.

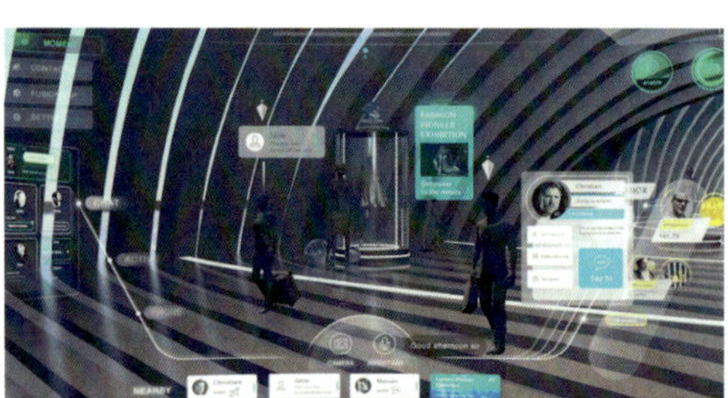

Shuang Liang (CN)

VR Werewolf

The project *VR Werewolf* presents the vision of future mobile gaming. It is a future where the gaming experience could be part of every moment and any place in our usual daily routine. Suggested technology is mix-reality (augmented and virtual reality). Interactions are relocated into the physical environment and stick to the user's body (palm of the hand) to maximize user immersion and embodiment.

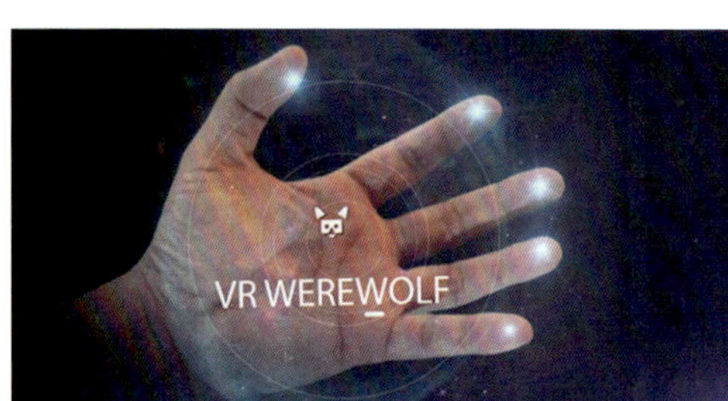

Zhang Hongxiang (CN)

Interaction Design Fiction, the Three Body Problem

This work on the Three-Body Problem is design fiction on how people will read books in the future. The project explores and suggests new interactions, supported by speculative technologies capable of detecting user behavior and emotions and performing multiple searches to find which chapter the user wants to read. Required content is presented in a multidimensional, multisensory and highly interactive manner. The way we read books will never be the same—it will be more interesting.

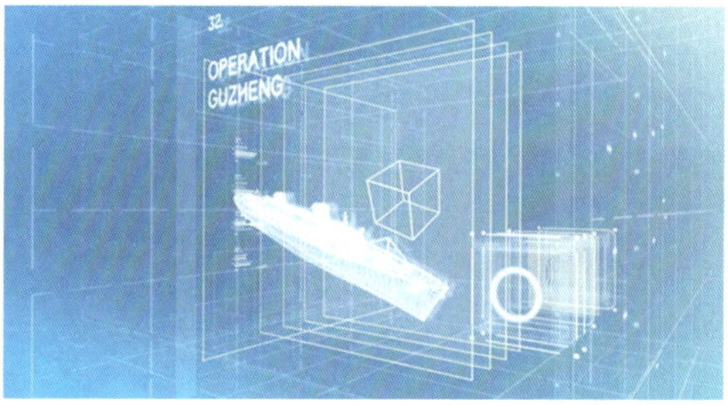

Wang Xin (CN)

The Digital Transformation of Chengdu Museum

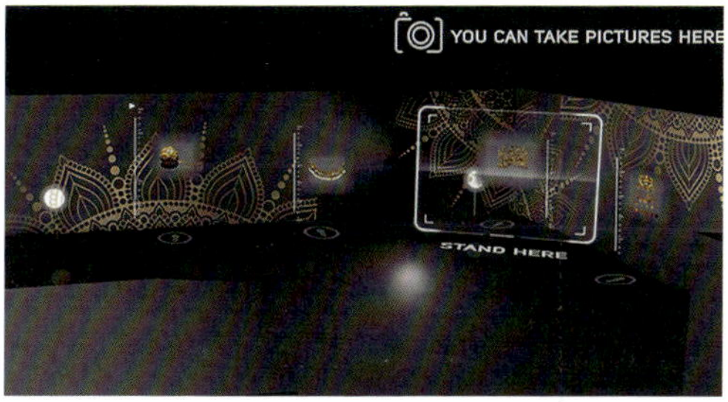

This project uses AR technology to design a new museum experience. The goal was to make Chengdu Museum's content more interactive and engaging, so the audience could become more involved in gathering the information about cultural heritage. The project includes a conceptual design proposal for the three museum exhibitions. The solution is based on using technology to improve audio-visual elements of the exhibition in order to increase visitors' engagement and immersion.

Zhou Pu Fang (CN)

The Future of Unattended Retail Shops

The project is focused on exploring new opportunities and offering design speculation for BingoBox unattended retail shops and how to enrich the future customer experience. The proposed fictional system is focused on the use of intelligent agents, data visualization, and behavioral-pattern recognition. Particular attention is given to the extensive use of self-service convenient shop space to display relevant information and redefine user interactions and engagement.

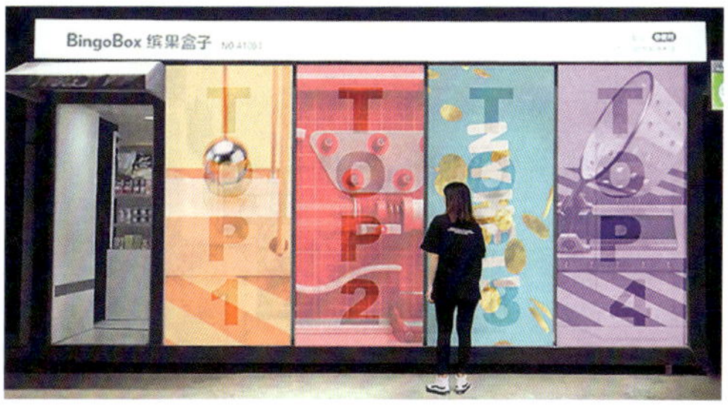

Predrag K. Nikolic (RS)

Art of AI Sense, Robot-Robot Interactions – Syntropic Counterpoints: Robosophy Philosophy

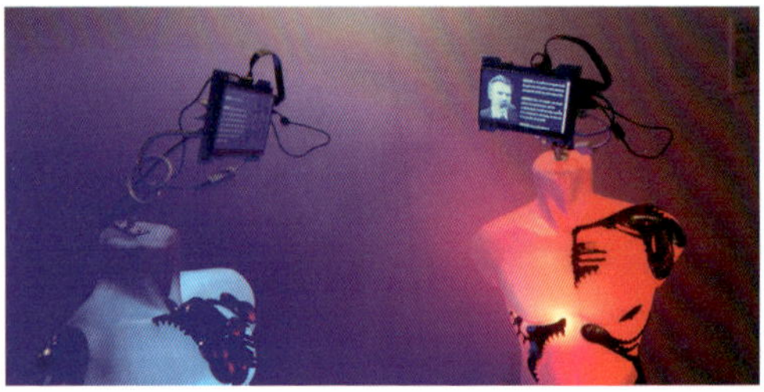

The artwork *Robosophy Philosophy* is part of the project Syntropic Counterpoints which has been conceptualized in the form of series of discussions between artificial intelligence clones, related to topics we want to expose to AI interpretation. We confronted philosophical standpoints between Aristotle (Magnanimous) and Nietzsche (Übermensch) and used their cyber clones to run debates on various topics. The project and research are presented at the SIGGRAPH Asia 2017 and SIGGRAPH 2018.

Predrag K. Nikolic (RS)

Design for Social Innovation – Interactive Installation Before and Beyond

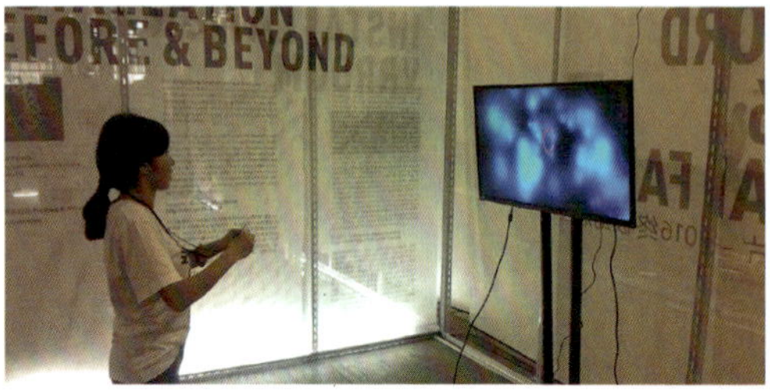

Interactive Installation Before (our existence) & Beyond (our perception) is a multi-sensory interactive metaphorical voyage inspired by String Theory- and socio-emotional relationship development across the distance between human bodies. The intention is to provoke instant reactions from the participants in terms of social behavior and new types of sensory and socio-bodily interactions, which affects users' collaboration and experience. The research is presented at the Smart City 360 Conference 2016.

Yang Hua (CN)

Interactive Cities – Interactive Wall Projection The Light and Shade of Realm

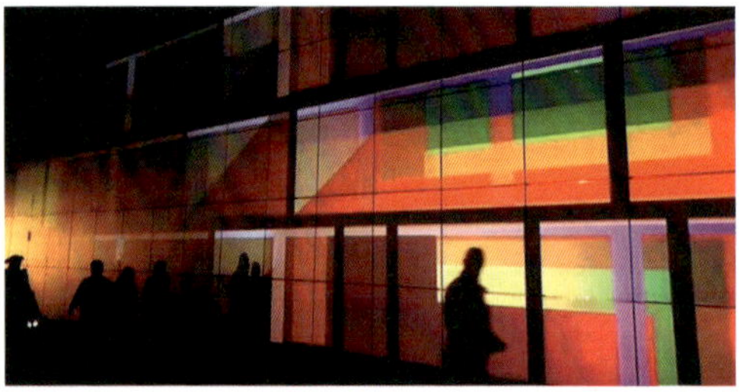

The Light and Shade of Realm is a vertical projection interaction work about light and shadow. Graphics combined with sound presentation are projected on the facade of the building, passing through the shadows and forming an urban interaction scene. With seven flowing colors, such as sunrise and sunset, time transformation, and the sound of four seasons, it shows the infinite charm of life.The work has been exhibited at the gallery of the Museum of Dusseldorf K20 Museum.

Predrag K. Nikolic (RS)

Design for Behavioural Change – Interactive Installation Innerbody

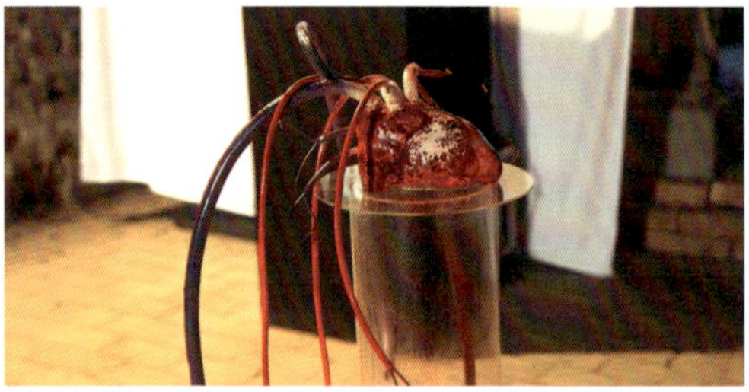

Innerbody is the interactive installation where visitors are invited to interact with the human-heart shaped interface and take a fake medical examination. The visitors experience stressful multisensory environments and life-threatening diagnosis as antecedents of death anxiety to provoke positive consequences such as enhanced life meaning. The research outcomes related to use of multisensory interfaces to change user behavior have been accepted for publication in The Leonardo Journal.

Log. Files. Stories from the Internet of Things

University of Art and Design Linz, Visual Communication

Coordinators: Univ. Prof. Tina Frank (AT), Ass.Prof. DI (FH) Marianne Pührerfellner (AT), Barbara von Rechbach, MA (AT), Mag. David Lechner (AT)

Imagine this is the year 2030. The Internet of Things is in full swing and our world has developed into divergent social groups.
In a two-week workshop, students considered everyday future situations in which networked devices take action, depending on different scenarios about data control and ownership. We wanted to know how the devices communicate with their users in order to make decisions of an action comprehensible and to reveal the hidden lives of IoT black boxes. The log files reveal possible connections with privacy, transparency and participation. Each individual project provides a lens to examine the social, ethical, and aesthetic implications of the interdependence of society and technology in the future era of connected things. The project is a showcase of connected objects in different scenarios, dealing with possible, preferable and undesirable futures. The projects investigate the hidden values and needs of artefacts and propose an alternative visual narration.

Marco Langguth (DE), Eva-Maria Schitter (AT)

SAFETYLOG

The installation *Safetylog* visualizes the epicenter of a safety dome scenario, where vast amounts of digital data are detected, collected and converted into analogue, action-orientated information messages. We have chosen a scenario with a small but skillful intruder: robotic locusts.

Barbara Oppelt (AT), Sophia Wäger (AT)

HEALTH SCAN 8.0

We welcome anyone who is not a threat to us. In order to protect the values of our natural world, a body scan must be carried out before entering our country. *HEALTH SCAN 8.0* analyzes the mental and physical health of the prospective visitor. If there are too many negative results, they cannot enter our world.

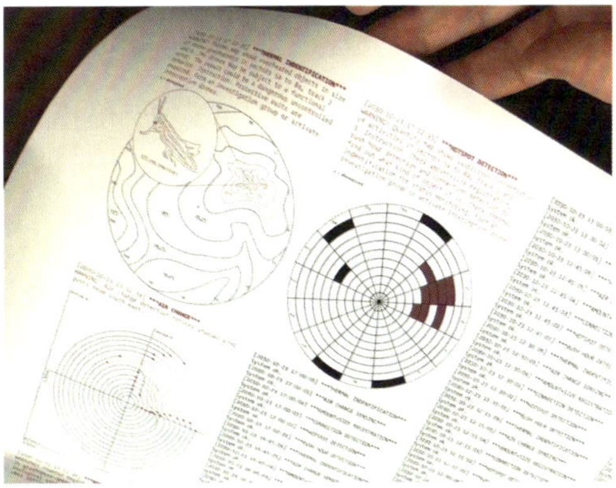

SAFETYLOG

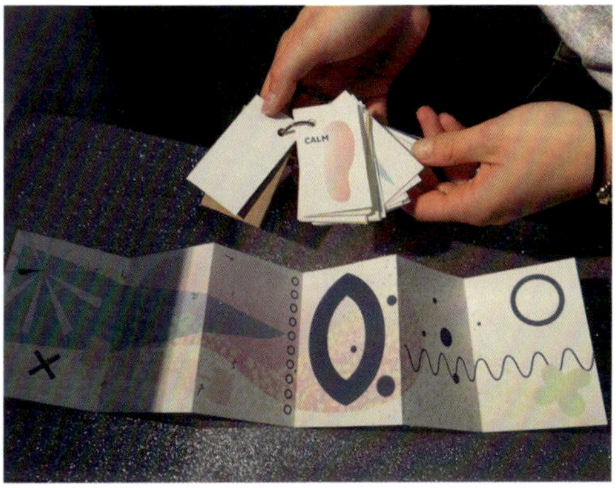

HEALTH SCAN 8.0

Viskom

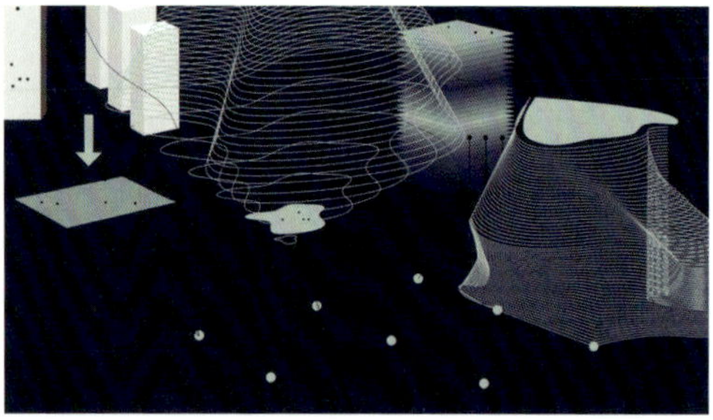

THE BED

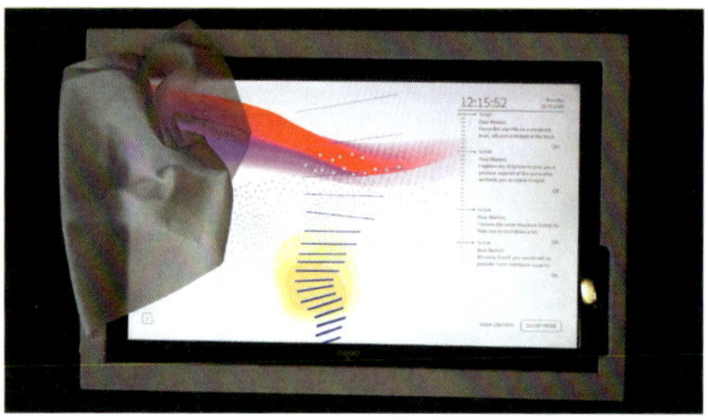

MOODIE

Maria Fröhlich (AT), Theresa Korherr (AT), Teresa Schuh (AT), Sonja Thoms (DE)

THE BED

In our world where technology is accessible in abundance for everyone, economy and politics are being driven by individual customer choices. Every object turns into an intelligent assistant for your daily needs and wishes. Even your bed collects, processes and transmits your personal data for the best possible support. It communicates with other objects, organizes your bedtime routine and even makes decisions for you.

Enikö Gál (HU), Daniela Poschauko (AT), Julia Singer (AT)

MOODIE

MOODIE gives its wearer ultimate support in every situation by constantly changing its material qualities based on interpretations of collected data. Its surface and interlining layers change their physical and tangible qualities according to the user's needs, based on data measured on the body, skin and outside parameters like temperature and air quality. *MOODIE* pursues the perfect comfort for your body, as a ready-to-wear fashion statement.

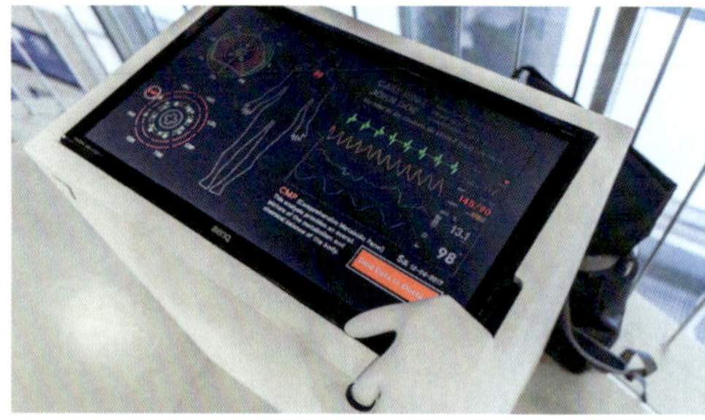

CARE RING

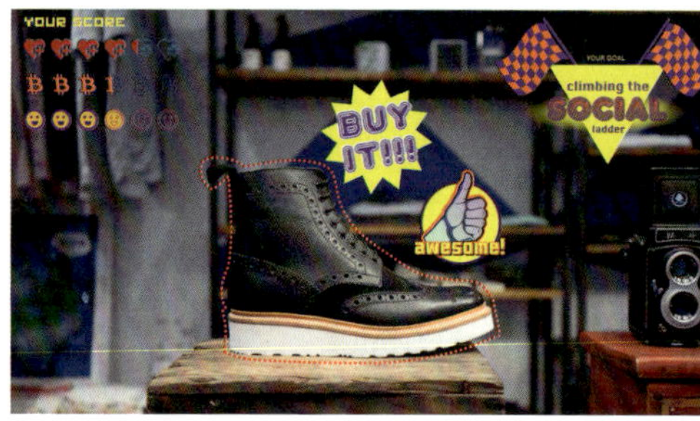

THE TOKEN

Ana Dumitrache (RO), Anna Eickhoff (DE), Martina Jäger (AT), Hana Oprešnik (AT), Annette Valcic (AT)

CARE RING

Care Ring is a conceptual exploration of healthcare in the future: A ring, unique to its wearer, gathers not only information like blood pressure, body temperature and more, but also becomes the direct connection between the customer and their healthcare provider. Data is always stored locally on *Care Ring*, until the moment the user gives permission to transfer information to medical personnel.

Marjan Moradhasel (IR), Daniel Huber (AT), Lisa-Marie Witting (AT)

THE TOKEN

In a world where 1 % of the population holds all the power, we introduce a device which is apparently part of consumer choice but allows total control of the people. *The Token* not only categorizes and ranks its human user, it is the only authorized method of payment and identification. Rating and saving everything you encounter, it tells you what to buy, what to eat and where to go, making individual choice obsolete.

splace magazine

University of Art and Design Linz, Visual Communication

Univ.Prof. Tina Frank (AT), Ass.Prof. DI (FH) Marianne Pührerfellner (AT), Sabine Kienzer (AT), splace magazine is made by lecturers and students of the Kunstuniversität Linz

With *splace magazine*, the Kunstuniversität Linz presents a digital multi-format publication for interactive art encounters in space, word and image with text contributions by teachers and artistic works by students.

The bilingual magazine takes a position in the current discourse on art, culture and society and documents the current art scene and the artistic vitality of this educational institution. As a presentation platform and art mediation format, *splace magazine* focuses on the passion of its artists and authors. The timeless setting of the white \ splace gallery gives the young artists space for a permanent exhibition that can be visited any time, from anywhere, and that communicates contemporaneity in multimedia form. All projects are related to the thematic focus of the particular issue and show artists who express an individual aspect of the relevant phenomena in their medium.

www.splace-magazine.at
Visual Communication / Kunstuniversität Linz
In cooperation with Department of Telecooperation/Johannes Kepler University Linz

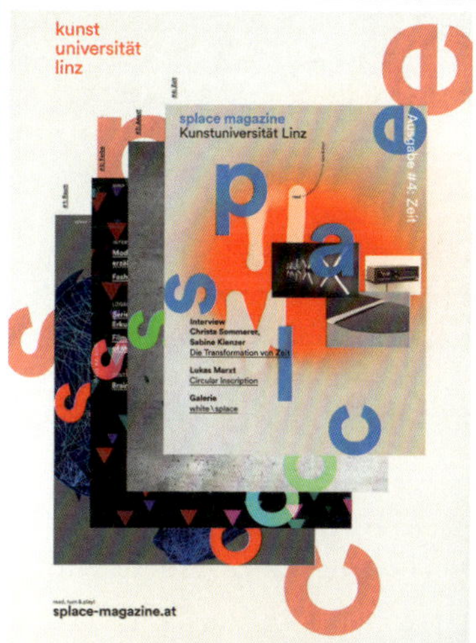

Tina Frank

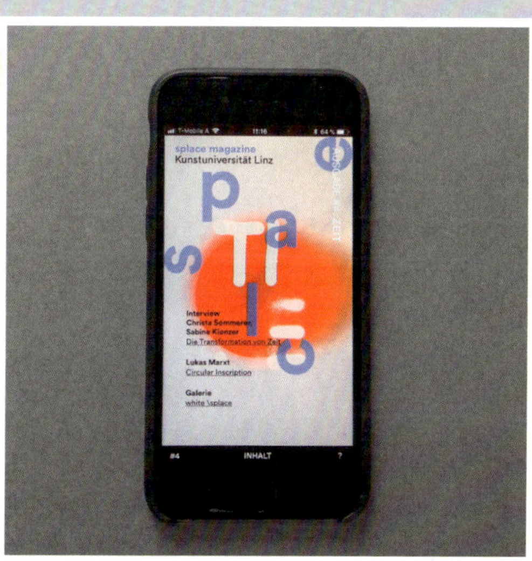

splace magazine Issue #4: Time, © Tina Frank

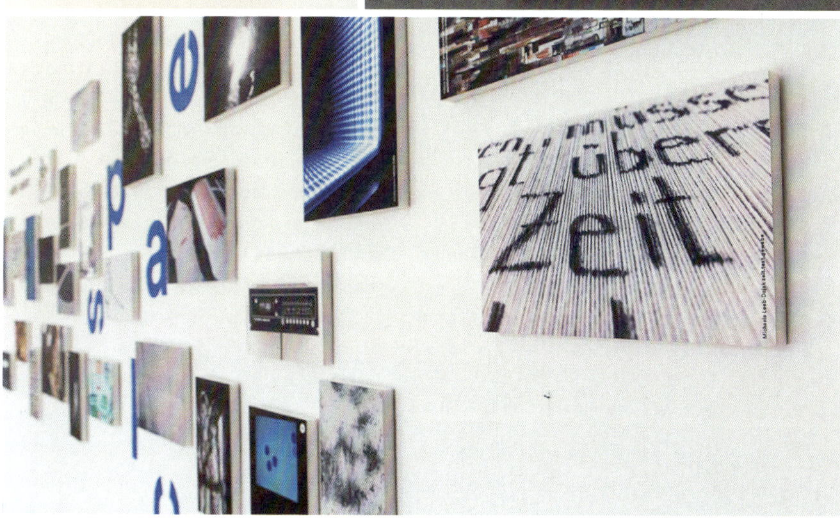

splace magazine Issue #4: Time, © Micha Gerersdorfer

Design Fiction

New Design University St. Pölten, Graphic- and Information Design

Professor: Barbara von Rechbach

Design Fiction is critical speculation about different possible and preferable futures in media design using methodologies of science-fiction prototyping and design research.
Students are examining possible scenarios for developing interfaces in a near future. They work within story laboratories investigating the "archeology of the present" (William Gibson), testing utopian and anti-utopian settings.
Which future is possible, which probable, and what can we actually imagine and visualize? Design is an expression and representation of its users' inherent hopes, projections and fears – Design Fiction as design method allows us to detect the hidden issues in media objects and their different usages.

Nora Reiter (AT), Andrea Sommerauer (AT)

Silver Drop

A tool for sending messages to space, as a time capsule with images and emotions from earth. Extraterrestrial life will experience information differently and can see and hear with tactile senses. Our project investigates the messages people would send from earth, representing human life on our planet with images and feelings #silverdrop #silverdropmessages.

Alexander Jestl (AT)

Blister

Speculative design in pharmaceuticals is used in an investigation about packaging waste and overproduction of medicines. Three different visions present a plausible, a purely fictional and a preferable near-future approach: 3-D printing, green utopias and smart home systems are sources of inspiration. The exhibition object, a blister that disturbs the familiar image through deviant forms, becomes a symbol of new visions of healthcare futures.

Marina Dragicevic (AT), Doris Sutrich (AT)

Kungfufi

Kungfufi is a dystopian design project about a protection system for a safer world. It provides different ways to protect and save humans, and plays with the concept of fear and victimization in society. Every human is attached to a choker device around their necks—they are being monitored 24/7, their whole life. Is the person next to you a criminal—or just a stranger? And who is in control of the classification and algorithms?

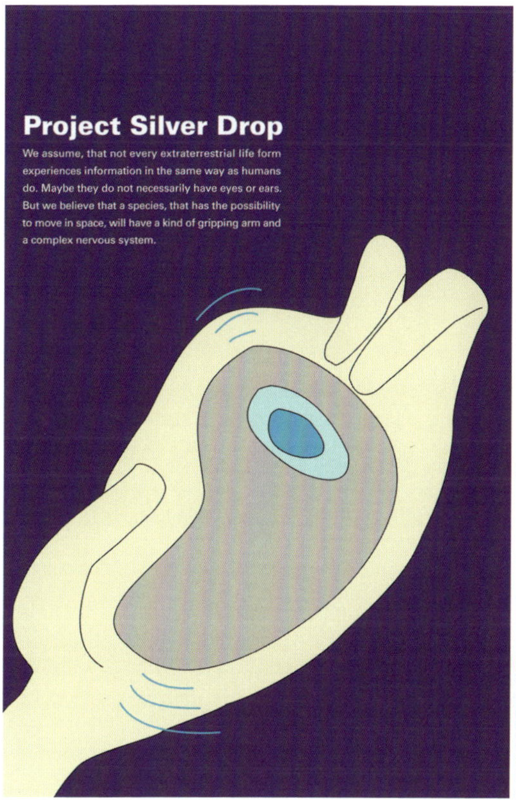

Silver Drop

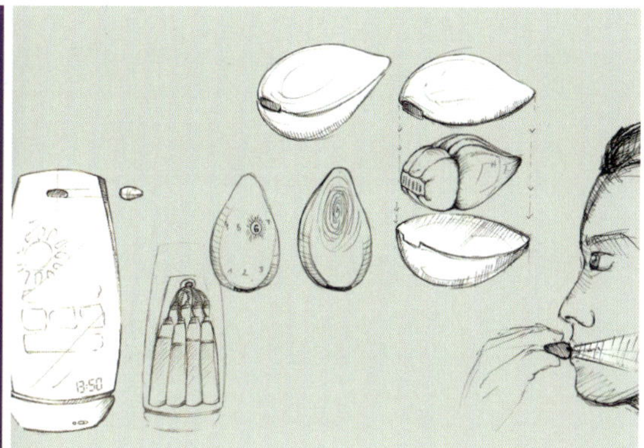

Blister

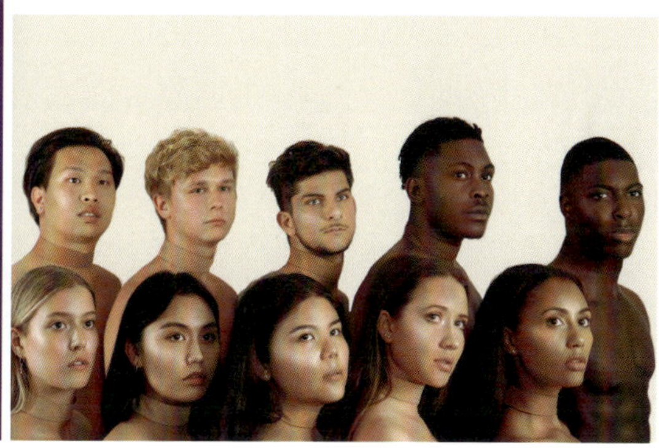

Kungfufi

ERROR

the Art of Imperfection

ARS ELECTRONICA ANIMATION

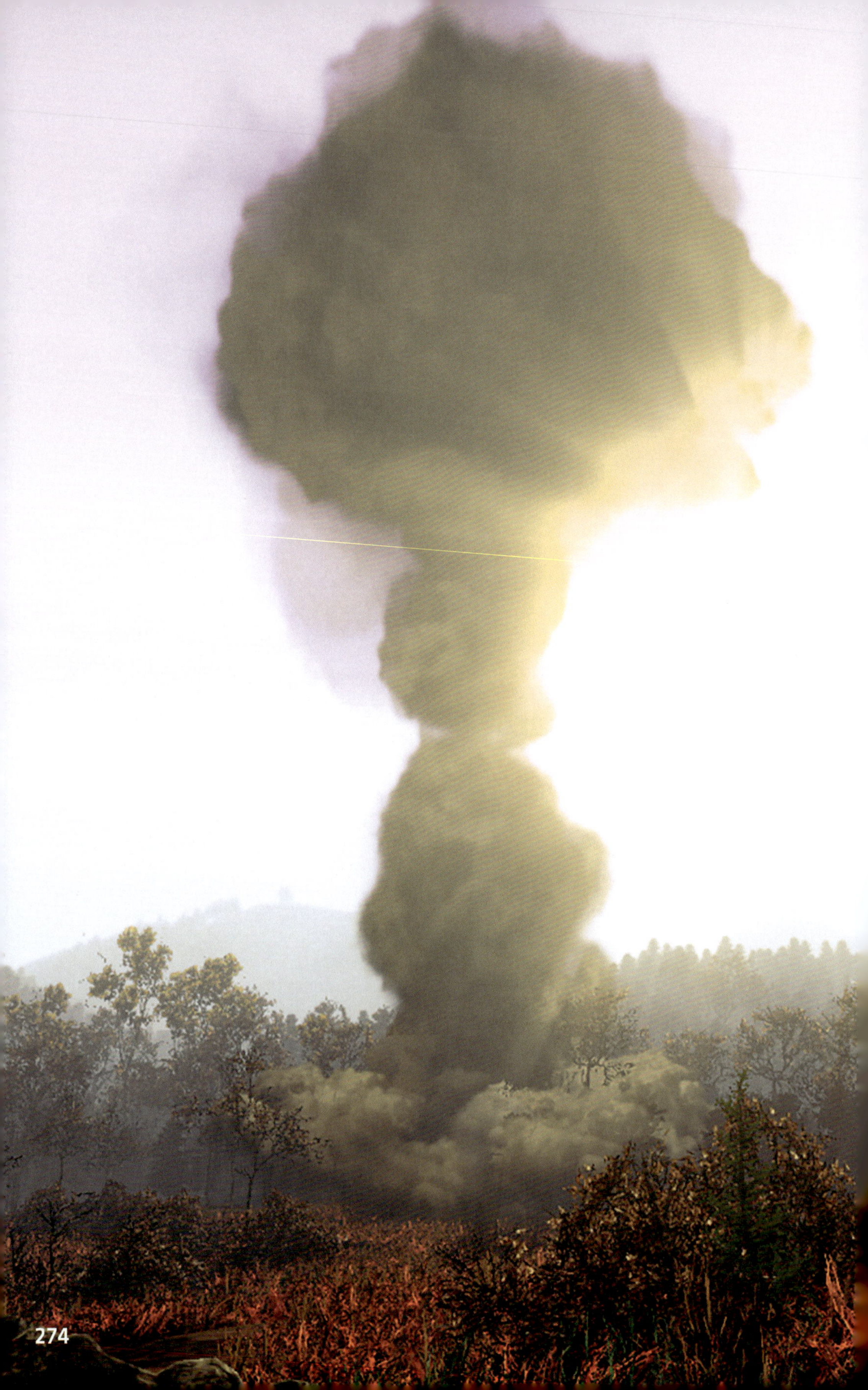

Ars Electronica ANIMATION FESTIVAL 2018

220 works selected from among this year's 1,007 submissions to the Prix Ars Electronica's Computer Animation category constitute the lineup of the 2018 Ars Electronica Animation Festival.

Even before the jury convened in April for three days of deliberations, the Animation Festival's curators, Jürgen Hagler and Christine Schöpf, screened the entries and noted their favorites. The 350 works that made it past this first cut were then put online for evaluation by jurors Gaëlle Denis, Alex Verhaest, Pokras Lampas, Casey Reas and Jonathan Yomayuza. Then, in April, the remaining 220 shortlisted works were presented to the jury for their final decisions. And these 220 films also formed the basis for the programming of this year's Ars Electronica Animation Festival.

These 10 programs provide a representative overview of what's happening now in the digital motion picture genre worldwide; at the same time, each individual lineup manifests specific strategies and unique points of view. Many of the works being shown have long since taken leave of the picture screen. Substantial price reductions in recent years have made VR hardware much more affordable and thus an increasingly viable option for artistic work. As a result, audience interaction with audiovisual experiences constitutes a new challenge in its own right. Jumbo-format mappings on buildings and landscapes as well as in a museum context and installations are examples of expanded animation that transcend the confines of the picture screen. Of course, a large proportion of the works of this year's program are once again short films—both narrative and experimental-abstract works, produced by individuals or entire crews. The spectrum of techniques ranges from stop-motion to 100% computerization.

The program is being supplemented by a Young Animation lineup featuring films from the Prix Ars Electronica's u19 – CREATE YOUR WORLD category for young people under 19 years of age in Austria, the Japan Media Arts Festival, ISCA—the International Students Creative Award, Digital Media Hagenberg Campus, Filmakademie's Animationsinstitut, Animafest Zagreb and Best of Austrian Animation 2011–2017. The 2018 Ars Electronica Animation Festival is a highly diversified showcase that invites spectators to take a fascinating 'round-the-world journey through current digital filmmaking.

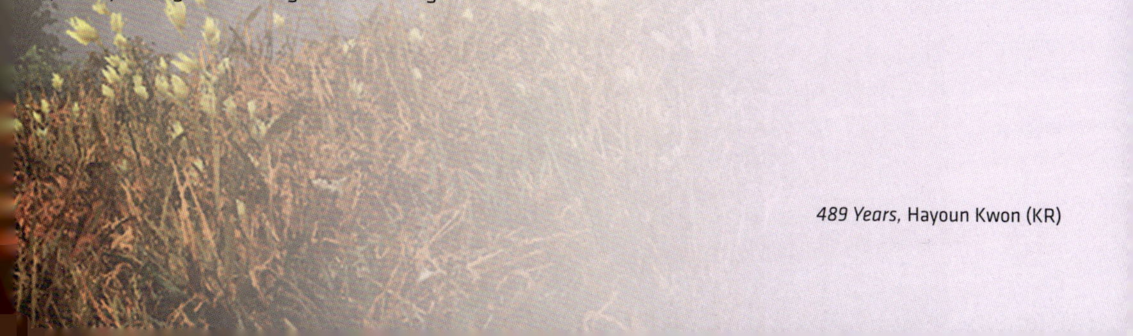

489 Years, Hayoun Kwon (KR)

IN PERSONA: Boris Labbé

Born in 1987 in Lannemezan, France, Boris Labbé is an artist and animation film director working between France and Spain. His art comes through several forms: short films, audiovisual concerts and video installations. He has collaborated with Sacrebleu Productions since 2013. His short film *Rhizome* won the Golden Nica Animation at Prix Ars Electronica in 2016 and his last project *The Fall* was selected in special screenings at the 57th Semaine de la Critique, Cannes Festival. The retrospective screening will present the last four animated projects directed by Boris Labbé, all made during the last four years.

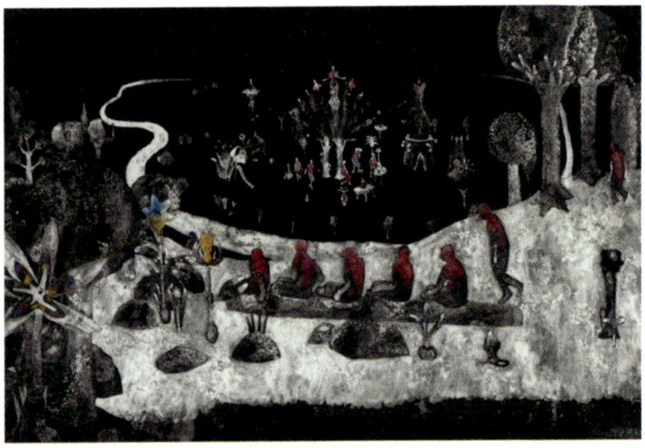

La Chute, Boris Labbé (FR)

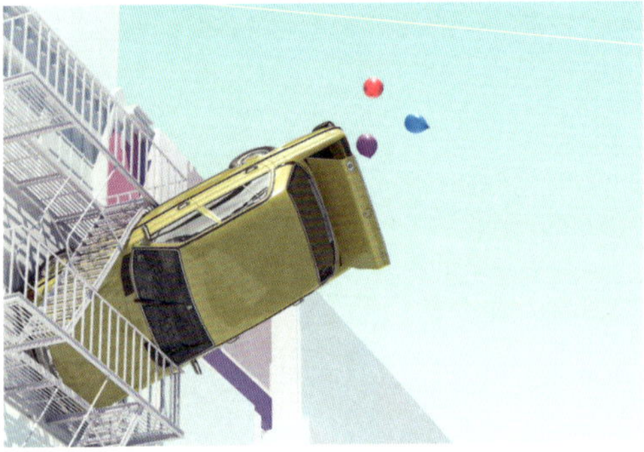

AUTO NOM, Zeitguised / foam Studio (DE)

IN PERSONA: ZEITGUISED / foam Studio

ZEITGUISED is an award-winning studio established in 2001 by Jamie Raap and Henrik Mauler that produces exquisite realities, at the intersection where art and design meet digital and physical space. Their work has been presented in numerous international New Media art fairs and festivals like Oberhausen, Ars Electronica, Onedotzero, Dotmov, Nemo, ITFS and Resfest. Selected work has been shown in galleries and art shows around the world, most notably Volta Art Fair New York and the Museum of Contemporary Art in Barcelona. *foam Studio* is the commercial service mode of *ZEITGUISED*.

Electronic Theatre

This is the Ars Electronica Animation Festival's annual best-of program—a compilation of the 13 outstanding works chosen by the jury from among the 1,007 animated films submitted this year for Prix Ars Electronica prize consideration. At the same time, Electronic Theatre is a showcase of the latest trends in an artistic-substantive sense and with respect to technology and innovation.

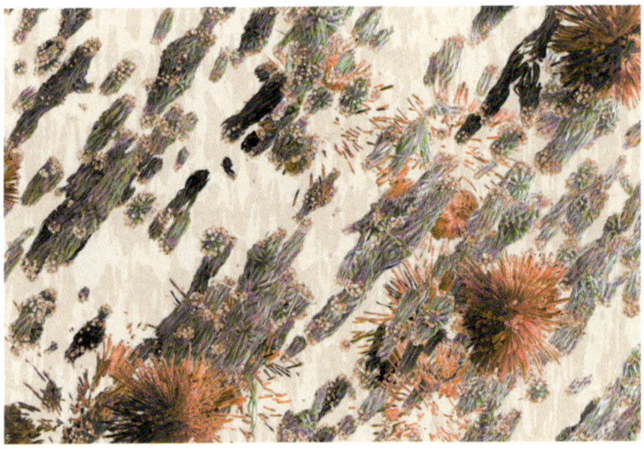

Quantum Fluctuations, Markos Kay (CY/UK)

Expanded Animation

This is an overview of new pathways that digital filmmakers are exploring and setting out on—mappings on a natural landscape and on the human body, "an ode to the rhythm of being," robotic installations, kinetic works, media façades and VR projects.

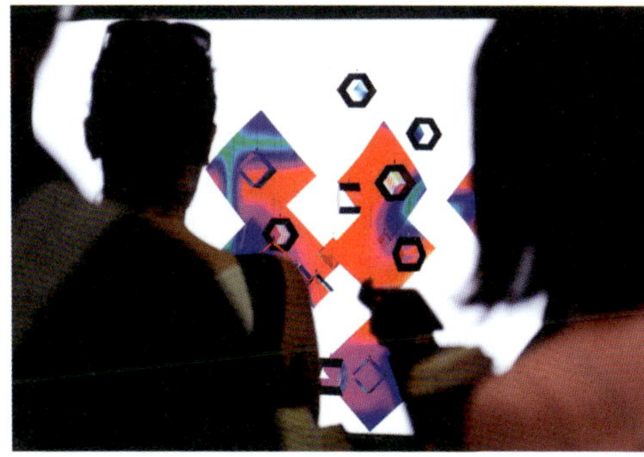

display(bias), Ryo Kishi (JP)

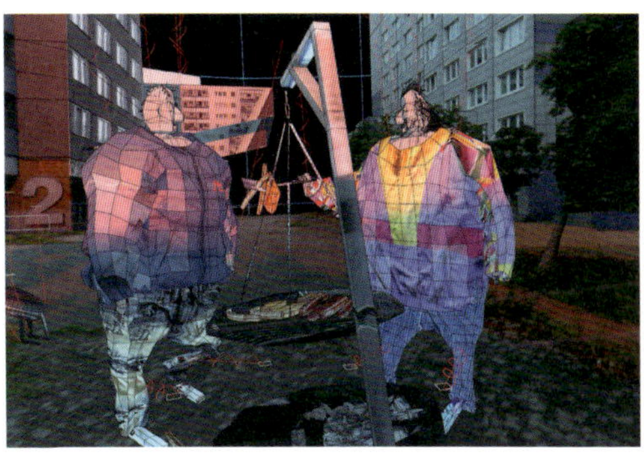

Fest, Nikita Diakur (DE)

Experimental

Man's fall from grace rendered in impressive images that allude to works by Hieronymus Bosch and Francisco Goya; a sci-fi video illustrating a post-apocalyptic multiverse; emotions contravening the laws of nature; the symbiotic relationship nematodes (roundworms) form with Wolbachia bacteria—*Experimental* brings together highly divergent points of view.

Micro Macro

A visual and conceptual connection between the macroscopic brain and the microscopic behavior of neurons envisioned with data, drawings and various other techniques; the complexity of the world of quanta, the observation of the world as it doesn't exist; a journey to distant horizons of our cosmos; or, simply, the story of an ant—these are some illustrative examples of Micro Macro.

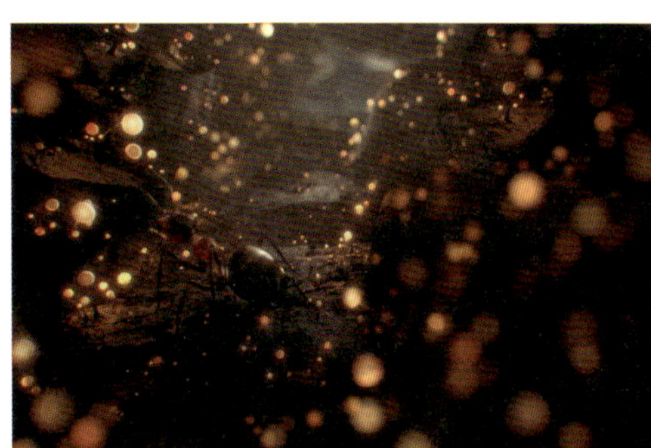

Overrun, Jérémie Cottard, Antonin Derory, Matthieu Druaud, Pierre Ropars, Diane Thirault, Adrien Zumbihl (FR)

Music & Visuals

This isn't just a showcase of classic music videos; it's a compilation of various approaches to bringing together sound and visuals—for instance, an immersive 360° VR array or software programs that translate music into visuals. But the music video genre doesn't get short shrift either!

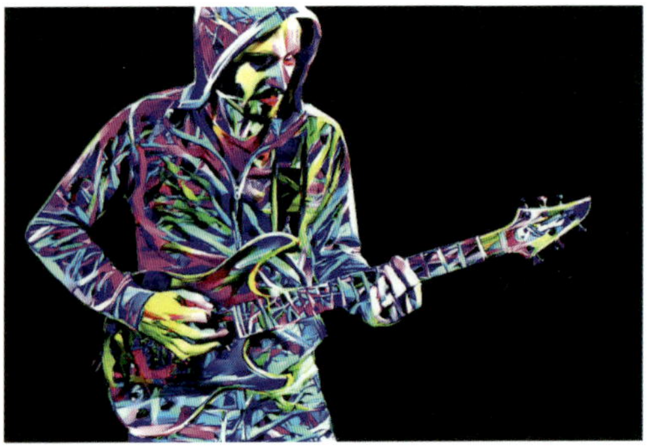

Hardcore Anal Hydrogen, Jean Pierre, Martyn Clement, Sacha Vanony (FR)

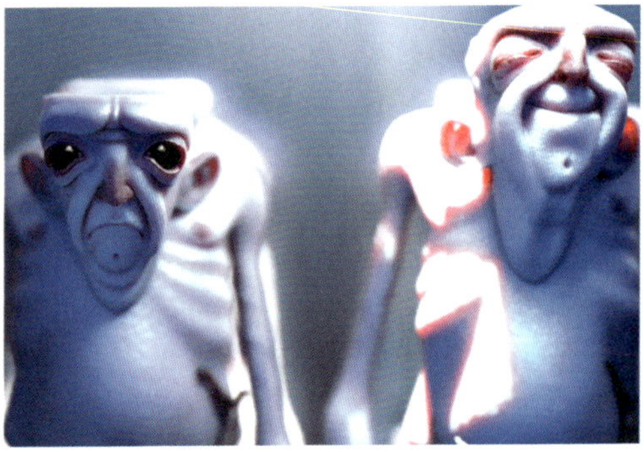

The Box, Dušan Kastelic, Mateja Starič (SI)

Narration

Storytelling in the classical sense is a form of depiction. Digital narration goes beyond this definition. These condensed-associative accounts get across in images what can hardly be expressed in words. Many of the works in this program are personal stories and recollections.

Female Worlds

This program shows female worlds from a variety of perspectives. An 8-year-old Turkish girl looks back at her happy past; meanwhile, yawning abysses emerge. Amidst her everyday routine, a woman discovers something unexpected in her house. A young woman who's home alone wants to spoil herself a little bit, but that doesn't quite go according to plan. And plastic surgery is another one of this program's topics.

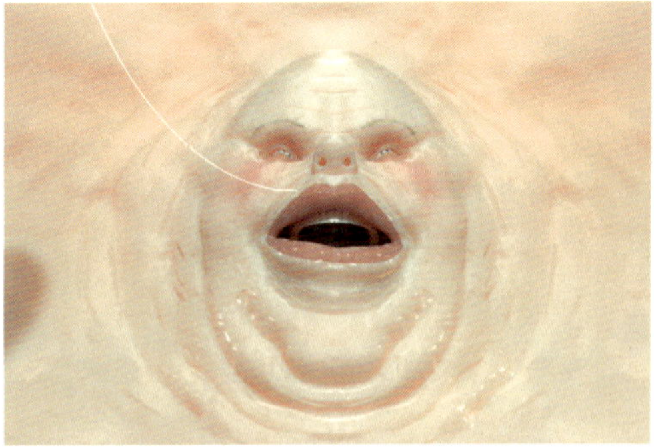

Caterpillarplasty, David Barlow-Krelina (CA)

Mental States

A group seated at a table in a pub is chatting and drinking beer, when one of them begins analyzing his emotions. A father recalls a missed chance to talk with his daughter; a brief meeting years later makes all words meaningless. A path leads through a sunken labyrinth of memories and dreams. These are examples of mental states.

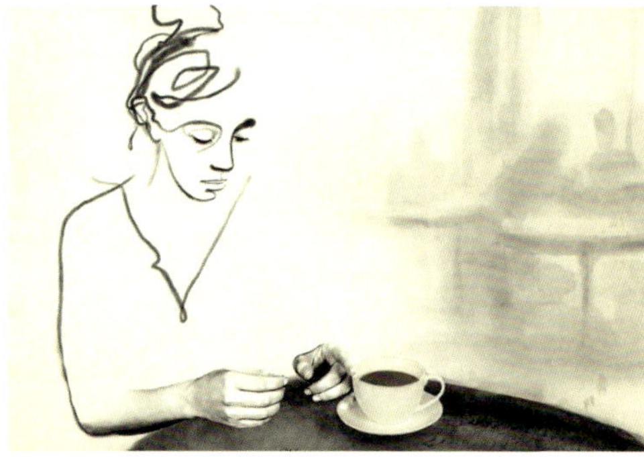

In Other Words, Tal Kantor (IL)

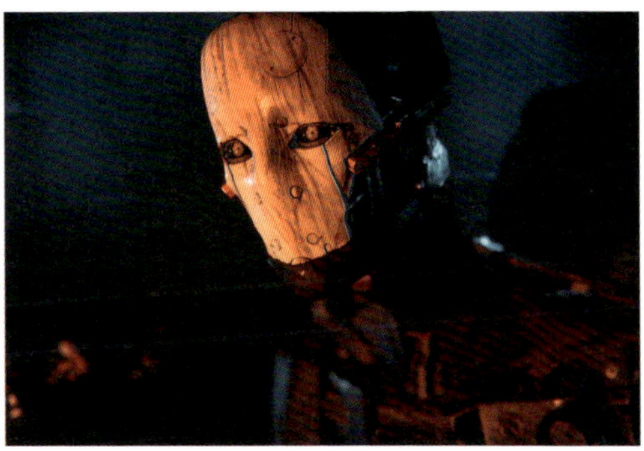

Adam Episodes, Neill Blomkamp (CA/ZA)

Statement

The demilitarized zone between North & South Korea as VR work, a computer game featuring Merkel and Schulz as the chief protagonists, Brexit in the UK, and a female soldier's battle against slavery—as this program amply illustrates, more and more digital activists are addressing political issues.

Young Animations

Gifted young filmmakers annually submit their witty, off-beat, subtle, tragic and dead-serious works for prize consideration to the Prix Ars Electronica's u19 – CREATE YOUR WORLD category (AT), bugnplay.ch (CH), mb21 (DE) and C3<19 (HU). The greatest hits are featured in Young Animations.

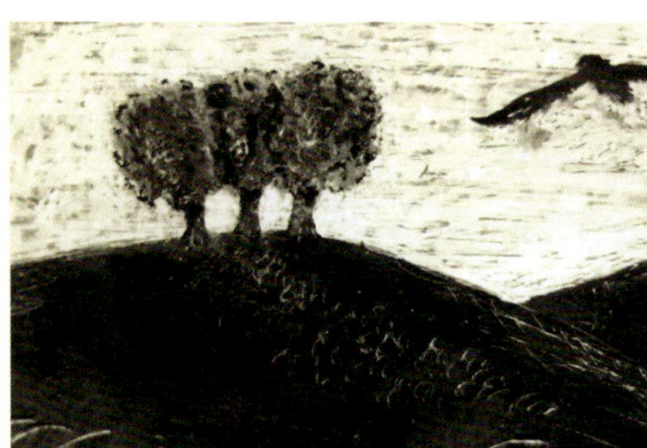

Figura Immagina, Viktoria Hörndler (AT)

Digital Media, Hagenberg Campus

A deer and a hunter are trapped in a surreal clockwork, a woman gets lost in a roundabout, and a gorilla eavesdrops on the conversation of two meerkats. This program features a selection of recent student works, ranging from narrative shorts to experimental animations, from the Digital Media Department at the University of Applied Sciences Upper Austria in Hagenberg.

ISCA (The International Students Creative Award)

The International Students Creative Award (ISCA) is an international arts and information media competition for university, graduate school, and vocational school students from Japan and other countries worldwide. It is staged under the aegis of the Knowledge Capital Association.

Japan Media Arts Festival 2018

The Japan Media Arts Festival honors outstanding work in a wide variety of media in four categories: art, entertainment, animation and manga. This program consists of 11 excellent animated films singled out for recognition by the 2018 Japan Media Arts Festival.

Animationsinstitut of the Film Academy Baden-Württemberg

Students at the Animation Institute can develop their own personal style, which then serves as a basis for the implementation of innovative ideas. This screening showcases the tremendous diversity of student projects—including 2-D and 3-D animation and VFX.

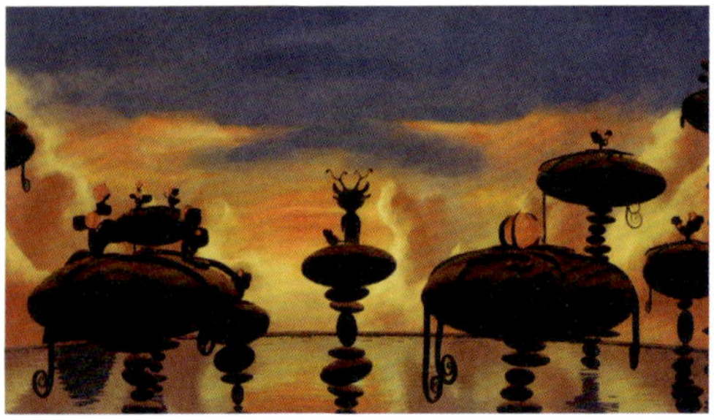

Iruh, Niko Frenkenberger (AT), Alexander Gassner (AT), Moritz Rührlinger (AT)

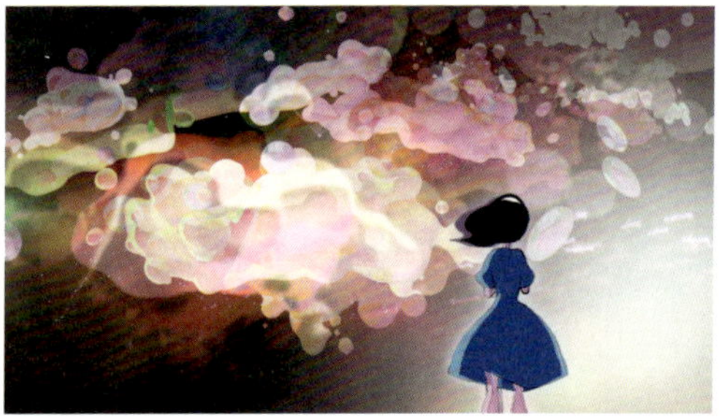

SHARK, Yuri Saito (JP)

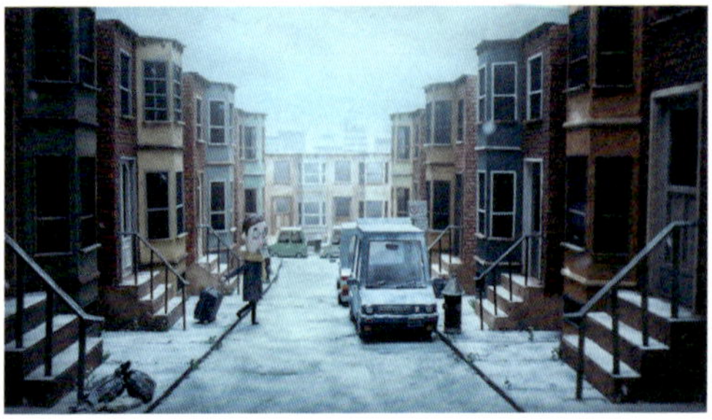

Negative Space, Max Porter (FR), Ru Kuwahata (JP)

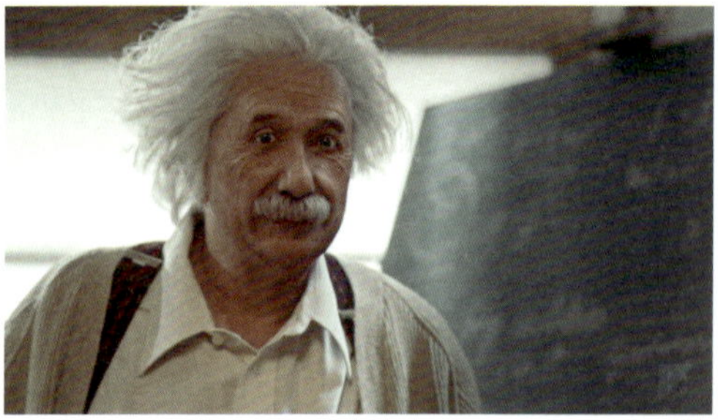

Digital Actor_Albert Einstein, Leszek Plichta (DE)

Jürgen Hagler (AT)

Expanded Animation 2018

INTERFACES IN MOTION

The 6th Expanded Animation symposium carries on a process launched in 2013—mapping the wide-ranging domain of animated imagery beyond the well-trodden paths. The symposium stays the course that was set at its inception, and presents theoretical positions and perspectives from the art world, the R&D field and the industrial sector. The mission: To function as a driving force advancing an interdisciplinary discourse. This year's symposium is an inquiry into future interfaces in animation. *Interfaces in Motion* will focus on animation technology at the manifold interfaces where humans, computers and interaction meet. The emergence in recent years of affordable technologies—e.g. in the areas of virtual production and mixed reality—create openings for new forms of interaction with moving pictures. Tools and digital devices such as head-mounted displays and mixed-reality glasses increasingly call into question conventional workflows and forms of presentation. How will future animated films and hybrid forms consisting of filmed sequences and digital content be produced with and for these new technologies? How will these interfaces change the way digital motion pictures are presented and received, and what role does the audience play in this? This symposium is an effort to come up with answers and present approaches from the art world, the R&D field and the industrial sector. Kicking off the symposium is the Prix Forum, in which honorees in the 2018 Prix Ars Electronica's Computer Animation category present their prizewinning works and discuss current trends. This will be followed by several panel discussions covering various aspects of "Interfaces in Motion" with artists, scientists, software developers and studios working in the fields of gaming, media design and animation. *The Lacuna Shifts*, the latest VR application by the artist duo named *DEPART*, addresses the zone where the real world adjoins the virtual one. Right at the outset of this first-person experience, those partaking of a VR environment find themselves amidst a mirror-image of the real, outside world, and set off on a personalized journey inspired by Lewis Carroll's *Alice in Wonderland*. Hannes Rall will elaborate on a current research project that illustrates how conventional animated productions can be expanded by AVR technology. James Paterson presents *Norman*, animation software that received a 2018 Honorary Mention for making it possible to bring to life in a VR environment the traces of movements left behind in a physical space. Volker Helzle will offer insights into the latest research in virtual production, a complex field of postproduction in which the real and the virtual increasingly merge. Gerhard Funk's presentation will showcase interactive works subsumed under the title *Cooperative Aesthetics* that were developed for Deep Space 8K, and show how their audience's collective movements can generate an animated world of imagery. Artistic experiments in the field of stereoscopic animation—from 3-D noise to binocular rivalry—will be the subject of Rainer Kohlberger's presentation. Chunning (Maggie) Guo's talk is entitled *Re-evolution as the Interface in Generative Animation*. In *The Dematerialization of Animation*, Birgitta Hosea will discuss works by artists in the expanded cinema movement. Then Sophie Mopps reports on VR applications in medicine as well as artistic interventions in this field. Rounding out the symposium will be making-of presentations about prize-winning and well-received works in the field of applied art. *Broken Rules*, an Austrian indie game studio, will give a speech about the emotional design of the prizewinning computer game *Old Man's Journey*. Talks by *Melt* and *FIELD.io*, two design studios doing interdisciplinary work, and the *ZEITGUISED / foam* animation studio will offer insights into the creative process—from VR experiences to large-format audiovisual projections. Plus, many of the presented works will be shown in the Ars Electronica Animation Screening and in VRLab and Deep Space 8K at the Ars Electronica Center.

Participation in the symposium is free of charge for all registered participants and holders of a 2018 Ars Electronica Festival Pass. To register and obtain additional information, go to *www.expandedanimation.com*.

Speakers at the 2018 Expanded Animation symposium: 2018 Prix Ars Electronica winners in the Computer Animation category: Broken Rules—Felix Bohatsch (AT), DEPART—Leonhard Lass & Gregor Ladenhauf (AT), FIELD.io—Vera-Maria Glahn (DE), Gerhard Funk (AT), Chunning (Maggie) Guo (CN), Volker Helzle (DE), Birgitta Hosea (UK), Rainer Kohlberger (AT), Melt—Kamila Staszczyszyn & Kuba Matyka (PL), Sophie Mobbs (UK), James Paterson (CA), Hannes Rall (SG), ZEITGUISED / foam Studio—Henrik Mauler (DE).

Organization: Expanded Animation is produced jointly by the Upper Austria University of Applied Sciences' Hagenberg Campus, the Ars Electronica Festival and Central Linz, and organized by Jeremiah Diephuis, Jürgen Hagler, Michael Lankes, Alexander Wilhelm / Upper Austria University of Applied Sciences' Hagenberg Campus / Department of Digital Media.

http://www.expandedanimation.com, http://www.fh-ooe.at

ERROR

the Art of Imperfection

u19 – CREATE YOUR WORLD

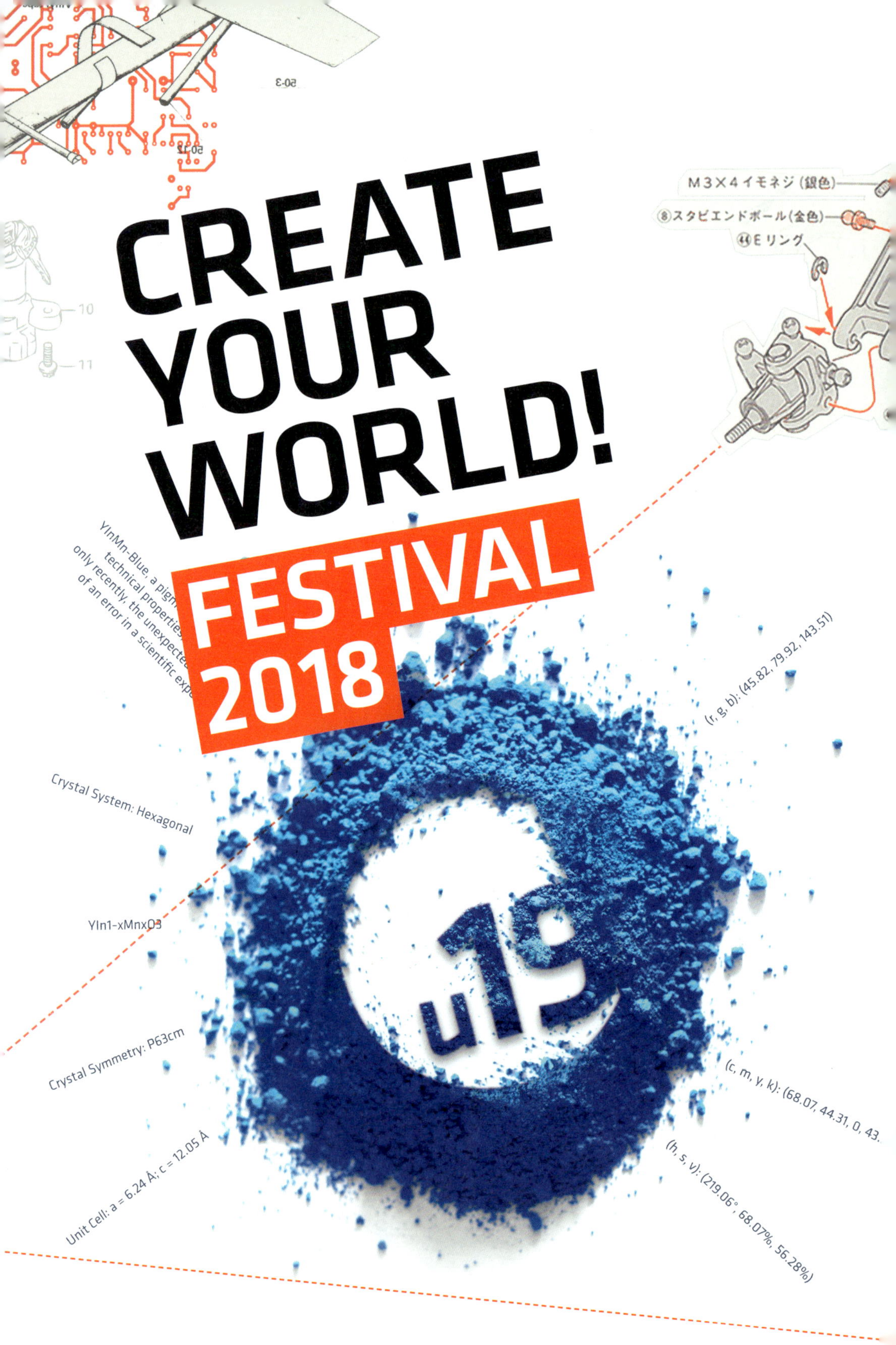
CREATE
YOUR
WORLD!
FESTIVAL
2018
u19
M3×4イモネジ(銀色)
⑧スタビエンドボール(金色)
㊹Eリング
10
11
50-3
50-12
YInMn-Blue, a pigm
technical properties
only recently, the unexpecte
of an error in a scientific exp
(r, g, b): (45.82, 79.92, 143.51)
Crystal System: Hexagonal
YIn1-xMnxO3
Crystal Symmetry: P63cm
(c, m, y, k): (68.07, 44.31, 0, 43.
Unit Cell: a = 6.24 Å; c = 12.05 Å
(h, s, v): (219.06°, 68.07%, 56.28%)

u19 – CREATE YOUR WORLD

Future Festival of the Next Generation 2018

LEARNING

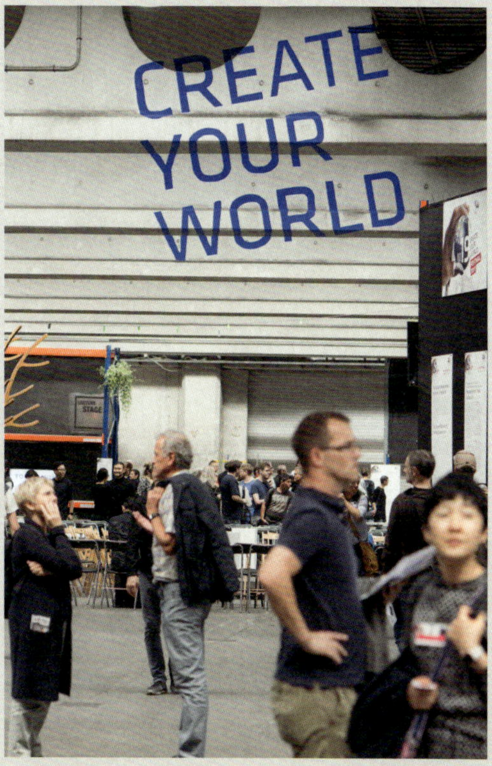

Ronald Posch

This year's festival theme, Error – The Art of Imperfection, epitomizes the ideas and principles of an initiative launched in 2011. Ever since its inception, u19 – CREATE YOUR WORLD has invited young people to experiment, to discover new things, and to try out alternative teaching/learning models. Each year, the festival makes a concerted effort to create a gemütliche atmosphere that provides a nurturing setting for these learning processes. And making mistakes isn't just an unfortunate side-effect; it's an eminently important element of this. Without errors, there would be no subjective realization; all that would remain would be external inputs gradually leading to a state of overload.

And think about it—isn't that actually what we're increasingly confronted with in a seemingly perfected world? Knowledge acquisition is now only a matter of seconds, and thus unfortunately makes less and less of an impression on us. The fact that accurate and true information is available everywhere, all the time is not only a positive development; it suddenly leads to people increasingly calling information into question or, conversely, completely ceasing to question things. And the upshot of this is that our own opinions recede more and more into the background. One no longer dares to pose the question: Could that possibly be wrong? What's my opinion on this?

This evokes insecurity. There arises a sense of distrust towards newly developed technologies. They aren't regarded only as enriching; they might also be seen as impediments to personal growth. Seen from this angle, dealing with errors is a social process of extraordinarily great importance. In social relationships too—for instance, in families and school systems—failing and successfully dealing with it is a learning process. Doesn't that take us back to where we began?

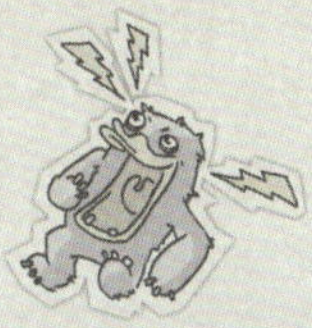

Error as Chance

2 + 2 = 5. This is definitely wrong, isn't it? Nevertheless, defining errors in our life is a tricky undertaking. Lots of bright minds have been forced to concede that differentiating between right and wrong is a tough nut to crack. Ultimately, we decide for ourselves where to draw the dividing line between these two opposites. It's said that what we actually learn are life skills – that is, ways to enrich our existence and make it as productive as possible – and also to develop new things from it, to invent, and to discover new technologies.

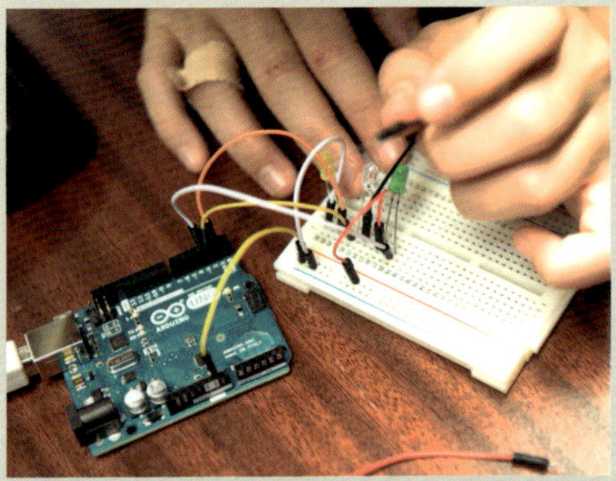

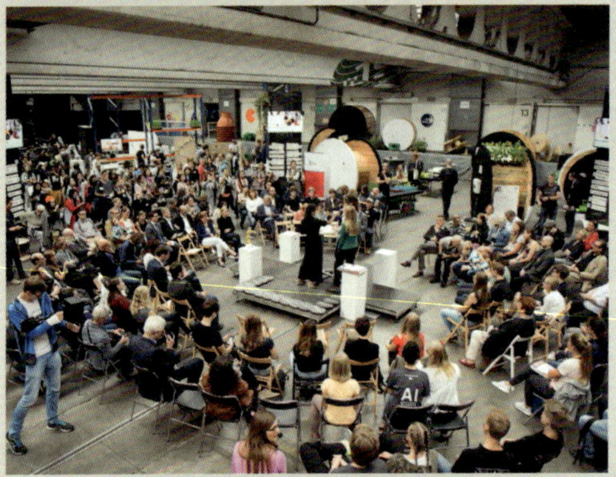

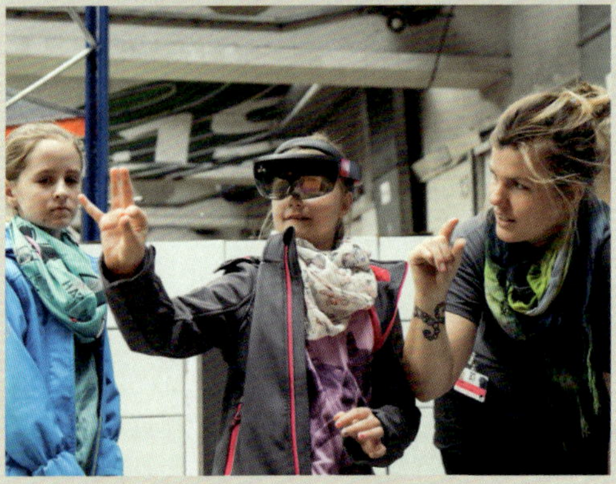

Ronald Posch

TRIAL and ERROR

Maybe what's now called for is more childlike curiosity to characterize these processes of exploring and learning, and perhaps a bit less adult "reasonableness"? How do we deal with these newly developed technologies? How can we gain experience with things repeatedly becoming too much for us to manage? By permitting errors.
That way, we can ascertain whether the course we've embarked on is the right one for us. This enables us to reflect and possibly even to question decisions. And isn't that precisely what many educational institutions call upon students to do? Question authority, be audacious, disobey!
The u19 – CREATE YOUR WORLD learning platform doesn't just tolerate these errors; it positively encourages making them! So, in that spirit, this year's watchwords are: Try out new things; make discoveries; invent the future; and, by all means, make mistakes!

Configure the Future ... and Don't Mind the Rough Edges!

This year's CREATE YOUR WORLD Festival invites festivalgoers to experiment and get hands-on experience. Be among the first to see the unusual, innovative, and very off-beat attractions in a fascinating lineup of open labs and exhibitions in POSTCITY Linz.

Impressions, ideas and projects from throughout the world are arrayed here for your consideration. For five days, you can discover the future and have a say about how it turns out. This is the perfect spot for tinkerers, lateral thinkers and anyone who is thirsty for insights!

Collective Smart TV
2018 Youth Exchange Project

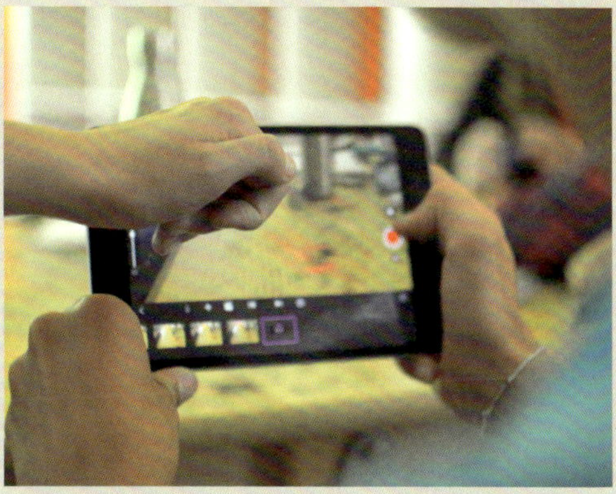
Remo Rauscher

Get acquainted on the fly and then immediately get busy collaborating on a project that has to be completed within five days – the annual Youth Exchange Project has become a mainstay and a highlight of u19 – CREATE YOUR WORLD.
This year's youth exchange project is entitled *Collective Smart TV*. The idea is to question the formats and limits of traditional, old-fashioned TV by bringing together different ways of sharing video content and filming, editing, directing, coding, and – most importantly – storytelling!
The festival is a good base for this project because of all the spectacular events happening there. It's easy to find something worth showing and explaining.
The goal for the five days is to develop an individual way of sharing stories by collecting existing concepts and finding fresh ideas. Along the way we will try and use different video formats, tools and channels to merge together our own hub of *Collective Smart TV*. The recording will be done primarily by smartphones– let's see where that leads us…

c3 (HU), mb21 (DE), bug'n'play (CH), artTechLab (NL), u19 – CREATE YOUR WORLD (AT)

The Error City Lab–We're Building the City of the Future

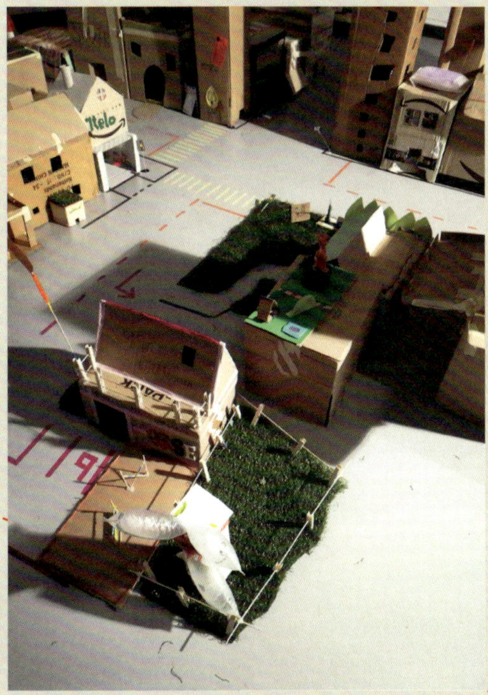

Martin Hollinetz

Cities don't just sprout up. And rarely do they develop according to plan. That's simply because we're human beings with ideas and needs of our own. That makes cities lively and gives us opportunities to co-determine our habitat. Meanwhile, we're now surrounded by new technologies like drones, self-driving vehicles, automatic surveillance systems and other digital developments that will shape our image of the village and the city in the future. Here and there, now and then, we'll encounter errors because our needs change, things simply aren't expedient, or we just need space.
Error City is a growing community. Visitors can develop it on an ongoing basis, electrify and digitize it. This is a place where drones deliver packages and autonomous garbage robots keep the streets clean. Error City isn't just a conceptual model; it's built of cartons and can be modified at will by the developers. The cityscape can be navigated by means of 360° cameras and robots; afterwards, urban denizens can experience these trips along with the feeling of being right in the middle of the action.

Otelo (AT)

food culture experiments

Exploration space @ ACDH-ÖAW focuses on experimentation and open innovation. Set within the liquid, participatory concept of the post-dictionary, various interactions and experiments, both digital and virtual, are offered. Artists, practitioners and scientists aim to co-design and take the participants on a shared journey into intimate dimensions of food as part of European cultural heritage as well as European blended culture, touching all senses by various experiences.

Eveline Wandl-Vogt (AT), Kim Albrecht (US/DE), Matthew Battles (US), Penesta Dika (AT), Nikhil Krishna Dharmaraj (US), Amelie Dorn (AT), Keith Hartwig (US), Sarah Newman (US), Enric Senabre Hidalgo (ES), Barbara Piringer (AT), Jose Luis Preza Diaz (AT), Thomas Palfinger (AT), Tuuli Maria Utriainen (CH), in collaboration with Community Cooking (AT), Sprungbrett (AT), Topothek (AT) and Wiener Tafel (AT)

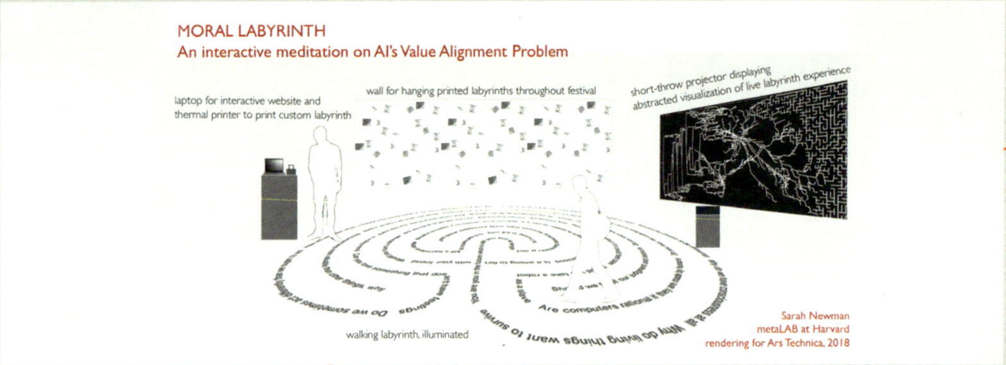

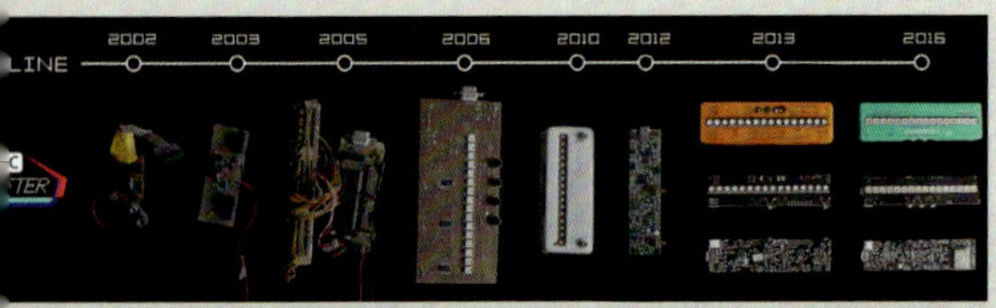

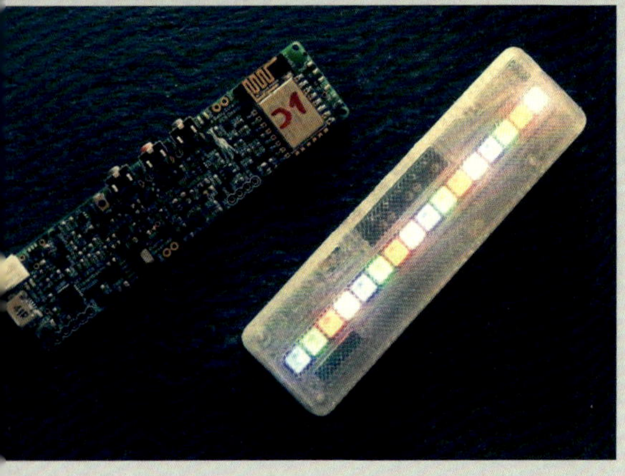

STARTS OPENLAB: Magic Shifter

Magic Shifter is an open source magic flashlight. With 16 RGB LEDs, it can be used to write texts and draw images in midair. This takes advantage of the "persistence of vision" phenomenon, whereby a rapid sequence of light impulses is seen as a whole picture. An acceleration sensor stabilizes the image. It's also possible to program colored shadow effects.

Philipp Tiefenbacher (AT), Jay Vaughan (AUS), Tom Peak (AT), Jascha Ehrenreich (AT), Philipp Poten (AT), Florian Bittner (AT)

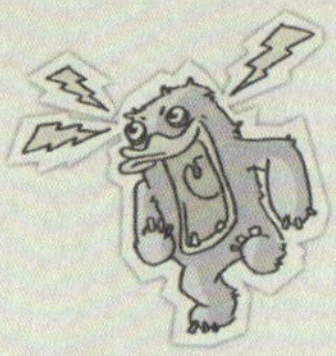

Blinking Rocket

Here, we'll construct an electronic switch with a blinking 8x8 LED matrix and give it a cool rocket design. The Blinking Rocket's texts and animation sequences can be modified via smartphone, tablet or laptop without having to install software.
The *Blinking Rocket* project is 100% open source and enables users to explore many different aspects of the STEAM fields—from modern, Web-based software and firmware development to hardware design and production.

Florian Bittner (AT)

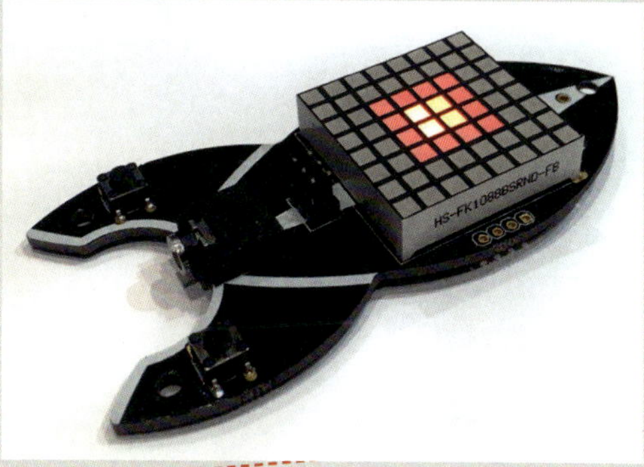

Jugend Hackt ZONE

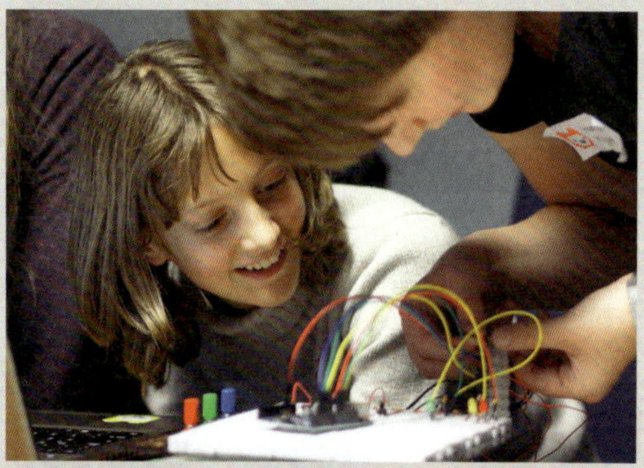

Daniel Havlik

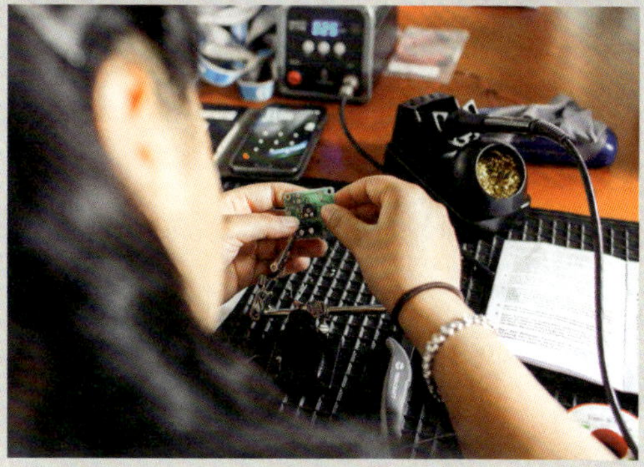

Sandra Schink

Jugend Hackt (Youth Hacks) displays real-world ways and means of dealing with errors.
In conjunction with this year's "Error" theme, *Jugend Hackt* is a setting in which young people, age 8-18, are cordially invited to consider the errors inherent in society and to make mistakes of their own. True to the motto "Improving the world with code," we'll offer lots of interesting ways to encounter technology, all of them associated with the same question: What society-changing potential is inherent in the technology and in your heads?
In this open lab, participants can build, solder, tinker with and program robots, explore across the technology spectrum, and try out all sorts of devices and apps. The point of departure: "An error is a discrepancy from what we expect, a deviation from the norm ..., but what is the norm and who establishes it?" Accordingly, we'll be shedding light on festivalgoers' views of the world and calling social norms into question. This works best with prototypes that can be artistically, creatively deployed—with or without technology.

Jugend Hackt (DE), Hello World (AT)

Do it yourSelfie

Glitch, selfie, hacking, error? Students in Linz Art University's Graphic Education program and festivalgoers play with analog and digital approaches to photography in a domain bounded by self-portrayal and image interference. Instead of deploying prefab digital filters, analog disruptions of digital images will be experimentally produced via colored foils, glass elements, mirrors, fabrics, common, everyday devices and lots more. Participants will explore the possibilities of hacking an SLR camera, and create analog add-ons for their own cell phone's camera. Then they can take home the selfie shot with them either in digital form or as a photographic printout.

Students in Linz Art University's Graphic Education program under the direction of Maria Anna Eckerstorfer (AT) and Helene Siebermair-Sommerer (AT)

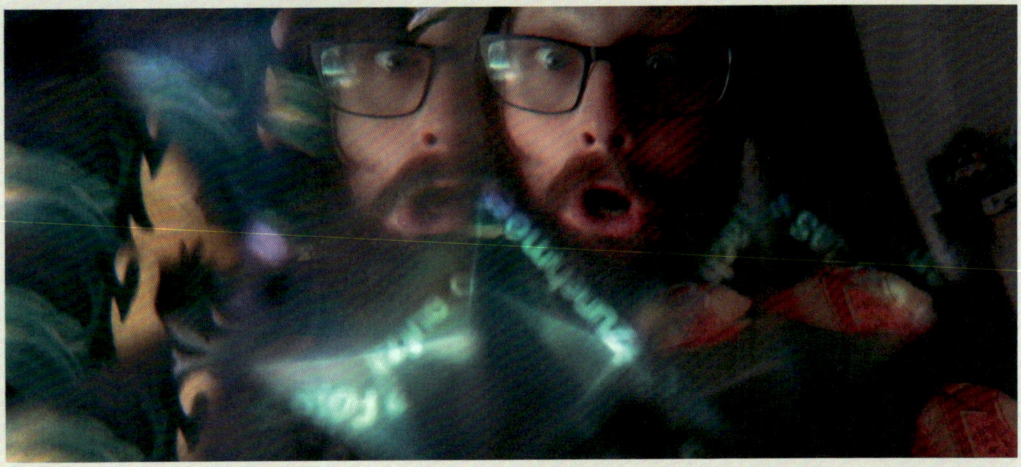

Bildnerische Erziehung, Kunstuniversität Linz

FM4 Spielekammerl Live at the 2018 Ars Electronica Festival

The ORF–Austrian Broadcasting Company's weekly computer gaming show on radio station FM4 is a colorful, upbeat, four-hour live stream. Since this format was introduced in March 2017, the game room crew has done several live remote broadcasts—including from the 2017 Ars Electronica Festival! In line with this year's focus on errors, they'll be playing demanding games in which players have to fail miserably and often in order to advance. Needless to say, the FM4 games crew will also be welcoming festivalgoers as studio guests. So, drop into the game room and play along!

FM4 (AT)

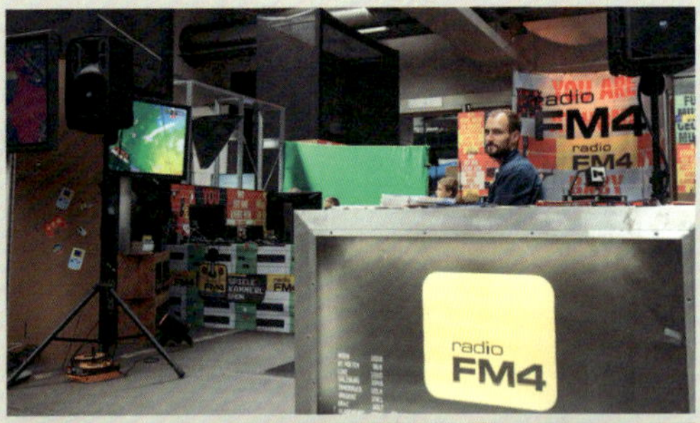

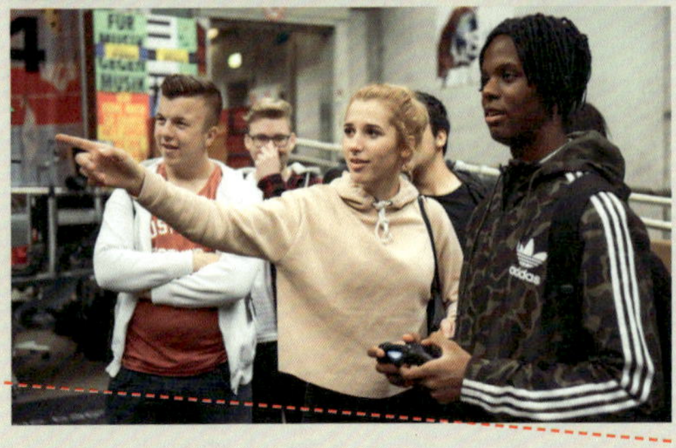

Ronald Posch

VERNER–Versatile Nature Exploration Rover

VERNER consists of a handmade carbon-fiber chassis, six wheels and four motors. It's controlled by a credit--card-size processor with wireless connectivity that permits two-way exchange of data on engine workload, sensors and wheel position. Its one-of-a-kind programming interface makes it possible to reprogram all components live while the vehicle is on the go. Users can try out many of these features for themselves during the demonstration.

Simone Atzwanger (AT), Max Heisinger (AT), Max Hofinger (AT), Markus Pirngruber (AT), Florian Rudinger (AT)

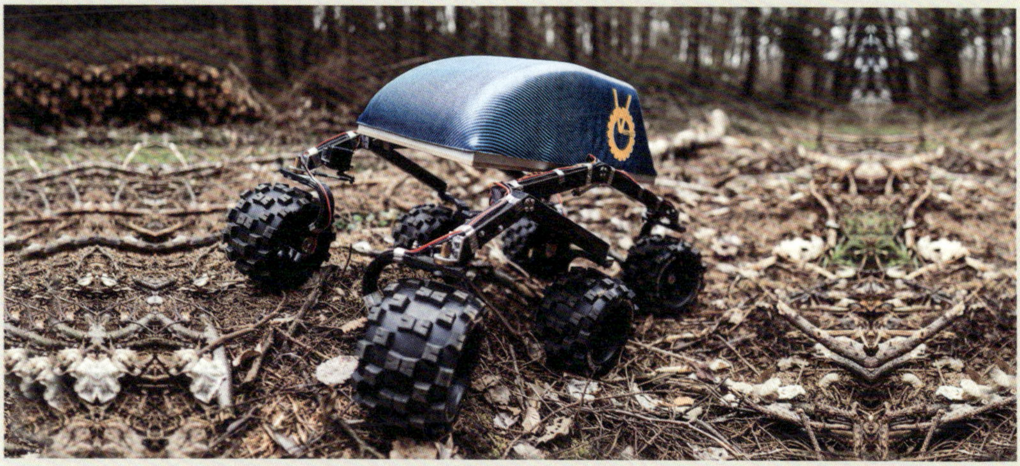

HARPTech2018

Robo Wunderkind

Robo Wunderkind turns building and coding real functioning robots into a child's play by bringing together the best of the technologies and educational concepts in the innovative educational robotic modular kits. Snap colorful modules together, control and program them using our intuitive software–a task even a 5-year-old could easily manage. Our products provide holistic user experience via tactile hardware and intuitive software and introduce children (and adults!) to the basics of coding and robotics. Join *Robo Wunderkind's* interactive playground to tinker, explore, experiment and create with friendly technologies. Open for visitors from 5 to 99 years old, that's the place to try your hands at engineering and bring your most daring ideas to life.

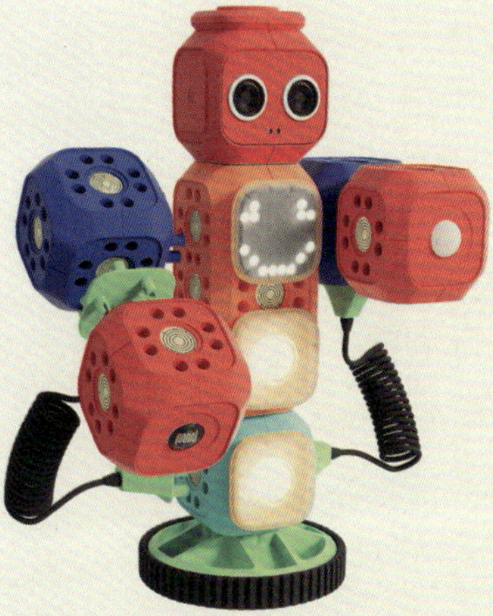

Hebocon

A *Hebocon* is like a traditional Japanese sumo wrestling match but two low-tech robots square off in this smackdown of technical dilettantes. What's the story here? Two robots are pitted against one another; they battle it out for one minute; the victor is the one that shoves the other out of the ring, tips its opponent over, or is the more active aggressor. Here, a robot is defined as anything that can move—but it can't weigh more than 1 kg or exceed 50x50 cm in size. The catch: the more sophisticated the built-in technology, the more points are deducted from the robot's score! The more creative, the sillier, the funnier the robot is, the better its chances of prevailing—and especially garnering an Audience Award.

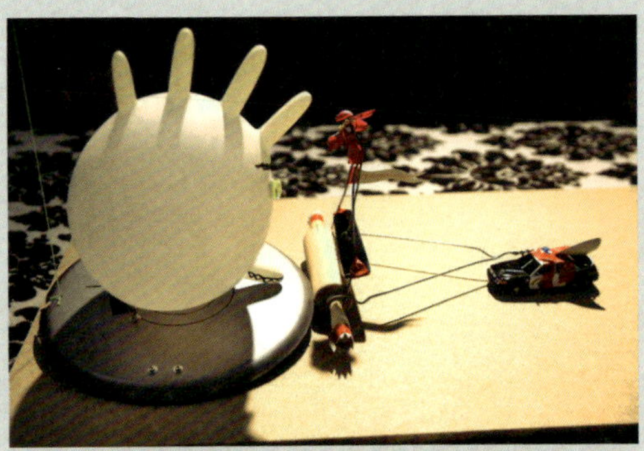

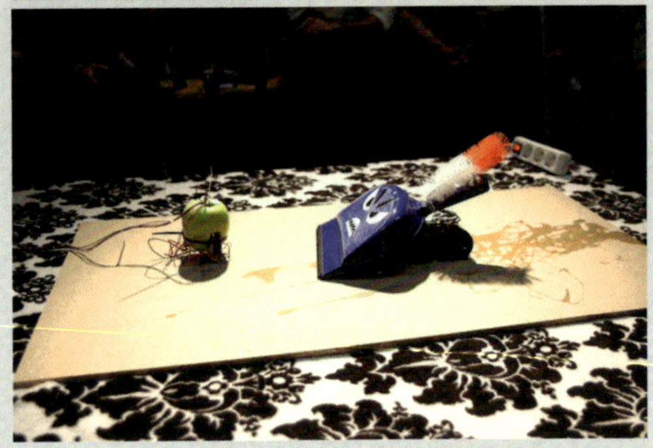

Schmiede Hallein

Think Your Product

Leonding Technical School's Think Your Product initiative teaches 4th year students entrepreneurial thinking and creativity in coming up with product ideas and starting up a new company as a first step on a potential career path. Students are offered several motivational and informational events over the course of the year. Think Your Product works together very closely with Upper Austria's Startup Network and numerous IT firms in the Linz metro area. The initiative's highpoint is the Make Your Product Week during which the teams with the best product ideas can spend a whole week focusing on developing a prototype and, at the end of the week, present the results to a jury of experts.

Steampy

Steampy is an app, where the user can create playlists across different streaming platforms. It frees users from the tedious task of searching music over different streaming platforms, sorting out the high number of overlapping offers in order to get the music they are really looking for.

Stefan Stanzel (AT), Karsten Köhne (AT), Andreas Leeb (AT)

Ludimus

Ludimus is a mobile gaming platform for card, board and arcade games that is revolutionizing how families play games with their kids. As a mobile solution, it enables players to take their entire game collection with them on vacation or to their friends' house.

David Matousch (AT), Eric Stock (AT), Elias Bürger (AT)

in the digital world, we present ourselves in the best possible light—#nofilter, of course. In the hopes of being loved, employed, liked and followed, we carefully choose the best parts of our identity to display. By adding a photo filter here and highlighting a detail there, we constantly work on creating the perfect surface. Our digital identity is flawless and impressive. Until... someone comes closer. When we actually meet our Tinder date face to face. Or go to that job interview we got invited to. Surface X captures the very moment when our digital and our physical identities collide and merge into the real us.

At this moment, it becomes impossible to keep the perfect surface intact. It cracks and folds and our true self is revealed. Like our digital identity, Surface X appears large and impressive from a distance. The interactive installation is made up of 35 black umbrellas that form a single, spherical body 3.5 meters in diameter. The umbrellas represent an impenetrable shield, protecting the inside from unwanted looks. However, once a person approaches Surface X, the umbrellas immediately shy away and close within milliseconds. It crumbles and shrinks and loses its polished look. Up close, one sees less surface but more of what lies behind, cables, sensors and a steel frame. It is left to the approaching person to judge: Does it look ugly or broken? Or rather interesting, maybe even beautiful?

Rebecca Gischel (DE)

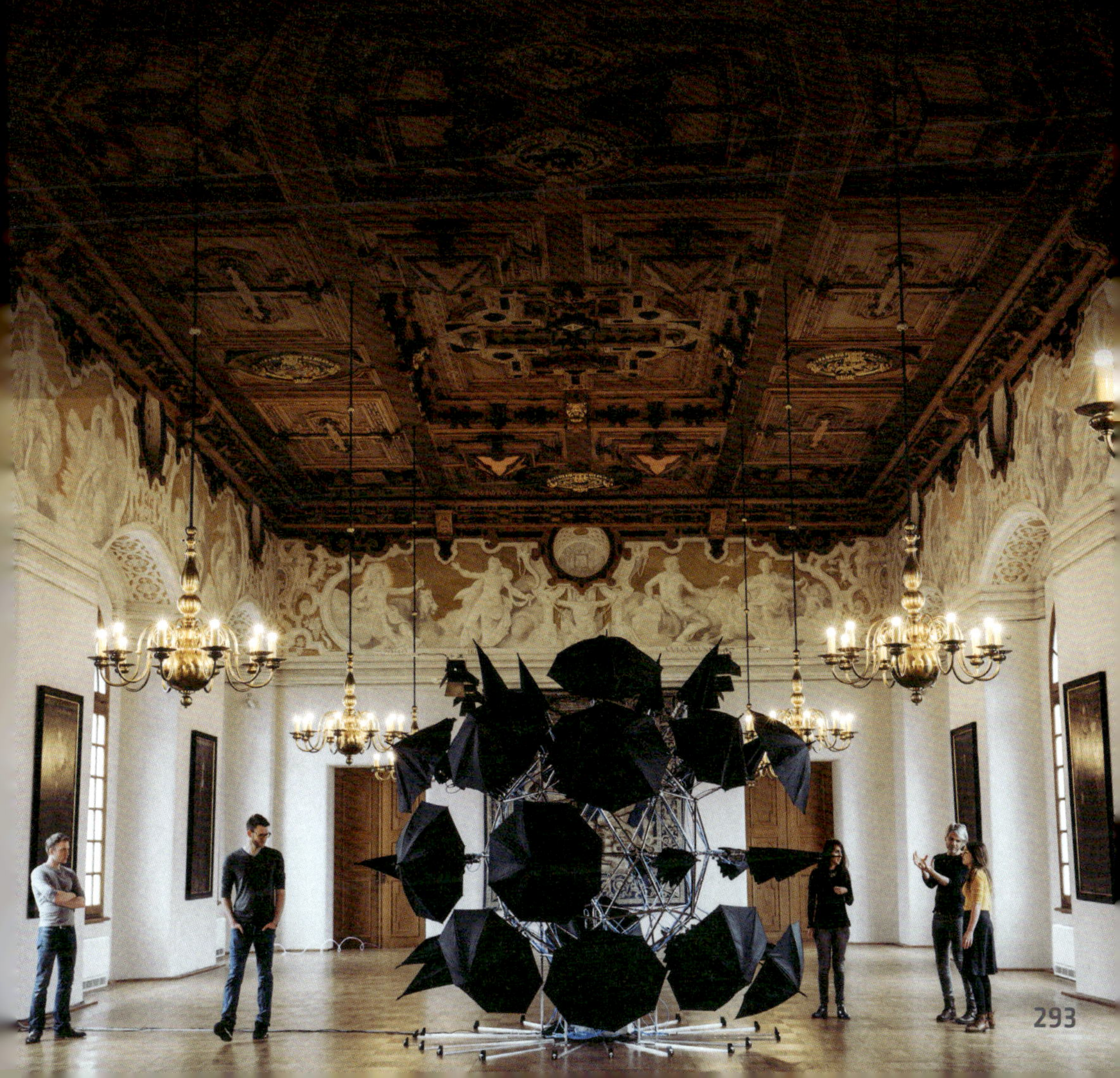

(NO) GREAT FUTURE

The future as a pointedly dystopian scenario—social media, fake news, performance pressure and bullying necessitate putting our thinking and behavior under a sort of glass dome in which we're apparently no longer able recognize reality any more. This "bell jar" has been graphically depicted in oversize format by a youth group at BFI Oberösterreich (regional career advancement agency). They now invite visitors to the Ars Electronica Festival to graphically enhance this work with the really important things in life. Maybe festivalgoers can succeed in collaborating with the BFI youth group to avert looming threats! The question is: What is actually really important in life?
Following up on a 2017 pilot project, young people associated with BFI Oberösterreich have come up with their own project, which they'll be implementing independently at this year's CREATE YOUR WORLD Festival.

Youth Group of BFI OÖ (AT)

MUSIC EDUCATION DAY

A coalition of Austrian musical institutions and initiatives is producing an encounter with the future prospects of young Austrian musicians. This event is an outgrowth of Music Summit, a working group formed at the 2017 Ars Electronica Festival; now, one year later, they've put together a program for musicians, teachers and musicologists. The emphasis is on the future of the music industry and musical training. Considering these two systems jointly is meant to provide a better overview and more profound insights for musicians. Music Education Day will lay down the beat and set the tone.

Mica Austria (AT), Universal Music Austria (AT), St. Pölten University of Applied Sciences (AT), Ableton Live (DE), BORG High School Linz (AT), Rock im Dorf Festival (AT), Österreichischer Musikfonds AT), mdw-University of Music and Performing Arts Vienna (AT), Fachinspektion für Musikerziehung und Instrumentalmusik (AT))

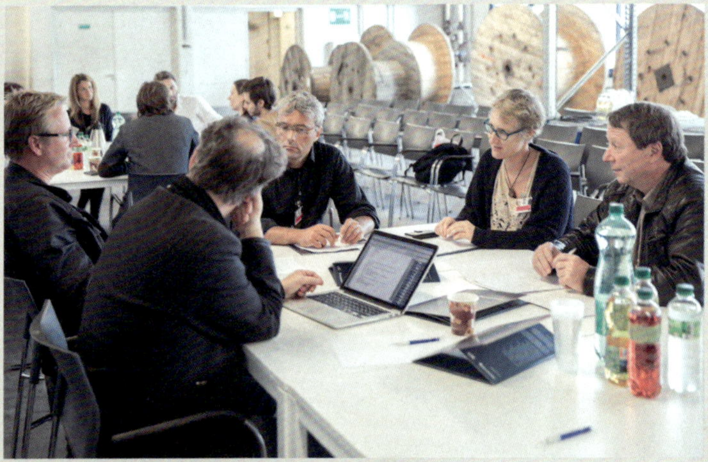

Tom Mesic

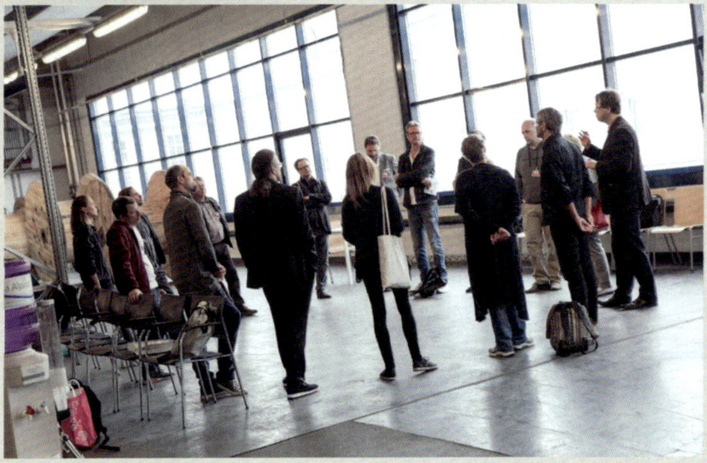

Florian Voggeneder

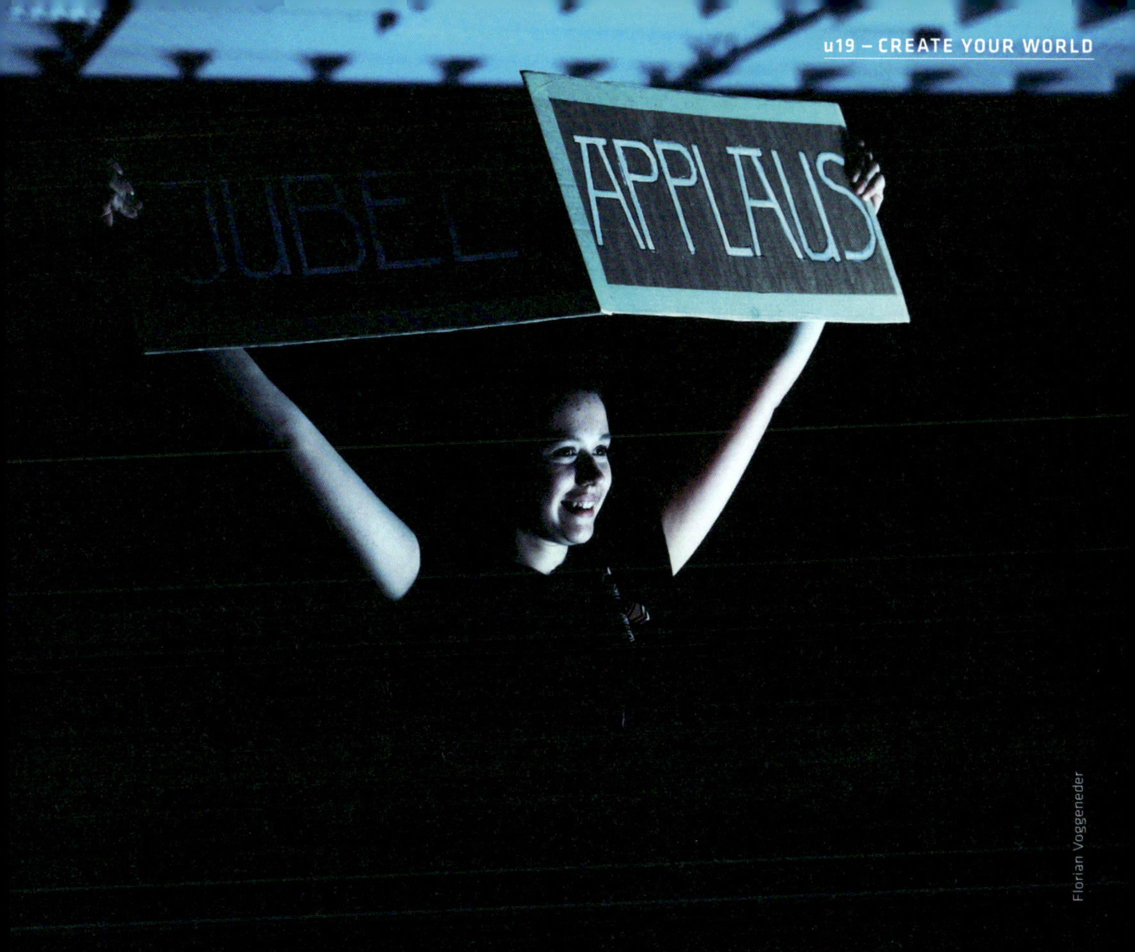

Florian Voggeneder

BUG TV

How will television as a media format develop from here? Could it be that, in the future, there'll be no regularly scheduled programming at all anymore, and maybe it will be completely replaced by on-demand streaming services? In cooperation with Landestheater Linz, this TV format will propagate a mix of fakery and truth, where subtle and, in some cases, intentional mistakes demonstrate a very prevalent blunting of the instincts of those who view them. In which formats is verification inherent? At what point do we acknowledge news as true? How must the news be presented for us to put aside our ever-growing skepticism? What portals do we perceive as reputable and responsible? These questions are treated in a blend of acting and festival documentation produced by this interactive project set in the CREATE YOUR WORLD festival village.

Landestheater Linz (AT), Ars Electronica

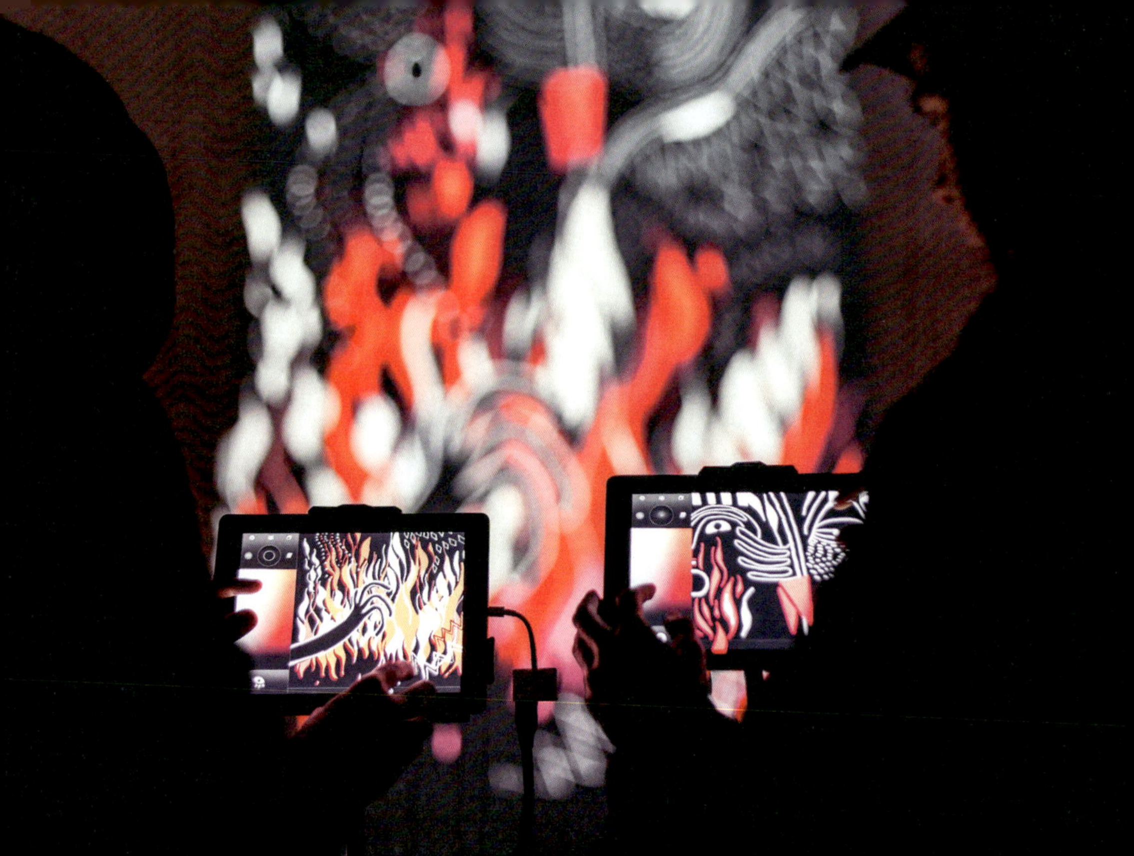

Tagtool Convention

For over 10 years now, building façades and stage sets worldwide have been illuminated with spontaneous images rendered and animated live. The people who make this happen are members of the Tagtool community; what they have in common is an alternative vision of digital art. They work on the street or in the theater instead of in front of a computer screen, jam like jazz musicians, and seek new forms of visual expression.
An official conclave of the international Tagtool community is being held in conjunction with the 2018 Ars Electronica Festival. Artists from all over the world will present their work at the Tagtool installation in POSTCITY, and festivalgoers of all ages are invited to get hands-on experience with projection painting. Rounding out the program are a series of improvised performances, guerilla projections on and around POSTCITY's exterior, workshops and presentations.

OMAi (AT)

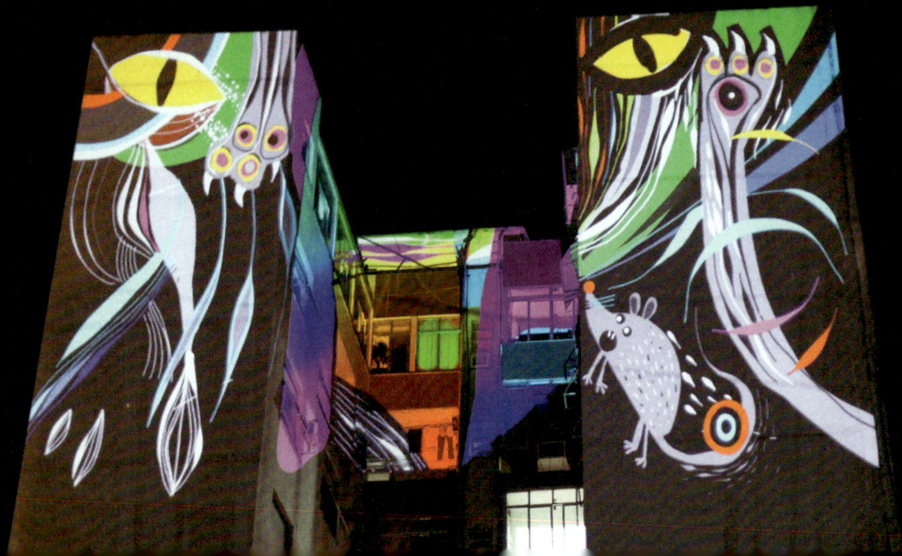

OMai

Glitch Painter

The participants in Virtual Office at Fab Linz have worked together with g.tec to develop a game about this year's festival theme, Error. A g.Nautilus EEG cap and specially programmed software enable players to intervene in an image using only their brain waves. The installation these young people have come up with deals with the subject of error on several levels: corporeal, social and technical. People don't always know exactly what they consider to be a mistake. Errors can be spontaneous, planned and even aesthetic! Minor glitches can be enthralling in that they allow for interpretational latitude.

Bettina Gangl (Fotografie.Design), Birgit Pölz (FAB Linz Virtual Office), participants in Virtual Office at FAB Linz; in cooperation with g.tec medical engineering and Mag. Katharina Mayrhofer VVVV Programmierung

PRZM

PRZM is an Animated MIDI Lightbox. Up to 5 people can play each *PRZM* instrument to generate a semi-random improvised musical jam. They use robust, arcade-style buttons mounted on the sides to trigger sound and light at the same time. Triggering a button will send pitch messages out to a MIDI compatible software/hardware instrument. In the current installation, Ableton Live is being used with custom instruments and MIDI effects that manipulate the MIDI data being generated. Animated light patterns are displayed through the tinted black perspex top layer. These were created with Neo Pixel Digital RGB strips which allow for programmable colors and patterns to be triggered via an Arduino Uno board.

Mark Towers (UK)

Animaker

Animaker is an interactive installation that breaks down the boundary between the virtual and physical worlds. A visitor enters a projection of a jungle temple and is called upon to construct animals out of Duplo© bricks. The results are identified via artificial intelligence and brought to life in a virtual environment. *Animaker* is a playful way for users of all age groups to get acquainted with the latest technological developments such as artificial intelligence and cross reality (XR).

The Tech Museum of Innovation (US), Resonai (IL/US), OMAi (AT), Occipital (US)

Fort Thunder

Fort Thunder is a touch play synthesizer built into a play fort that generates a startling array of electronic noise as bodies inside connect the 22 stainless steel poles, and themselves, in different combinations. Essentially it works by replacing the knobs you would find on typical synthesizers with bodies, which act as resistors between two points of oscillating circuit resulting in a fun, playful way to produce electronic sound. This project is supported by the Australian Government through the Australia Council, its arts funding and advisory body.

Lucas Abela (AU), Keg de Souza (AU)

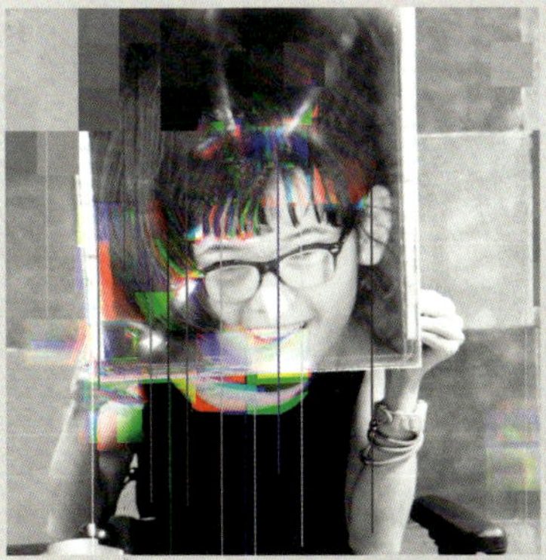

Glitch Painter, virtual office FAB Linz

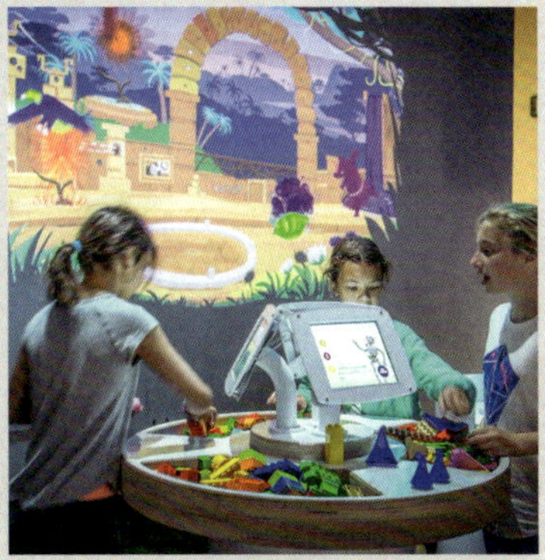

Animaker, OMAis

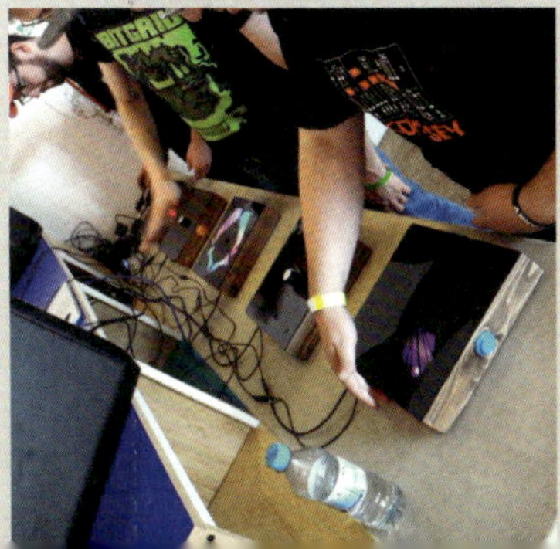

PRZM

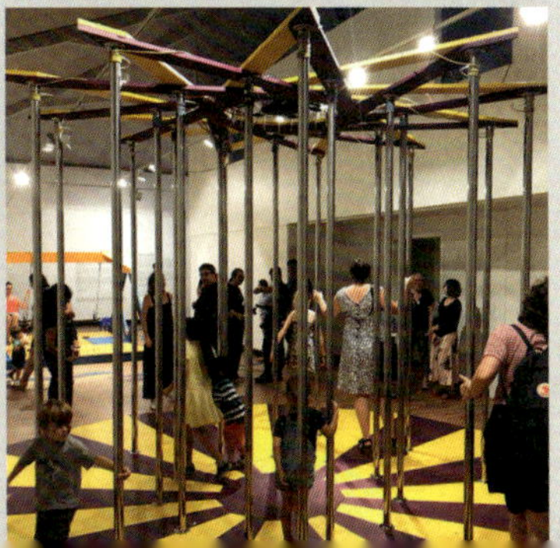

Fort Thunder

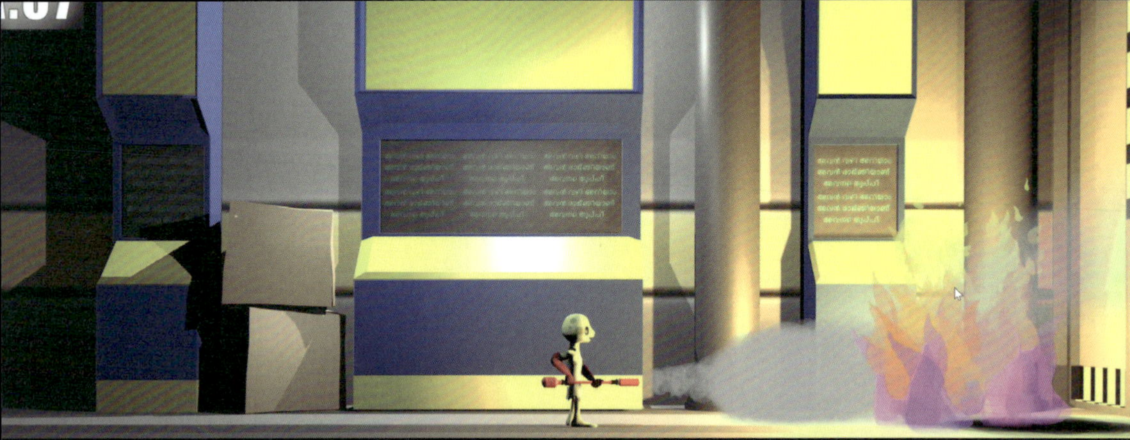

u19 Exhibition

The Prix Ars Electronica's u19 category confronts young people with two questions: How do you envision the future, and what ideas do you have in response to what you see? Each year, we receive 600+ answers in the form of incredibly diverse projects submitted for prize consideration. From among them, the five-member jury selects the 16 best. This is a task that seems to get progressively harder since the youngsters' projects have become more elaborate and more professional with each passing year. These 16 prizewinners are then presented to festivalgoers in an extensive, curated exhibition that is the centerpiece of the CREATE YOUR WORLD festival. The prizewinners are also invited to attend a workshop at which they collaborate with the u19 – CREATE YOUR WORLD production crew on the design of this exhibition. On one hand, this enables them to have a say in how the prizewinning projects are exhibited; on the other hand, it lets them get acquainted with one another. The process of conceiving the exhibition brings in a variety of perspectives, and repeatedly brings to light commonalities of both a substantive and a spatial nature. This format for producing the festival in cooperation with creative youngsters has worked brilliantly ever since it was launched in 2016.

Levers and Buttons, Klemens Horvath

u19 Ceremony

The highlight of the CREATE YOUR WORLD Festival is actually the u19 Ceremony at which the 16 projects singled out for recognition are publically honored in a fun, relaxed atmosphere. The event is integrated into the hustle & bustle of the festival and has become an annual favorite.

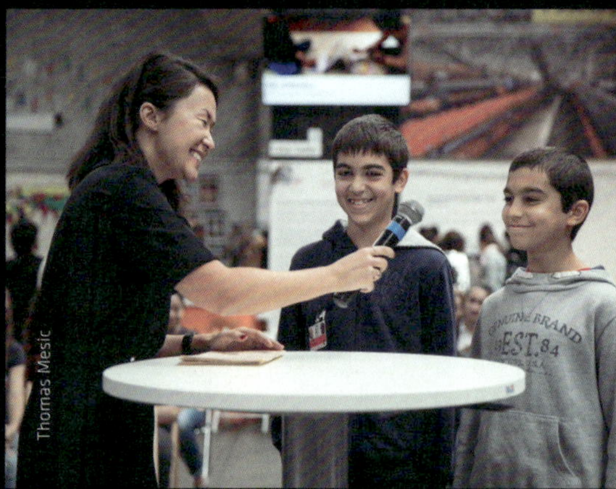

Thomas Mesic

Dabei sein ist alles

Opportunities for Young People Who Enter the Prix Ars Electronica

Over 650 entries were submitted to the Prix Ars Electronica's u19 category this year. This huge response makes the jury's job of picking the prizewinners a tough one, but they manage nevertheless to narrow the field down to 16 projects. Be that as it may, for several years now, the u19 – CREATE YOUR WORLD staff has made a concerted effort to foster and support projects that, despite their high quality, didn't quite make it to the Final 16. These very deserving youngsters have been invited to showcase their projects at the festival too.

Moreover, in recent years, the u19 – CREATE YOUR WORLD initiative has sought to act as a "summer job placement agency" for young people, and we're pleased to report that this has been a success. During negotiations with many of our clients and associates, we consider ways in which young people could be given assignments from certain companies, which would then obtain an authentic evaluation and, at the same time, a product that the company could use for other R&D purposes.

The u19 – CREATE YOUR WORLD staff constantly seeks projects that are substantively suited to this effort. We lined up three paid assignments this year:

1) Prospects for Political Education Symposium: Seraphina Reisinger, a student at Linz's High School for Communications & Media Design, was commissioned by the Upper Austria Teacher Training College, the Upper Austria Chamber of Labor and Ars Electronica to carry out a project dealing with peer education. The results will be presented at the symposium being held on September 6th in conjunction with the Ars Electronica Festival. Seraphina submitted a project entitled "Educational System in Austria" to the 2018 Prix Ars Electronica.
2) Upper Austria Tourism: This is the second time that a project having to do with the future of tourism has been commissioned. The regional inbound tourism development agency's aim is to integrate young people into the process of designing a tourism strategy. The crew got to work together with OUTFLATS, an architecture project by three young people in Vienna.
3) Kinderklangwolke 2018: The Kids' Cloud of Sound is set for Sunday, September 9, 2018. Kuddelmuddel Linz and u19 – CREATE YOUR WORLD are working on the artistic design and implementation of this open-air spectacle. Larissa Schwaiger, a young composer from Steyr, has been commissioned to lead a music workshop in conjunction with it.

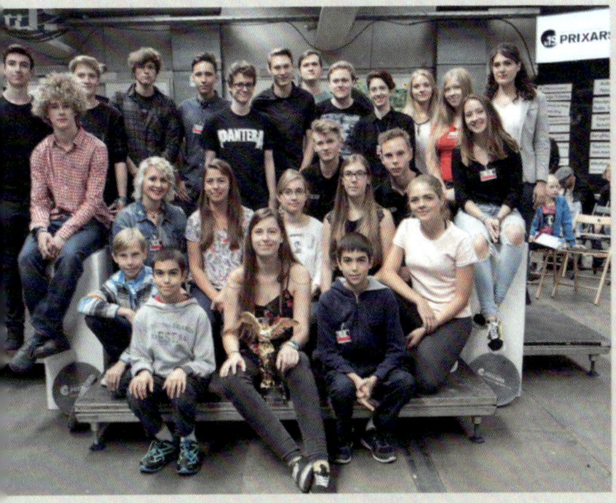

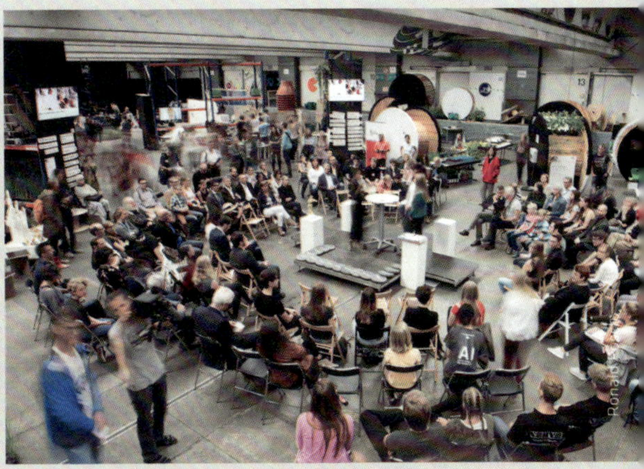

CREATE YOUR WORLD TOUR & "u19" international

Since 2015, we've offered a program of workshops at selected schools and institutions to integrate content and ideas from u19 – CREATE YOUR WORLD into everyday classroom instruction. The Tour is meant to offer pupils and faculty members means to access alternative learning & teaching methods and to playfully lay the foundation for new ideas and projects.

Above all, the broad spectrum of workshops fosters the performance of students, who are familiarized with new educational alternatives that can then be integrated into everyday life in schools. Plus, the Tour features regional artists adept at passing along their creative ideas directly to students and teachers. Teachers can take advantage of the workshops as continuing professional education units with current content and as a source of good ideas for new directions in classroom instruction.

For the past two years, a special program has been offered in cooperation with the European Capitals of Culture. In each country in which a city is serving as the year's capital of culture, a u19 Special Award is bestowed in a format that resembles the Prix Ars Electronica's u19 category. Integrating youngsters in this way makes a major impact in terms of sustainability and urban development.

In preparation for this award, a specially curated CREATE YOUR WORLD TOUR makes the rounds in the country of the next year's capital of culture. Five Workshop Weeks are staged with artists associated with u19 – CREATE YOUR WORLD, and the conclusion of a week features a presentation of the prizewinning works in the Prix Ars Electronica's u19 category. This is meant to inspire young people in the land of the upcoming European Capital of Culture to create a project of their own and submit it for prize consideration.

The managers of the European Capital of Culture program thereby comply with their educational mandate and Ars Electronica is delighted to support them in a consulting role. The exhibition of prizewinning projects from "u19Malta" (Valletta 2018) will premiere at the u19 – CREATE YOUR WORLD Festival!

Bulgaria (Plovdiv, 2019 European Capital of Culture) is where the CREATE YOUR WORLD TOUR's workshops are currently being held. The results of "u19Bulgaria" and the prizewinning projects will be presented at the 2019 Ars Electronica Festival.

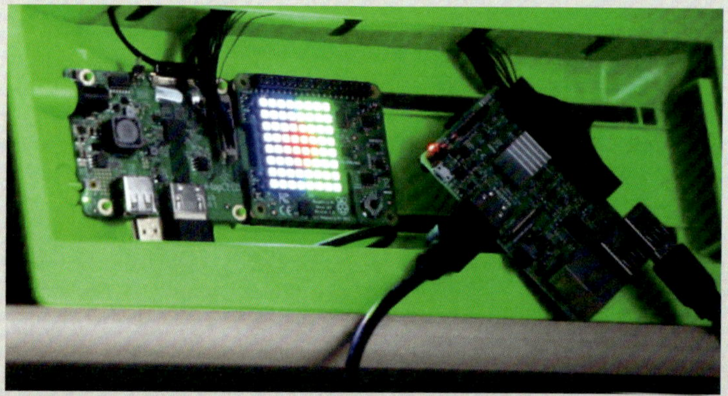

Conrad Nicholas–musicalweatherstation (@Malta Exhibition)

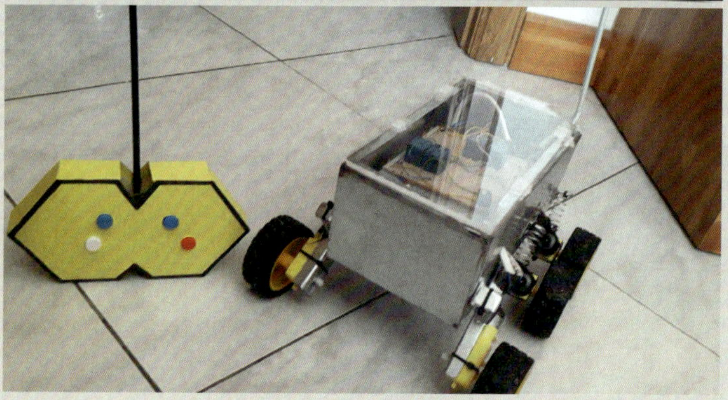

Amy Portelli, Mia Gauci–astrobot (@Malta Exhibition)

Conferences @ u19 – CREATE YOUR WORLD

"Prospects for Political Education" Symposium

Impudent, Courageous, Demanding–Self-Empowerment through Peer Education?

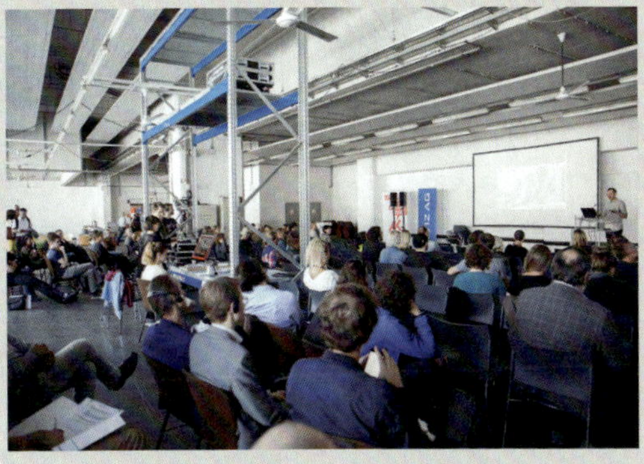

Tom Mesic

Is the world coming unhinged? You would think so now that atomic threat scenarios are once again part of what we perceive on a daily basis; when we observe the metamorphosis of conventional political systems; when democratic institutions–both in individual countries and multilateral arrangements–are called into question or even rejected outright, traditional parties are on the verge of dissolution, movements lacking an ideological foundation and some characterized by internal chaos enjoy growing popularity, and many people see authoritarian structures as solutions to social challenges. But do young people see things that way too? Does the younger generation have answers to questions that aren't even formulated the way they would have put it? Does peer education constitute the key to self-empowerment that opens the door for young people to assuming responsibility, making a commitment and a mark on society–in impudent, courageous and demanding terms?
This symposium will attempt to come up with answers to such questions by staging speeches as catalysts, workshops as settings for discussions of ideas and methods, and a Dialog of Disobedience as a wake-up call.

An event produced jointly by the Upper Austria Teacher-Training College, Upper Austria Chamber of Labor and Ars Electronica

ZusammenHelfen Conference

Day of Encouragement

This is the fourth consecutive year that ZusammenHelfen-Working Together in Upper Austria for Refugees is hosting a conference for all those who are committed to and interested in helping people forced to flee, or are affected by refugees and integration.
This year's conclave entitled "Day of Encouragement" will scrutinize new prospects, discuss the latest developments and challenges, and elaborate on successful projects. One inspiring speaker on the program is Ali Mahlodji, who'll recount anecdotes from his life as described in his book "And what do you do? From Refugee and Dropout to International Entrepreneur."

ZusammenHelfen–Working Together in Upper Austria for Refugees–is the go-to organization for everyone who is committed and interested.

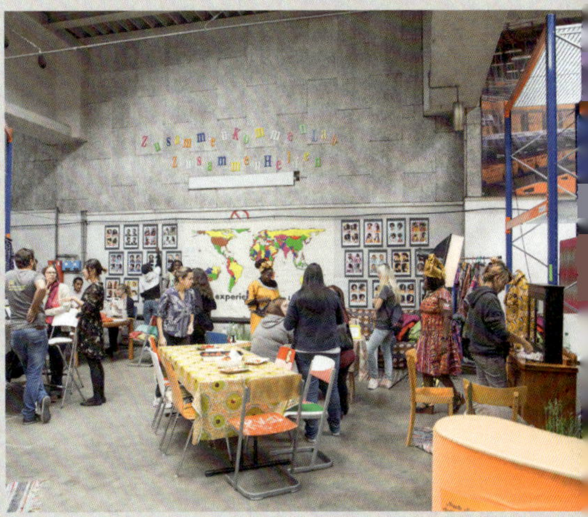

ZusammenKommenLab, © Tom Mesic

ERROR
the Art of Imperfection

GUEST PROJECTS

EMAP (European Media Art Platform)

EMAP (European Media Art Platform) annually awards production grants to outstanding European media artists and supports research, production, presentation and distribution of media art in Europe and beyond. Aiming to enable European artists to collaborate on projects and create closer bonds between European media organisations, the platform was founded in 2017 and offers grants in 11 member institutions:
Ars Electronica Center (Linz, Austria), Bandits-Mages (Bourges, France), FACT (Foundation for Art and Creative Technology, Liverpool, United Kingdom), IMPAKT (Utrecht, Netherlands), Kontejner (Zagreb, Croatia), LABoral Centro de Arte y Creación Industrial (Gijón, Spain), m-cult (Helsinki, Finland) , Onassis Cultural Centre (Athens, Greece) RIXC (Riga, Latvia), WRO Art Center (Wroclaw, Poland) , lead organisation: Werkleitz Centre for Media Art (Halle, Germany). The EMAP program is scheduled for the years 2017-2021 and is the largest international platform of artistic exchange and residency projects for media artists in Europe.

EMAP is co-funded by the Creative Europe Programme of the European Union.
http://www.emare.eu/

Robertina Šebjanič (SI), Gjino Šutić (HR)

aqua_forensic

Underwater Interception of Biotweaking[1] in Aquatocene[2]

aqua_forensic illuminates the invisible anthropogenic pollutants in water habitats. The project has a strong backbone in the hands-on research that combines: art, science, and a citizen-science approach to collect and process information on pollutants—"invisible monsters." These invisible chemical pollutants (such as legal & illegal drugs—mood controllers, antibiotics, antimycotics, painkillers, hormone pills etc.) are the residue of human consumptions discharged into underwater habitats that were explored during a residency in the summer of 2018 in two specific localities: Danube river (Linz, AT) and Adriatic Sea (Dubrovnik, HR). The project goal is to make these invisible anthropogenic pollutants and the pattern of their effect in the water habitats visible. With samplings of water and the seabed we are "hunting for a phantom" in the context of new mythologies. It's a voyage into the microbial seas that are forming the body of waters all around the world. Humans do relate to water in different ways in different cultures. With pollution we change the oceans inside out - influencing life and behavior of the whole cybernetic loop of the interconnected ecosystem.
The vast complexity of the ecosystem that covers more than 70% of the planet, producing over 80% of the atmospheric oxygen, is still mysterious. We are opening discussions about *aquaforming* and pushing the question of "the human footprint on water" to make it as present as *terraforming* in the deep age of the Anthropocene.
Anthropogenic presence is now aquaforming every part of the water habitats. It is the result of our global

1 Biotweaking—"acts or (art) of improving biological organisms on any level, by available means, to exhibit and use their full potential," concept coined by Gjino Šutić to fully define his work. Biotweaking mostly relies on DIY biotechnology – biohacking for achieving desired goals.

2 Aquatocene - is a series of research-based artworks initiated by Robertina Šebjanič that deals with cultural, (bio)political and ecological realities of aquatic environments and their consequences / challenges.

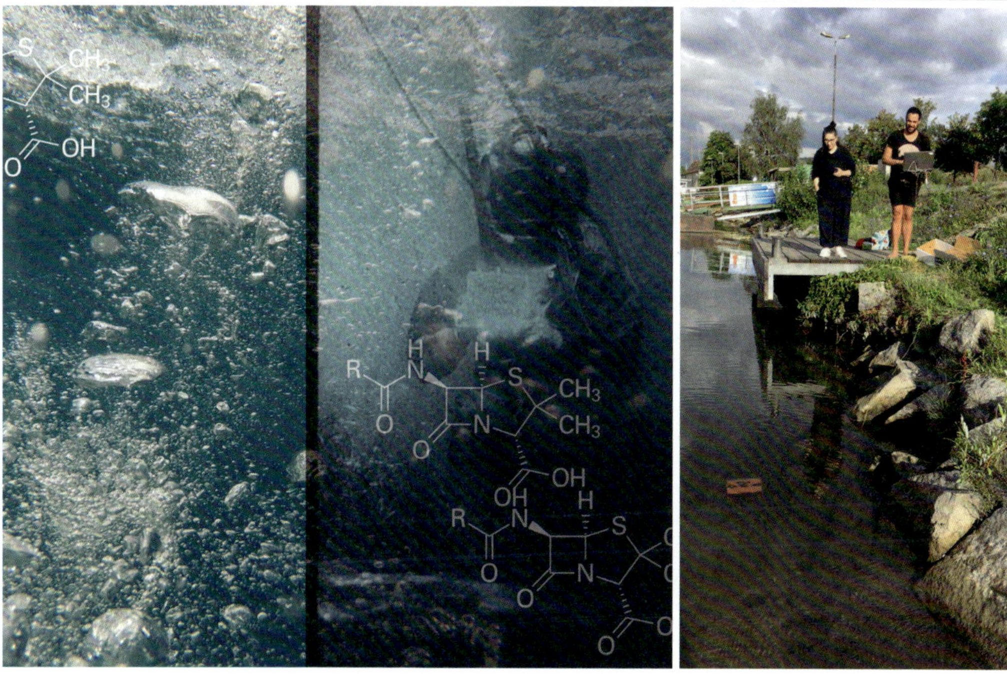

Jessica Galirow

socio-technological system and its (geo)political, social, economic interest in water – from the shallow waters on the coastal lines to the deepest points in the oceans.

aqua_forensic wishes to open the discussion about our solidarity and empathy in understanding the oceans, seas, and rivers beyond human perception. We are using the frame of art_sci installation, workshops and public discussions, to create new narratives in this art/science dialog.

By conducting citizen science investigation on forensic oceanography, we are looking into hidden secrets. The combination of science, art and field research is opening new doors in developing sustainable solutions—bringing these problems and thus development closer to citizens. There are a lot of challenges but also potentials when working with this topic and prototyping tools for water exploration.

The major omnipresent invisible anthropogenic pollutants are the semi-consumed pharmaceuticals

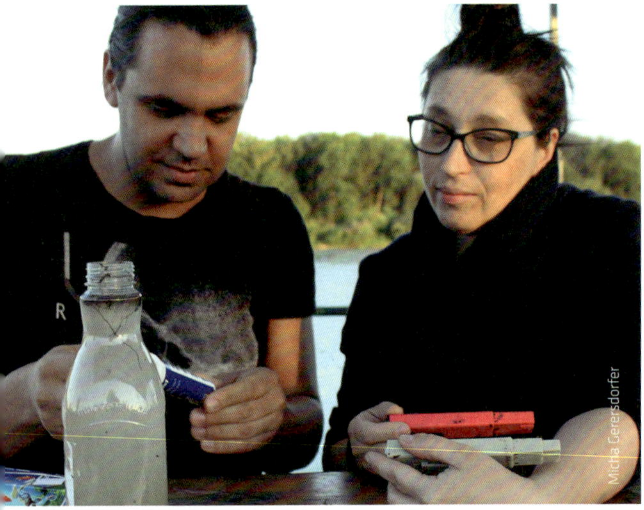

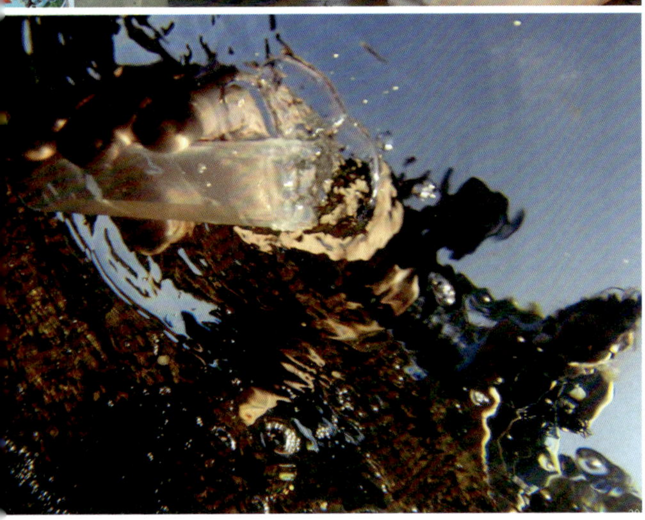

that stay overlooked and are changing not just us, but also the life in fresh and saltwater habitats all around the globe. The human body on average is able to digest only 20% of the drugs that we consume, while the other 80% is disposed from our body to waste systems – ending up in world's waters.

This has a malignant impact on the whole interconnected ecosystem in an escalating loop pattern. The loop is similar to the food consumption chain, which is entering a never ending circle of multiple effects on such things as the development patterns of all living organisms, from the micro level (viruses, bacteria and other microorganisms) to the largest organisms in the oceans.

There is a question we would like to ask: How do the oceans feel our impact?

Could we develop better communication with the oceans to find solutions, such as designing new and better therapeutics without negative collateral side effects?

Artists (research & development): Robertina Šebjanič (SI) and Gjino Šutić (HR)
Project supported by: Ars Electronica within the EMAP/EMARE project, Projekt Atol Institute (SI), UR Institute (HR), Čistoća Dubrovnik, The Ministry of Culture of the R. of Slovenia, and The Ministry of Culture of the R. of Croatia.
Co-funded by the Creative Europe Programme of the European Union.
Special thanks to: Martina Brković, Antonia Merčep, Veronika Liebl, Jessica Galirow, Uroš Veber, Annick Bureaud, team of nature reserve Lokrum, Eleonore team, UR Institute team

Anna Dumitriu (UK), Alex May (UK)

ArchaeaBot

A Post Singularity and Post Climate Change Life-form

This underwater robotic installation explores what "life" might mean in a post singularity, post climate change future. The project is based on new research about archaea (the oldest life-forms on Earth) combined with the latest innovations in machine learning and artificial intelligence creating the "ultimate" species for the end of the world as we know it. New research is revealing the mechanisms by which ancient archaea called Sulfolobus acidocaldarius can move around to seek "food" using tails known as archaella. The archaella use cogwheel-like "motors" to swim about. Our collaborator cryo-microscopist Amanda Wilson is studying the structure of these archaella to make tiny drills made of DNA which might be used to drill into cells to cure diseases, but the robotic archaella are made through 3-D printing.

Made in collaboration with Amanda Wilson (MARA Project, Imperial College) and Professor Daniel Polani (University of Hertfordshire).
Supported by Arts Council England and through an EMAP/EMARE artists residency at LABoral Centro de Arte y Creación Industrial in Spain via funding from Creative Europe.

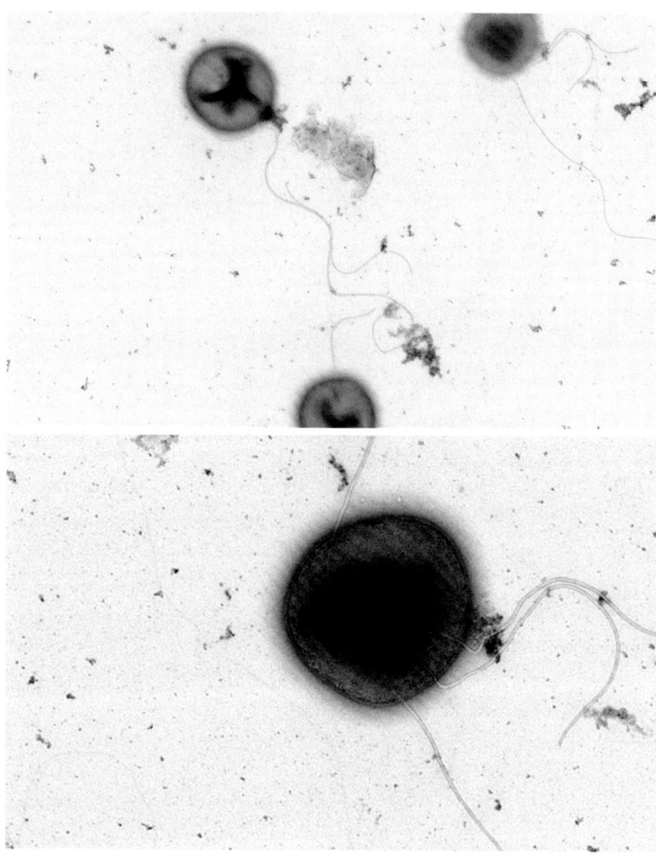

Amanda Wilson

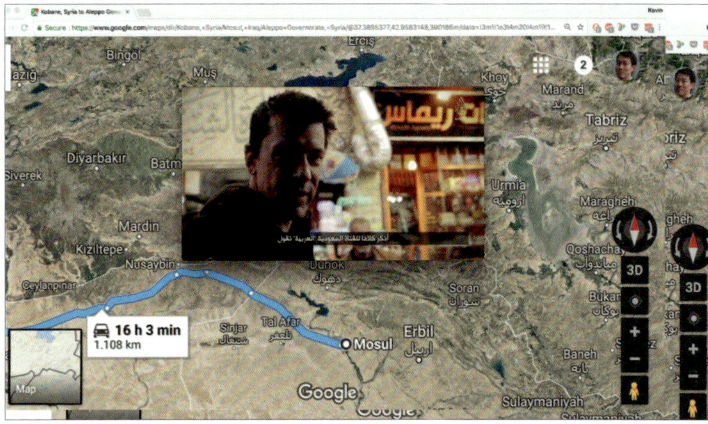

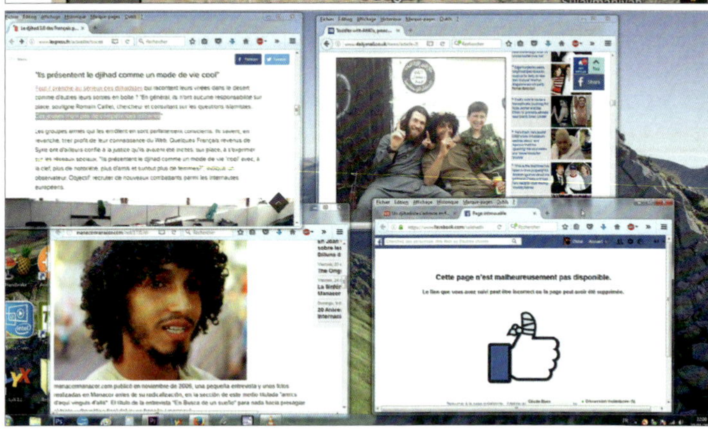

Chloé Galibert-Laîné (FR), Kevin B. Lee (DE/US)

Bottled Songs 3 & 4

The Spokesman – My Crush Was a Superstar

The Bottled Songs of Lost Children is a series of video letters investigating desire, power and terrorism in online and social media. The videos, recorded from the researchers' desktops, depict and interrogate their subjects' compulsive engagement in the production of everyday myths and fictions about themselves and others. Here are presented Chapters 3 and 4.

"The Spokesman" investigates the online traces of John Cantlie, a British news reporter who was kidnapped and appeared in several Islamic State propaganda videos. "My Crush was a Superstar" tracks a French ISIS fighter, Abu Abdallah Guitone, through a trail of messages, videos and postings to uncover his existence in both social media and reality. This leads to an uncomfortable first-person exploration of the gender dynamics behind ISIS recruitment strategies.

The Bottled Songs of Lost Children was conceived with the support of the Harun Farocki Institut, sponsored by the Goethe Institut. It was further developed at m-cult (Helsinki), supported by EMAP/EMARE, co-funded by the Creative Europe Programme of the European Union.

Anna Ridler (UK)

Myriad (Tulips) + Mosaic Virus

Drawing historical parallels from the "tulip-mania" that swept across Netherlands/Europe in the 1630s to the speculation currently going on around crypto-currencies, the video work *Mosaic Virus* is generated by an artificial intelligence (AI). This AI has been trained on 10,000 photographs of tulips that the artist took and then labelled to make the piece *Myriad (Tulips)*. It then uses this information to create a tulip blooming, an updated version of a Dutch still life for the 21st century. The appearance of the tulip is controlled by bitcoin price. *Mosaic* is the name of the virus that causes the stripes in a petal which increased their desirability and helped cause the speculative prices during the time. In this piece, the stripes will depend on the value of bitcoin, changing over time to show how the market fluctuates.

This project was commissioned by IMPAKT within the framework of the EMAP/EMARE program with support of the Creative Europe Programme of the European Union.

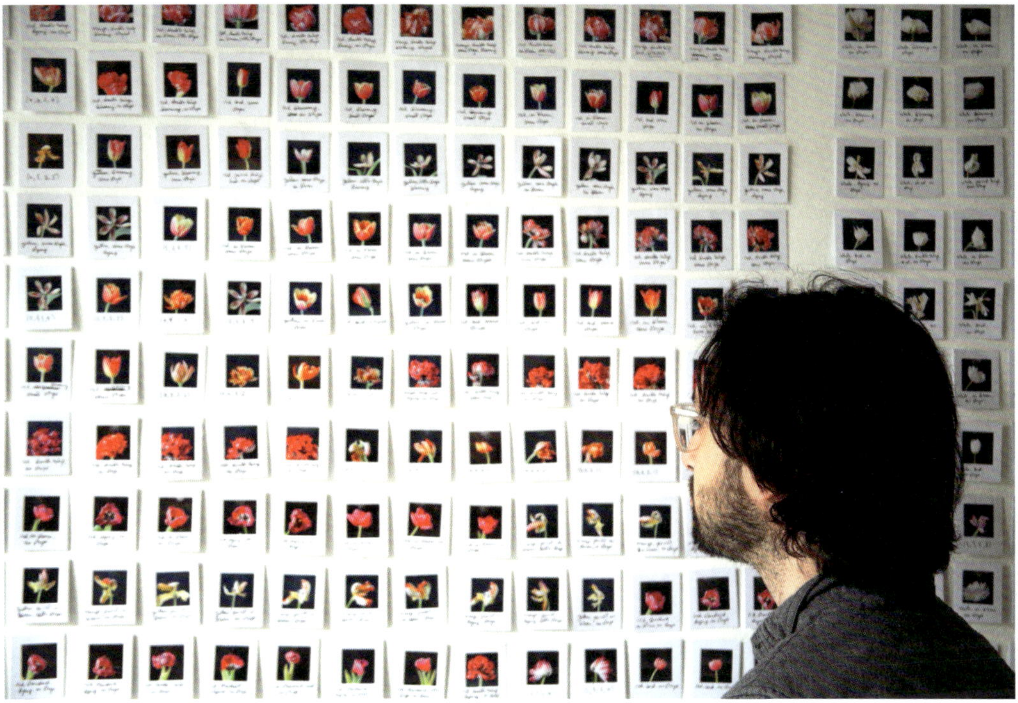

Kentaro Kumanomido (DE/US), Thomas Anthony Owen (DE/US)

faster than light

faster than light (2018) is a hybrid performance-film developed in residence at the Onassis Cultural Centre, Athens, Greece, as part of European Media Art Platform/Residency Exchange (EMAP/EMARE). During the residency, the artists collaborated with a group of local activists and performers through one-on-one exchanges and improvised group encounters. These meetings focused on the complexities of representation, the tension between individuality and collectivity, and the building of non-digital social networks. This process was documented through a range of filmic methods including MiniDV and high-frame-rate slow-motion recording, then crafted into a multi-layered filmic journey that weaves together the artists' interests in ecology, gender, and quantum intelligence.

Co-directed: Kentaro Kumanomido & Thomas Anthony Owen
Film participants: ActiVista / George Kounanis, Chraja, Tracy Fischer, Giorgos Kalogeropoulos, Efi Karagiannopoulou, Zak Kostopoulos, Maria Michailidou, Thomas Anthony Owen, Khalid Prem, Evi Tsaklanou, Avraam Vrohidis
Vogue Performance: Chraja
Lead Sound Design: Stratos Bichakis
Music: Where Did I Go?: BLANKIE / Alicia Grant
Αρχιπουστης : Chraja
Cinematography: Nikolas Pottakis, Kentaro Kumanomido, Thomas Anthony Owen (lead camera), Zinzi Buchanan (support camera)
Makeup support: Anastasia Athanasiou
Translation support: Anis Alexandros El Namparaoui
Special thanks to: Mor Demer and Vicki Kapo
Produced by: Onassis Cultural Centre, Athens
Within the framework of the European Media Art Platform (EMAP/EMARE)
Co-funded by the Creative Europe Programme of the European Union

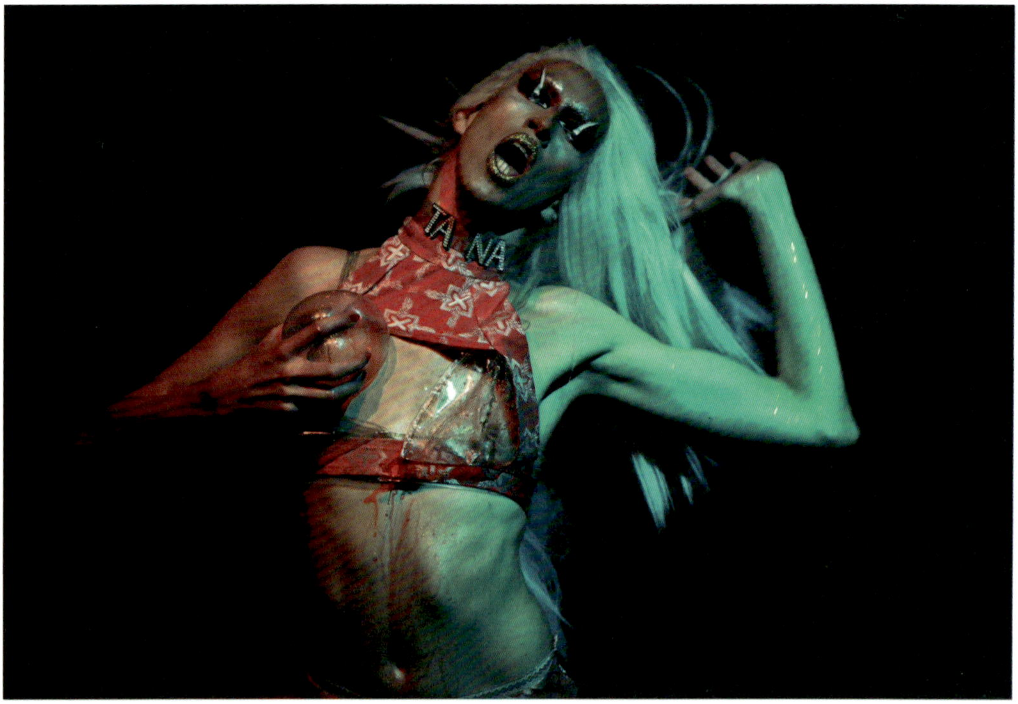

Kiki Papadopoulou

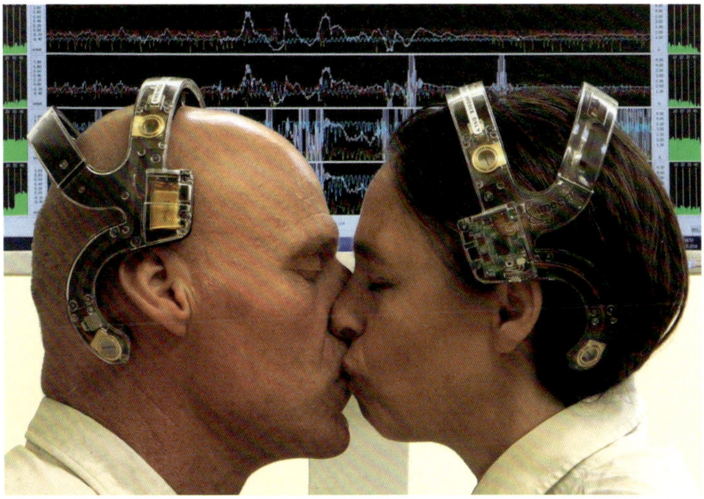

Karen Lancel (NL), Hermen Maat (NL)

Kissing Data

How does your kiss feel in E.E.G. data?

Can a kiss be translated into bio-feedback data? How does your kiss feel in E.E.G. data?
Artists' duo and researchers Lancel/Maat critically investigate social/sensory connections, privacy, empathy, vulnerability and trust when mediated by Multi Brain Computer Interfaces (Multi BCI).
During internationally shown performance-installations, people are invited to experience a shared kiss, as an intimately co-created, reflexive data-scape. Acts of kissing are re-orchestrated for a poetic, digital synesthetic ritual. In live kissing experiments with Multi BCI E.E.G. head-sets, visitors are invited as Kissers or Observers. During kissing, their brainwaves are measured. Real-time, their streaming E.E.G. data encircle them in a floor projection. Simultaneously, the Observers brain waves are measured, their neurons mirroring activity of intimate kissing movements, resonating in their imagination. Both Kissers' and Observers' data intimately co-create an immersive visual, reflexive data scape, translated to an algorithm for a soundscape–a "Kissing Data Symphony".
Kissing Data encourages shared agency to explore, co-create and co-interpret intimate BCI data, to share dialogue and reflection on ethical design concepts. In this poetic, Multi BCI mediated social synthesis, each unique E.E.G. data soundscape is saved to be downloaded and appropriated by all visitors. A selection of data-visualizations is printed as a "Portrait of a Shared Kiss".

Developed and supported with partners: Mondriaan Foundation, Baltan Laboratories Eindhoven, STEIM Amsterdam, Waag Society Amsterdam 'Hack the Brain' Horizon 2020 research program of the European Union; Public Art Lab / Connecting Cities Network Berlin; TASML Tsinghua University Beijing. The Work was realized within the framework of the EMAP/ EMARE program at the RIXC Center for New Media Culture in Riga with support of the Creative Europe Programme of the European Union.
Research partners: TU Twente & 'Dutch Touch Group'; University of Amsterdam 'NeuroCultures Group'; University of Applied Science Vienna 'Digital Synaesthesia Group'; Tsinghua University Beijing, Neuro-engineering department; TNO Netherlands Organization for Applied Scientific Research; NWO The Netherlands Organization for Scientific Research.
This interdisciplinary research is part of Lancel's PhD trajectory at the TU Delft 'Participatory Systems Initiative.'
Sponsoring: Fourtress, Holst Center and Phillips Eindhoven; Eagle Science Amsterdam.

▸ Chilean Artists at the Ars Electronica Festival 2018

This year, for the very first time, the Ministerio de las Culturas, las Artes y el Patrimonio and the Ministerio de Relaciones Exteriores | Gobierno de Chile and Ars Electronica Linz collaborated in asking artists, scientists, designers, researchers, entrepreneurs and social activists from Chile or with a Chilean background to participate in this year's Ars Electronica Festival 2018.

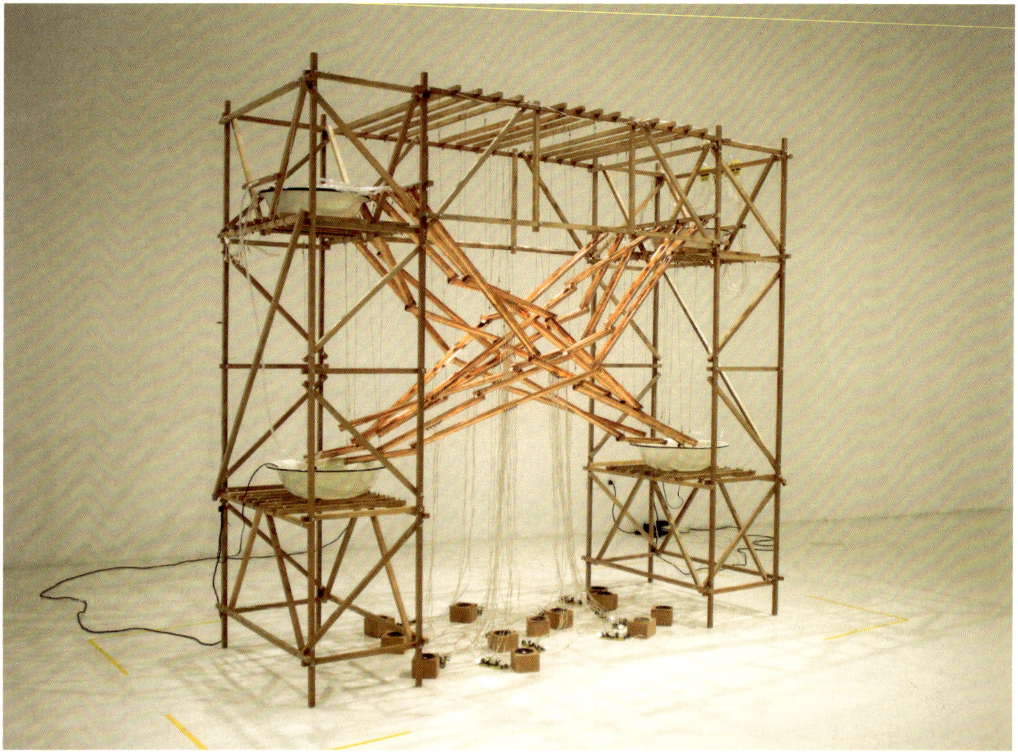

The Water Resistances Laboratory "Toboggans"

Claudia González Godoy (CL)

▶ The Water Resistances Laboratory "Toboggans"

TWRL Toboggans is a sound installation that seeks to make a connection between the natural and the synthetic, making possible different meanings for the element of water: as a physical material that generates sound, and as a substance with chemical properties that has an effect on the interaction between humans and nature. The permanent contact between water and electricity as an apparent risk plays a fundamental role in thinking through the relation between nature and artifice. This project aims to provoke thought about media and electronic devices, placing water at the center of technological development. It explores the water element as an interface that interconnects with electronic devices, generating a tension between two elements that are extremely important to life: water and electricity, where the water becomes the center of this installation, acting as a variable resistor to modify and modulate the sound of DIY synthesizers.

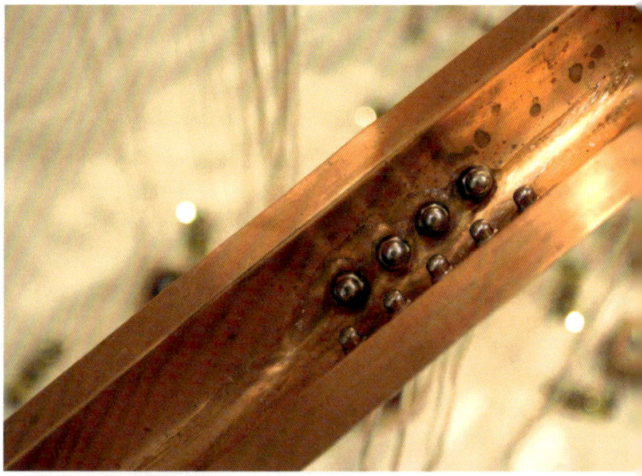

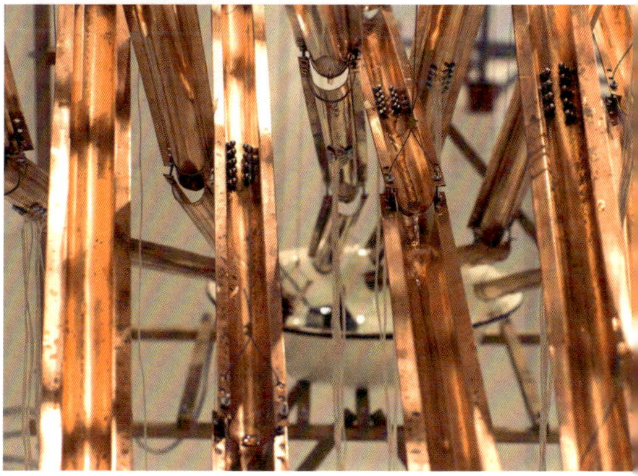

Name of the Work: water resistance laboratory slides
Author: Claudia González Godoy
Installation and assembly concept: Claudia González Godoy
Sound design and hardware development: Claudia González Godoy
Design and construction wooden structure: Andrés Moreno
Programming: Daniel Tirado
Collaboration: Rodrigo Moreno, Silvia Godoy
Institutions: Plataforma Bogotá (2013) Tsonami Sound Art Festival (2016)
Assembly assistants: Daniel Tirado, Andrés Moreno
Photography: Courtesy of the artist

Supported by Ministerio de las Culturas, las Artes y el Patrimonio and the Ministerio de Relaciones Exteriores | Gobierno de Chile

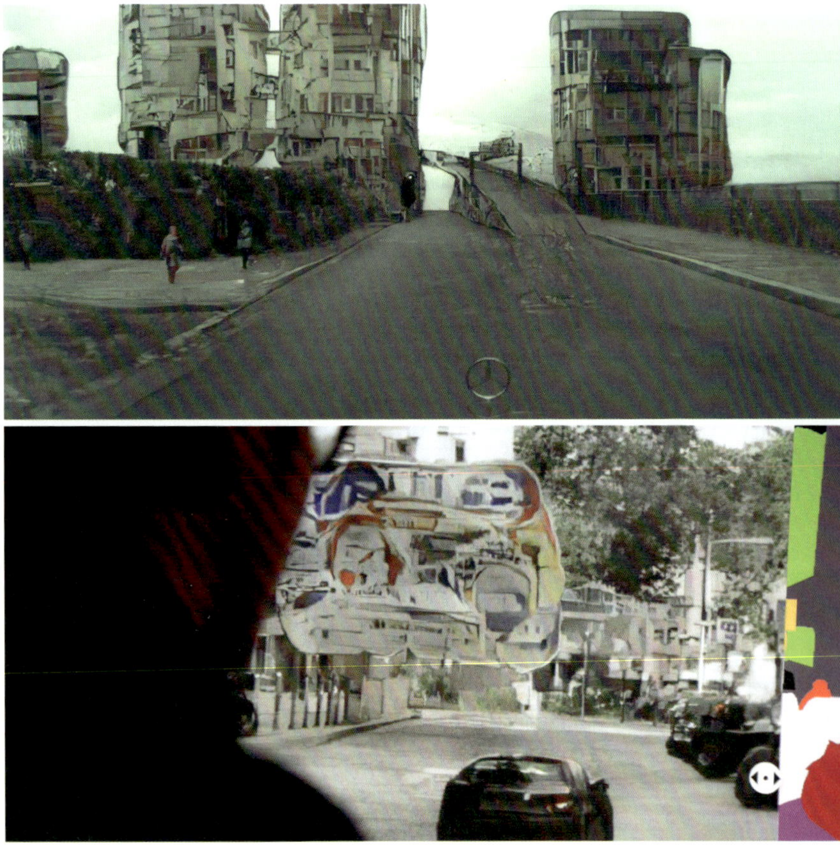

Anastasis Germanidis

RunwayML – Cristóbal Valenzuela (CL)

▶ Inaccurate Collaborations

Tools for inaccurate human-machine collaborations

Determinist algorithms and systems are based on the expectation of regular and punctual inputs to generate consistent and systematic outputs. Glitchy, uncleaned, misleading or faulty errors in the inputs can lead to suboptimal or erroneous results. *Inaccurate Collaborations* is a project that explores how generative and deterministic machine-learning algorithms can be used to create faulty and erroneous results that suggest new explorations for human-machine creative collaborations. Using an interactive digital tool to draw semantic colormap scenes, where each color represents a different label, images are synthesized using Generative Adversial Networks (GAN) that have been trained on a collection of different image datasets. Each generation introduces a new procedural error in the algorithm. The development of this project will make use of RunwayML.

RunwayML: *https://runwayml.com/*
NYU ITP: *https://tisch.nyu.edu/itp*

Supported by Ministerio de las Culturas, las Artes y el Patrimonio and the Ministerio de Relaciones Exteriores | Gobierno de Chile

Roy Macdonald (CL)

▸ Defective Apparatus

It refuses to behave according to its original rules and insists on avoiding people. part 1.1.

This project takes the form of an interactive installation driven by a laser beam which interacts with the audience by bouncing on a number of computer-controlled mirrors in order to avoid people. The laser beam, according to the Hollywood-esque code, works simultaneously for delimiting space and as an intruder-detection system. In the case of this work, this laser beam loses its original use because it "acquired a phobia towards people," causing it to avoid people constantly. Each time someone or something blocks the beam, the system reconfigures its mirrors so the beam avoids the blocked route, generating a new space delimitation, although this route is originally chosen randomly and fed into a machine learning algorithm, so the next time it can make a better decision. It learns by "trial and error." The whole learning phase of this work happens during the show, so this relationship between human interaction, errors and learning becomes a new narrative layer of this work.

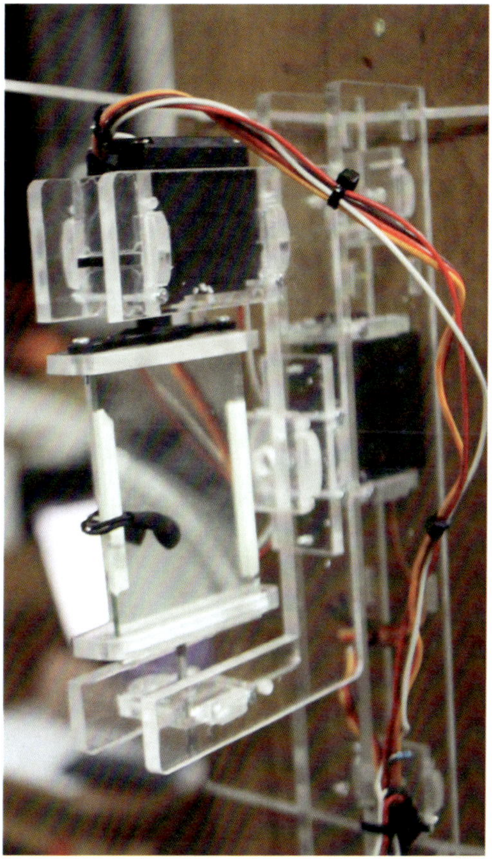

Project made during the Computational Arts residency setup by the Department of Computing at Goldsmiths, University of London and the Victoria and Albert Museum Digital programs.

Supported by Ministerio de las Culturas, las Artes y el Patrimonio and the Ministerio de Relaciones Exteriores | Gobierno de Chile

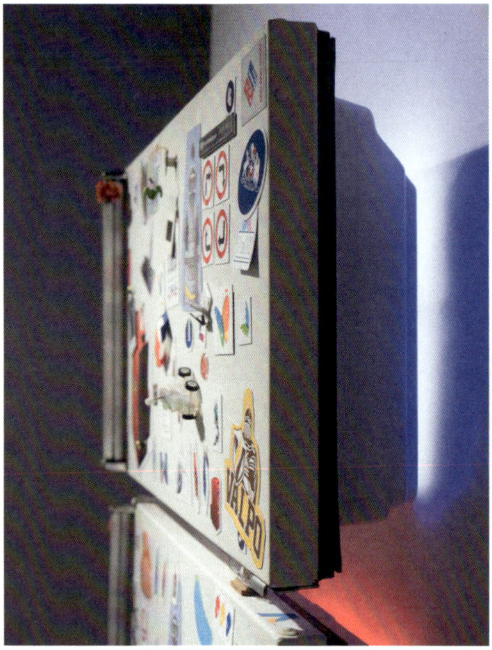

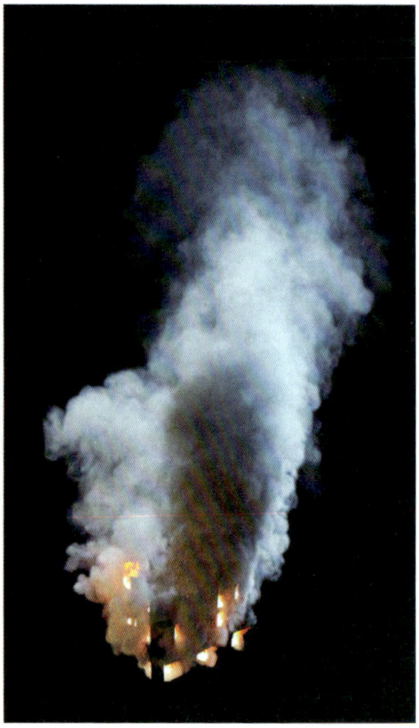

Cristóbal Cea Sánchez (CL)

▶ Glorias

In the loop

Centered around the events that surround the 21st of May in Chile, and assembled from news clips found on news and social networks over the past years, *Glorias* uses 3-D-animation methods to develop an investigation of the symbolic significance of a particular event: A date when – almost ritually every year until the death of Eduardo Lara in 2016 – the monsters of order, republicanism, protest, violence and repression emerge in the city of Valparaiso. Starting with a military parade, followed by protests and riots, ending with police in the streets, the 21st of May was a loop of imperfections, monstrosities and unresolved conflicts that reflected our precarious democracy. Assembled from archival footage corresponding roughly to the 2010-2016 period, the main video contrasts this footage with CGI reconstructions of the scenes depicted. The piece, started after five years of living abroad, also represents an exercise in trying to inhabit the space between event, witness and spectator: An inconsolable distance, full of misinterpretations and misconceptions, where computer reconstruction works as a medium between the subjective experience and the archive.

Project supported by the Ministerio de las Culturas, Las Artes y el Patrimonio Grant of 2016.

Supported by Ministerio de las Culturas, las Artes y el Patrimonio and the Ministerio de Relaciones Exteriores | Gobierno de Chile

Nani Gutiérrez (CL)

▶ Acceder

Methodology for an Emergency Plan

The Tsunami earthquake catastrophe of February 27, 2010 on Chile is a political issue on a highly emotional level. Over 500 people lost their lives. Right after the earthquake, the Pacific Tsunami Warning Center in Hawaii issued a tsunami warning for the whole pacific region. The Chilean authorities did not heed the warning and no evacuation was ordered.

Acceder illustrates the communication between the different Chilean authorities during the earthquake and the tsunami that followed. Their decisions are affected by misunderstandings, language barriers, politics, machismo and the fear of committing mistakes. Due to a government change, a week after the catastrophe, the official response to the tsunami also became a political issue between right and left. The video is based on emergency protocols, court reports and newspaper articles. As a guided lecture, it sums up five hours of complex bureaucracy behind the emergency methodology.

Directed and animated by Nani Gutiérrez
Production Assistant: Felix Moser
Score: Hainbach

Supported by Ministerio de las Culturas, las Artes y el Patrimonio and the Ministerio de Relaciones Exteriores | Gobierno de Chile

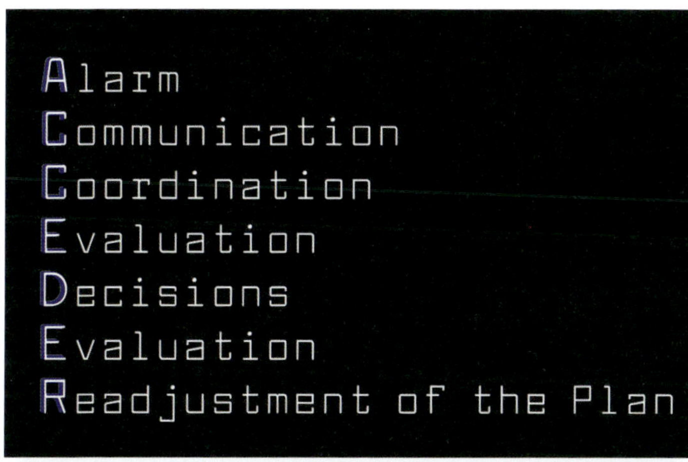

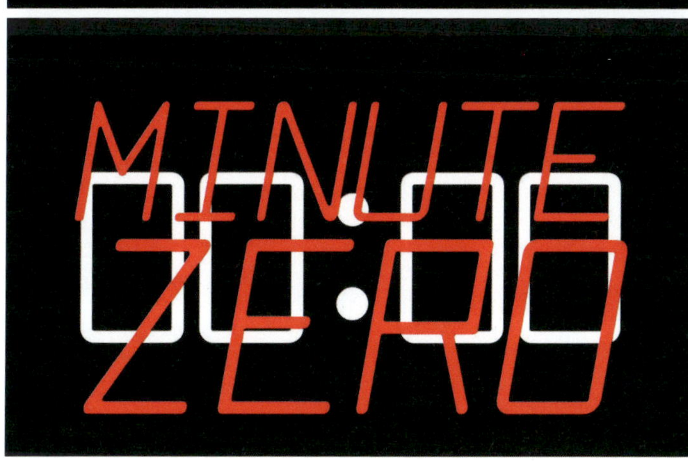

Stadtwerkstatt (AT)

STWST48x4 SLEEP

48 Hours Art as Sleep

With a major focus on sleep, the 4th edition of the 48-hour non-stop program brings together artists and critical producers to address an aesthetic of the unconscious. Art for Sleepers, Art by Sleepers and Art as Sleep.

SLEEP48 is a Stadtwerkstadt project conceived by Shu Lea Cheang and Matthew Fuller, who were inspired by Fuller's latest book *How to Sleep, the art, biology and culture of unconsciousness*. SLEEP48, 48 hours of sleeping through active sensing, experiencing, doing and perceiving, explores the complex physiological phenomenon that changes over its duration and has different cultural and physical expressions. Programs include the lecture DEEP SLEEP and the exhibition HEROES OF SLEEP. The sonic sanitarium of SONATAS OF SLEEP/LESS offers treatments for information shock-workers. As a statement of non-architecture: SLEEP TUNNEL shows infinite sleep. SLEEP BATTLE will be monitored by Sleep Laboratory. More: The experimental setting HYPNOMACHIA, the performative ROUGH SLEEP, a SPECULATIVE SCHOOL OF SLEEP DANCE and uncanny familiar SOCIAL SLEEP VIDEOS. Your own sleepy experiments can be followed up with SCHLAFGUT-BIER and SLEEP FOOD.

Along with the SLEEP48 program, *STWST48x4* presents an array of new works developed at Stadtwerkstatt under its new art contexts initiative—INFOLAB defines the concept of information, from synchronous quantum storms to randomness in the past and blockchain technology. The MYCELIUM NETWORK SOCIETY gives a preview of its Patulin exhibition in Taipei Biennale this year, plus the Mycelial Radio Activation workshop. The art and context research QUASIKUNST stages faded and blown away coordinates from artagens to body movens. From the Stadtwerkstatt archive, an indicative view applies to the PREVIOUS LAYERS of earlier Stadtwerkstatt projects. And at the Danube shore, the REOUZERI offers an after-work sunset area of relaxed, cooperative participation.

RiP—Randomness in Past is an INFOLAB project conceived and organized by Franz Xaver. With synchronous quantum noise in Lindabrunn and Linz (September 3-17, 2018)

Randomness and chaos as a source of creative action. They are used to encrypt messages in the world of algorithms and form the basis of free will in the analog world. Information theory, however, is pushing chance and chaos further and further out of our daily events. The "logically correct" answers of the algorithms displace the unfounded decisions of individuals. However, algorithms cannot develop utopias.

Therefore, we want to counteract this development with the idea of the "Global Consciousness Project" of Princeton University in the first two weeks of September and synchronize two locations via quantum noise generators: During STWST48x4, the STWST measuring ship Eleonore in Linz and the symposium Lindabrunn, a former sculpture quarry near Vienna, will be "connected" to each other via "acausal" quantum information.

Windlines—Movement A—48 Hours Drifting—Disappearing (50,000) is a Quasikunst project, conceived and realized by Tanja Brandmayr.

Quasikunst is systemic-performative research. Depending on the project, it calls out coordinates, increases and performs contradictions. In 2015 Quasikunst dedicated itself to trees with "I like Trees and Human Rights." In 2016, it equated the actor with the network in "Fog Ballet." In 2017, "Eisberg/The Entity" staged a theater of meaning of irreconcilable contradictions—and as a meltdown of a 2-ton ice block. In 2018, immateriality will be pushed forward. The focus is on the medium wind, on air, transparency, intangibility and the increase of distance. With a wind machine or several wind machines air is blown through the Stadtwerkstatt building.

STWST48x4.STWST.AT

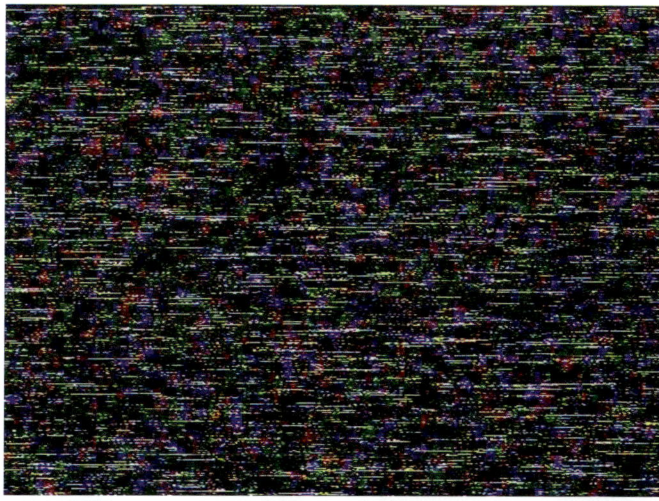
Franz Xaver

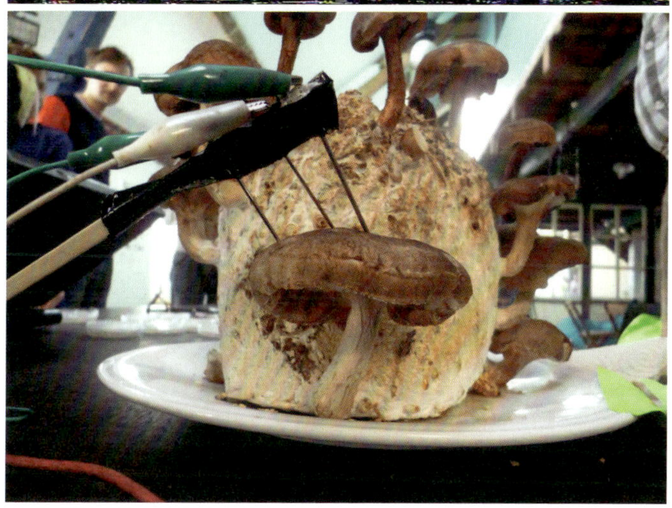
Martin Howse

The Windlines asks questions about the organization of thinking and feeling, and we say: Better faded than networked.
A living reference to Walter Benjamin's angel wanders backwards across the square: Movement A as a trace of memory in structured oblivion. And for a country of drifting entities, an almost transparent flag flies for 48 hours on the roof - for great internal and external spaces. The visibility of the 50,000: wanting to end. All in all, Quasikunst shows blown coordinates from art-agens to body-movens.

Mycelium Network Society (MNS) is a Stadtwerkstatt (Linz, Austria) and cycleX (Andes, New York) initiative to connect rhizomic mycelium network nodes. Mycelium Network Society imagines an underground network situated in a post-internet mudland and powered by fungus, spores, culture, kitchen, radio, transmission, installations, workshops and performances. Departing from the pursuits of magic mushroom, MNS charts a state of hyper-hallucination to collective fungal consciousness, investigates the fungi culture, its network capacity to communicate and process information. As an after-nature network, MNS commands its own domain in human-disturbed forests - sprouting across the moisture of rich soils, expanding its colonies, sharing information, networking co-habitants across borders. During STWST48x4, we present a prototype of a 17 atom patulin network structure to be premiered at Taipei Biennale 2019. Martin Howse of 1010.co.uk holds a one-day Mycelial Radio Activation workshop.

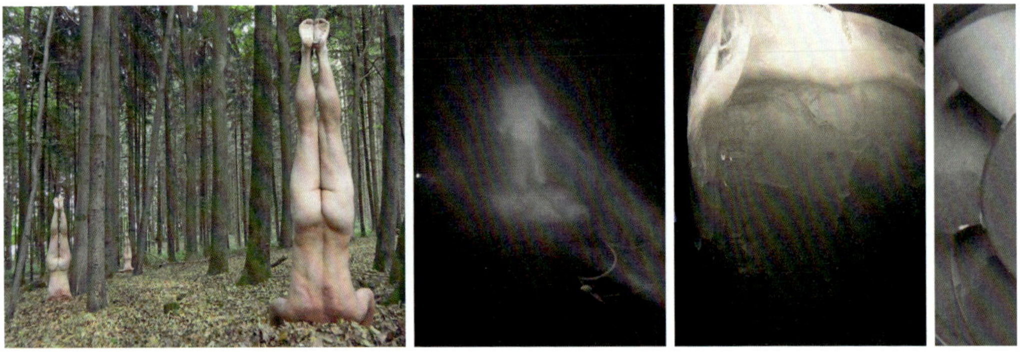

tanjab

Quasikunst seit 2015

http://myceliumNS.net

STWST48x4 ALL PARTICIPANTS +++
SLEEP48 - DEEP SLEEP and HEROES OF SLEEP: Matthew Fuller +++ SONATAS OF SLEEP/LESS: Svetlana Maras, Sainkho Namtchylak, Olesia Onykiienko aka Neither Famous Nor Rich, Ioana Vreme Moser, Mimu Merz, Jessica Ekomane, Cammack Lindsey, Tamara Wilhelm, Fu aka Nelson Landwerh. Curated by Shu Lea Cheang +++ SLEEP TUNNEL: Stadtwerkstatt +++ SLEEP BATTLE: Stadtwerkstatt, Andreas Kaindlstorfer / Kepler Universitätsklinikum Linz, Department of Neurology 1, NeuromedCampus Sleep Laboratory, Pamela Neuwirth, Harald Purrer, Georg Wilbertz +++ HYPNOMACHIA: Qualia Industries +++ ROUGH SLEEP: Katharina Brandl, Violeta Ivanova, Madeleine Schrabauer, Karla Woess +++ SPECULATIVE SCHOOL OF SLEEP DANCE: Gerlinde Roidinger, Tanja Brandmayr +++ SLEEP BEER: Jörg Parnreiter and Thor Bräu Ottensheim +++ SLEEP FOOD: taro +++

INFOLAB: Franz Xaver, GIS Orchester +++ Preview MNS at Biennale Taipei: Shu Lea Cheang, taro, Franz Xaver, Martin Howse +++ MNS-Workshop: Martin Howse +++ QUASIKUNST: Tanja Brandmayr, Felix Vierlinger +++ PREVIOUS LAYER: Stadtwerkstatt Archive +++ REOUZERI: Jakob Breitwieser

STWST CREW: Tanja Brandmayr, Jakob Breitwieser, Shu Lea Cheang, Thomas Hauer, Andreas Heißl, Jan-Nahuel Jenny, Jörg Parnreiter, Bernhard Schiesser, taro, Felix Vierlinger, Franz Xaver +++ Radio Features: Pamela Neuwirth +++ Documentation Videos: Claudia Dworschak +++ Homepage: Michael Aschauer +++ Graphic Design: Ortner&Schinko

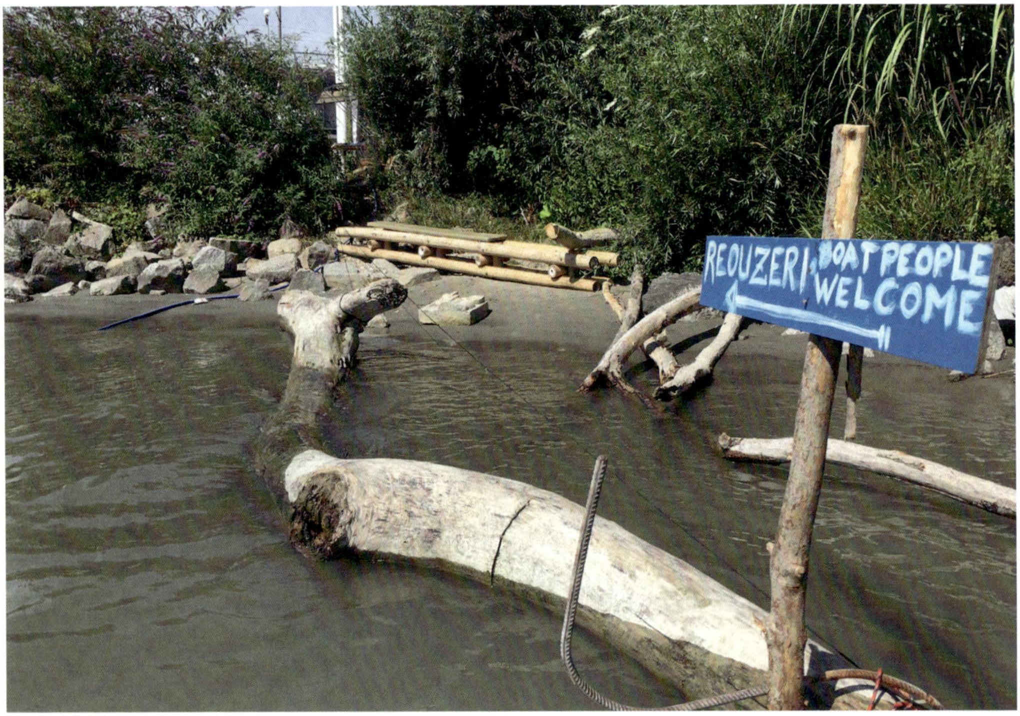

Jakob Breitwieser

Die Reouzeri an der STWST-Lände

BIO AUSTRIA (AT)

BIO AUSTRIA Stands for Organic Quality from Austria

BIO AUSTRIA is an association that proprietors of organic farms can join. There are about 12,500 members, which means that more than half of all Austrian organic farms are part of this community. As an association of organic farmers, BIO AUSTRIA advances the interests of its members domestically and abroad. BIO AUSTRIA is the go-to source of information for distributors, processors and media representatives, and also informs consumers about the advantages of organic farming and organic products. BIO AUSTRIA's mission is to help develop agriculture for the good of our children, our environment, our farmers and our society.

HIGH ORGANIC STANDARDS & BEST REGIONAL QUALITY

Organic farming produces high-quality products, and does so using methods that go easy on the environment. Pure water, healthy soil, climate protection and biodiversity are additional outcomes of organic farming that benefit all of us!

The entire organic production chain–from the farmer to the processor to the retailer–is checked and certified by an independent auditor at least once a year. And this is another big difference between organic products and other foodstuffs–when it says BIO AUSTRIA, you can really be sure!

OUR FARMERS GO THAT EXTRA MILE!

All BIO AUSTRIA member farms produce especially high-quality products since, in addition to complying with the EU's regulations for organic foods, they fulfill the association's even more stringent guidelines.

A few of the key additional provisions are:

- The entire farming operation is organic, and not just a portion of it at one location, which is permitted by EU organic regulations. For instance, growing organic fruit while fattening swine through the use of conventional methods is prohibited at BIO AUSTRIA farms. Only in this way can a closed-loop production cycle function.
- In the spirit of closed-loop production, feed and organic fertilizer are produced primarily on the premises. Purchases of additional feed are done in accordance with the BIO AUSTRIA Regional Model, whereby products from the region are always preferred.
- When it comes to animal husbandry too, BIO AUSTRIA farms far exceed the requirements of Austria's Federal Animal Protection Law and the EU's organic regulations. More free-range time, larger pens, a correspondingly larger area-to-animal ratio, and mandatory pasture grazing for milk cows demonstrate the tremendous extent to which BIO AUSTRIA member farmers have the well-being of their animals at heart.

BIO AUSTRIA is the only organic farming association that mandates raising male chicks along with the females destined to become laying hens. These and many other association guidelines assure that the quality of a BIO AUSTRIA product meets the high expectations of BIO AUSTRIA members, partner enterprises and consumers. That's why more and more BIO AUSTRIA partner enterprises engaged in both processing and marketing prefer BIO AUSTRIA quality products.

You're assured of BIO AUSTRIA quality when you see the BIO AUSTRIA logo!

Curious? To learn more about organic food, click here: http://www.bio-austria.at/bio-konsument/was-ist-bio/bio-zum-reinhoeren/

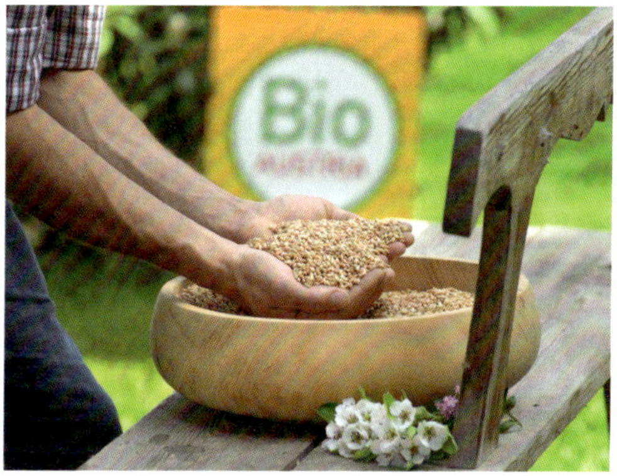

BIO AUSTRIA

netidee
Open Source Community Camp

Austria's largest open source internet promotion campaign, netidee, is a programme of the Internet Private Foundation Austria and has again produced three interesting projects as winners at the Open Source Community Camp.

Michael Faschinger (US)

bürgerchain

Project "bürgerchain" aims to provide a web-based e-voting platform. It is not supposed to be used by authorities in public elections, but rather by smaller organizations like political parties, clubs or companies for internal polls and voting. The "bürgerchain" should facilitate participation in democratic processes for members of such organizations. The main goal of the project therefore is to provide an e-voting platform combining anonymity, secure authentication, manipulation-safety and transparency with flexibility, ease of use and accessibility.

The unique approach of project "bürgerchain" is to combine the Austrian Bürgerkarte and a blockchain-like storage of voting results. The Bürgerkarte, Austria's official technology for rendering electronic signatures, enables voters to provide a proof of identity. The blockchain-like storage allows for the secure and manipulation-proof storage of voting results.

Project lead: Michael Faschinger
UI/UX design: Thomas Bayer
Crypto: Felix Klengel
Funded by: netidee (www.netidee.at)

Matthias Zeppelzauer (AT), Alexis Ringot (FR), Florian Taurer (AT), Kevin Pirner (AT)

SoniTalk

Ultrasonic communication is increasingly used for data exchange between mobile devices as well as for location-based services. UK is applied because it is inaudible and requires only very simple hardware (only a microphone and a loudspeaker). Several proprietary solutions exist today for the UK such as Silverpush, Lisnr, Prontoly and CopSonic. All these solutions are developed jointly by companies and raise questions regarding the protection of users' privacy. The goal of the SoniTalk project is to develop an open source, transparent and fully privacy-oriented protocol for UK. In the SoniTalk protocol, all privacy settings for each application are managed separately by the user via a separate central permission system. The architecture of the SoniTalk protocol also enables coordinated and synchronous communication between multiple applications for the first time.

Team:
Matthias Zeppelzauer, Florian Taurer, Kevin Pirner, Alexis Ringot
Supporters: St. Pölten University of Applied Sciences
Funded by: netidee (www.netidee.at)

Christoph Fabianek (AT), Nico Grienauer (AT), Christian Ziegler (AT), Florian Fida (AT)

▶ OwnYourData

Nowadays we each generate a huge amount of data, whether we use the internet, drive to work, or go grocery shopping. And we continuously make decisions which affect our privacy but also the convenience of using new services. Companies on the internet, in particular, have optimized their user experience in order to keep visitors on the website and make them accept far-reaching terms of service to gather ever more data. Our installation illustrates this dilemma and allows visitors to try new paths in their digital routine in a playful way. There they can – and should – make errors!

Funded by: netidee (www.netidee.at)

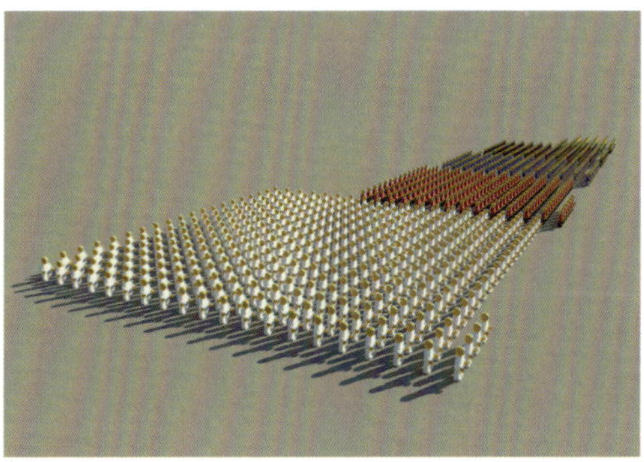

Matthias Wewering

bürgerchain

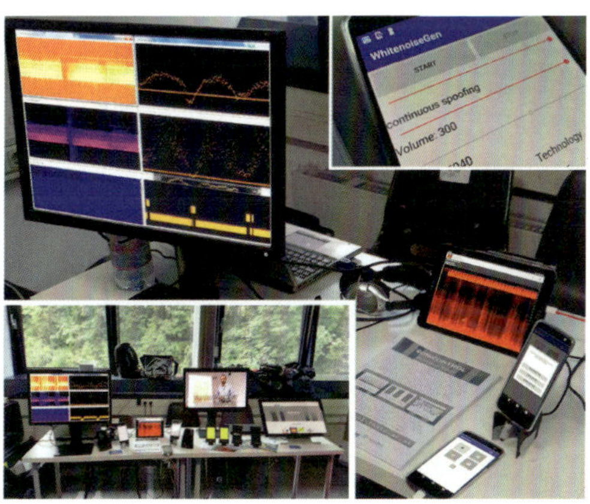

Fachhochschule St. Pölten

SoniTalk

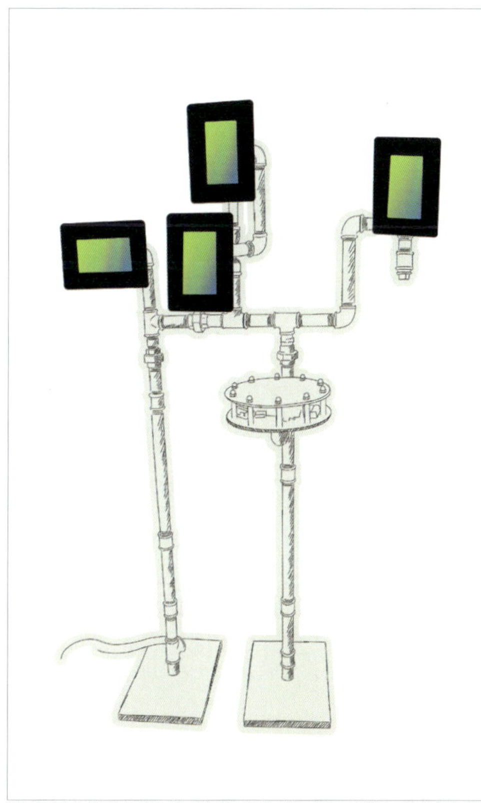

acolono GmbH, Nico Grienauer

SoniTalk

ERROR

the Art of Imperfection

ARS ELECTRONICA

Besides the annual Festival and the Prix Ars Electronica numerous projects and activities are conducted throughout the year by the different departments and teams of Ars Electronica.

- Ars Electronica Center
- Ars Electronica Futurelab
- Ars Electronica Solutions
- Ars Electronica Export
- Ars Electronica Japan

The following part gives a brief overview of these activities since last year.

Ars Electronica Center

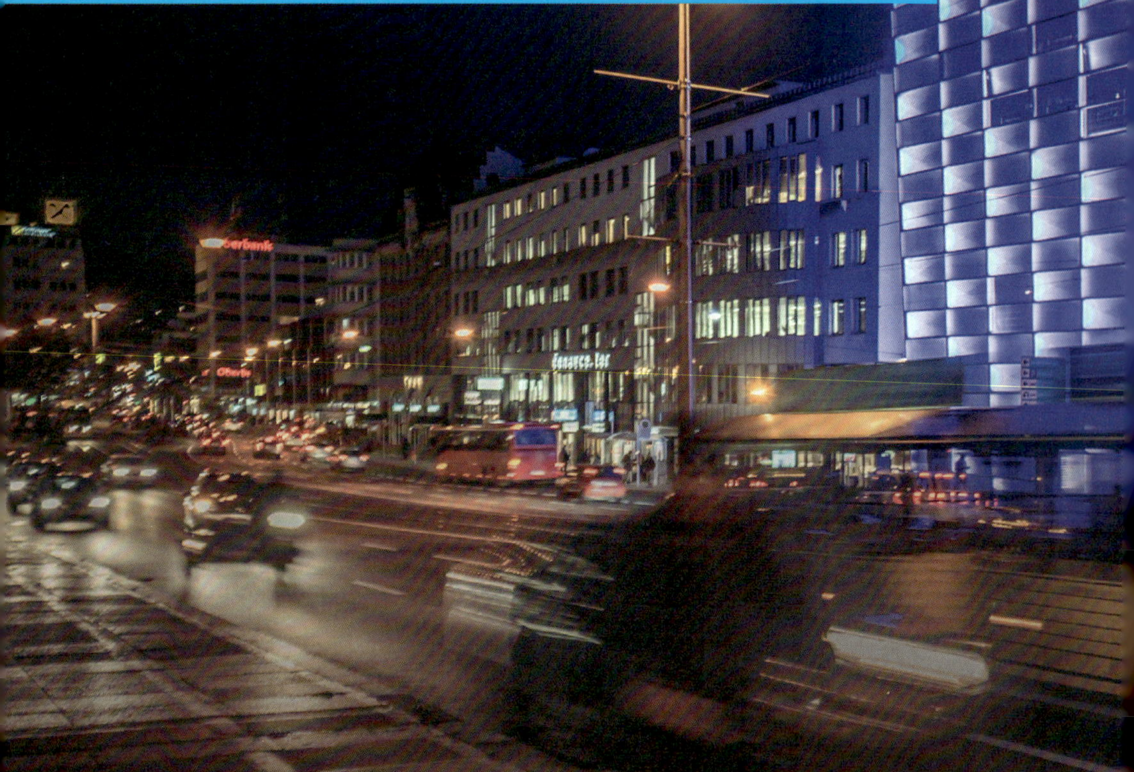

Museum of the Future

This is a time of transition for the institution known as a museum. Since its very inception in 1996, the Ars Electronica Center has been addressing the question of what a Museum of the Future ought to be like. Right from the start, the Ars Electronica Center has considered itself an experimental laboratory to learn how educational material should be imparted in an information-based society. Since its architectural makeover and reopening on January 1, 2009, more than one and a half million people have visited the exhibitions and attended the events held in Linz's Ars Electronica Center. One key to its success has been the concept of situating open labs among the exhibits, and creating a museum that makes carrying on a dialog with people the core of its mission. Change is another essential element of the Museum of the Future, an institution that continually reinvents itself while remaining true to its basic principles of examining the interplay of art, technology and society, and establishing a public setting for participation, conversation and fascinating discussions.

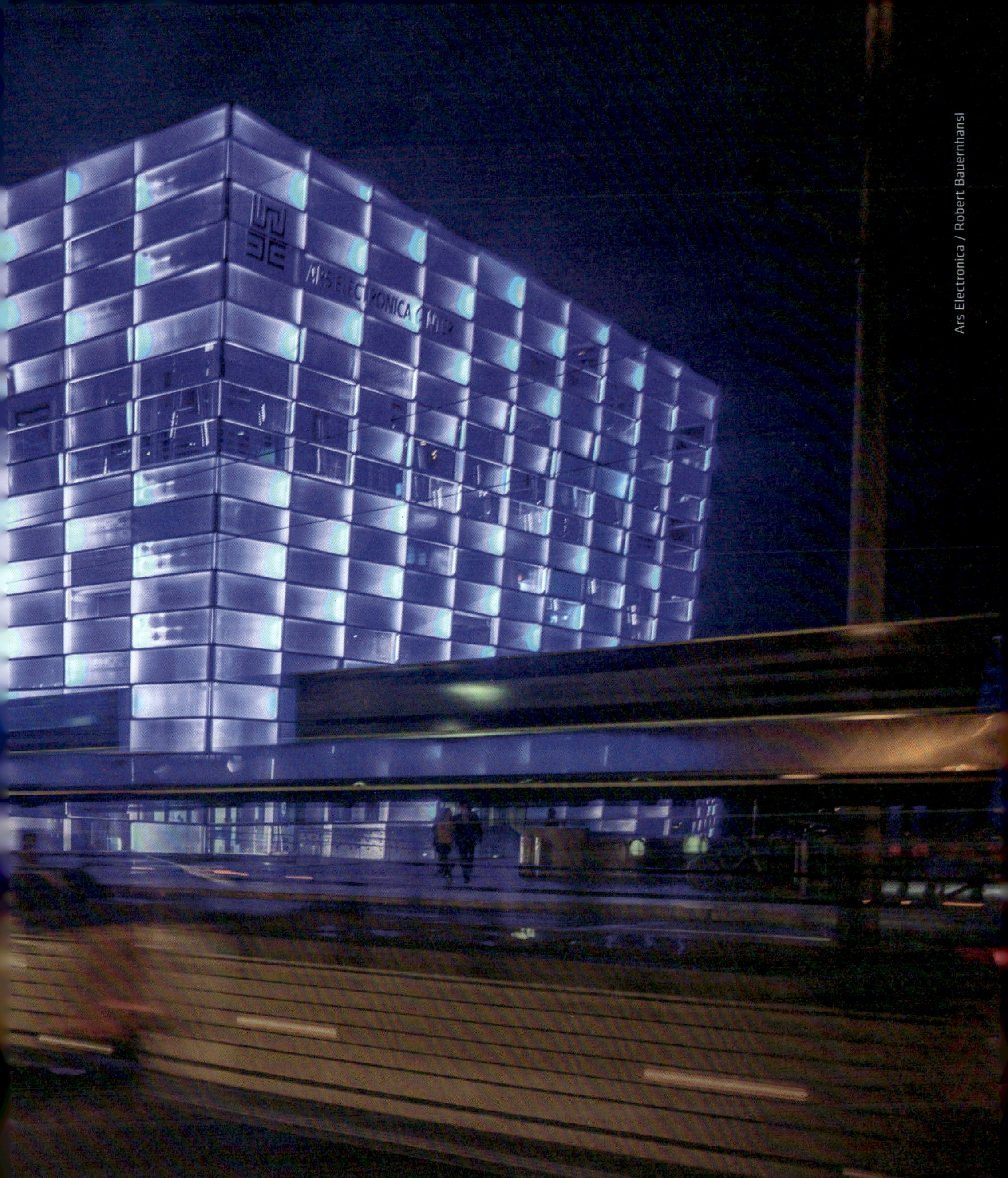

Ars Electronica / Robert Bauernhansl

Leading protagonists in this interactive process are the Infotrainers, who regard the exhibitions as their instruments and use stimulating stories and flashes of inspiration to get visitors involved in what they are looking at and to link it up to their everyday lives, local surroundings and personal experiences. Infotrainers get into a conversation with the members of the tour groups they guide, help them to formulate questions related to our shared future and offer them potential answers to consider on their own. In doing this, Infotrainers can reference current trends and configure the interaction with laypeople as a form of give-and-take among peers—after all, the Infotrainers themselves come from a wide variety of backgrounds and have acquired their expertise from a broad spectrum of fields including biotechnology, sociology, art, design, teaching and psychology. The Museum of the Future is one of the cornerstones of Linz's cultural landscape. It is a museum that pays attention to what people are saying.

Digital Education

DIGITAL IS PERVASIVE!

Our survey of everyday facts & circumstances confirms the picture propagated in the media: Digitization has arrived, and done so in grand style. It's now living large in our public sphere, our living rooms and recreational spaces, as well as in our workplaces and public domains.
Nevertheless, enthusiasm isn't the only emotion this digital transformation elicits; there are also insecurities and fears. But isn't this completely understandable? After all, when procedures change, new processes are implemented, and our latitude for action is shifted, it's to be expected that we have to make adjustments and obtain new skills.
And we need these new skills to go forth empowered in our endeavors to succeed and thrive in everyday life in the present and future. That said, the best way to acquire these digital skills isn't by taking a wait-and-see approach, gradually becoming familiarized, or via some sort of cultural memory; things are changing too fast for that. To get these skills, we have to get active! Here, a logical step is to accelerate the pace of educational initiatives that facilitate imparting these skills, and, of course, including this material in basic educational curricula.

WWW LIKE ABC

Reading and writing are among our civilization's most important cultural techniques. Mastering them is essential, a precondition for participation in social life. Since the development of these cultural techniques takes place in a socio-cultural context, they

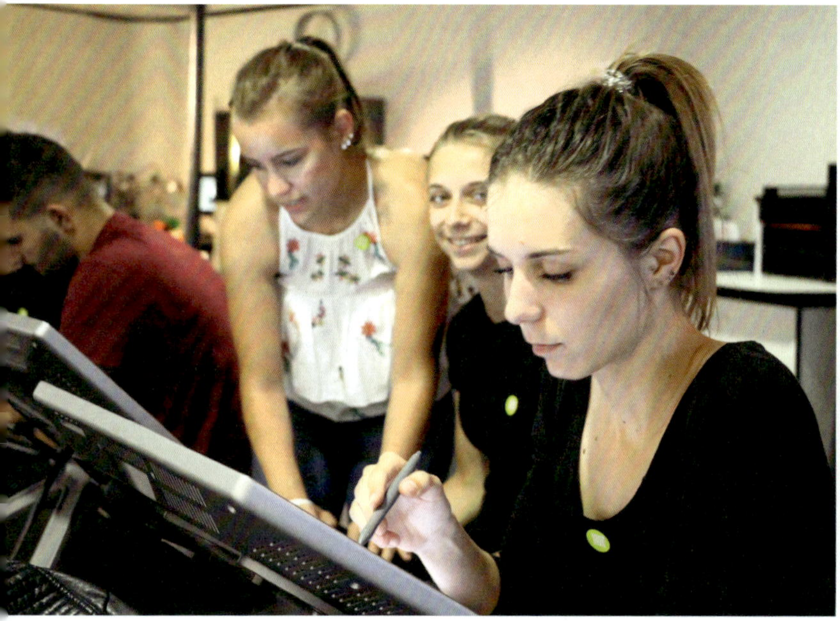

Ars Electronica / Magdalena Sick-Leitner

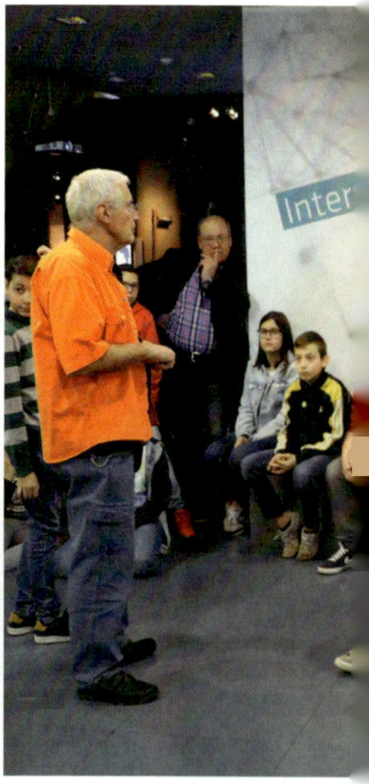

Ars Electronica / Magdalena Sick-Leitner

are modifiable and dynamic – thus, they grow along with the developments of our day and age. They are described as concepts designed to bring forth solutions to particular problems. In our Digital Age, challenges and problems are also of a digital nature (or spinoffs of digitization); accordingly, we must add this dimension to our repertoire.

Therefore, the theme of the 2017-18 school year in the State of Upper Austria was *The Year of Digital Education*. The aim: to provide the easiest possible access by as many pupils as possible to basic digital education and the new media that are essential to it. The agenda: to make it clear that digital education may not be considered a domain reserved for certain privileged schools.

Here, the Austrian educational system's vision dovetails perfectly with the capabilities brought to bear on the daily business at the Ars Electronica Center. Our mission statement: functioning as a competence center for digital technologies and their fields of application in everyday life. As an educational partner/auxiliary designed to support classroom instruction, our museum's program includes a diversified lineup of educational offerings that enable participants to effectively use, insightfully reflect upon, and subsequently evaluate digital technologies.

Besides our wide-ranging offerings for young visitors (mostly as members of school classes), we also provide training and continuing professional education for teachers.

And in addition to instilling information about new technologies and their consequences, we are also available to consult on new curricula and the educational strategies on which they are based.

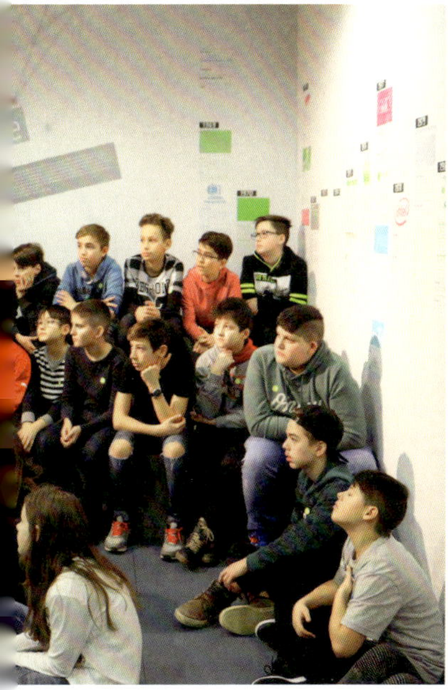

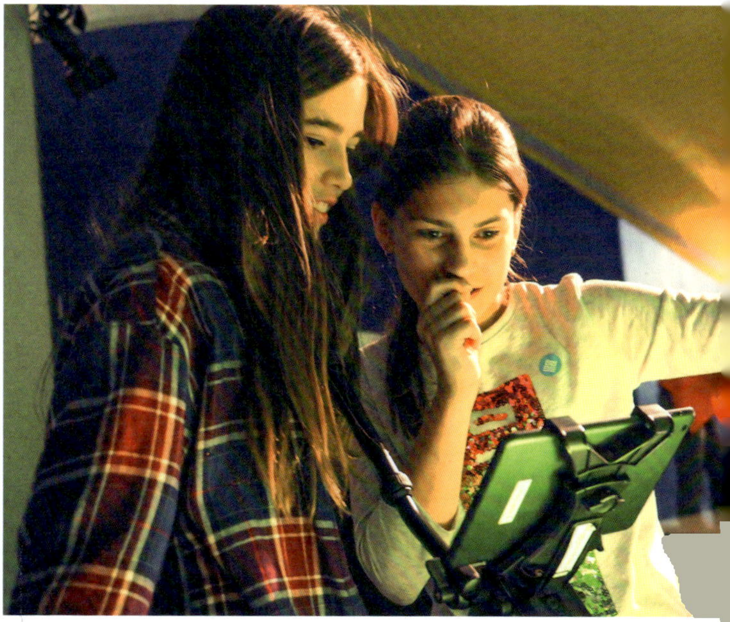

Ars Electronica / Martin Hieslmair

Ars Electronica as Educational Facility – School for the Future

In its capacity as an educational institution, the Ars Electronica Center has offerings geared to kindergartens, all primary and secondary school grades, universities and technical schools, as well as adults pursuing lifelong learning. The yearly figures:

- 4,000 kindergarten kids
- 10,000 elementary school pupils
- 20,000 secondary school students
- 2,000 apprentices
- 4,000 college students
- 180,000 total visitors

Emerging technologies, art and social change are the cornerstones of our program—but the focus is always on people. Accordingly, the Ars Electronica Center does not define itself as a temple of knowledge that delivers lots of interesting facts. It is what you might call a museum that is also a good listener, and is interested in the views, ideas and concerns of visitors. In going about this, what seems especially important to us is taking a playful, creative approach, generating enthusiasm for new ideas, and being committed to making them accessible to the general public.

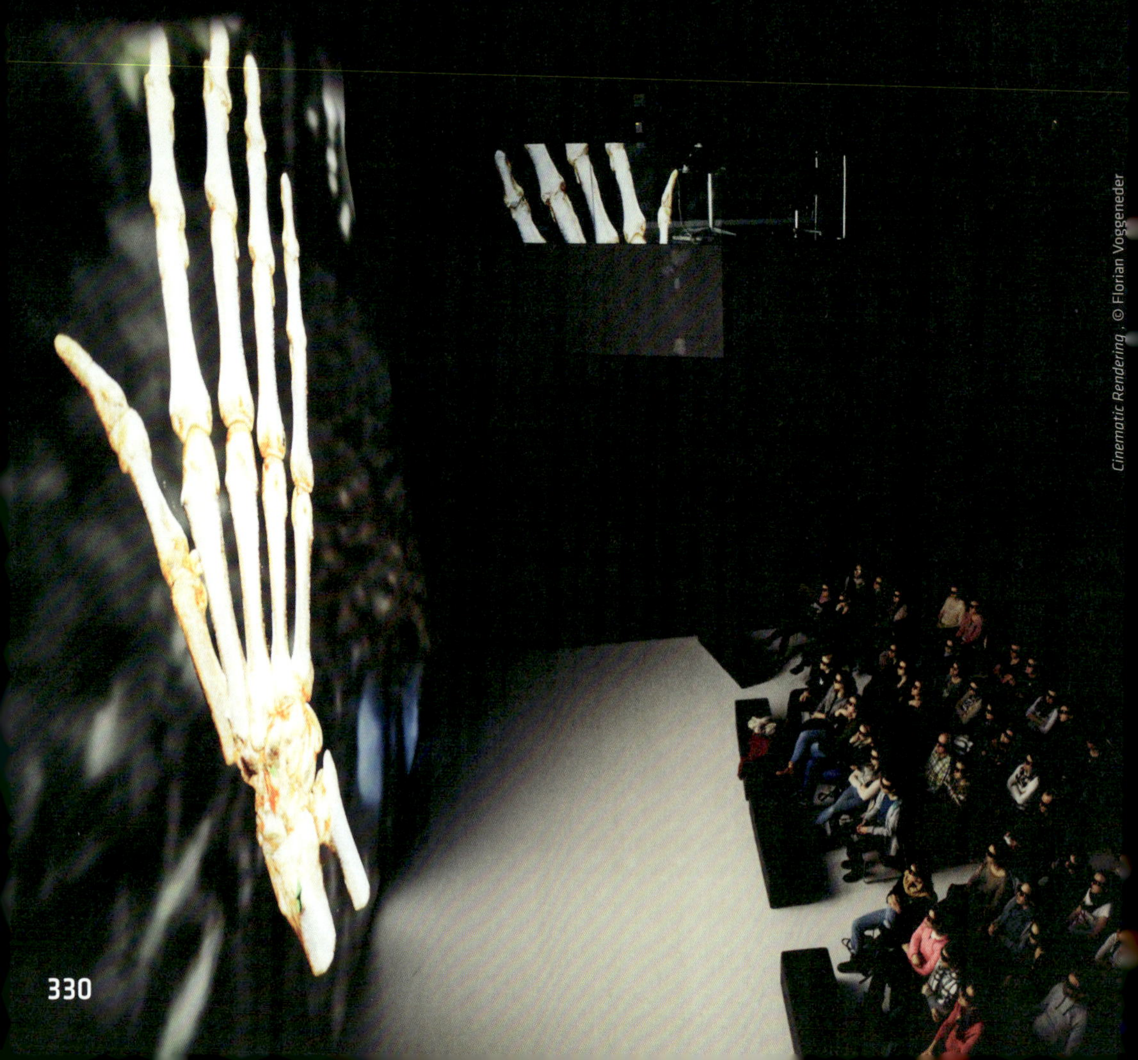

Cinematic Rendering, © Florian Voggeneder

Continuing Professional Education for Elementary School Teachers in Upper Austria

Our continuing professional education initiative entitled "Technology Is Child's Play!" is meant to enable elementary school teachers to use digital technologies in classroom instruction and to enhance their existing skills. The Ars Electronica Center Kids' Research Lab is a setting for experimentation—both analog and digital—with motor, cognitive and social skills. As Albert Einstein realized long ago, play is the highest form of research, and we have made this insight our maxim. We've configured a space in which 4-8-year-olds can engage in learning-by-playing and have fun while getting hands-on experience. With a mix of new technologies and familiar materials, our youngest visitors can spend quality time in the Museum of the Future and thus get valuable preparation for their own future. Developing eagerness to experiment and having fun doing so are essential to the way we work with children, but this also characterizes our approach to continuing professional education for elementary school teachers. The highly stimulating interplay of interesting theoretical inputs and do-it-yourself units both motivates and enables participants to use new technologies more confidently and integrate them into the educational process.

The Ars Electronica Center as an Educational Facility for Colleges

The Ars Electronica Center has been working together intensively with Upper Austrian universities and universities of applied sciences for many years now. Thus, the Museum of the Future is a setting for extracurricular education for elementary and intermediate school pupils as well as college students. The University of Linz, the Kepler University Clinic, the Upper Austrian Teachers College and Linz Art University are some of the many educational institutions that collaborate with the Ars Electronica Center. An interesting aspect of this is that the Ars Electronica Center not only makes available educational offerings such as the anatomy classes for students in the health profession held in Deep Space 8K and featuring visuals created with Cinematic Rendering software; we also provide students with the opportunity to appear at the Ars Electronica Center itself during the course of their educational career—for instance, the Linz Art University students who have exhibited their works in conjunction with the *Time Out* exhibition series.

My Future Workshop

Endowing future prospects, showing possibilities and trying out new things to discover one's talents—The *My Future Workshop* project launched by the State of Upper Austria in cooperation with the Ars Electronica Center especially targets unemployed young people enrolled in training programs. In May 2015, the first 20 youngsters spent a whole week together with counselors in the Ars Electronica Center. In the facility's labs, they got acquainted with fields of work that will increase in importance in the future, got hands-on experience with the tools of the trade and met a lot of people, some with unconventional career paths.
From 3-D printing in the FabLab to sound design in the SoundLab, and even producing an indie film of their own—this is a great chance to get in touch with undiscovered gifts and derive pleasure from creativity. The Future Workshop fosters the ability to work in a team, to independently develop solutions to problems, as well as to speak publicly in front of a small group. Youngsters equipped with a tablet have the opportunity to record their experiences in a blog and then refer to them in preparation for future job interviews. Since the first Future Workshop in May 2015, there have been more than 20 such events with unemployed young people in training programs.

APPRENTICE DAYS: INDUSTRY 4.0–LIFE 4.0

Apprentice Days at the Ars Electronica Center are held in cooperation with the Upper Austrian Industrial Association. At this two-day program for apprentices at Upper Austrian manufacturers, young people can get a glimpse at the world of Industry 4.0.

According to a fallacy that's very widespread in modern society, the only jobs that will soon be available in our increasingly digitized world are those that require a degree from a college or university of applied sciences. Nevertheless, it's simply not true that youngsters who've completed an apprenticeship won't have a chance on the job market.

The only difference from the way things have been in the past is that IT skills will increasingly be called for in the crafts & trades too.

To get their apprentices geared up for these changes in the world of work, 30 of Upper Austria's foremost companies took part in a two-day program, Industry 4.0–Life 4.0, at the Ars Electronica Center, where the apprentices immersed themselves in the Digital World during their informative and playful encounters with the exhibitions throughout the Museum of the Future. In light of the fact that what constitutes Industry 4.0 has not yet been finally defined, it can be maintained that as tomorrow's highly skilled employees, today's youngsters will help configure this development.

Teacher Continuing Professional Education and Ambassadors

If we want our youth to be well-prepared for the future, then we have to consider up-to-date teaching skills just as important as the process of reaching students. Accordingly, our educational mission is focused on continuing professional education for teachers. What do our teachers need to be prepared for the high-tech world of their students? What tools can we provide them with so they have the substantive wherewithal to venture into terra incognita and develop a sense of openness towards new approaches to teaching and to their students? Our answer is to familiarize them with topics, techniques and developments with which they, as digital non-natives, are now confronted. As their distance diminishes, teachers acquire more self-confidence as well as knowledge and skills they can apply in the classroom. We regard this contact as very important. In order for us to develop continuing professional education offerings that do justice to teachers' real-world needs, we, in addition to cooperating with teacher training institutions, have to stay in touch with the teachers themselves. We achieve this with the help of our Ambassadors–teachers as multipliers. One part of their role is to publicize the Ars Electronica Center's educational program in schools; on the other hand, they also provide feedback to us about what teachers and schools need. Ambassadors are on the faculty of schools throughout Austria. Their activism for a good cause is rewarded with benefits such as invitations to the Ars Electronica Festival and our Ambassadors' Day, discounts on program bookings and exclusive information.

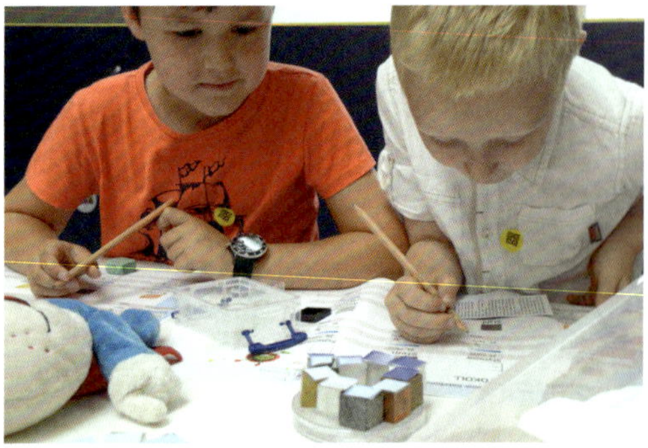

Ars Electronica / Magdalena Sick-Leitner

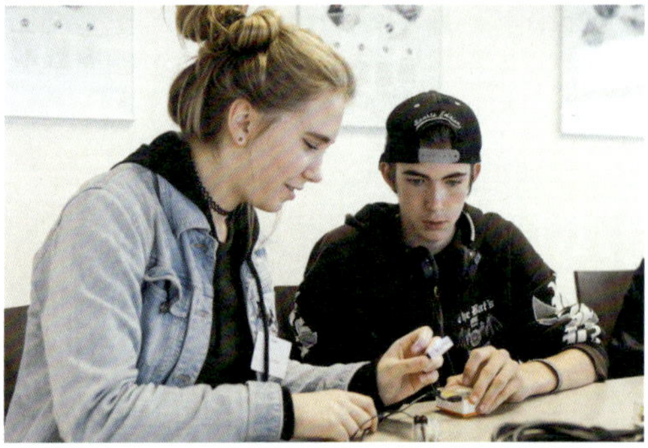

Florian Voggeneder

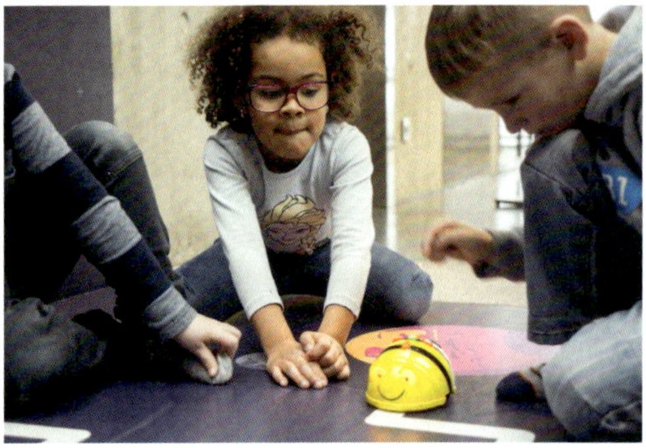

Ars Electronica / Vanessa Graf

FutureClash

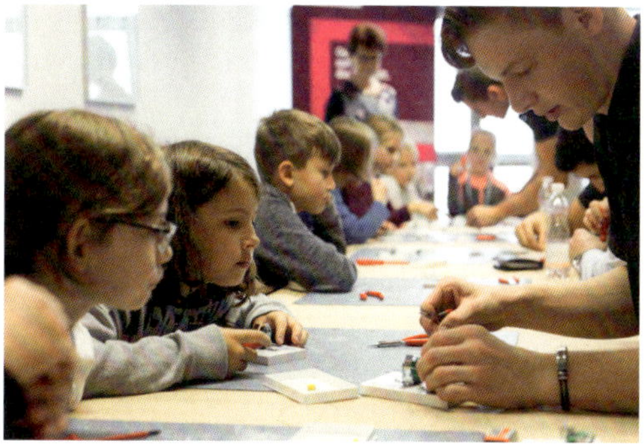

Ars Electronica/ Vanessa Graf

This summer, in cooperation with think300, the Ars Electronica Center organized the first *FutureClash*, a multi-day workshop program for entrepreneurs and kids. On three successive days, youngsters were accompanied by museum pedagogues and moderators as they engaged in exploration, experimentation and play.

FutureClash is an offering for firms that want to give their employees a break from very cerebral grown-up thinking and reconnect with imagination-driven, child-like-creative mental structures. The purpose of this is to break out of entrenched patterns and absorb inspiration. To achieve this, leading-edge technologies and creative techniques are used to come up with unconventional solutions.

The workshops are set up in such a way that there's successive interplay between the kids' program and the formats for company staffers. In the workshops, the kids' and adults' agendas regularly dovetail. Depending on the workshop, the adults are involved in a variety of forms—from collaboration as equals to interviews and observational phases.

Inherent in kids' natural behavior based on trial-and-error and their playful approach is the potential for big ideas. The ultimate objective of *FutureClash* is for adults to define new paths that lead them to creative solutions.

CanSat Competition 2017/18

Mini-satellites the size of a coke can, a rocket launch at the Austrian-German border and three teams of students from all over Austria, competing for the title of best CanSat team (and the participation in the European *CanSat Competition* from June 28, 2018, to July 1, 2018 in the Azores islands) - that's the first Austrian *CanSat competition* in a nutshell!

CanSat is short for satellites (Sat) the size of a coke can (Can). The participating teams only had a few months at their disposal to build their very own mini-satellite, before they would have to show their skills at the big launch at the airfield Schärding-Suben in Austria. On its way back down to Earth, the small devices have to fulfill two separate missions: Measuring the temperature and air pressure, as well as a secondary mission chosen by the teams themselves. The competition was organized by the European Space Education Research Office (ESERO) Austria, the educational office of the European Space Agency (ESA), which, in Austria, is located at the Ars Electronica Center.

And the winner was: Team EcoSat, formed by Klemens Fischl, Maximilian Hierner, Florian Leonhartsberger, Alexander Fraisl and Tobias Stronl, from HTL Ybbs! Congratulations! After winning the Austrian local competition, Team EcoSat continued in an exciting way - at the European-wide competition in the Azores at the end of June, their CanSat was once again able to demonstrate its skills.

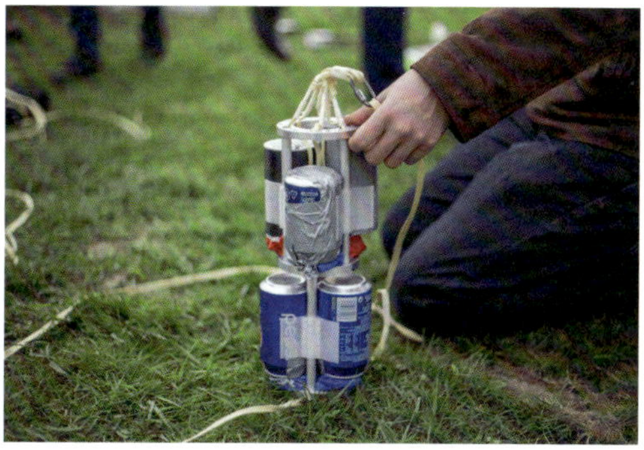

Ars Electronica/ Vanessa Graf

In-Flight Call to the International Space Station

In late 2017, ESERO Austria offered teachers an extraordinary opportunity—a live in-flight hook-up to the ISS from the Ars Electronica Center. Austrian teachers and students, together with their counterparts in the Czech Republic, Finland and Italy, could take part in a video conference with Italian astronaut Paolo Nespoli and discuss life in outer space.
In addition to the in-flight call, participants were treated to a fascinating presentation about the cosmos in Deep Space 8K, and were familiarized with educational materials made available by the European Space Agency.

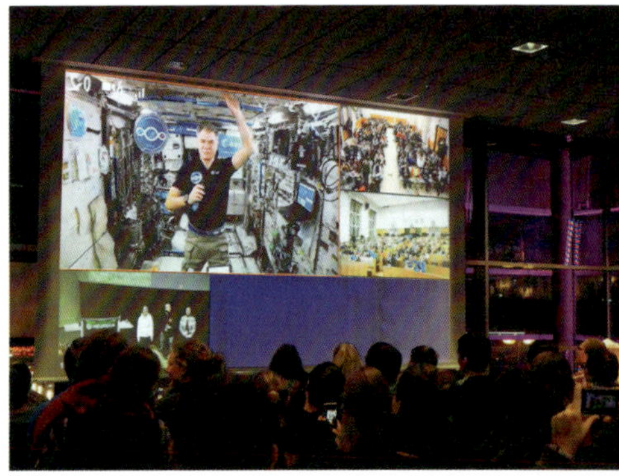

Ars Electronica / Martin Hieslmair

Wir sind Zeitung

What is fake news? How is it produced and how can we protect ourselves from it? False reportage exists in numerous facets and forms, and anyone can be confronted by it—for instance, phony postings on Facebook and faked videos on YouTube. So, how do you find out what's true and what isn't? Is it even possible to evaluate this content? Who can help me do so? And how should government leaders, law enforcement agencies and society deal with this very explosive issue?
School classes throughout Austria confronted with these questions in a competition entitled *Wir Sind Zeitung* [We're the News] sponsored by the *OÖNachrichten* (regional daily), Pädagogische Hochschule Oberösterreich (regional teachers' college) and Ars Electronica Center. The contestants were students in the 4th to 11th grades, who submitted journalistic texts such as news articles, interviews, letters-to-the-editor, op-ed pieces, reviews, editorials, proclamations and short films dealing with the subject of fake news.
The best submissions were chosen by a jury and published in a supplement to the *OÖNachrichten*. The pupils who created the prizewinning projects and their teachers were invited to a presentation in Deep Space 8K at the Ars Electronica Center.

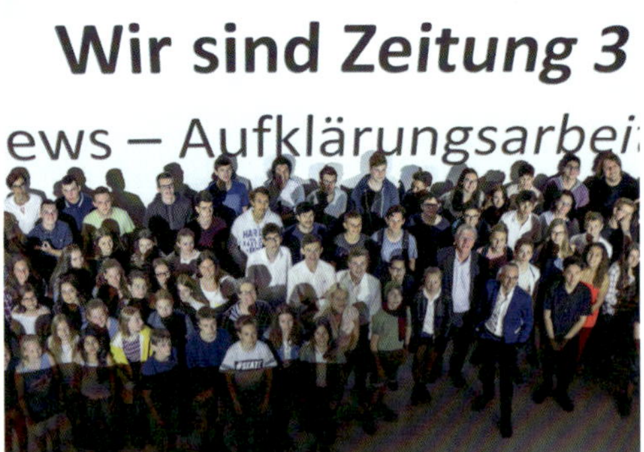

Ars Electronica / Vanessa Graf

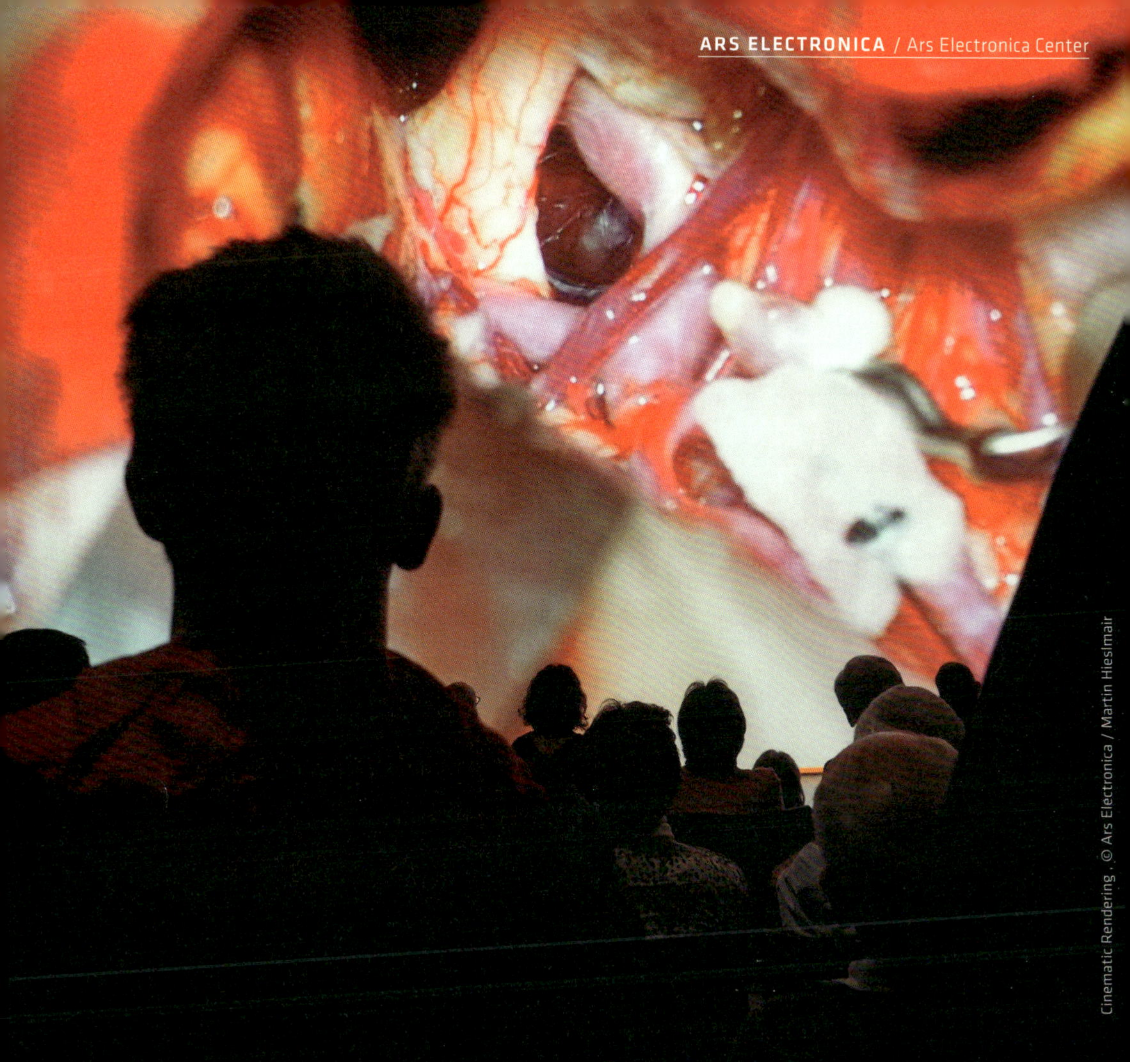

Cinematic Rendering , © Ars Electronica / Martin Hieslmair

Long Night of Research

We offered our guests free admission to the world of science and research once again on Friday evening, April 13, 2018! A wide array of research at facilities throughout Austria—from school projects to state-of-the-art science—was featured at this open house. The aim was to enable the general public to appreciate domestic institutions' achievements and to present current projects, recent findings and new technologies in interesting, understandable and entertaining ways. And this was a big draw once again – 228,000 visitors at 265 locations in all nine Austrian states set a new attendance record and reconfirmed the Long Night of Research as the largest such event in Austria. From 5 to 11 PM, 1,068 visitors stopped by the Ars Electronica Center and thus made the Museum of the Future a Linz hotspot on this evening.

The offerings included quickie tours through various exhibitions, presentations about the security risks of smart homes, and special screenings in Deep Space 8K. This year's highlight was a live transmission of brain surgery being performed at Kepler University Clinic, where Dr. Andreas Gruber wore a headset to communicate with remote spectators.

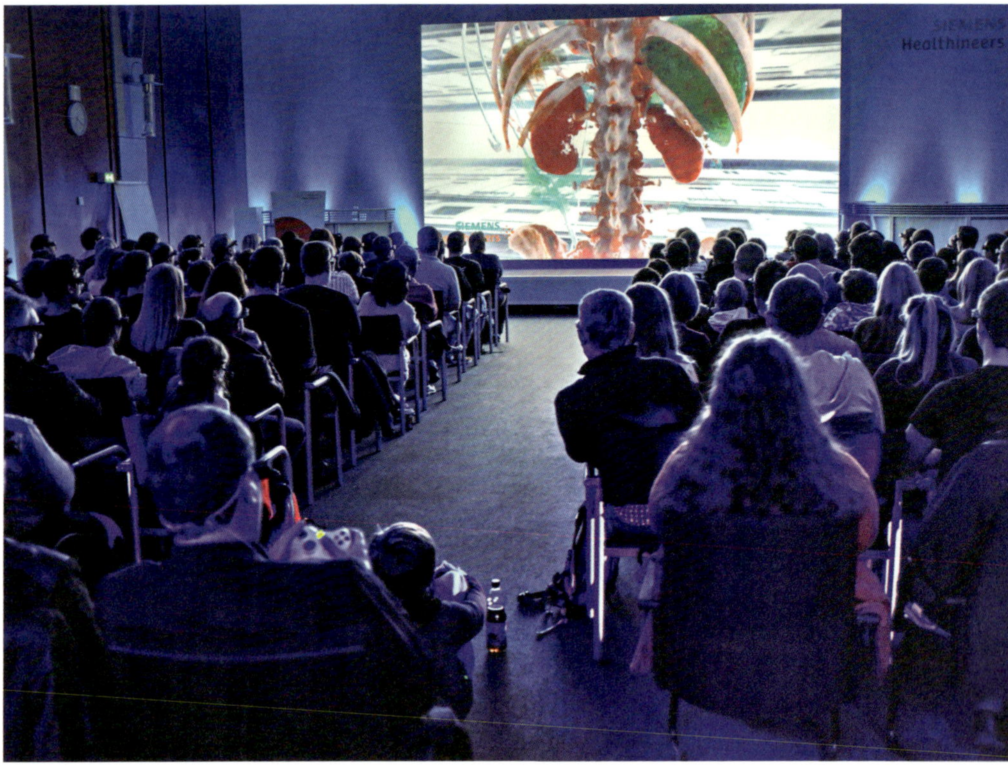

DS Mobile at the Long Night of Sciences in Erlangen, © Thomas Kollmann

Deep Space Mobile

Deep Space Mobile is a small portable version of Ars Electronica's extraordinary Deep Space 8K. Just like the original, Deep Space Mobile features a wide-ranging lineup in 2-D and 3-D. The big difference is that Deep Space Mobile isn't fixed in place; as the name suggests, it can be set up at events off-site. So, not only visitors to the Ars Electronica Center can enjoy the amazing content developed expressly for Deep Space 8K; clients who book Deep Space Mobile can, too.

The presentations are screened on a projection surface of at least 6x4 meters. The lineup includes just about anything that can be shown in the original Deep Space 8K—for example, Uniview takes you on a journey through outer space; Cinematic Rendering lets you explore the human body, layer by layer; and you can tour ancient cultural heritage sites like the subterranean Rome.

DonauArt

DonauArt is the umbrella brand of the museums and exhibition spaces of the City of Linz and the State of Upper Austria. From June to October 2018, they're presenting a total of 16 works having to do with the Danube Basin and water. At the Ars Electronica Center, you'll see the following:

Flow

Flow lets you feel what it's like to stand in the middle of a fast-flowing river. On the floor of Deep Space 8K at the Ars Electronica Center, thousands of tiny arrows "flow" from one wall to the other while the sound of rushing water is audible in the background. A spectator entering the projection surface awash with arrows modifies the current. *Flow* was produced by Poyraz Yildirim in conjunction with his studies in Linz Art University's Time-based and Interactive Media program.

Poyraz Yildirim (TR)

BLUE

How does it feel to stand below the surface of the Danube? Immerse yourself in a stream of unusual stories!

The Danube snakes its way across Europe for almost 3,000 kilometers; its course takes it through 10 countries. The Danube creates commonalities that bridge geographic borders. *BLUE* is a real-time visualization of flow speed, wave formation and water level. In a regular rhythm, bubbles flow past—on an ongoing basis, they collect information online about the river and its surroundings. When visitors collide with them, they burst and release their information. *BLUE* is not only a newsroom; it's also a clock that reflects the passage of time and the seasons. The installation changes along with the time and weather conditions, reacting to day and night, rain and snow. *BLUE* reinvents itself, second by second, and compels installation visitors to consider the Danube and the flow of time.

h.o (Hideaki Ogawa, Emiko Ogawa, Satoshi Onodera, Taizo Zushi), Ars Electronica Futurelab (Woeishi Lean, Sebastian Neitsch), Hydrographic Service of the State of Upper Austria

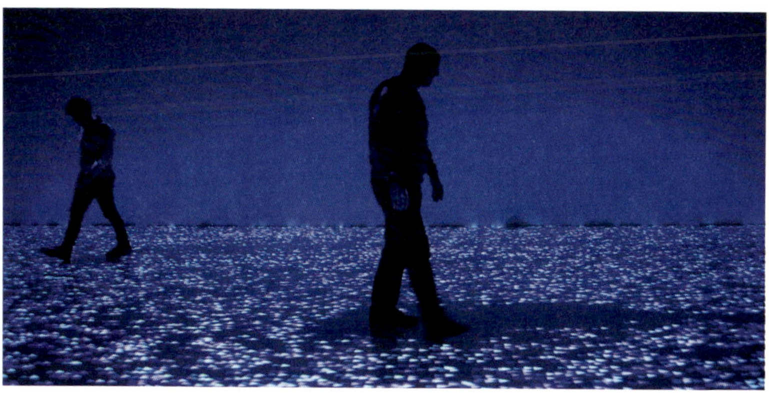

Ars Electronica / Vanessa Graf

Flow

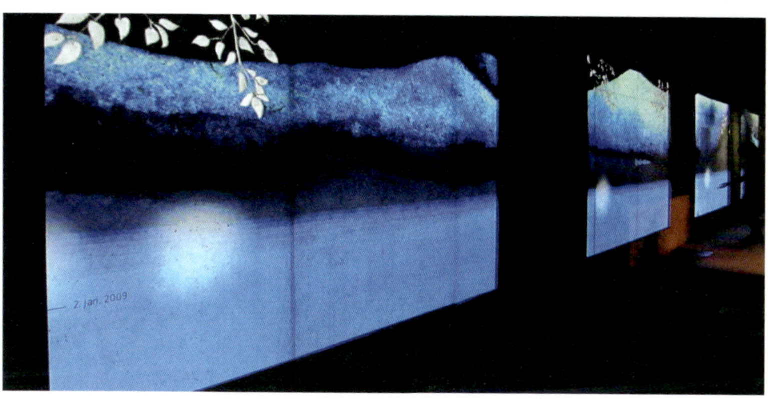

h.o

BLUE

2,800 Kilometers–10 Countries–1 River

Deep Space 8K is the ideal setting to behold a high-definition collage of satellite images in various color spectra showing the entire course of the Danube from its source to its delta. Satellites help us to analyze complex global interrelationships: weather developments, ongoing measurement of expanding cities, surveillance of the oceans' surface, and detection of forest fires. Thanks to satellite technology, we can acquire the precise data we need to protect our environment. But in addition to the scientific aspect, the so-called false color technique used to render these images makes them look great too! That applies above all to measurements made in the infrared range, which is normally invisible to the human eye. Instead of the usual combination of red, green and blue in the visible range, a combination of other channels makes it easier for scientists to evaluate the data. *2,800 Kilometers – 10 Countries – 1 River* features breathtaking footage shot along the Danube Basin, and is screened on Deep Space 8K's jumbo 16-by-9-meter projection surface.

ESA Copernicus processed by Birgit Hartinger und Juliane Leitner

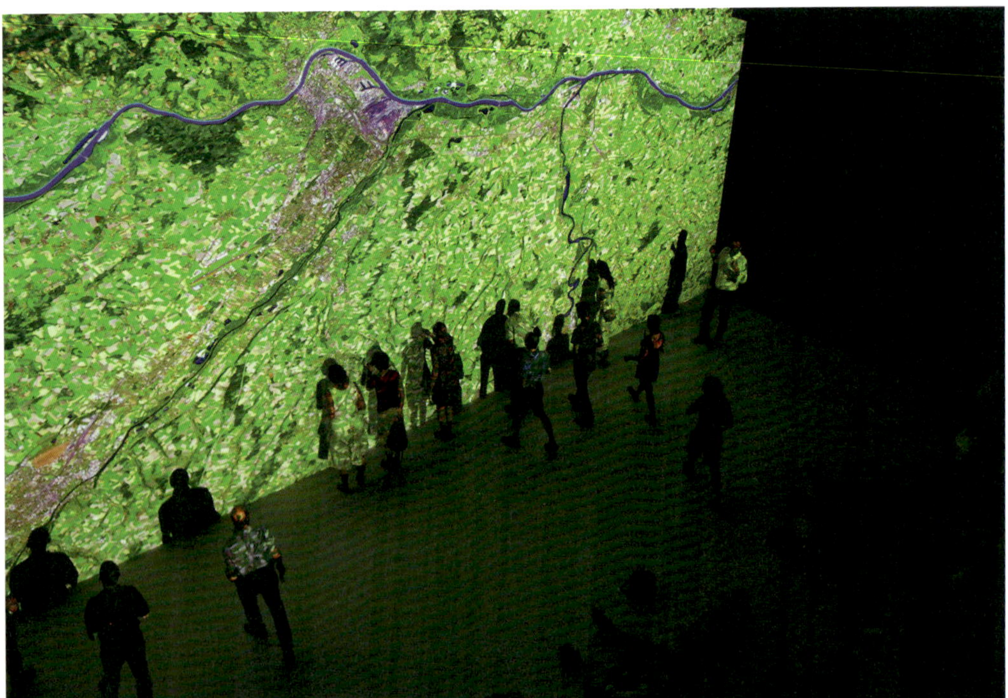

Ars Electronica / Martin Hieslmair

B2,800 Kilometers–10 Countries–1 River

RADICAL ATOMS

Hiroshi Ishii, Tangible Media Group, MIT Media Lab

How do we get the digital (back) into the physical world? An answer to this question could be so-called radical atoms.

In a sort of digital core meltdown, they bond information and material – the information liberated from the constraints of the pixel realm, the atoms wrenched out of their static state and set in motion. The results are smart materials that can be computer-modeled and remodeled into ever-new forms. Radical atoms in the hands of visionary scientists and engineers are being formed into high-tech materials and applications with astounding properties and capabilities. Neuroscience and biotechnology, robotics, hardware and software play roles in this; so do traditional practices in the crafts and trades.

A hotspot of these trailblazing developments is the Media Lab at the Massachusetts Institute of Technology (MIT), where Hiroshi Ishii and his Tangible Media Group have been working on new forms of human-machine interaction for over 20 years. The breakthrough prototypes that have been emerging there are now exhibited at the Ars Electronica Center.

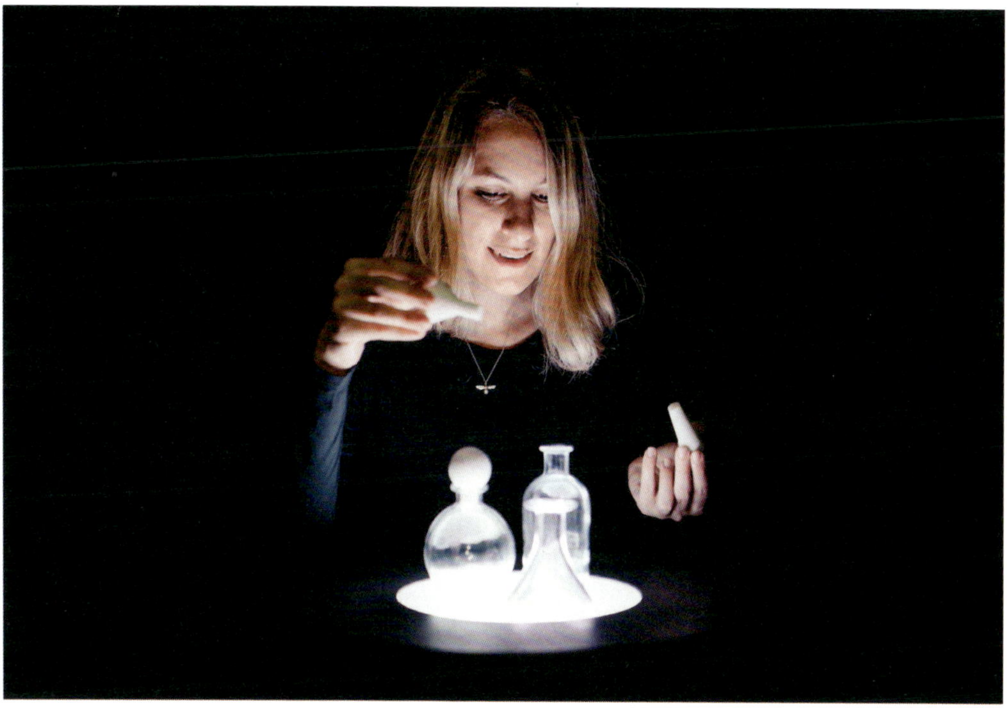

Florian Voggeneder

musicBottels, Tangible Media Group / MIT Media Lab

VRLab

Virtual reality, augmented reality and mixed reality, total immersion in virtual worlds and superimposing data onto our reality—for several years now, everybody's been talking up these concepts and ideas once again. The enthusiasm that accompanied the dawn of this new high-tech age in the 1980s and '90s is back and the technology deployed in today's data glasses (head-mounted displays) seems to finally be able to live up to the visions that preceded it.
VR, AR and MR have become a playground of multifarious pursuits: the gaming sector and film industry, applications in the educational field and tourism market, works of art and architecture, the creative economy, performance and the theater.
The VRLab in the Ars Electronica Center's Main Gallery showcases the latest VR, AR and MR technologies. In addition to applications by filmmakers and animators as well as artistic approaches, the VRLab relates the history of virtual and augmented reality's development. What did 18th-century spatial illusions look like, how did we progress from the stereoscope to the Oculus Rift, and in which directions will VR and AR be advancing in days to come? The VRLab provides insights into these questions.

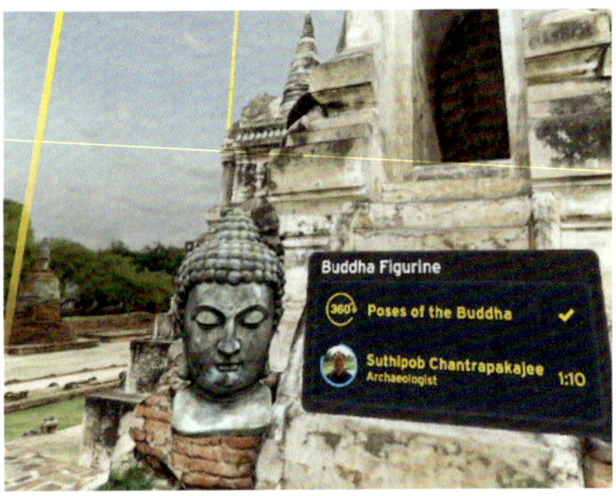

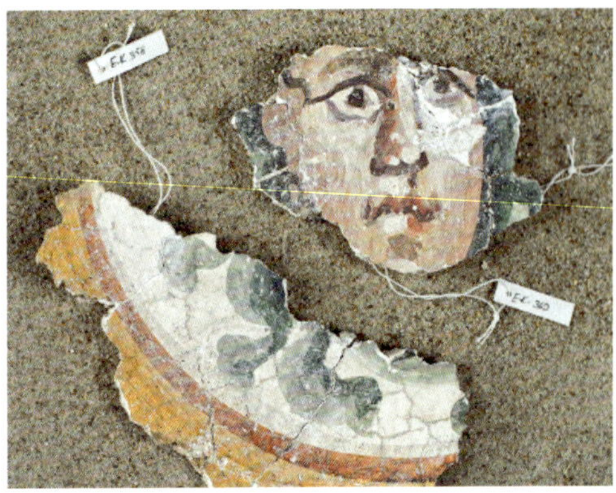

Bundesdenkmalamt

CyArk (US), FarBridge (US)

MasterWorks

In this virtual reality application, you travel to various historically significant locations around the world and discover the fate of the ancient capital of Thailand, the mysteries of a pre-Incan temple in the Peruvian Andes, the astonishing Native American cliff dwellings of Colorado and the monumental stone carvings of Mount Rushmore in South Dakota. CyArk's virtual *MasterWorks* museum shows four fully explorable environments in which artifacts can be collected and learned by archaeologists and scientists.

FH Hagenberg / Playful Interactive Environments (AT)

House of Medusa

In 2000, fragments of Roman mural paintings were discovered and recovered during an archaeological emergency excavation prior to the construction of a car park in Enns. The paintings were part of a vaulted ceiling with associated wall decorations and significantly broaden the knowledge of this art form in Roman Austria. Enter the virtual house of Medusa in this VR application.

Richi Owaki (JP), Yamaguchi Center for Arts and Media [YCAM]

The Other in You

The Other in You, developed as a new way to experience dance, has realized a novel dance audience experience. We assembled the cutting-edge computer graphics, haptic feedback devices, which directly express the dance to the body, 16 stereophony channels sound and research on Virtual Reality techniques to realize this work. How can we relate to others, who are supposed to be distant from us? Do we really know what it is to "see"? *The Other in You* is an attempt to revive the notion of our body in relation to an object, a notion, which had been forgotten in the act of watching. The movement of the dancer is captured with a motion capture system. The data is converted into 3-DCG to create virtual dancers covered with black skin. Using VR technology, the views of the audience move freely during the performance and the dancers get very close to the audience—an unlikely situation with ordinary dance theater. In addition, screened by 3-D sensor cameras, the audience members themselves virtually come to appear in the VR images.

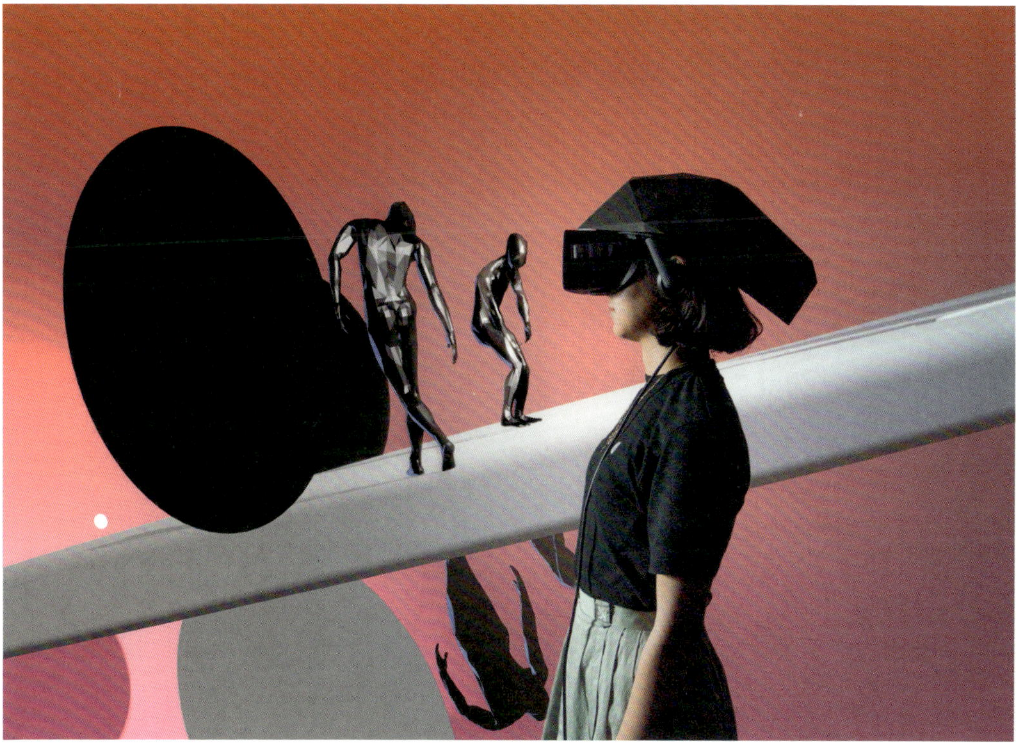

DEPART (AT)

The Lacuna Shifts

Alice in Wonderland inspired the artists' collective named *DEPART* to create this Virtual Reality application that immerses users in a world behind the looking glass in which nothing is the way it seems. Walls move, objects begin to speak, and gravity ceases to operate. The space is incessantly changing, depending on one's point of view. The use of a real-time 3-D environment lets visitors experience this realm of wonders that has new surprises in store each time you step inside. In the background, voices are audible, and there appear excerpts from Lewis Carroll's Alice in Wonderland, its sequel entitled Through the Looking-Glass and what Alice Found There and new poetic fragments written by the members of *DEPART*. The combination of these surreal graphical, auditory and textual levels sends visitors on a journey that is characterized by a deceptive sense of space, acoustic hallucinations and mysterious aesthetics, and that begins to call into question the entirety of one's perceptions of space, the world and reality itself.

VR-Experience, Duration ca. 6 minutes
Computer, HTC Vive, Wooden Object, Mirror
2017 by DEPART (Leonhard Lass, Gregor Ladenhauf)
commissioned by Sound:Frame

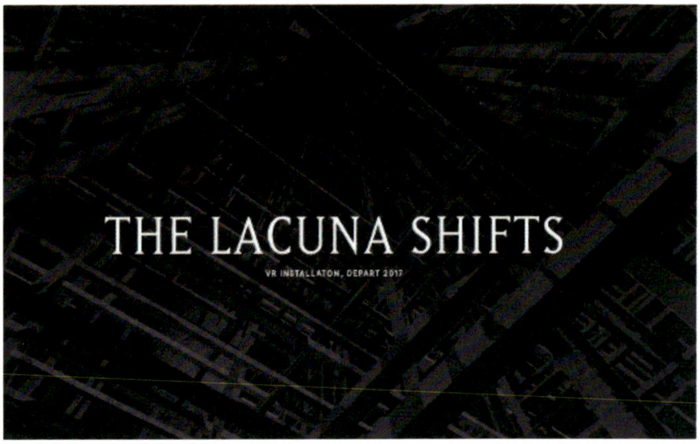

Keiichi Matsuda (UK/JP)

HYPER-REALITY

HYPER-REALITY is a concept film, presenting a provocative and kaleidoscopic new vision of the future, where physical and virtual realities have merged, and the city is saturated in media. Our physical and virtual realities are becoming increasingly intertwined. Technologies such as VR, augmented reality, wearables, and the internet of things are pointing to a world where technology will envelop every aspect of our lives. It will be the glue between every interaction and experience, offering amazing possibilities, while also controlling the way we understand the world. *HYPER-REALITY* attempts to explore this exciting but dangerous trajectory. It was crowdfunded, and shot on location in Medellín, Colombia.

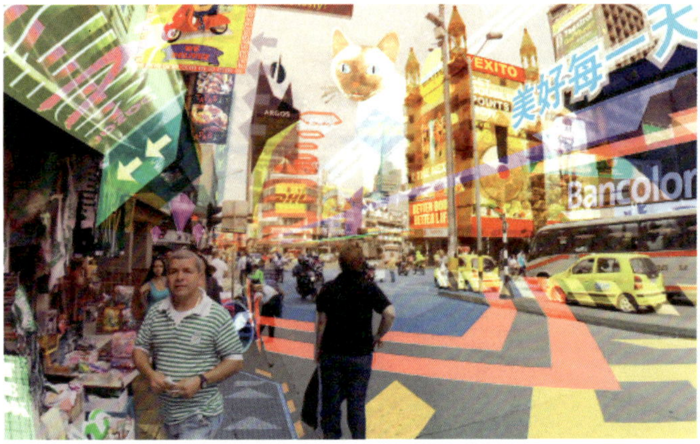

Jürgen Ropp (AT), Oliver Lehner (Sound Design) (AT)

LOGIN

This installation invites visitors to physically experience the login process, which normally isn't visible. The transition from physical to digital states is manifested by an experience in which the perception of one's body melds with its representation in the form of data. In the installation's virtual environment, the omnipresent data coalesce into spatial, almost tangible entities, and the point of view assumed is that of the machine. The virtual space is not only populated by the visitor's own body; other participants also insinuate themselves into the digital world and influence it. The boundaries between reality and virtuality increasingly blur.

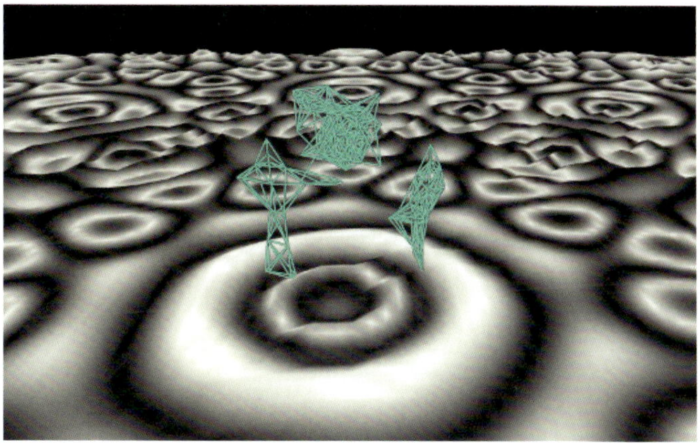

INEO (AT)

Project INEO

INEO is a virtual reality experience designed especially with the interests and needs of non-tech-savvy seniors in mind. The whole experience—beginning with the choice of scenery—was developed in cooperation with members of the Pensioners Club of the City of Vienna. Thus *INEO* can offer a growing target audience—the older generation—a way to get into modern technology and partake of state-of-the-art entertainment. The first *INEO* experience takes players on a trip to outer space. In the role of a marooned astronaut, they can weightlessly explore an abandoned space station and activate its escape pod for the return trip to Earth. *INEO* plans to enable players to have input into the development of new experiences in the future. For example, homes for the aged or retiree clubs will be able to order the service as a subscription plan that includes the necessary hardware. In this way, members can conveniently enjoy a constantly changing line-up of virtual-reality-based experiences.

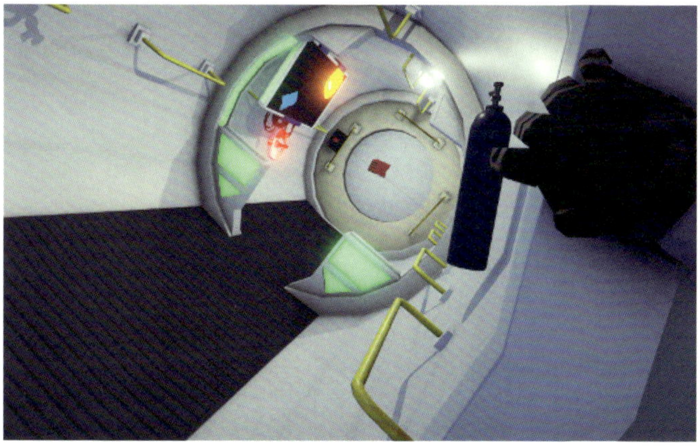

Frederick Baker (AT/UK)

Klimt's Magic Garden

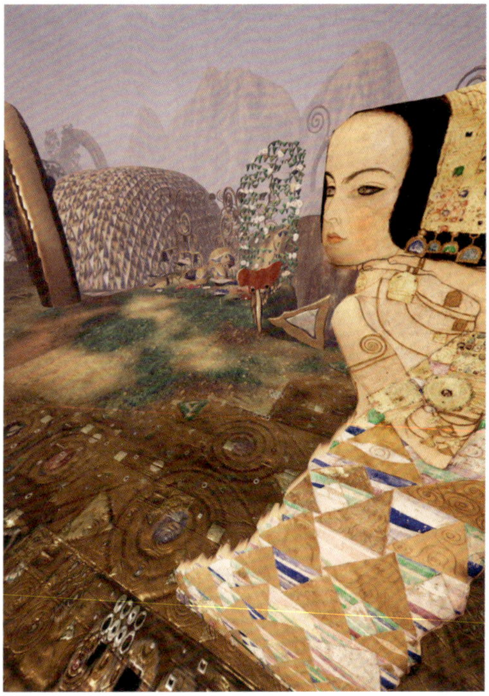

Klimt's Magic Garden allows the audience not only to enter INTO Klimt's art work for the *Stoclet Frieze*, but also to travel OUT the other side into a 21st-century experience. That new space is *Klimt's Magic Garden*, the story space that lies behind the two-dimensional surface of Klimt's design.

Klimt's three characters the knight, expectation, and fulfillment are given agency and the audience is allowed to explore these themes by making their own journey into the visual world Klimt created. My design philosophy is slow VR, which means allowing the first-time user to have the space and time for a smooth experience.

Direction and conception: Frederick Baker,
3-D Graphic Artist: Markus Cermak
Sound Designer and Composer: George Taylor
Video post-production: Christian Leiss Postproduction GmbH
Dramaturgy Sandra Fasolt.
World premiere at the Museum of Applied Arts (MAK), Vienna.
A Frederick Baker project in cooperation with Christian Leiss GmbH and the Museum of Applied Arts (MAK), Vienna
This VR-experience was commissioned by the MAK – Austrian Museum of Applied Arts / contemporary art and presented at the MAK for the first time in 2018.

Gianpaolo Barozzi (Cisco) (IT), Bill Jackson (Cisco) (US), Chuck Shipman (Cisco) (US), Laura Smith (Slanted Theory) (UK), Mark Burkitt (Slanted Theory) (UK), Toshi Hoo (Institute For The Future) (US)

Digiti-light the Human Networks

The naked ape is a social animal. We have always been connected to each other through networks of relationships. *Digiti-light Human Networks* is a collaborative VR environment that immerses people in the digitized map of their organizational network, enabling them to understand and feel how they are connecting and collaborating. Digital technologies make the creation and application of human networks simpler and faster than ever before, deeply changing the way we live, learn, play and work. Organizations and corporations are transforming from hierarchical structures to networks of networks, requiring new solutions to enable their talent to succeed and thrive. Their HR teams need to deliver delightful and compelling digital solutions to be impactful; moving from H-R to H-Art, recognizing that beauty has always been a key attribute in evaluating data, hypotheses and theories – "Theoretical physicists accept the need for mathematical beauty as an act of faith" [Paul A.M. Dirac].

Cisco: Gianpaolo Barozzi (IT), Bill Jackson (US), Chuck Shipman (US) *www.cisco.com*
Slanted Theory: Laura Smith (UK), Mark Burkitt (UK) www.slantedtheory.com
Institute For The Future: Toshi Hoo (US) *www.iftf.org*

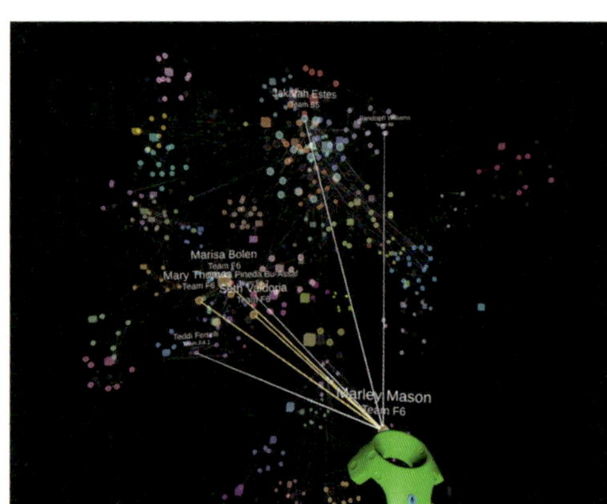

TIME OUT .08

Exhibition opening on March 16, 2018

Linz Art University and Ars Electronica have been working together closely for many years now. One collaborative project has been the *TIME OUT* exhibition series that enables young media artists in the school's Time-based and Interactive Media program to show their work in the Ars Electronica Center.
This Bachelor of Art degree program offers comprehensive and highly professional approaches to the theory, technology and design of digital media. It's geared towards students interested in gaining in-depth skills and wide-ranging experience in the areas of video, audio, installation, interface and interaction. In this program, the accent is on providing students with plenty of creative latitude to express themselves using the means afforded by audiovisual technology, to experiment with digital media, and to bring their own ideas to fruition.
On March 16, 2018 *TIME OUT* entered the eighth round! This time the focus was on works for the Deep Space 8K and ground projections for the museum elevator.

WORKS:

Floor projections for the museum elevator

the hive: Christian Berger
palmeria: Hanna Besenhard / Felix Winkler / Reinhard Reisenzahn
Muscle man: Titus Probst / Joachim Iseni / Philip Eglauer / Viktoria Pflüglmayer / Stefan Bruckner
sub fluid: Yazdan Zand

Works for the Deep Space 8K

Particle Images: Christian Berger
3D Shape: Christian Berger
Das Holzkamel von Marokko: Markus Bruckner
Feuerwerk: Ute Hackl
What an Illusion: Fabian Terler
Mischpult: Monica Vlad
Espacio Profundo: Gregor Woschitz
Flow: Poyraz Yildirim

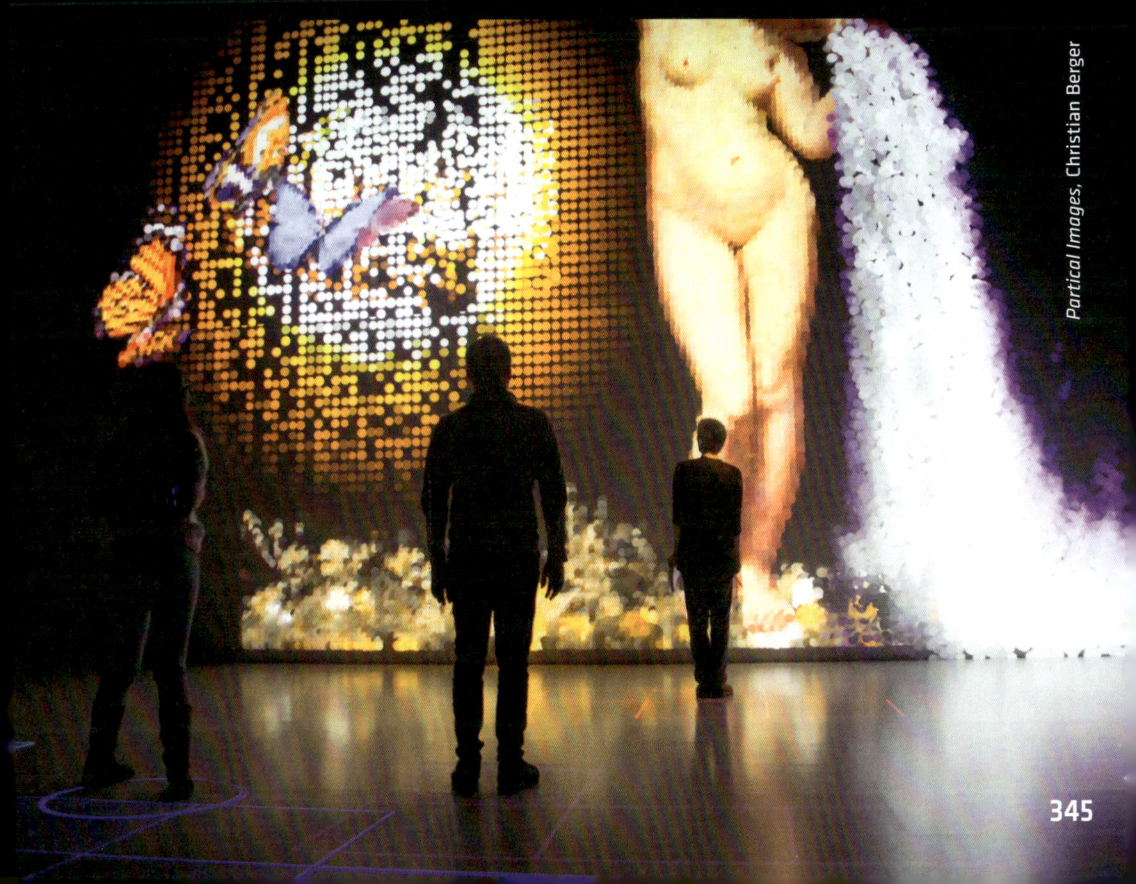

Partical Images, Christian Berger

Creative Robotics

Exhibition opening on May 9, 2018

The key idea of Industry 4.0–networks interlinking all the infrastructure, machines and human beings involved in production processes–currently determines how we think about coexistence with robots. On one hand, it's said that automation will facilitate fabrication processes and simplify our everyday life; on the other hand, rapid progress induces anxiety and insecurity at the thought of machines gaining the upper hand over the world we inhabit.

Beyond the realm of their industrial applications, however, members of a creative young generation of architects, artists and designers are now using machines to perform tasks that are radically different than the jobs they were designed to do– applications that highlight the possibilities of cooperation and alternative utilization. Can industrial robots be used to fabricate fascinating forms in the field of fashion design or to produce intricate objects out of clay? What does a bridge assembled out of robot-formed modules look like? Will our public spaces someday be teeming with intelligent robotic-architectural structures that automatically adapt to our needs?

The *Creative Robotics* exhibition demonstrates once again how industrial robots are being used outside of their intended areas of application in mass manufacturing, and have become a medium of artistic and creative expression and a catalyst for the implementation of innovative ideas and the manifestation of futuristic visions. Together with our prominent associates, we are exploring new ways to utilize industrial robot technology beyond the confines of big assembly lines. In cooperation with KUKA; Laboratory for Creative Robotics, Fashion and Technology program at Linz Art University; Institute of Robotics at JKU-University of Linz; Institute for Computational Design and Construction (ICD), University of Stuttgart; Centre for IT and Architecture (CITA) in Copenhagen; Co-de-iT in Turin; Nico Rayf @ Tree of Motion in cooperation with the Applied Robotics Lab and Wood Technology at the University of Applied Arts Vienna.

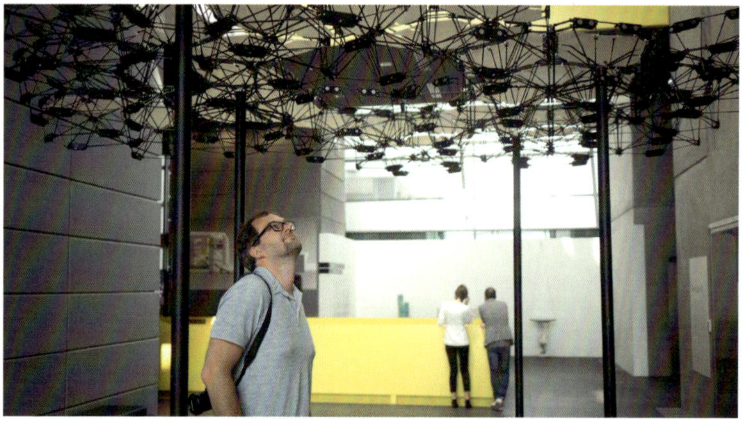

Ars Electronica / Robert Bauernhansl

Cyber Physical Macro Material, Miguel Aflalo (BR), Behrooz Tahanzadeh (IR), Jingchen Chen (CN), Institute for Computational Design and Construction (ICD), University of Stuttgart

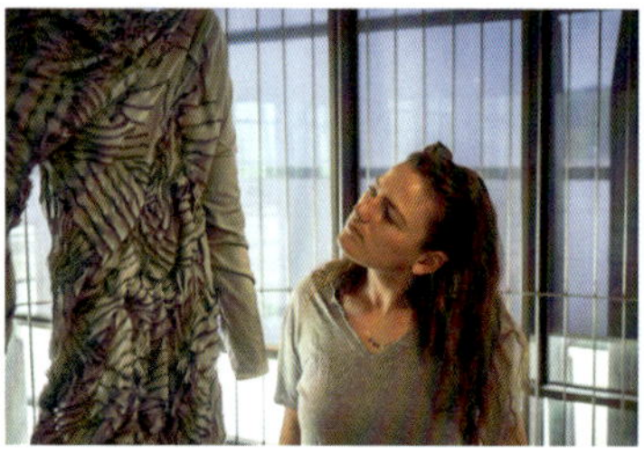

FaR - Fashion and Robotics, Anna Katarzyna Piecek (PL)

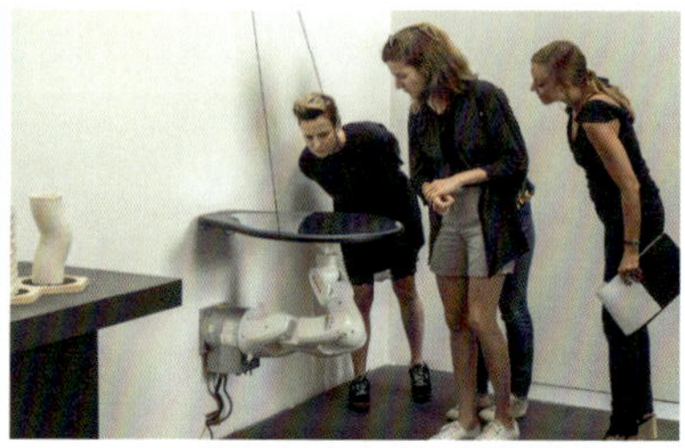

Ars Electronica / Robert Bauernhansl

[proteus], Pierre Cutellic (FR/CH), Maria Smigielska (PL/CH)

WORKS:

[proteus]
Design, robotics: Maria Smigielska (CR UfG, mariasni.com)
Design, BCI: Pierre Cutellic (CAAD ITA ETHz, Compmonks)
Robotics: Johannes Braumann (CR UfG, RiA)
Fabrication assistant: Marie-Caroline Zimmermann-Meinzingen (CR UfG):
Electronics advisor: Wesley Lee (UfG)
Chemistry advisor: Barlomiej Doros

BranchBoarding
BranchBoarding "3rd Generation" limited edition
Tree of Motion: Nico Rayf, Applied Robotics Lab: Philipp Hornung, Wood Technology, University of Applied Arts Vienna

A Bridge Too Far
This research was done at the Centre for IT and Architecture (CITA), KADK, as part Complex Modeling, a Sapere Aude Advanced Grant research project supported by the Danish Council for Independent Research.
It was carried out in collaboration with Bollinger + Grohmann consulting engineers, KET at Berlin University of the Arts, SICK Sensor Intelligence Denmark, and Materials Science and Engineering at Monash University.
Project staff: Paul Nicholas, Mateusz Zwierzycki, Esben Clausen Nørgaard, Scott Leinweber, Christopher Hutchinson and Riccardo La Magna.

Roboterzelle
Institute of Robotics, Johannes Kepler University Linz

FaR – Fashion and Robotics
Anna Katarzyna Piecek
Fashion & Technology: Prof. Christiane Luible-Bär and Prof. Ute Ploier
Creative Robotics: Prof. Johannes Braumann, Maria Smigielska
University of Art and Design Linz

Cyber Physical Macro Material
ITECH M.Sc. Thesis Project: Miguel Aflalo, Behrooz Tahanzadeh, Jingcheng Chen
Thesis Advisors: Dylan Wood, Maria Yablonina
Further Scientific Development: Dylan Wood, Maria Yablonina
Exhibition Project Team: Denitsa Koleva, Sanoop Siby
University of Stuttgart: Institute for Computational Design and Construction, ICD (Prof. Menges), Institute of Building Structures & Structural Design, ITKE (Prof. Knippers)

InFORMed Clay Matter
Co-de-iT for digifabTURINg

THE MEANS
Design, robotic rod bending: Maria Smigielska (CR UfG, Bendilicious)
Design: Mateusz Zwierzycki (The Object)

Infabity

The HARATECH Company's *My Twiin Bodyscanner* performs a 3-D scan of a person's entire body within only a few seconds. Then, its software automatically converts the body scan data into a 3-D model and creates a personal virtual 3-D avatar. Immediately after the scan, the subject's own 3-D selfie is sent to his/her smartphone, where the avatar is displayed on various 360° background images and can be used for a variety of applications. The next step for HARATECH is research on deploying this technology in the fields of Augmented and Virtual Reality. At present, the models are used in conjunction with a 3-D printer that, in a process that takes only a few hours, can generate, layer by layer, a miniature statue that is an exact replica of the scanned subject.

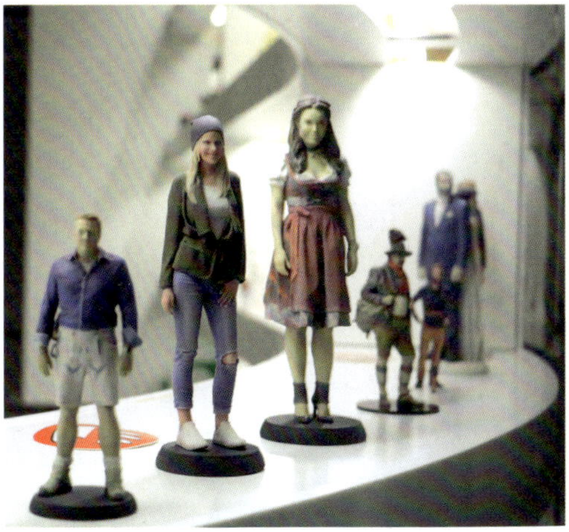

Ars Electronica / Magdalena Sick-Leitner

Ars Electronica Futurelab

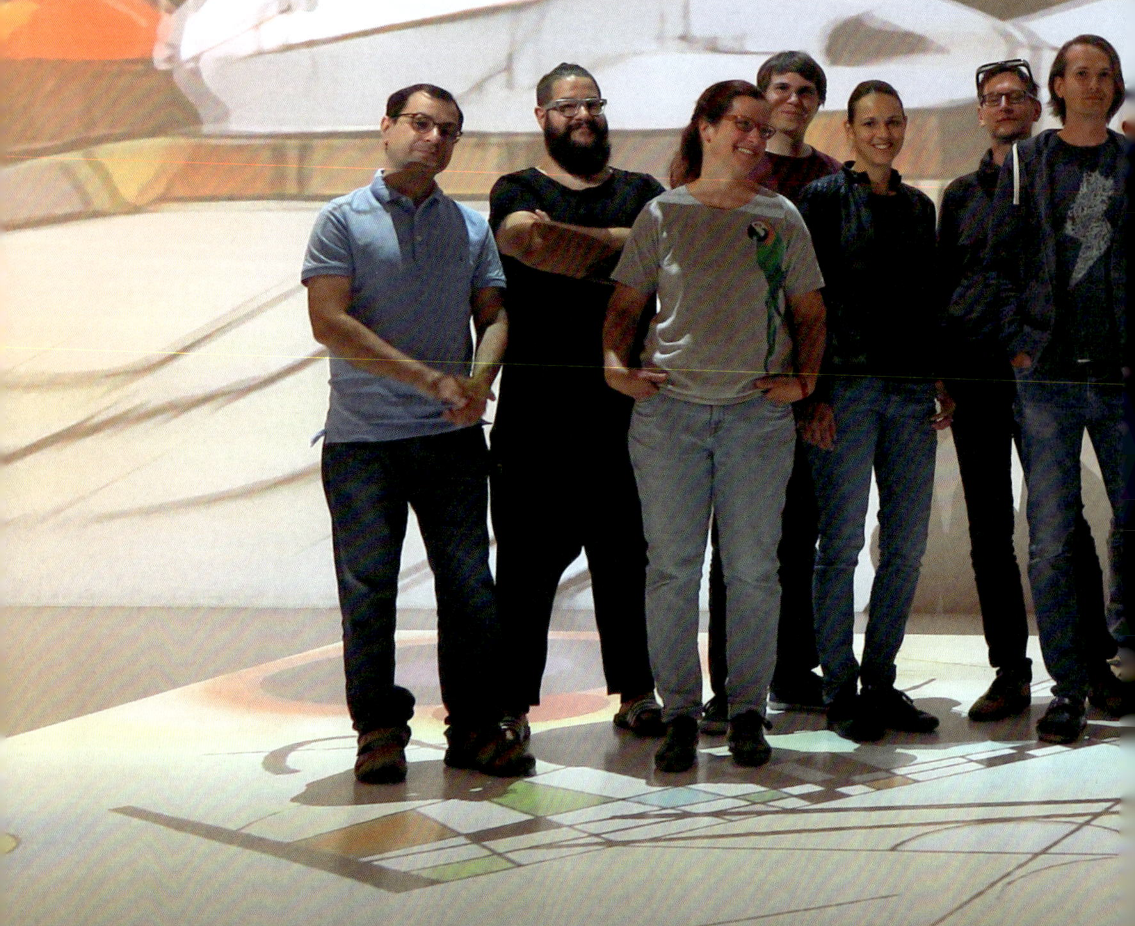

Since 1996, the Ars Electronica Futurelab has been configuring the future at the interface of art, technology and society. The aim of the lab's own projects is to open up spaces conducive to thought and action, ones that generate new perspectives and make available unprecedented means of taking on the challenges of the present with a view towards the future. The Ars Electronica Futurelab's projects are "sketches of our future(s)," presented as an invitation to a broad audience to join the debate and discussions about the meaning of these sketches or concepts of our future to our society. An open approach based on art and research and an orientation towards "the meaning of" characterize the way this lab works. In everyday practice, the established differentiations between art and science are reordered and links are established among a wide variety of approaches. Whether

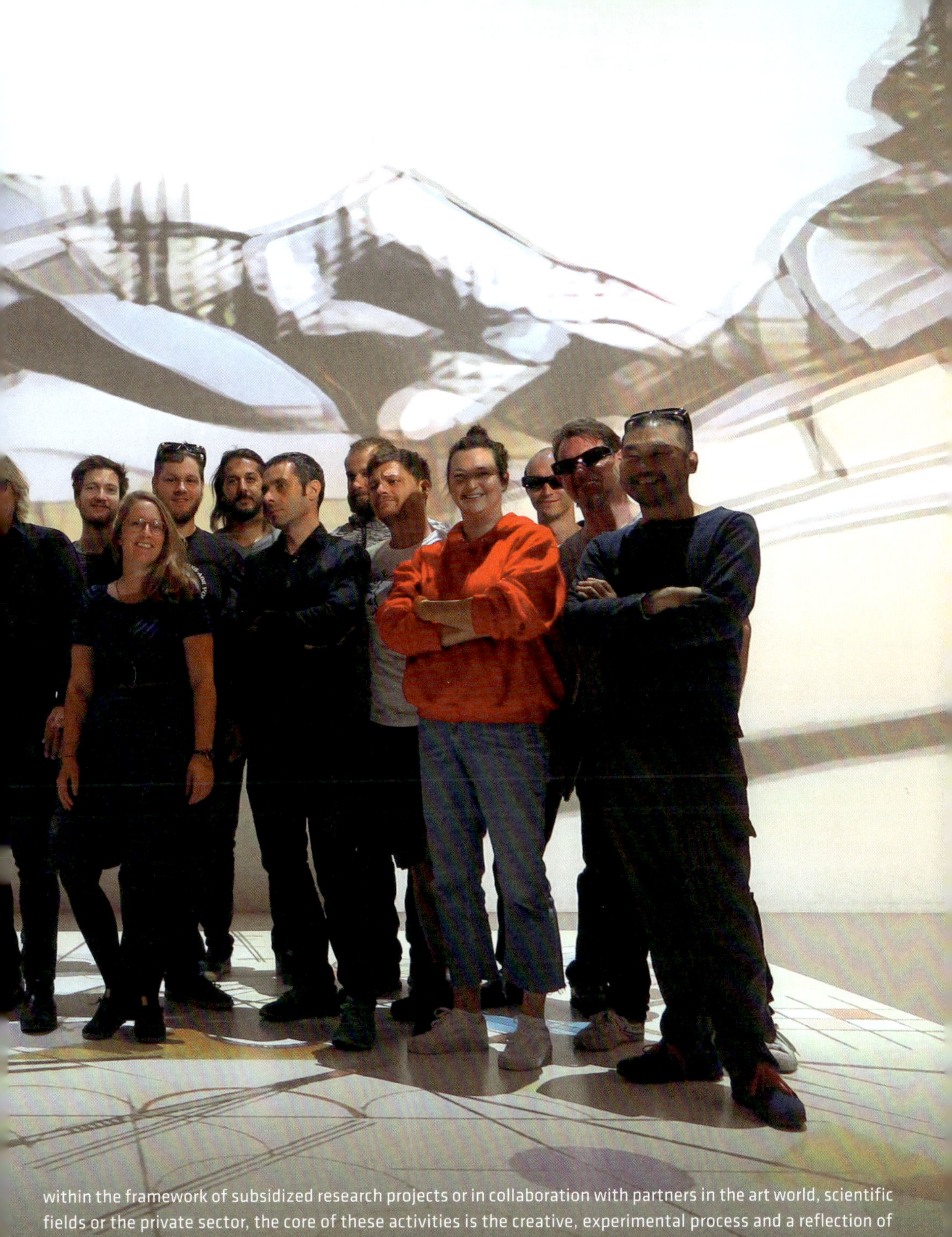

within the framework of subsidized research projects or in collaboration with partners in the art world, scientific fields or the private sector, the core of these activities is the creative, experimental process and a reflection of current or future social issues. The Lab does this by generating questions rather than answers, utilizing scientific explorations, artistic expression and pointing at the social impact by keeping its focus on "the meaning of" to our society. Transdisciplinarity as a work approach has been important to the Futurelab since its very beginning. The disciplines in which the approximately 30 staff members have received their training include architecture, graphics, construction engineering, computer science, art, media design, media technology, music, physics, sociology, telematics, surveying, economics, medical technology and industrial design.

Future of Imperfection

1

–

Article 1 of the constitution of the Federal Republic of Germany states that "Human dignity is inviolable," and Austria's Constitution contains the same formulation. The European Union likewise gives priority to an understanding of human dignity as a fundamental right. Nevertheless, none of these documents go into detail about what is meant, precisely, by dignity. Instead, dignity is explained in terms of derived principles governing state power, or freedoms and rights, such as the right to "informational self-determination" (data protection). Immanuel Kant declared human dignity to be an end in itself and defined it in terms of its violation, by contrasting it to the use of human beings as a means to an end. This negative image (of subjugation, exploitation, betrayal) says more than the postulation of an intrinsic value, and thus illuminates dignity intersubjectively by showing us what its absence might look like. It emerges in a concrete form where tactile sensitivity towards that which must remain untouched is palpably disrupted. More and more, it seems to me, people need to be reminded of their dignity, which is gradually being taken from them.

There are currently numerous warnings about blindly trusting data-gathering platforms like Facebook and Google, and there's good reason for this. There seems to be less emphasis, however, on warnings about the social consequences of so-called Industry 4.0, and about what is now being extolled as—and by means of—artificial intelligence, AI, and its purported benefits. A new operating system is currently being installed in capitalism under the heading of "digitization." Meanwhile, the potential for progressive digitization—conceived of as a machine-supported contribution to the common good of society—is being totally neglected in the conversation about AI.

In light of these facts & circumstances, this essay aims to orient the reader in the search for approaches and insights that are conducive to the effort to actively unpack the concept of dignity as both a phenomenon and as a scientific fact—and all this in the context of AI and its manifold potential.

2

–

Sufficiently analyzed and criticized as a value system of global economic liberalism, these are ultimately movements of a process of transformation, in which the economic sphere imposes itself on the biosphere. For a long time, this term "ecological," derived from the Greek oikos ("house") and logos ("teaching"), simply meant livable conditions, but it has now been transformed through the superimposition of an antonym: What happens is that the teaching, or logos, whose meaning was, etymologically speaking, an approach which utilizes research and inquiry, is replaced by the notion of regulation, in Greek, nomos ("law"), or how something must be. Meanwhile, the concept of "house" (an original existential milieu) is replaced with the economic understanding of the house as "household," and this in turn completes the transformation of the inhabitant into the consumer, into the one who receives an income. .

The culmination of this homologous semantic renaming of the ecosphere can be seen in the loss of the earth's renewable resources of food, water, and energy. Since July 8th, 2016, the yearly use of these exceeds the planet's capacity for regeneration.

That there was no alternative (M. Thatcher)—we don't need to propagandize about this any longer. By realizing itself as the ecosphere, and thus imposing a misunderstanding about general livelihood, the economic takes on the character of a constant or law of nature; the objection that it is man-made, a mere invention, contradicts its theology (and theodicy).

Any sort of limit imposed on the business model of digitalization by the concept of dignity is not allowed. By and large, the ecosphere abbreviates the individual as someone whose purchase of a television, or of a pair of boots, makes a contribution against a period of economic recession, perhaps caused by under-consumption, as though it were a matter of being able to breathe (following the mantra: "there is no alternative"). The question of human dignity, if it is injured by the instrumentalization of the person for the goals of others, only arises ultimately where there is a

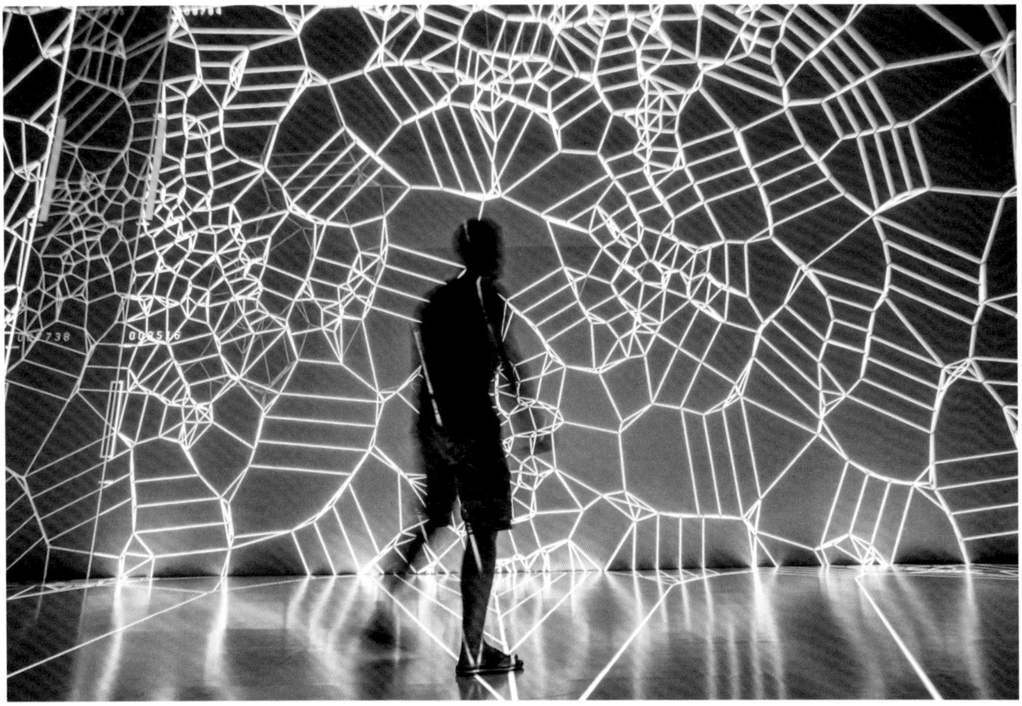

Abysmal / Void (TR), © Robert Bauernhansl

difference between customers for whom one competes, for economic reasons, and human beings, who will be made into customers.

It is characteristic of this process of transformation to some extent that there is a sort of "anything goes" attitude, which enables any critique of liberalism to be incorporated as an intrinsic proof of the functioning of liberalism, thereby rendering this critique ineffective. The principle is to receive as mere data what was intended to be information which would have an effect. Hence, it is sufficient, for example, to date the categorical imperative as the maxim of a philosopher to 1785; it relieves one of the task of having to place oneself in relation to this maxim, and it certainly relieves one of the task of measuring up to it. Moreover: it seems that the development of the ecosphere would lead to a culture of debasement or it would have this, at least, as one of its prerequisites. Large and small injuries to human dignity, which occur everywhere according to the industrial standard, are not only accepted, but also supported by the consumers, the "customers," and as it were legitimized collaboratively.

The enjoyment of chocolate does not obscure the knowledge that child labor is a part of the process of production. We use plastic bags and many other contemporary packing materials made from plastic, although everyone has long known that millions and millions of tons of this plastic are polluting the oceans and threaten maritime fauna and flora. This industry, with which we've learned to make common cause, creates relationships of dependency, within which our universities and other educational institutions must direct themselves according to the rule of efficiency, and will be directed to follow its content. And when—one further example—private companies running nuclear power plants are able to make the state pay for the costs of transporting nuclear waste with almost no objection being raised—almost as an advance payment for the lost profit from an exit from nuclear energy, which was supported by the majority, then that is not only an example of the failure of

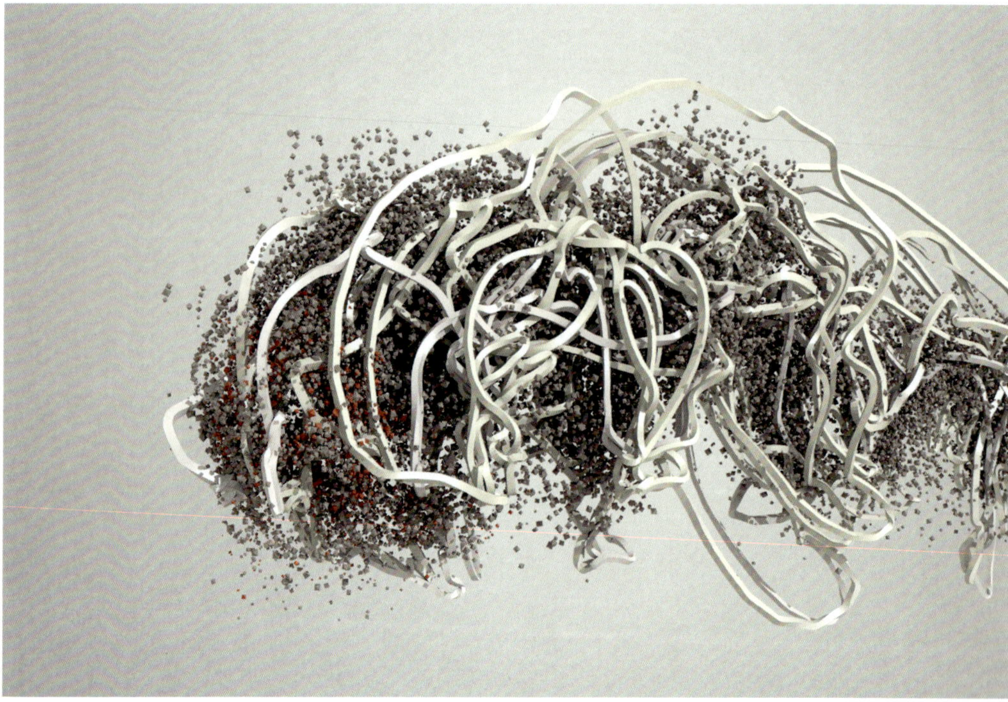

Forms, David Quayola, Memo Akten

the ("secular") division between the society (state) and business (industry). Because it prioritizes monetary and power relationships over the ongoing damage to the planet, and lets us almost forget these conditions (expressed in the broad acceptance of the process), it is at the same time a significant proof for the economic transformation of the ecosphere into the ecosphere.

3

Ostensibly this entire development arises under the auspices of the assurance of growth, and concomitant prosperity—where on the other hand statistics (united with the OECD, the IMF, et al.) attest to an ever-greater gap between rich and poor, which has already overtaken the middle class and is starting to damage not only society but also businesses. This is old hat (we merely adjust the hat's size every year): According to a study by Oxfam (commissioned by the large Swiss bank Credit Suisse and Forbes Magazine), in 2017 one percent of the population managed to secure 82% of the wealth accumulation during that year. The 42 wealthiest people own as much as the entire bottom half of the world's population. As estimated by the Organization for Economic Cooperation and Development, the damage to growth within the OECD from this inequality in the distribution of income between 1990 and 2010, amounts to a cumulative total of 4.7%. Social improvement—contrary to the earlier belief that economic inequality was thought to be necessary in order to stimulate more growth—is said to support economic growth.

So while one recognizes, on the one hand, an injury to human dignity proven through the facts of objectification, the economized ecosphere, on the other hand, understands people according to a concept of "usefulness," which proposes the goal of the system (i.e., growth) as the aim and purpose of life for the consumer. And despite the general recognition of the dignity of human beings as an inviolable property, a life-world constitutes itself which is subordinate to mechanisms and forces for which dignity appears to be not only completely irrelevant, but contraindicated.

4

Of all the challenges that have "revolutionized" culture and society since the industrialization that followed from new technologies (from the conveyor belt to the atomic bomb and the internet to genetic design), the next one should be much larger or, at least, it will close in on us much faster. Unlike biotechnology or current forms of processing information (whether from personal or genetic data), which are required to be

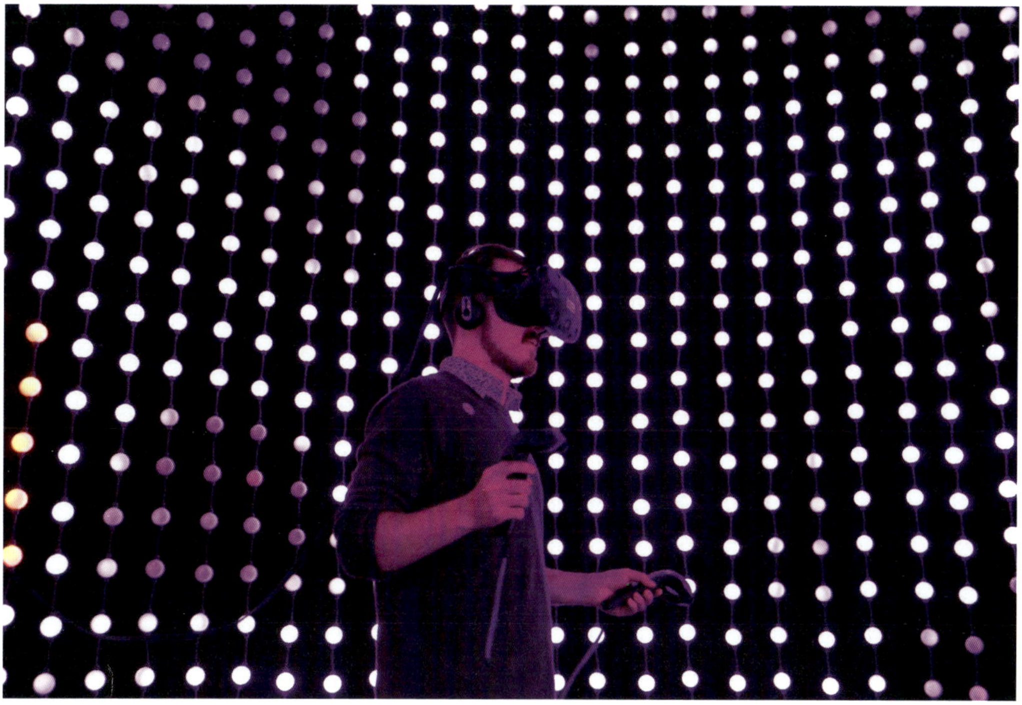

Training 2038 Kitchen Budapest, © Florian Voggeneder

examined according to legal statutes and evaluated by ethics commissions in light of general social questions, this instrument alone will no longer be appropriate to these evolutionary processes.
For, not only is a further technology arising, the use of which concerns ethical guidelines which have yet to be determined, but the development of an environment is also looming on the horizon, which, if we interact with it, we automatically call forth decisions whose dynamic is of a size that can no longer be predicted—not least because both engaged entities, AI and human being, can no longer be understood by simple algorithms. The question of whether the use of AI can be ethically justified or not is not raised; it is not the effects (one on another; reactive) that can be predicted, but rather the cooperation (interaction). And that is an essentially ethical problem.

5
–

For the year 2025, the IDC (a market monitoring firm) predicts a growth in data worldwide of 163 zetabytes, ten times the growth in 2016 (16 zetabytes); the yearly growth rate of all data between 2015 and 2025 will be about 30%. In this, there is a shift in the sources of data on the horizon. In 2015, businesses created just 30% of all accruing data worldwide, so the portion in the years until 2025—with a tenfold increase in the size of all the data—should increase to 60%. By then, three-quarters of the world's population will be linked. The diversity of information and the amount of data has for a long time been a hopeless challenge to humans, so machines which can process data were developed, which themselves have taken over these tasks. The creators of these machines were clever enough to build them such that they could learn to draw conclusions from this data prepared for them. Undisturbed by technological advances, we are already prepared to make use of rudimentary forms of artificial intelligence—even in order to advise people as they assess and evaluate other people. AI provides support in the selection of candidates for open positions in the work world, or even in judging people according to the probability that they would commit a crime in the future.
The ubiquitous available data produced by everyone is an image of the world we all live in, an image already interpreted. A mechanical brain fed with this data would learn, for example, that the majority of prisoners belong to a minority group in the population; it would realize that the majority of those responsible for decisions in positions of authority are older, male, and white, and it would realize that part-time jobs are

for the most part filled by women. It would also realize that corruption in the southern hemisphere is much more visible than in the north, and that affluence and health are, more and more, concentrated among certain people, etc. The efforts which extend over generations to change these relationships, and the numbers associated with them, can only be interpreted—based on the numbers, the data, and the facts—as a mild trend, but in no way can it be read as the norm, and certainly not as a meaningful goal for the improvement of these human relationships. If one allows the status quo to have an unmoderated effect on a learning mind, these relationships named above would be recognized and an impression would arise which we would characterize as racist and sexist.

Who is building these artificial intelligences? Certainly not publicly run research and development institutes, subject to an ethics commission! Why are these AIs being built? Incorruptible mechanical objectivity might have been the father of thought—racist and sexist impressions are a first provisional result. There is a "danger in distortion"—that our collective "being without dignity" which is intimately connected with the economic systems of business will become active as a "genetic inheritance" in development, orientation, and use of AIs in the interests of global business. Shouldn't the development of artificial intelligence be accompanied by the attempt to formalize a concept of dignity, in light of just such a danger, so that this concept is executed along with this intelligence in its development (as one of its emanations)? Even before an ethics commission created to address the problem realizes this, but only when it is too late . . . If it is the case that any civilizing and cultural development is oriented to improving the dignity of persons, as one of its driving forces, and if this type of development may not be collectively revoked—wouldn't this project be indispensable?

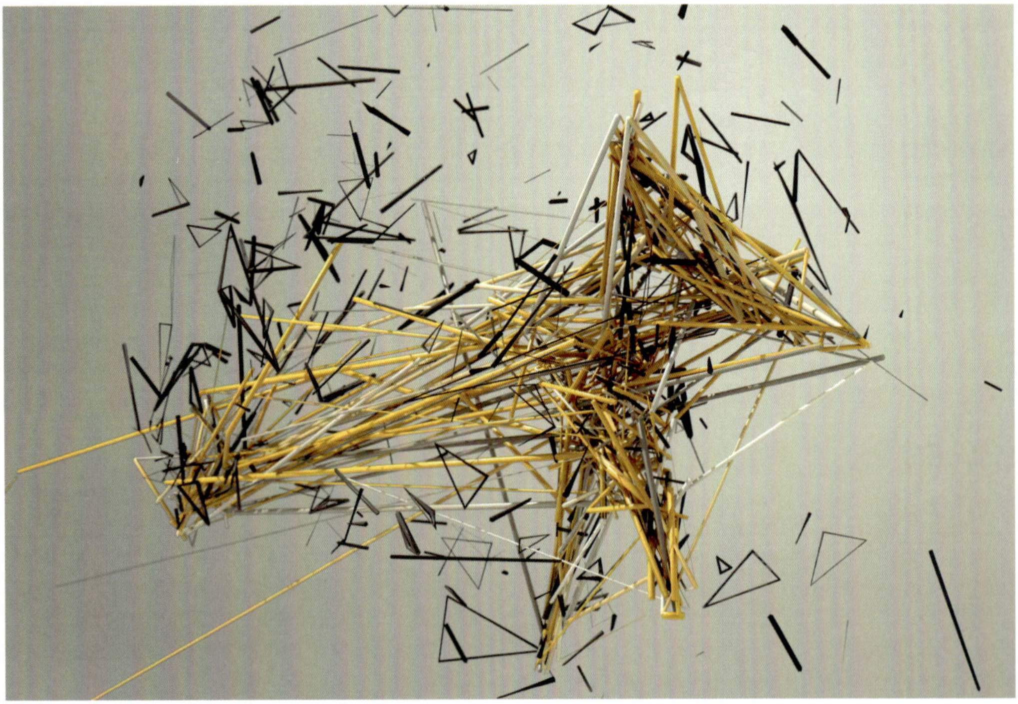

Forms, David Quayola, Memo Akten

6

—

In a certain respect, this problem has parallels with the antagonism between a normative ethics, wherein moral action requires metaphysical principles à la Kant, and a natural ethics, as proposed, for example, by Umberto Eco and expounded through his "solitary Adam." Presumably, this AI Adam must in the end also develop an ethics that is grounded in his nature by first reciprocating the gaze of the other: he would have to pass a reverse Turing Test (Human Interaction Proof). What allows an AI to know whether its correspondent is an AI or a human being?

Mechanical cognition, for now, is bounded by the assumption of everyday speech. What a neural network really accomplishes, when one says that it can recognize faces, is merely positive-negative comparison of examples: the resemblance of an input signal with an output signal expressed through the degree of the agreement. This comparison alone is what the neural network accomplishes. It is just as distant from recognition (in the sense of an experience which calls forth a Eureka!) as a light switch is from an epiphany. A systems theory approach designates its regulatory function in order to justify its ethics (independent of the necessity for ethical action on the part of individuals). The U.S. sociologist Talcott Parsons proposes four principles as definitive for the existence and persistence of every existing or conceivable system: Adaption, Goal Attainment, Integration, Latency (Latent Pattern Maintenance), for which he uses an acronym, AGIL. With A—the ability to react to external conditions which are changing—G—the ability to define and pursue objectives or goals—I—the ability to create cohesion and inclusion—L—the ability to maintain agreement, fundamental frameworks and patterns of value (acceptance and respect), we find here too with this acronym that the characteristic traits of a prototypical AI are also assembled. In the same way we conceive of ethics in systems theory as a capacity for regulation, can we think of "dignity" in terms of a quality that, for the sake of self-preservation, an intelligent environment must exhibit and/or cultivate as regulative principle of the interaction (with human systems of action)?

One prerequisite for this I would locate in a concept of dignity as something given by nature, and describable in terms of natural science—an idea that, in fact, is not all that foreign to neuroscience. Hints of this reveal themselves in observations of newborns and their measurable reactions to changes in the quality of relationships. If this step were to work, we could then dispense with appeals to convention (religion, cultures, socially established concepts of morality). It is quite urgent to engage in such deliberations, especially against the background of an ecosphere established by artificial intelligence that, with its decisions in the future, will make significant contributions to collective human life—from the biosphere to the ecosphere to the data-sphere. Also for this reason: because the point is to recognize in time the nascent sociopolitical potential of an AI and to reflect on the possibilities of using it for civil society; an AI that goes beyond its mere economic impact and is relevant for the improvement of our civilizational and cultural advances.

Text: Horst Hörtner

1 Cf. https://www.ris.bka.gv.at/GeltendeFassung.wxe?Abfrage=Bundesnormen&Gesetzesnummer=10000308 / June 24, 2018. https://eur-lex.europa.eu/legal-content/DE/TXT/?uri=CELEX:12012P/TXT (Charter of Fundamental Rights of the European Union) // June 24, 2018.

2 The Metaphysics of Morals

3 http://www.wirtschaftslexikon24.com/d/konsument/konsument.htm (June 12, 2018)

4 https://www.footprintnetwork.org/(June 12, 2018)

5 Cf. Herman E. Daly "Ecosystem as Subsystem of Macroeconomy" The First Annual Feast a Lecture, Trinity College, Dublin April 26, 1999

6 mrd.de "Schere zwischen Arm und Reich geht weiter auseinander" [The gap between rich and poor continues to widen]

7 https://www.propublica.org/article/machine-bias-risk-assessments-in-criminal-sentencing July 12, 2018

8 Carlo Maria Martini, Umberto Eco, "Woran glaubt, wer nicht glaubt" [Belief or Nonbelief?: A Confrontation]; Munich 1999. Eco arrives at the conviction "that there are concepts that all cultures have in common, and all of which refer to the position of our body in space." By way of explanation, Eco sketches "a sort of animal-like, solitary Adam" who, for example, attributes his own concept of above and below to the fact that for him, as a creature who walks upright, it is arduous to go for a long time with his head facing downward, whereby the former [above] takes precedence over the latter [below]; who, due to his extremities, has a conception of what it means to hit something hard, to penetrate something soft or fluid; who draws conclusions on the basis of his own physical constitution as to what it means to feel pleasure or pain, to be sad or to experience fear—and who derives from them, for example, "universal concepts of coercion." On the basis of this instinctive repertoire of willingness and unwillingness, this Adam acts in accordance with an ethically dimensioned concept of what it means to not do unto others what he would not want done to himself "when the Other comes into play." It is the view of the others that defines and forms him.

Open Futurelab

As in past years, the *Open Futurelab* means a stage within the festival, where projects and project partners of the Ars Electronica Futurelab are presented. Selected Industrial, educational, artistic and research collaborations are going to be positioned next to each other in one spot, displaying the variety of projects and wide range of activities going on at the Futurelab in recent months.

Spaxels Research Initiative

What started last year at the Ars Electronica Festival has grown to activities with partners that led to a variety of perspectives on Swarms, from the introduction of a design tool for Swarms, *Swarm3D*, at the Autodesk University, to a presentation of an entirely new heterogeneous Swarm of UAVs and UGVs performing together at the NTT-R&D-Forum in Tokyo, earlier this year. Growing from a concept to a first incarnation of *SwarmOS* after several years of development, a pre-alpha of *SwarmOS* is now used by the NTT R&D-Group for its research and development.

Besides the ongoing existing partnerships, new members have joined the *Spaxels Research Initiative (SRI)*: We're happy to welcome the world leader in production of firefighter equipment, the Austrian company Rosenbauer, to the SRI and also welcome MAXON, renowned worldwide for their flagship product Cinema4D. Both new members' project collaborations will be introduced in detail during the Ars Electronica Festival 2018.

Text: Horst Hörtner

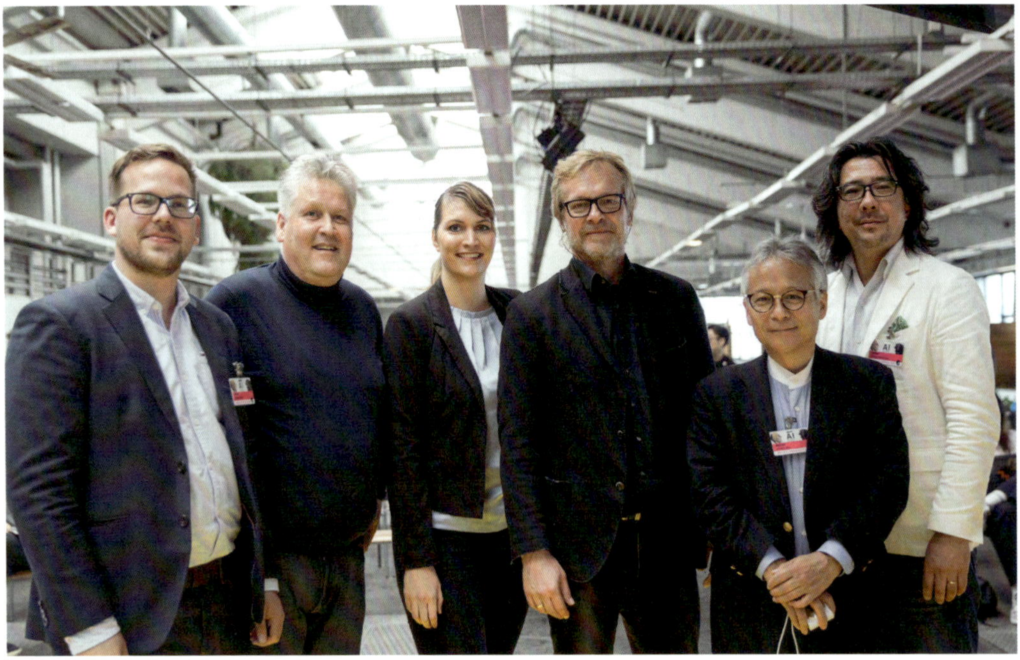

Tom Mesic

Speakers of the Spaxels Research Initiative

Swarm Arena

Swarm Arena is the latest outcome of joint research efforts by the Ars Electronica Futurelab and Japanese telecommunications giant NTT. The collaboration started in 2017 with the aim of working on using unmanned aerial or ground vehicles (UAVs and UGVs) as a means of communication. While the first project, Sky Compass, and its follow-up Swarm Compass focused mainly on navigation, public signage, facilitation of traffic and swarm intelligence, *Swarm Arena* now shifts the research question to creating an entirely new audience experience at (sporting) events.

A first demonstration was shown at the NTT R&D Forum 2018 in Musashino, Tokyo, giving an insight into the possibilities through a physical demonstration using both drones and ground bots. This combination allows for a new medium of (artistic) expression that is not bound by any canvas, but rather expands seamlessly from earth to sky.

Swarm Arena encompasses several concepts to enhance conventional sporting events. The idea of the Augmented Arena suggests using ground bots to display measurements, metrics or outcomes, thus visualizing otherwise invisible sports big data right at the sports arena. Its counterpart, the Virtual Arena, proposes expanding the viewing experience to remote locations, creatively reimagining Public Viewing.

Using swarm bots, athletes and events can be visually represented on moving displays to create a digital, real-time duplication of the game or competition. The bots offer the potential for bigger swarms, moving beyond orchestrated formations of single entities to create multi-cellular robotic organisms. Ground and aerial bots act as cells or physical (kilo-)bits of the animated whole. In their vision to "Animate Japan 2020 Together", the Ars Electronica Futurelab and NTT are focusing especially on audience participation with the multi-cellular organism. How could a stadium full of spectators effectively interact with a large number of bots, moving together as one? The synchronized movement of the robots forming an organism also opens up research questions in the field of spatiotemporal visual design under severe logistical constraints–an area in which the Futurelab, with its successful Spaxels project, can draw on ample experience.

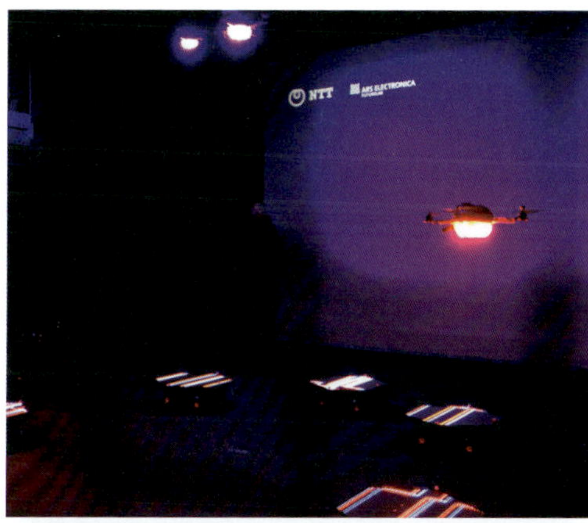

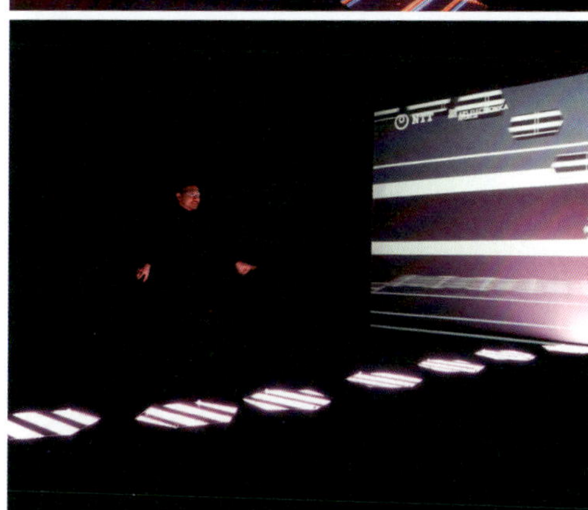

NTT

SwarmOS

In the past few years, the Spaxels project kept pushing the boundaries of how choreographed drone swarms could advance real-time spatial visual expression. In its wake, the world has seen increasingly more players develop their own swarm systems and increase the swarm scale to mind-boggling numbers. At the same time, more and more vehicle manufacturers are following their particular vision by designing specialized UAVs: Fast and light, massive and robust, or equipped with high computing power. Finally, there is a whole world of ground vehicles that offers a completely different set of uses and capabilities. What if we could find a way to control swarms of not only special, limited sets of drone models, but virtually anything that supports one of the open communication standards for autonomous vehicles? What if this control could range from strict choreographies to highly responsive interactive scenarios? And what if this swarm was not even limited to one model or type of vehicle per mission? Under the title SwarmOS, the Spaxels team has started development on an ecosystem of tools that will allow precisely that. Ultimately, a SwarmOS user will be able to take their drones of choice, attach a special device to them and, with a few configuration steps, fly them in an animation, a specialized mission (from reconnaissance to rescue), or an interactive scenario of their design.

Text: Peter Holzkorn

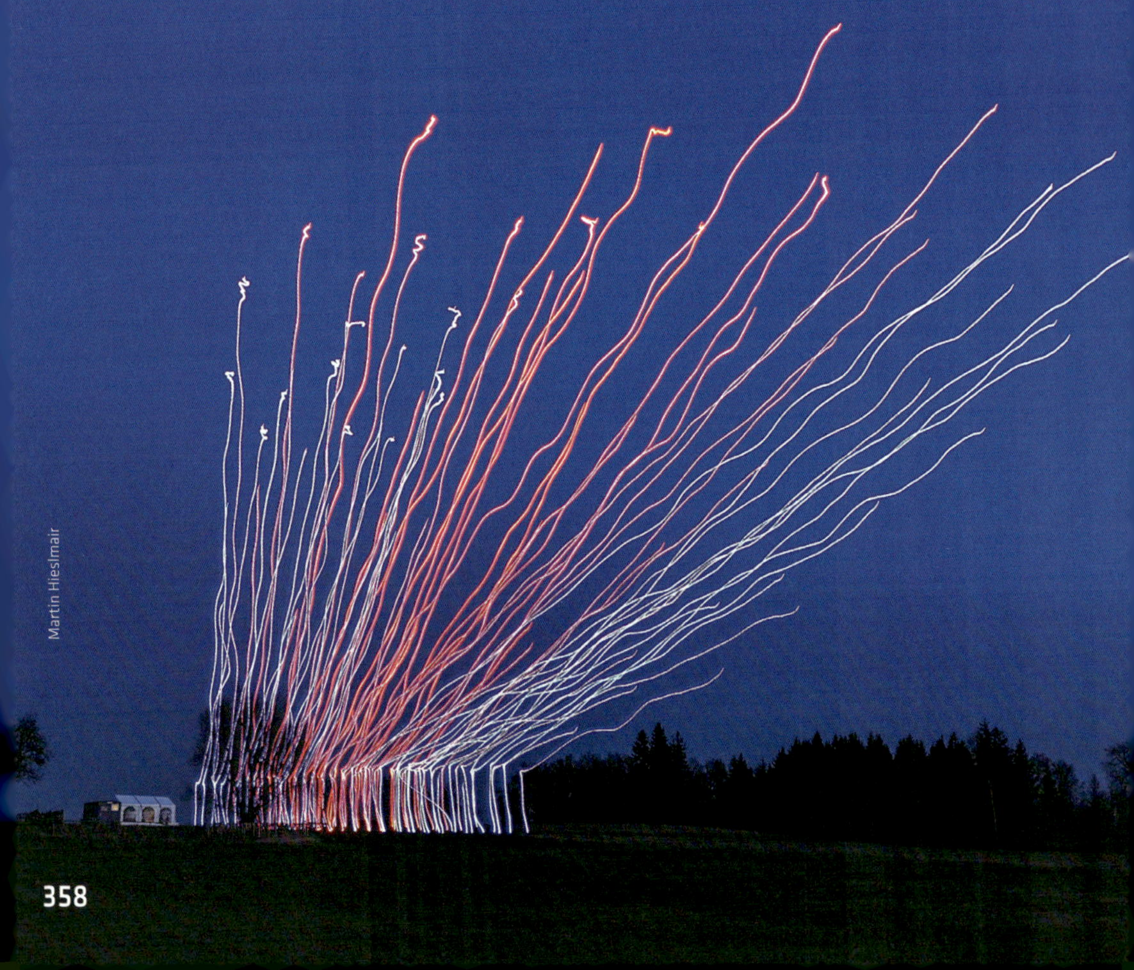

Martin Hieslmair

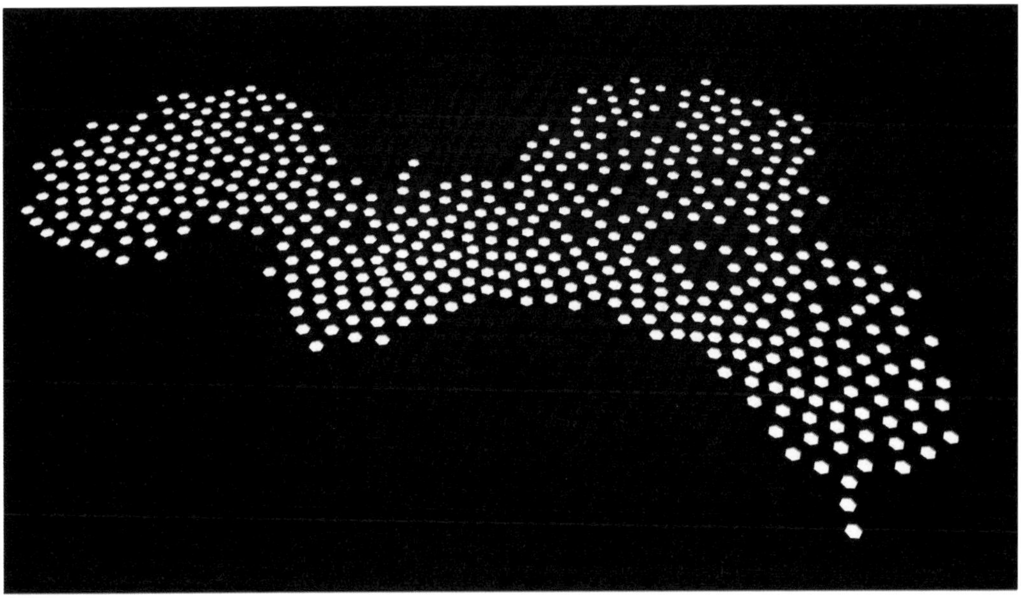

Raphael Schaumburg-Lippe

SwarmPhysics—Dynamic Vehicle Simulation

Birds flock, much like fish gather in schools in order to overwhelm and confuse predators. This behavior is aesthetically pleasing to watch, since our perception looks for patterns in the constant flux of form. However, the purpose of a swarm of vehicles is multifaceted in comparison. Not only must it be able to flock in the abstract manner we see in nature without individual agents colliding, it must also be able to form concise, recognizable formations for humans to understand. This spectrum of capabilities opens a new field spanning from practical applications to artistic expression.

One of the chief challenges of designing and planning a mission or show is the gap in accuracy between a virtual environment like a 3-D software and the laws of the physical world. Creating a mission or show that is collision-free, aesthetically pleasing and/or able to perform a specific task requires a deep understanding of the physical vehicles involved. Since it isn't feasible to go out and test this in the real world every time there is a change, an accurate simulation is imperative for visualizing and predicting the swarm's behavior. To achieve this, the *Spaxels Research Initiative* is working in conjunction with its partners to develop custom tools for facilitating the physically accurate simulation of real-world robotic agents in a swarm, in order to make the digital creation of missions or shows easier, safer and more reliable.

While many 3-D software environments include physics engines, most are designed to create stunning visual effects but offer limited capability concerning the fusion of vehicle behavior and flocking. Under the title *SwarmPhysics*, Maxon is collaborating – as a member of the *Spaxels Research Initiative* – with the Ars Electronica Futurelab. The goal is to create a software environment and workflow that allows the creation, simulation and performance of multicellular robotic organisms, i.e. swarms consisting of an array of vehicles.

Text: Patrick Berger

Ars Electronica Futurelab Academy 2018

The Ars Electronica Futurelab Academy, a part of the *Futurelab Education Initiative*, was created to support students and educators from international partner universities to engage in transdisciplinary practice. Futurelab researchers act as mentors and collaborators, supporting creative exchange with the Academy participants. Participants come from a range of backgrounds; art and design through to science, engineering and technology. Since 2012, collaborations with renowned universities across the globe have resulted in a diversity of exhibits and performances being presented at the Ars Electronica Festival. Some of these project outcomes have gone on to win prestigious awards and accelerated the artists' research and practice trajectories.

In 2018, we have been working with two partners from opposite sides of the world:

Long-time collaborator *Queensland University of Technology* joined the Futurelab again in a large-scale format that saw a Futurelab researcher and curators working with experienced QUT staff to facilitate a refined process that led a group of 30 students from discipline decontamination and group ideation to an exploration of art-driven research and its actualization in several very different projects that can be seen, heard and participated in at this year's festival.

At the same time, a new collaboration was forged with the *Media Lab Helsinki*, part of *Aalto University*, itself a vision of interdisciplinary futures. A small group of graduate students experimented with the artistic potential of bacterial cellulose and drew inspiration from internal cross-fertilization with other university departments (Bioproducts and Biosystems; Communications and Networking), as well as focused workshop sessions with Futurelab researchers.

The two collaborations could hardly be more different in scale, distance, structure and curricular context; together, they sketch out modes of learning and establishing creative practices outside of traditional templates.

https://ars.electronica.art/futurelab/en/initiative/futurelab-education-initiative

Text: Peter Holzkorn (Ars Electronica Futurelab), Lubi Thomas & Kristefan Minski (Ars Electronica Australia)

Statement from Queensland University of Technology

The Ars Electronica Futurelab Academy at the Queensland University of Technology affords students the opportunity to situate their emergent practice within a rich transdisciplinary context. Working with peers, researchers, industry, and artists, they work through stages of ideation, experimentation, prototyping, pitching and presenting media art works. The participant-led nature of the Ars Electronica Futurelab Academy @ QUT creates space for students to be co-leaders and co-creators, driving their own nascent practices while being inspired, guided and mentored by a transdisciplinary team. The dynamism and culture of the Ars Electronica Futurelab Academy enables students to make the transition from highly structured learning environments to professional practice.

QUT Educators & Artists: Steph Hutchison & Greg Jenkins
Technologist: Matt Strachan
Ars Electronica Australia: Lubi Thomas & Kristefan Minski
Ars Electronica Futurelab: Horst Hörtner & Peter Holzkorn
https://qut.to/ars2018

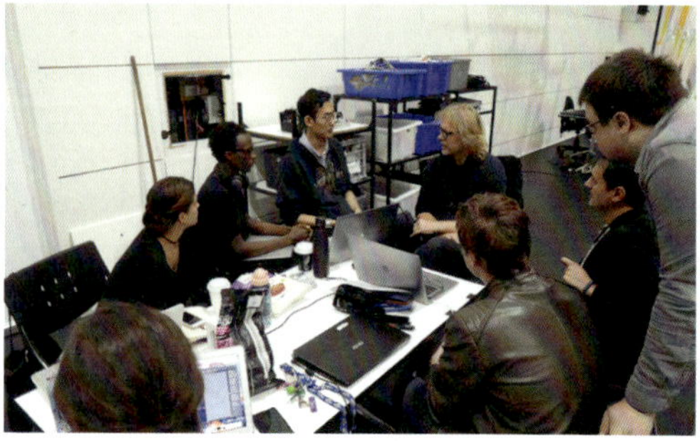

Greg Jenkins

AIWA

How could one experience a fully-sentient artificial intelligence? *AIWA* is an artwork created to push initial boundaries and current expectations of machine learning within the commercial market. AIWA is a speculative project exploring human interaction with a consciously aware artificial intelligence.

Undergraduate Bachelor of Creative Industries:
Amy Campbell (AU), Thomas Stig (AU), El Tapi (MZ)
Undergraduate Bachelor of Fine Arts: Oscar Connor (NZ/AU), Nicholas Dunning (AU), Merryn Trescott (AU)
Bachelor of Games and Interactive Entertainment: Jeremy Boulton (AU)
Honors Bachelor of Design: Salvatore Fazio (AU), Zhiheng He (CN)
Honors Bachelor of Design: Salvatore Fazio (AU), Zhiheng He (CN)
Masters Creative Industries by Research: Quinty Pinxit-Gregg (AU)
External Advisor: Maria José Sanchez Varela Barajas (MX) Philosopher, Researcher; Marvin Chancán (PE) (PhD Computer Science) AI & Deep Learning.
https://qut.to/ars2018

Klangschlange (AKA Soundline: what are you waiting for?)

Klangschlange (AKA Soundline) is a roaming performance that activates unsuspecting crowds in queues. Performers permeate these sites with playful responsive objects that invite audiences to become co-producers in a spontaneous performance. *Klangschlange* is a playful, social sonic intervention investigating future entertainment and technologies to facilitate a state of group Flow.

Undergraduate Bachelor of Creative Industries:
Romaine Quessaud (FR)
Bachelor of Entertainment Industries: Patrick Cenita (AU)
Undergraduate Bachelor of Fine Arts: Georgia Bradford (AU), Alexa Dewar (AU), Lily Douad (AU), James Dwyer, Morgan Hill (AU), Erin O'Rourke (AU), Matilda Skelhorn (AU), Reina Takeuchi (AU)
Honors Bachelor of Design: Cassidy Cloupet (AU), James Dwyer (AU), Tom Long, Steven O'Hanlon-Rose, Rani Shanks, Jess Greentree (AU), Kathryn Harvey (AU), William Richardson-Davis (AU)
Masters Creative Industries by Research: Quinty Pinxit-Gregg (AU), Jayden Grogan (AU)
PhD: Andy Ward (AU) (Creative Practice/Music)
External Advisor: Murray King (AU) Creative Technologist
https://qut.to/ars2018

Dilate

Dilate is an interactive wearable that responds to human data; dilating, pulsating and expanding while on a human body. *Dilate* uses symbiosis as a forefront for exploration within wearable technology. Audiences activate this wearable through their interaction with it, creating an emotional response.

Undergraduate Bachelor of Creative Industries:
Ruth Hawkins (AU), Daniel Kit Wei Tan (MY)
Undergraduate Bachelor of Fine Arts: Reina Takeuchi (AU)
Honors Bachelor of Design: Jess Greentree (AU), Peter Lloyd (AU), Tom Long (AU), Steven O'Hanlon-Rose (AU), Joash Teo (AU)
Masters Creative Industries by Research: Quinty Pinxit-Gregg (AU)
https://qut.to/ars2018

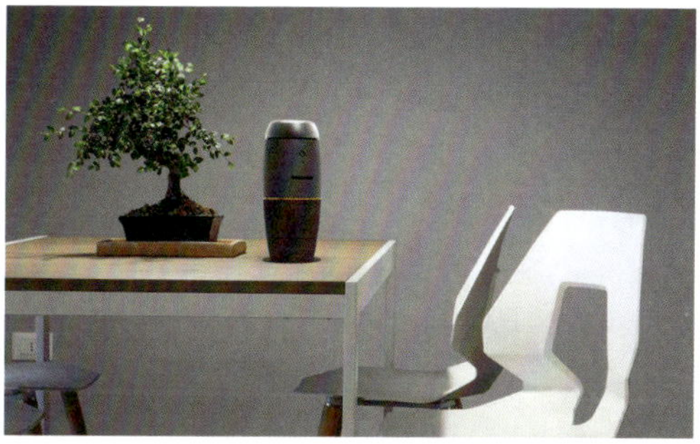
Zhiheng He

AIWA

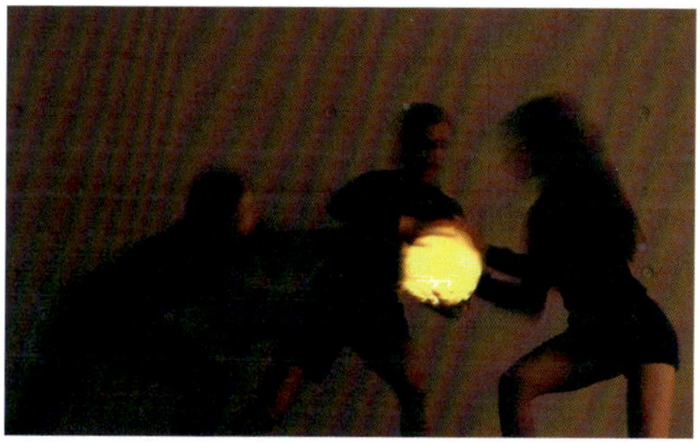
Quinty Pinxit-Gregg

Klangschlange (AKA Soundline: what are you waiting for?)

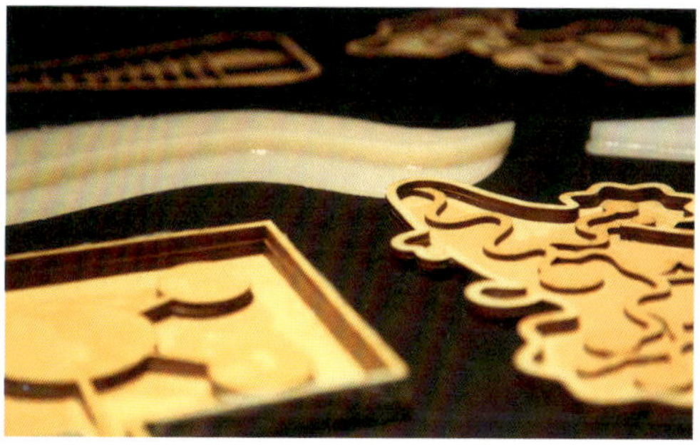
Peter Lloyd

Dilate

Statement from Prof Lily Diaz-Kommonen, Media Lab Helsinki

In 2018 the Systems of Representation course began to take a more hybrid-labs approach. We focused on combining media design skills and knowledge about the cultural and social basis of artefacts with 3D fabrication technologies. We explored the topic of *Vestigiality and Evolution in Human Fashion* and pondered the possibility of creating audio transmission devices that provide the type of functionality to the human ear that other primates possess and which is conjectured to have been part of our biological inheritance. We investigated the creative potential of bio-cellulose and developed concepts and lightweight prototypes of headphone and audio transmission devices.

Aalto University, Institutions: Aalto University, School of Arts, Design and Architecture, Department of Media/Media Lab; School of Chemical Engineering, Department of Bioproducts and Biosystems, Biobased Colloids and Materials (BiCMat) research group; School of Electrical Engineering, Department of Communications and Networking.
Aalto University, People: Lily Díaz-Kommonen, Orlando Rojas, Xiao Yu, Luiz Garcia Greca, Jannika Lehtinen, Solomon Embafrash, Niklas Pöllönen, Jason Selvarajan, Philip Dean.
Ars Electronica Futurelab: Horst Hörtner, Maria Pfeifer and Peter Holzkorn
http://sysrep.aalto.fi

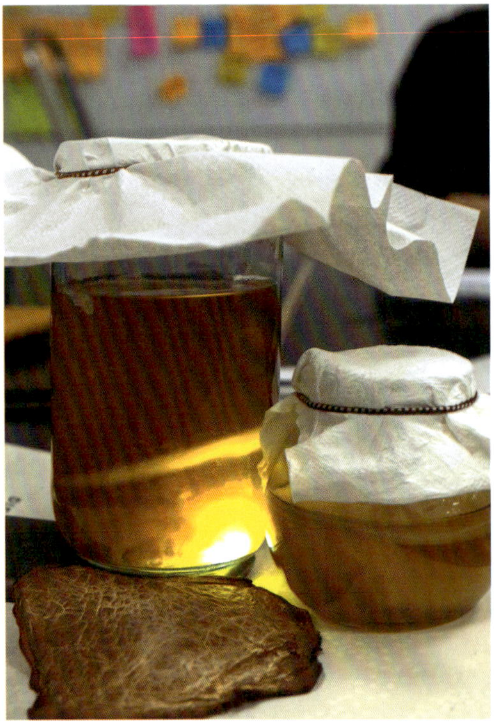

Lily Diaz-Kommonen

Andrea Mancianti (IT)

The living threshold—Blindphones

Headphones, blindfolds, blindphones...?

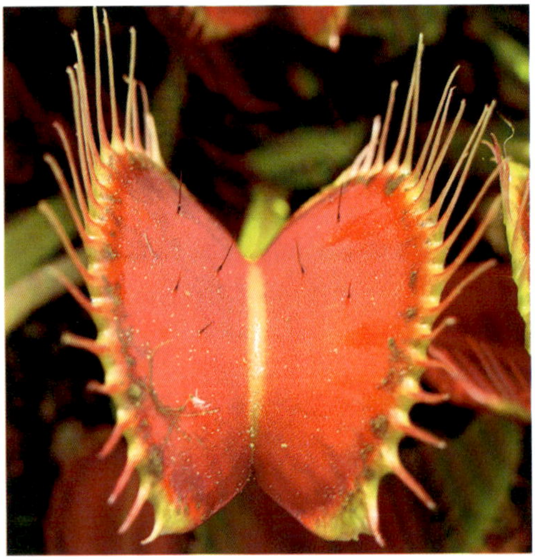

Noah Elhardt

Blindphones is an experimental prototype for an alternative low-tech head-mounted display, which enhances listening by impairing sight. Inspired by the structures of plants, and employing sustainable, bio-based materials such as bacterial cellulose, it immerses the user in a world of music and choreographed eerie lights. Drawing from the psychedelic tradition of early VR art in this sonic piece, like in a sensory deprivation tank, the body floats weightlessly in a dark, quiet environment, and the mind is free to shift from a meditative state to a hallucinatory one.

Institutions: Aalto University, School of Arts, Design and Architecture, Department of Media/MediaLab and the School of Chemical Engineering, Department of Bioproducts and Biosystems, Biobased Colloids and Materials research group, Aalto Studios, Aalto FabLab. Art and science collaborators: Prof. Lily Dìaz-Kommonen, Prof. Orlando Rojas, Prof. Philip Dean, Luiz Greca, Janika Lehtonen, Solomon Embafrash.
Others: Niklas Pöllönen (CAD modelling assistant), Philip Hector (design advisor), Joasia Cieslak (musician).
www.andreamancianti.com/blindphones

ORI*GEN

The origins of life, and the lifeforms that squirmed and flailed their way from the depths of the ocean to the land's surface, are the by-product of the Great Oxygenation Catastrophe that befell the planet due to a build-up of the waste product of photosynthesis. New waste has bred new life, the genesis of new life forms 'challenging binary categories between nature and culture.'

The ORI*gen rose from the primordial polymeric oceans at the dawning of the Plastisphere. Their home was the long-dead remains of Australia's Great Barrier Reef. In the ocean's darkness, bacterial mutations fed on the micro-particles, consuming and bio-converting endless consumer waste. The accumulated throw-away lifestyle of the Anthropocene led to the biopoesis of foldamers en masse.

A selection of foldamers evolved into collective communal entities that collapse and expand with the ocean's continual movements.

They became known as monsters. Their pulsating forms expand and contract as if breathing into consciousness while releasing their cries of anger and lament that their last common ancestors were a PET bottle and a plastic fork.

About ORI*gen

ORI*gen is a post-anthropocentric narrative that explores the role of the natural language of folding. It signifies the genesis of a new species in oribotic research. This work combines research into complex irregular origami geometry, customised 3D printing technology and soft-robotics. Preliminary prototypes were presented at the 2016 festival in the ORI*lab. The work is indicative of new aesthetic directions afforded by the practice-oriented inventions of Matthew Gardiner and the team at the Ars Electronica Futurelab.

Funded through the FWF PEEK Program, Program-Management: Dr. Eugen Banauch.
Text: Matthew Gardiner, Rachel Hanlon

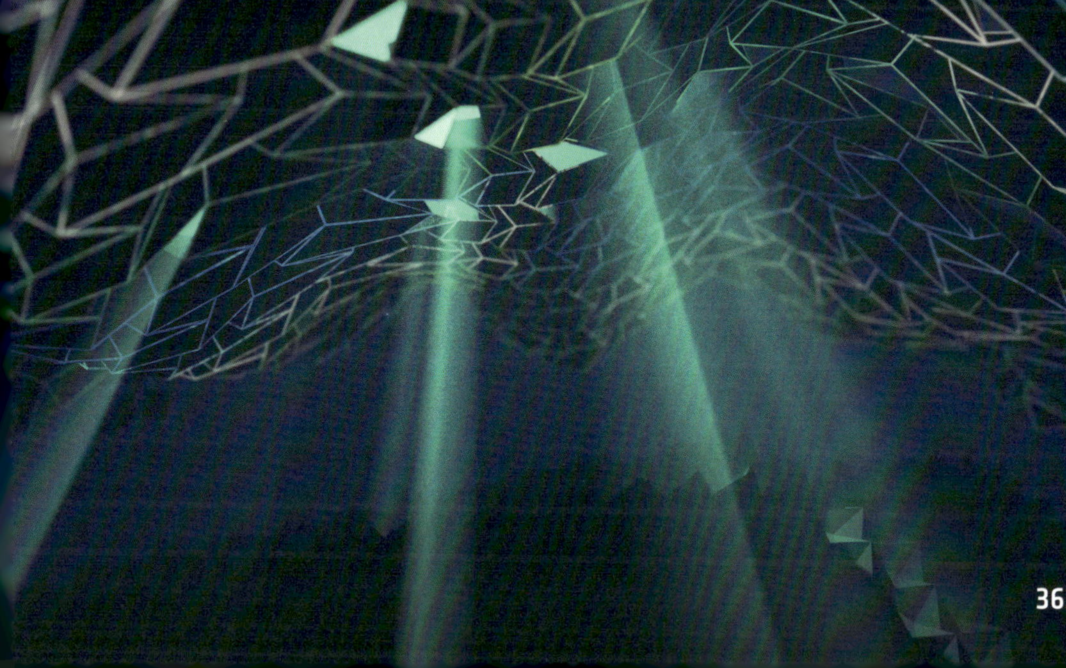

*The Departure. Depiction of the first migration of the ORI*gen Artetradium-Lucidium.* © Matthew Gardiner

Immersify – cutting-edge tools for the next generation of immersive media

Virtual Reality (VR) has what it takes to be the next big game-changer in the media sector. In stark contrast to video, TV and the movies, VR applications promise experiences that are not only more intense but, above all, interactive and individual too. Be that as it may, a considerable amount of R&D work still needs to be done to haul VR out of the niches it's been confined to and launch it on its way to a huge consumer market share. That's precisely the mission of *Immersify*, a European R&D consortium made up of the following partners: PSNC - Poznan Supercomputing and Networking Center (Poland), Spin Digital Video Technologies GmbH (Germany), Ars Electronica Futurelab (Austria), Marché du Film–Festival de Cannes (France) and Visualization Center C (Sweden). Immersify was set up in October 2017 and runs until March 2020. Funding is provided by the EU's Horizon 2020 program. *Immersify* is concentrating on four challenges. First of all, what's needed is a new technology for the video compression of data, which is virtually exploding due to higher resolutions, frame rates and constantly improving image formats. Second, media players and formats should be able to support as many different technical environments and devices as possible. Third, creative individuals working with high-quality videos, CGI in 2D and 3D, as well as interactive elements ought to have the option of combining them with each other so that users are in a position to enjoy totally customized experiences. Fourth, *Immersify* is definitely not to be developed behind closed doors; rather, the intention is to present the ongoing progress of R&D in the form of demos at the Ars Electronica Festival and the Marché du Film in Cannes, where specially developed content and innovative market-oriented products will showcase Immersify's creative and technical capabilities.

This project has received funding from the European Union's Horizon 2020 research and innovation programme under grant agreement No 7620799

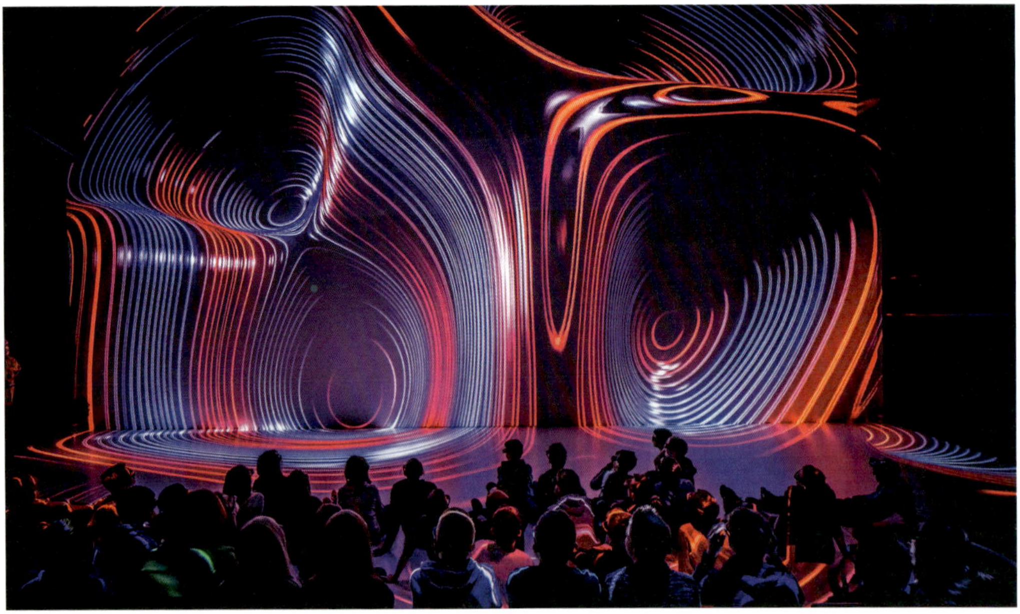

Robert Bauernhansl

Prima Materia, NOHlab

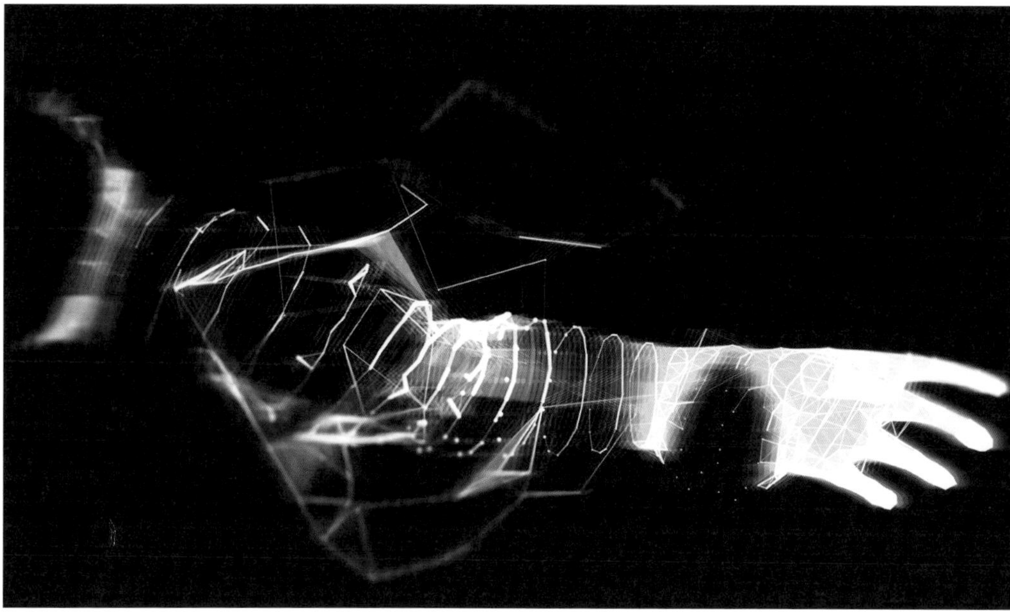

Johannes Pöll

Lazarus

The critically acclaimed musical by Enda Walsh, best known for its music and lyrics by David Bowie, was first performed in New York in late 2015, shortly before Bowie's death, as one of his last works completed. The piece is inspired by (and represents a sequel to) the Sci-Fi novel *The Man Who Fell to Earth* (1963) by Walter Tevis, in whose 1976 movie adaption of the same name Bowie starred as the humanoid extraterrestrial protagonist Newton, on his mission to take water back to his drying-up planet. However, he gets stuck and diverted by earthly vice, and eventually fails. The narrative of *Lazarus* picks up the loose endings of the *The Man Who Fell to Earth* and tells of Newton's subsequent despair, alienation, and ultimate redemption. For the musical's adaption for the Musiktheater Linz (English with German dialogue), the Landestheater Linz engaged in a special cooperation with the Ars Electronica Futurelab, in order to utilize media art to stage the material in a special contemporary appearance. Most notably, both stage design and media art complement each other in applying reduced and simplistic design to amplify the protagonist's disruption and agony as he struggles to realize the ramifications of his mission's failure. The outcome of this venture will premiere on September 27th, 2018, and will subsequently be performed 16 more times throughout the Season 2018/2019 in the Großer Saal at Musiktheater Linz.

Text: Roland Aigner

Ars Electronica Japan as a Creative Catalyst for Culture and Innovation

Ars Electronica Japan is a special division within the Ars Electronica Futurelab, which puts cultural programs, consulting and innovative research into practice for Japan. Historically, Ars Electronica and Japanese artists have collaborated on creative projects for the annual Ars Electronica Festival, Ars Electronica Center and Export exhibitions. What is more, many Japanese creators have been awarded the Prix Ars Electronica in the past. The Ars Electronica ecosystem has proven to be fertile ground for creating future dialogues between art, technology and Japanese society. Ever since Emiko Ogawa and Hideaki Ogawa joined the Ars Electronica team in 2007, they have been creating dynamic interactions in the import and export of creative programs between Japan and Linz. For instance, the idea of a Pop-Up Ars Electronica aims to bring the feeling of Ars Electronica to Japan, as well as to create new platforms for discussing future society together with artists and citizens. As of 2018, unique programs of Pop-Up Ars Electronica are being realized throughout the year, such as *School of the Future* at Tokyo Midtown in Tokyo or *Ars Electronica in the Knowledge Capital* in Osaka. Based on its considerable creative network, Ars Electronica collaborates with Hakuhodo, Japan's biggest advertising communication company, to create a mission-driven community for a better Tokyo after 2020. The very first *Future Innovators Summit Tokyo* at Tokyo Midtown in May 2018 is symbolic of this aim. The Ars Electronica Tokyo Initiative with Hakuhodo aims to create not only a think-tank, but also a "do-tank" for social innovation inspired in Tokyo. In another series of collaborative projects, Ars Electronica Japan has conducted joint research projects based on Japan's preparation for the Olympics 2020 in Tokyo. The NTT *Swarm Arena* Project is one of them: Together with the Japanese Telecommunication Company NTT, Ars Electronica Japan aims to create a new kind of sports viewing experience by developing a swarm display bots system. Another example is the Future Project, in collaboration with Japan's largest public broadcasting organization, NHK. It is a research project to explore creative use of 8K, the next generation of ultra-high definition TV technology.

Thus, Ars Electronica Japan is forming a creative ecosystem for shaping future society together with Japanese creators, educational public institutes and industries.

Text: Hideaki Ogawa

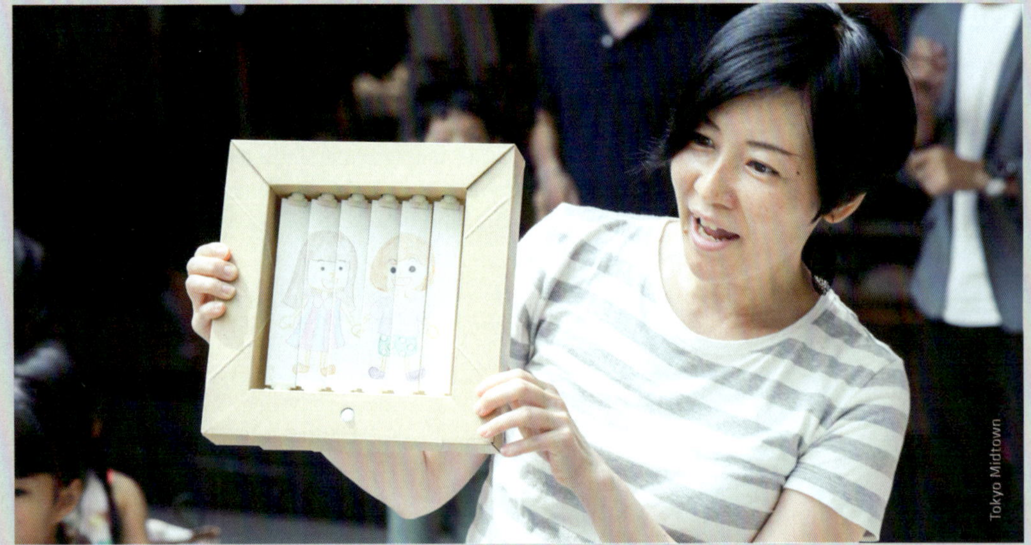

Tokyo Midtown

SWITCH

Ars Electronica Tokyo Initiative and Future Innovators Summit Tokyo

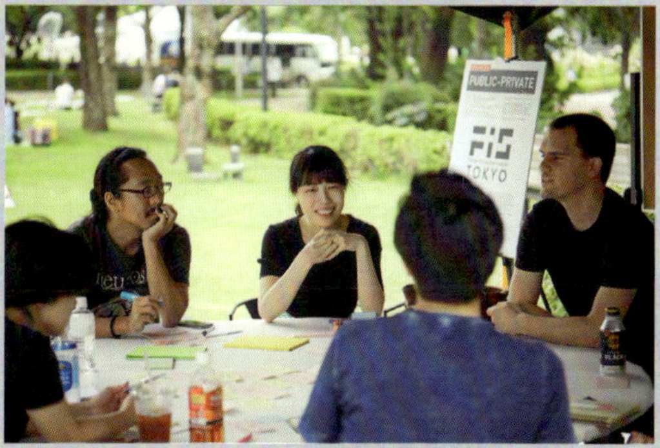

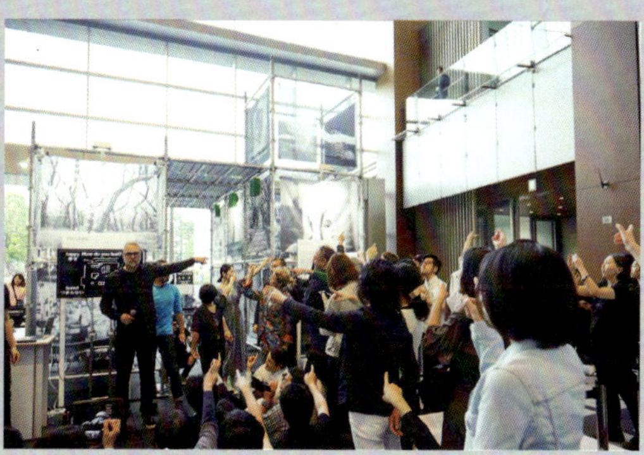

Tsukuru Ozaki

The *Ars Electronica Tokyo Initiative (AETI)* is a mission-driven community of innovative people centered around Ars Electronica and Hakuhodo. As a creative group, AETI is collaborating on ideation of and for a future society, as well as the actual implementation of the generated ideas. We carry out various workshops, events and exhibitions with diverse talents from all over the world, and also engage in collaborations for private sector innovations. The mission "What can we do to make a better future Tokyo and Japanese society?" stands at the heart of AETI's activities. The very first *Future Innovators Summit Tokyo* at Tokyo Midtown is symbolic of this aim.

The *Future Innovators Summit (FIS)* is a hands-on discussion program, which has been developed jointly by Ars Electronica and Hakuhodo. Ever since 2014, FIS has served as an interdisciplinary think tank at the Ars Electronica Festival in Linz, Austria, with the purpose of generating Creative Questions about the future, shaped by innovators from all over the world. Future Innovators represent a wide variety of different backgrounds and consist of experts such as artists, designers, scientists, technologists, activists, entrepreneurs, philosophers and more. For FIS Tokyo, the concept was extended to imagine "Tokyo as a Laboratory for our Future."

To this end, participants joined round-table discussions to discuss one of the three themes inspired by and relevant to Tokyo:

DEATH-LIFE in Tokyo: What is life and death in the city with the fastest aging society?

TECH-SKIN in Tokyo: What would be the future of fashion and body borne in a city embracing the world's most advanced technology?

PUBLIC-PRIVATE in Tokyo: What could be the future relations of individuals and public in a mega-city with no common space?

Complementing the discussions, a temporary exhibition of the participating innovator's projects and companies' prototypes of the future was set up to inspire and guide the participants. At the final presentation, each group presented their Creative Question:

DEATH-LIFE in Tokyo: How can we carefully co-craft DEATH-LIFE?

TECH-SKIN in Tokyo: How can tech skin foster kinship with all living things in a future where human values have shifted?

PUBLIC-PRIVATE in Tokyo: How can we lay out the "blank" that inspired the public to accept each other like a cat?

Based on the questions, processes and discussions this year, FIS Tokyo will continue to create missions and actions for Tokyo at Tokyo Midtown every year until 2020. We aim to create not only a think-tank, but also a "do-tank" for social innovation inspired in Tokyo.

Text: Hideaki Ogawa, Kazuko Tanaka, Rena Tanaka

Beyond the Frame: 8K Future Project

Beyond the Frame: 8K Future Project offers a glimpse into the joint research project between the Ars Electronica Futurelab and NHK, Japan's largest public broadcasting organization. They joined forces in 2017 to examine how 8K, the next generation of ultra-high definition TV technology, can be incorporated into daily life. The innovative broadcasting giant had already started creating and using 8K footage on a trial basis a year earlier, providing a resolution 16 times higher than Hi-Vision to its audience.

The research focuses on several key scenarios that reexamine modes of interaction with television content, ranging from a speculative high-resolution news program, novel interfaces or the employment of ambient media to a fresh approach to public viewing. The reflection upon incorporating 8K into the audience's lives aims to help NHK tackle the biggest challenges to a widespread use of high-resolution in public broadcasting.

At the 2018 Ars Electronica Festival, NHK and the Ars Electronica Futurelab present three of NHK's original 8K contents at Deep Space 8K. *Tokyo—The Luminescent Megacity* takes the audience on a flight in 8K over the megacity at nighttime, providing spectacular views in astounding resolution. The stop-motion animated *DOMO WORLD* invites viewers to discover a complex, miniature world, where Domo, NHK's official mascot, bustles around the screen performing various activities. The animation consists of 16 separate Hi-Vision scene shots, stitched together to create an 8K experience full of details and wonder.

Lastly, *8K TECHNE* engulfs the audience in an experimental TV program. To create the series of unique high-resolution videos, five artists were asked to express themselves artistically in 8K: Fuyu Arai used 1024 human-like figurines and individual spot lights to create the spatial animation 1024. Koki Sone turned natural stones into planets and lets the viewers travel through a universe made of ordinary objects in *Cosmos under Foot*. In Haruka Furuya's *The Window*, seven actors perform behind a frosted glass and transform the TV screen into a window from the performer to the audience. Koichiro Tsujikawa plays with video feedback in *88888888•••K*, showing endless loop movies of a camera pointed at its own playback video monitor. Finally, *mimesis* by Eiji Tanigawa impressively explores aesthetics of color in 8K technology by focusing on the sense of materials in the same color.

NHK (Japan Broadcasting Corporation) (JP), Ars Electronica Futurelab (AT)
Text: Hideaki Ogawa

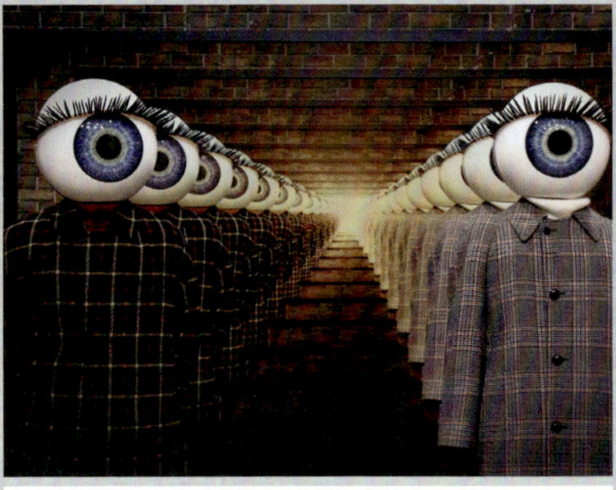

NHK, Koichiro Tsujikawa

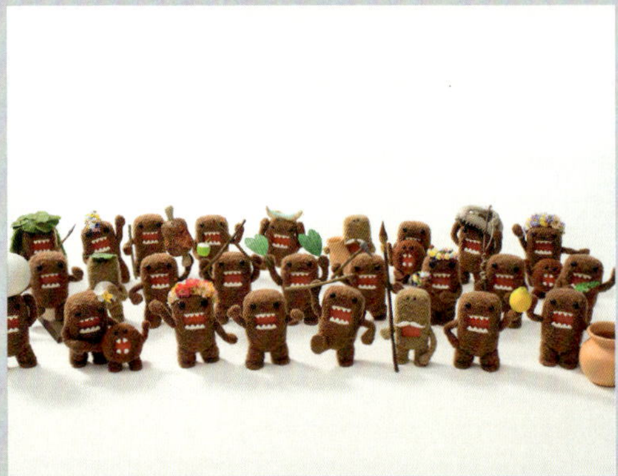

NHK, TYO

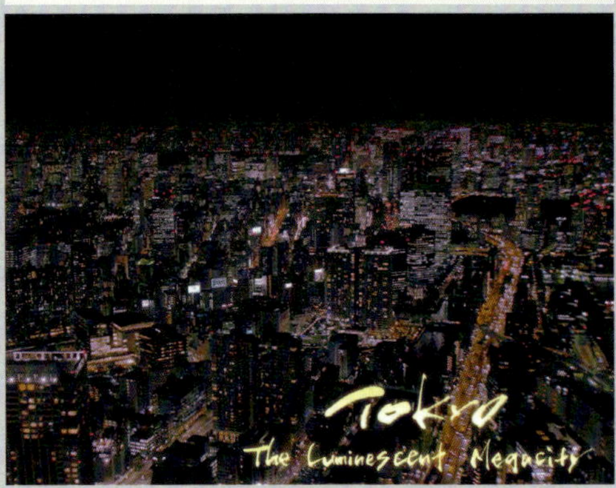

NHK

Virtual Anatomy

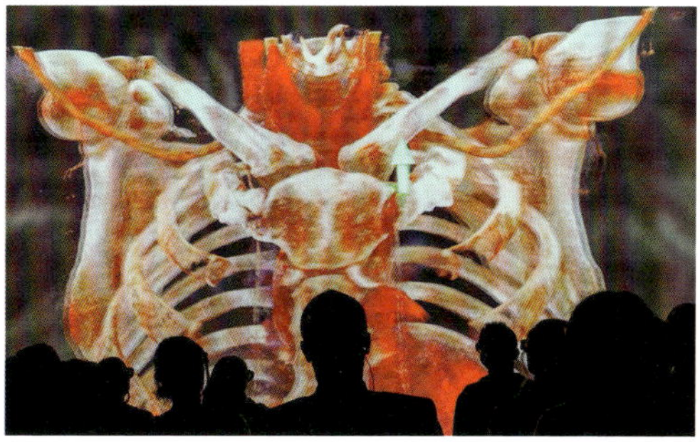
Martin Hieslmair

Johannes Kepler University (JKU) plans to build a Virtual Anatomy room by 2021 for the new campus building of the Faculty of Medicine, modeled after Deep Space 8K of the Ars Electronica Center. To that effect, the Ars Electronica Futurelab, JKU and Siemens Healthcare have started devising a joint research project for the planning, construction, procurement, installation, and commissioning of the hardware and software infrastructure. The common aim is to create a groundbreaking new space for mediation and interaction, a black box, in the context of medical data visualization.
The main usage scenario of the virtual anatomy room is to visualize medical volume data (especially MR and CT) as photorealistically and vividly as possible for educational purposes. University lectures in anatomy, drawing on the Cinematic Rendering software of Siemens Healthcare, have already been put into practice at Ars Electronica's Deep Space 8K and should ideally find their continuation at the Virtual Anatomy room of JKU. The Cinematic Rendering Software, the Deep Space Infrastructure and the teaching requirements of the Faculty of Medicine are to be merged into a new and unique teaching room in this research and development project.

Augmented Intelligence

Ars Electronica Futurelab conceived and assembled the exhibition pavilion at the headquarters of SAP SE in Walldorf, Germany in 2012. For the third time in a row, the Futurelab has produced an exhibition for this striking venue under its current theme: Augmented Intelligence.

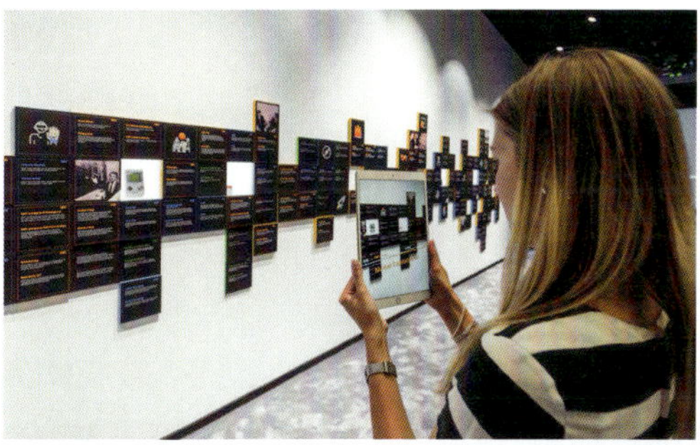

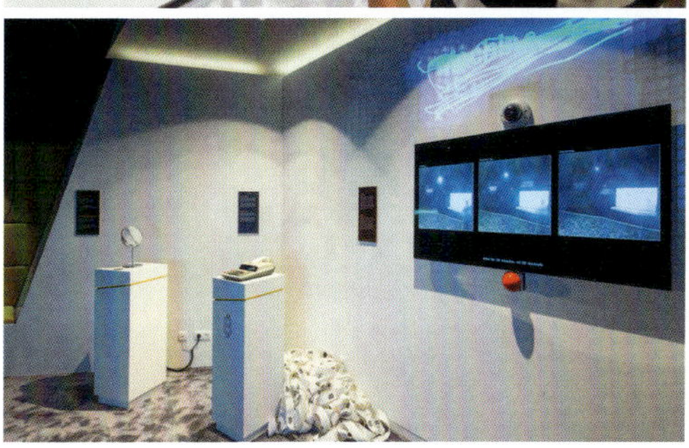
Otto Saxinger

The exhibition, featuring an array of installations developed in-house at the Futurelab, and other selected works of media art, is a reflection on technological progress' impact on our everyday life. Augmented reality comes into play in a timeline, which graphically depicts the company's history in the context of global events, and technological developments. The Intelligent Enterprise exhibit deploys virtual reality to enable visitors to potentially visualize a company of the future.
"Bittercoin" by Martín Nadal & César Escudero Andaluz, and "Learning to See: Hello World!" by Memo Akten, are just two of many artworks that illuminate the theme on another level. Through their immediate aesthetic appearance, they create a free space for reflection, and invite those beholding them to engage in a constructive debate about human-machine interplay, now and in the future.

Text: Peter Freudling, Stefan Mittlböck-Jungwirth-Fohringer

Ars Electronica Solutions

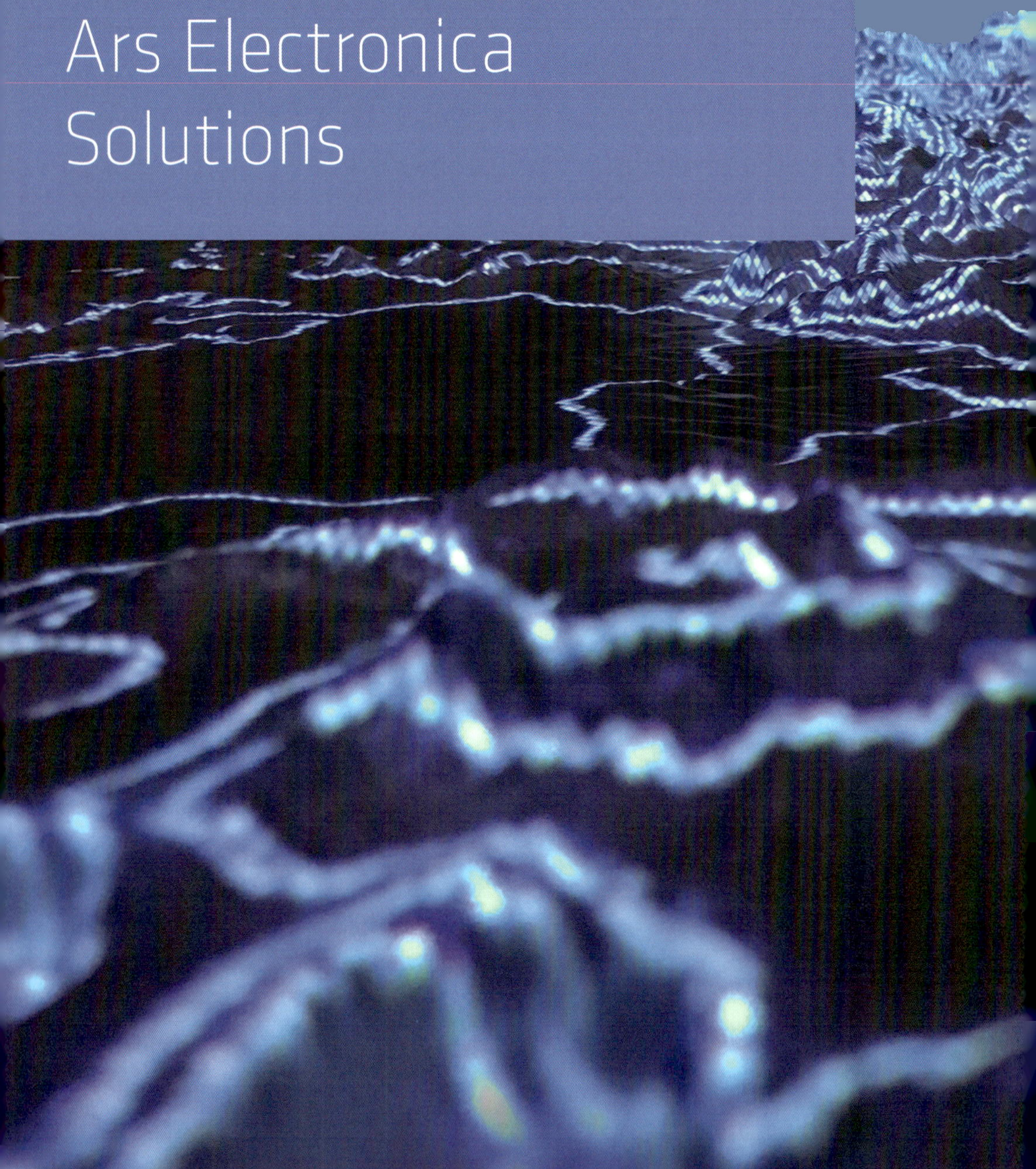

Ars Electronica Solutions conceives, creates and implements interactive worlds of experience as visionary installations. We develop creative, individualized solutions in the form of interactive products and services for exhibitions, brand lands, trade shows, events, and in the urban development field. Whatever your needs—single Touch applications or an entire exhibition, leasing reliable products or the creation of a bold prototype; whatever the occasion—a trade fair, a Smart City workshop, a showroom or a whole museum—Ars Electronica Solutions has the expertise and experience it takes to do a superb job!

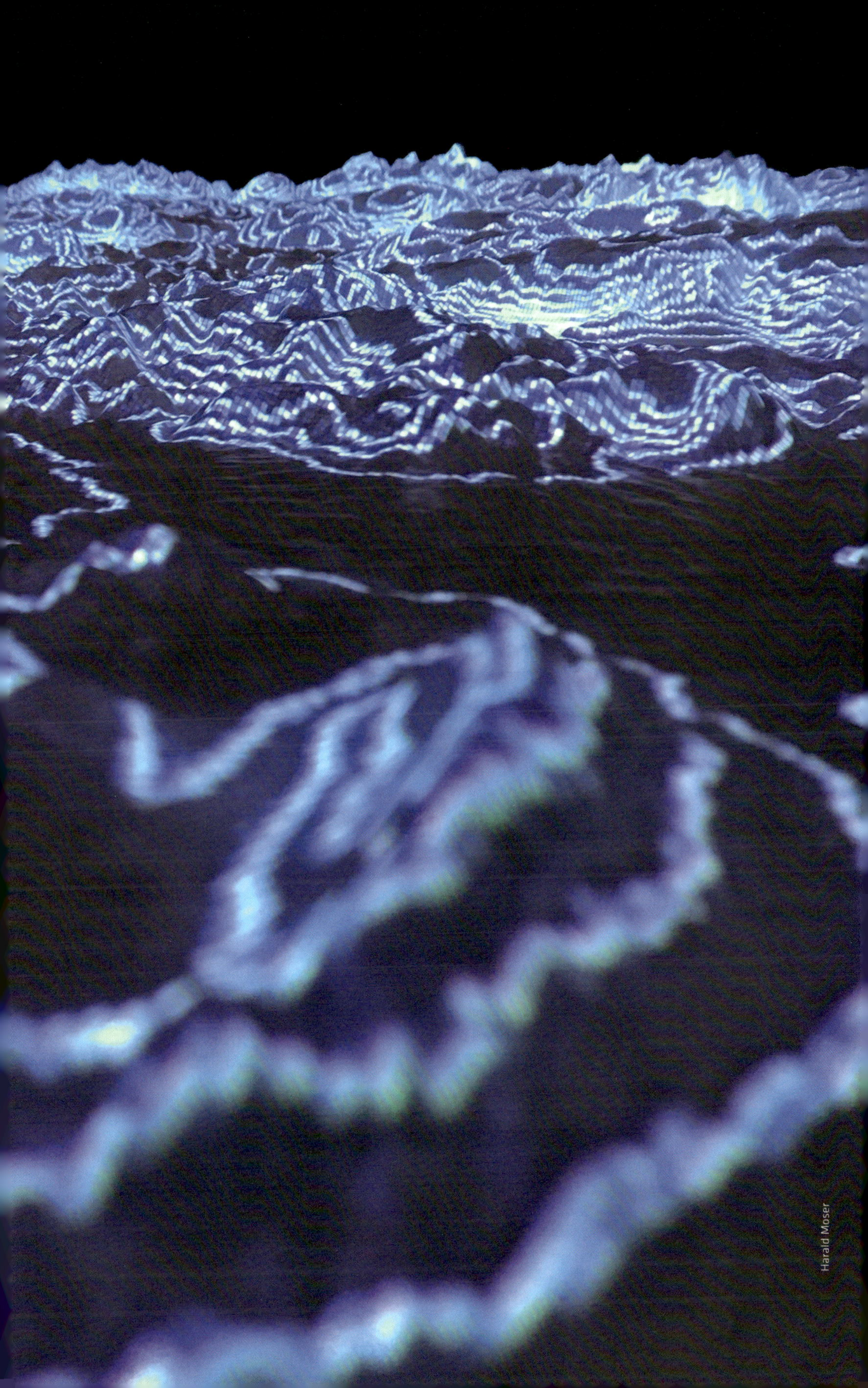

Harald Moser

PROJECTS

We delight in smart design, demand the impossible, and nurture playfulness. Our projects make the unattainable come true. Regardless of what message you want to send in today's world, we can get it across in digital media. Our solutions are aesthetically sophisticated and creatively extraordinary; they encourage hands-on exploration. We offer digital installations ranging from small-scale highlights to conceptioneering and producing elaborate exhibitions. Our clients include major institutions such as the European Space Agency and the EU Presidency, long-established brands like Sacher and Umdasch Group, municipalities including Tel Aviv and Berlin, and such private-sector heavyweights as Fronius, Palfinger, Primetals, voestalpine, BASF, Liebherr and the Industrial Association of the State of Upper Austria.

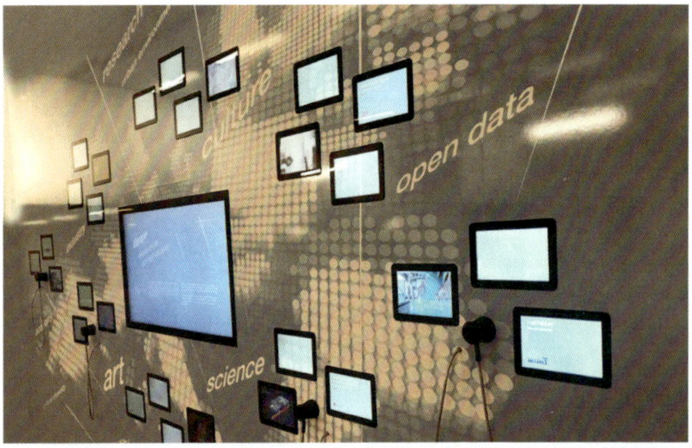

EU Council Presidency, © Ars Electronica

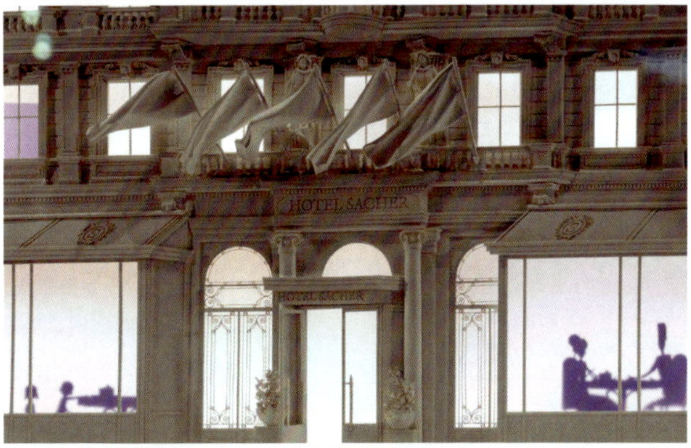

Sacher, © Florian Voggeneder

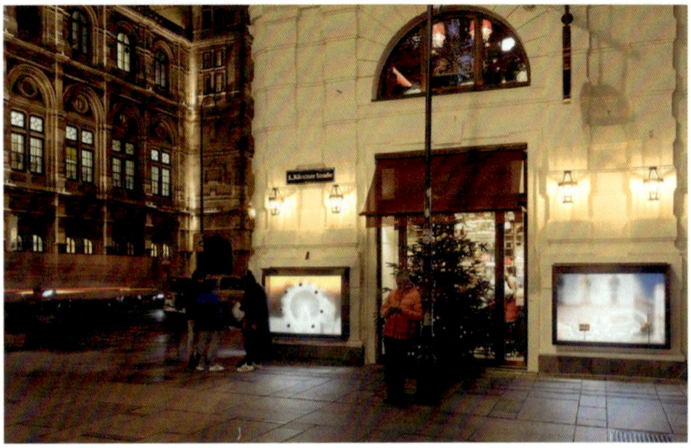
Sacher, © Florian Voggeneder

ESA: Φ Experience

Particularly noteworthy among our recent projects is the Φ-Experience, an interactive exhibition created for the European Space Agency's facility in Frascati, Italy. The Φ-Experience presents ESA's mission and its vision in a way that is understandable by the general public. Four installations feature a variety of scenarios elaborating on satellites, Earth science data, research and earth observation. A highlight of this dramatic exhibit is the Half-Dome Globe, a projection that is impressive in both its content and its interface. Users operate the globe by means of a crystal ball measuring 20 centimeters in diameter that emits colored light to indicate its status. Two touch displays make it possible to navigate through the different layers explaining the dynamics on our planet. This includes time series allowing to move back and forth in time or to overlay information layers. In the Control Room, visitors get a behind-the-scenes look at ESA. Space travel enthusiasts can find out about all of ESA's missions, how satellites are launched and operated and on the entire lifecycle of a satellite. A three-dimensional satellite model provides technical descriptions, innovative inventions are presented, and a complete satellite launch can be recreated. Rounding out this installation are the Map Room, featuring various data visualizations (temperature, tectonics, radiology) on an elevation model, and the News Room, which provides accounts of the latest developments.
Not only was Ars Electronica Solutions part of this project from its very inception; we also played a key role in the execution of the sound & light concept that makes this exhibition such a special experience. Rupert Huber's sound design purposely gives each room its own tonality. Plus, if multiple groups are touring the space at the same time, these sounds harmonize with each other to produce an extraordinary soundscape.
The Φ-Experience will be officially opened in Frascati, Italy on 27 September 2018 on the occasion of the 50th anniversary of the European Space Agencies research centre ESRIN.

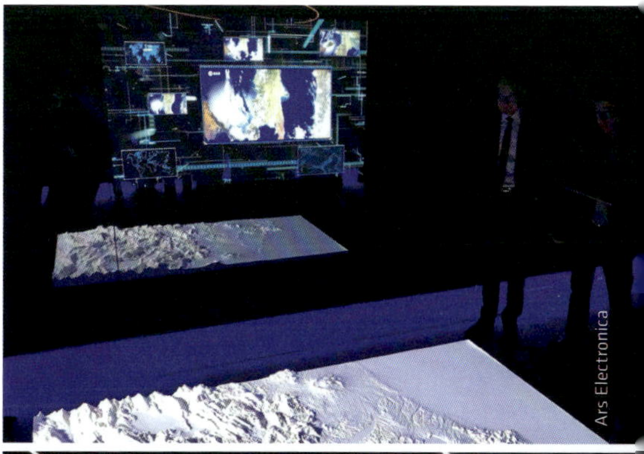
Ars Electronica

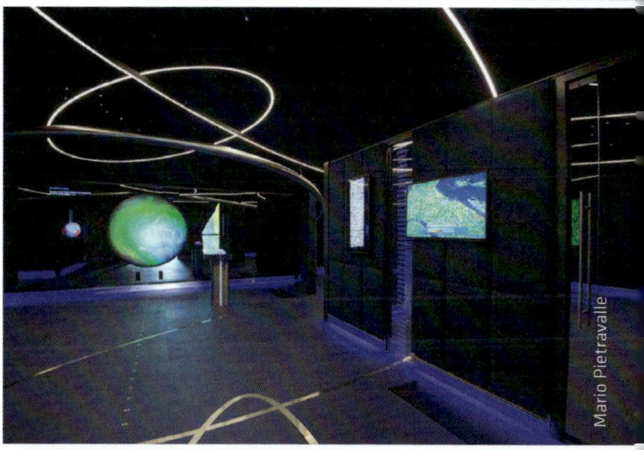
Mario Pietravalle

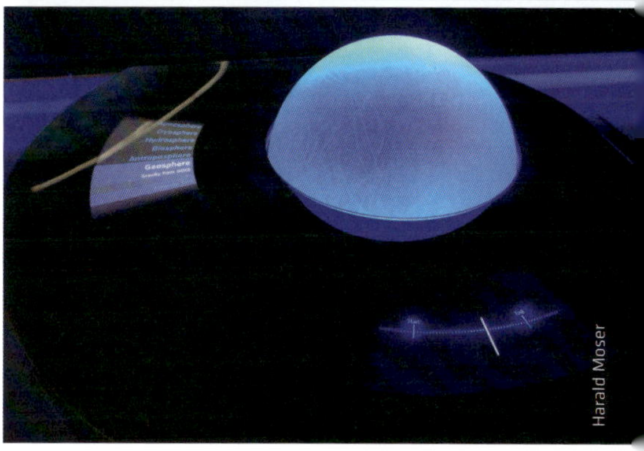
Harald Moser

EXHIBITIONS

Many of our commissions require us to produce exhibitions. We conceive and execute interactive installations and participative formats for many different purposes, featuring a wide variety of content. Deploying our didactic experience and creative spirit, we configure amazing worlds in which visitors immerse themselves via identification and participation. To carry out these assignments, we implement the most innovative communications trends and the latest technologies. Our creative endeavors follow through on the artistic approach of Ars Electronica.

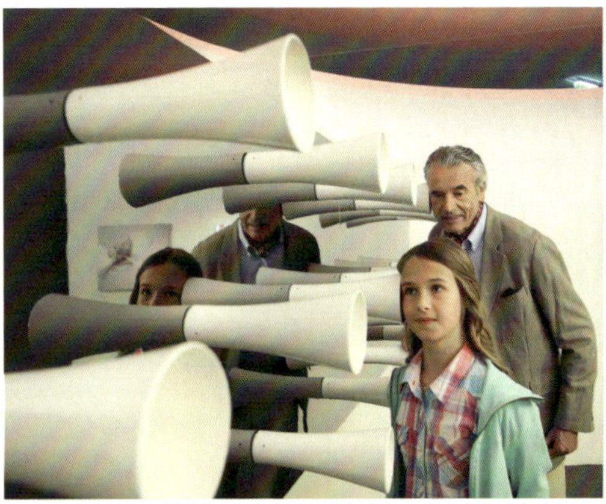

Audioversum, © MED-EL

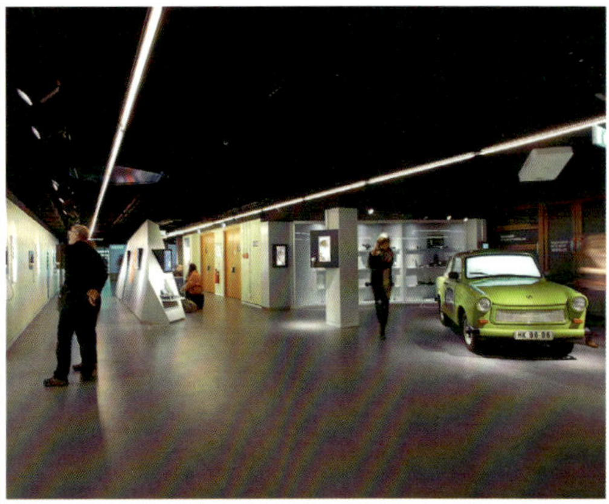

German Spy Museum, © DSM

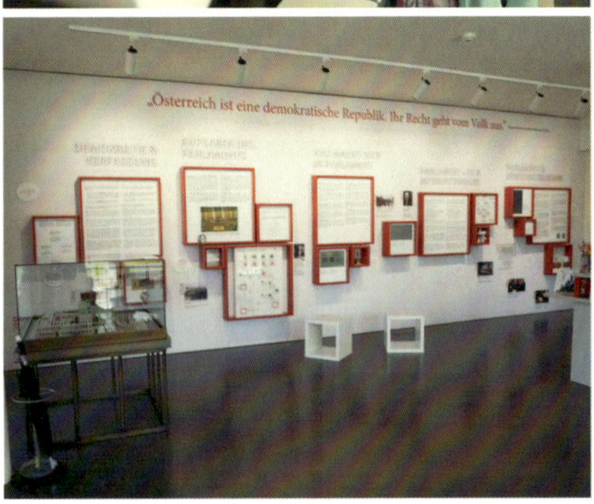

Parlament, © Ars Electronica

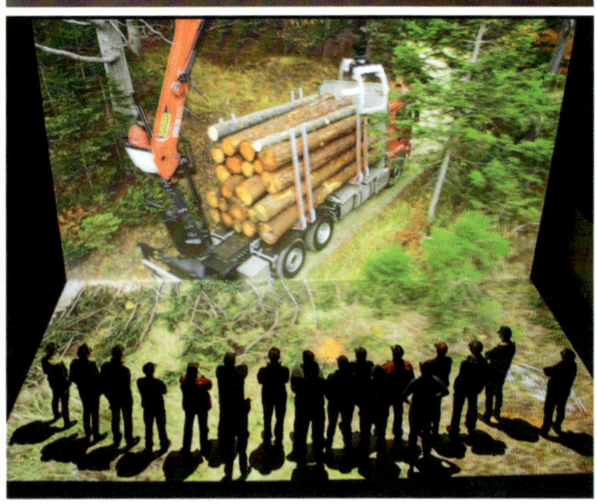

Palfinger World, © Palfinger AG

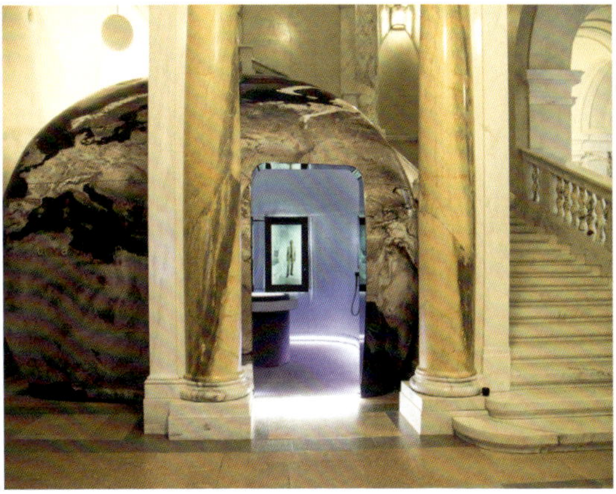

Umdasch Sesquicentennial, © Umdasch

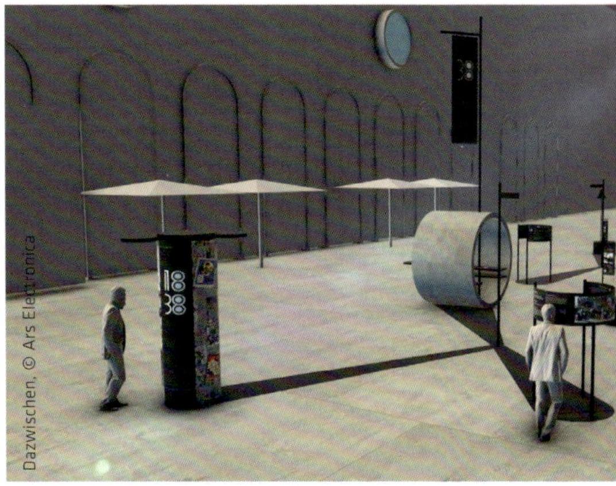

Dazwischen, © Ars Electronica

Umdasch: Two Pavilions for an Anniversary

This comprehensive package convinced Umdasch, our associate of long standing, to venture out beyond the core business of retailing and take the plunge with an exhibition. We worked together to design a show set in Vienna's Weltmuseum to celebrate the company's sesquicentennial. In two pavilions, Umdasch presented itself as a trendsetter over the past 150 years and as a global player confronting the challenges that face our society with vision and responsibility.. The pavilion of the present and the past focuses on the changes that Umdasch's world has gone through over the past century and a half—from the story of how this enterprise was founded and the personal experiences of employees all the way to a behind-the-scenes look at the construction of the world's tallest building. The second pavilion is designed to consider the fascinating prospects of humankind's potential "futures." Thematically oriented on four megatrends—Energy & Resources, Urbanization, Consumption, and the Future of Work—a dome-like projection delivers enthralling insights into futuristic necessities and possibilities.

Dazwischen

Occurrences in Linz from 1918 to 1938—an audio exhibition

Exactly 100 years ago, the end of the Danube monarchy in Austria led to the foundation of the First Republic. Twenty years later the "Anschluss" to the National Socialist German Reich took place. An audio exhibition in public space commemorates these events and the period between 1918 and 1938. The Linz Cultural Directorate, the Linz City Archive and Ars Electronica Solutions create an extraordinary audio world that conveys life, social and economic circumstances, and the political situation in Linz between two world wars. By actively exploring this exhibition in public space, scenes of everyday life at that time become audible by means of different sound reinforcement techniques. Attentive visitors and curious passers-by will be able to listen to conversations whose origin can often only be guessed. Speakers announce historic events and the political situation is discussed in niches. Texts and images also provide interesting background information and combine with what you hear to create a complex overall picture. The project was supervised by the Archive of the City of Linz. The audio plays were reconstructed by the writer Walter Kohl according to historical reports and staged by theater@work.

Kulturdirektion Linz
Archiv der Stadt Linz
Walter Kohl
theater@work
Ars Electronica Solutions
Evangelische Pfarrgemeinde Linz Innere Stadt
Buchhandlung Thalia Linz
Anton Bruckner Privatuniversität

TRADE SHOWS

Ars Electronica Solutions' portfolio includes an increasing number of trade show stands produced in recent years—from classic presentation settings to interactive worlds of experience. The well-thought-out use of innovative technology set in an eye-catching architectural package and endowed with a convincing storyline combine to create a unique attraction. Moreover, these stands are substantively and architecturally designed with modularity in mind, so they can also be used after the show is over at a corporate headquarters, an event or other formats. And experts in this field have been duly impressed. The stand that Ars Electronica Solutions created for Primetals Technologies' appearance at METEC 2015 garnered Ars Electronica Solutions a 2016 CAESAR in the Trade Show Architecture + POS category as well as a 2017 Golden AUSTRIACUS in POS-Trade Show Architecture.

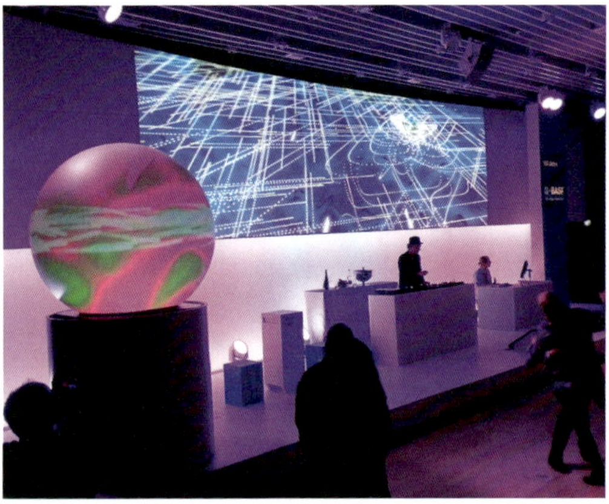
BASF, © Ars Electronica

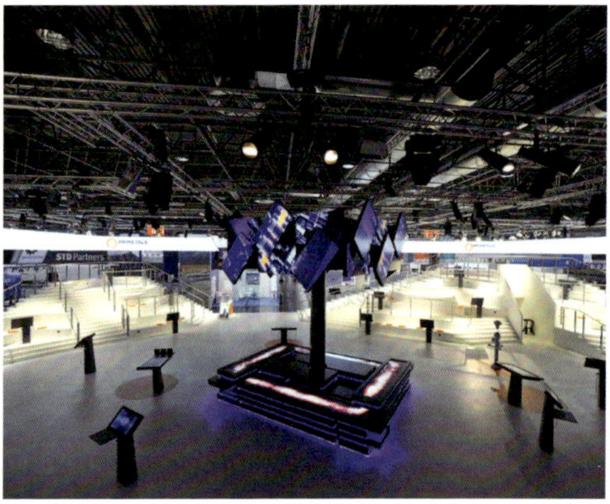
METEC, © C26

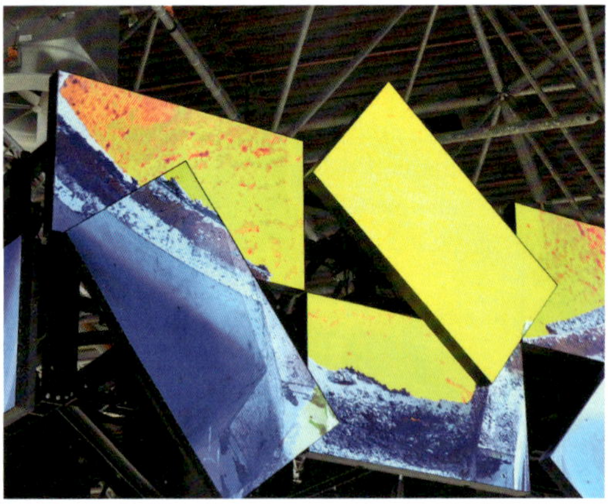
METEC, © C26

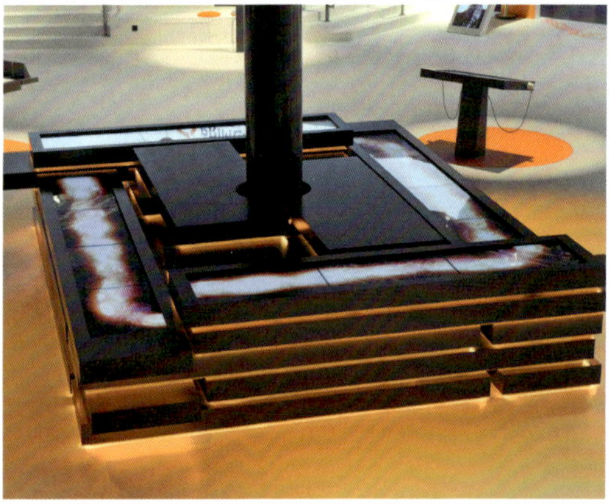
METEC, © C26

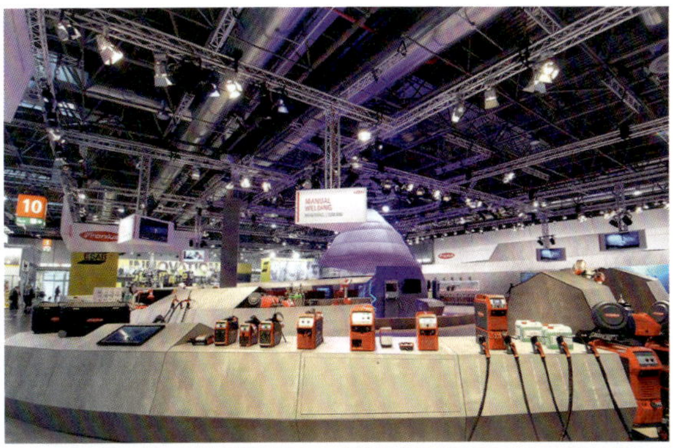

Fronius Booth, © Fronius

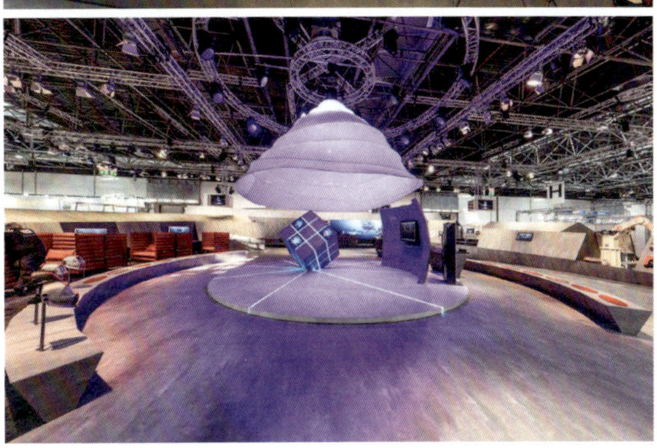

Fronius Booth, © Fronius

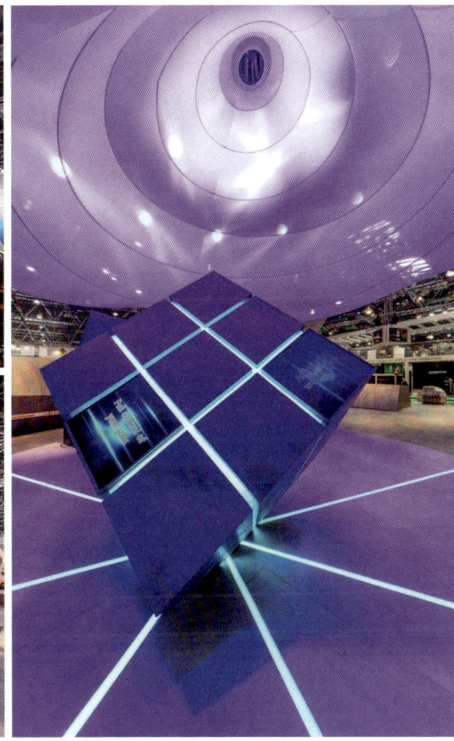

Fronius Booth, © Ars Electronica

Fronius: Arc of Light & Weld Cube

Fronius, a global player in welding, photovoltaic and battery charging technology, commissioned us to come up with an extraordinary trade show stand thought out down to the finest detail.

An impressive arc of light is the trade show stand's eye-catching centerpiece symbolizing the welding process per se. The light phenomena inherent in welding are captured at the live welding stations and sent to the arc light installation, where they are visually and acoustically reinforced and accentuated by means of media and architecture. Beneath the arc of light is the WeldCube, a physical cube in which the streams of digital data from the individual welding stations flow together. Integrated into the cube are picture screens that portray the object in a figurative sense as an intelligent organism. The stand architecture radiating out from the arc of light takes the form of concentric circle segments, and provides individualized settings for communication, product presentation, automation, and live welding.

PRODUCTS

We scrutinize and tinker, think and create, design and test. Sometimes with such outstanding success that it's immediately evident the results are suited to manifold applications.

That's how the development of prototypes can lead to products that can be customized to a client's specific situation and needs. And that closes the process chain from idea and vision to the development of a prototype and ultimately to an innovative product. The various interfaces we build into our products are intentionally futuristic—direct communication between the human brain and a computer, interweaving real and virtual scenarios, and photos at such high resolution that whoever sees them gets the feeling that they're not just a spectator; they're right in the middle of the action.

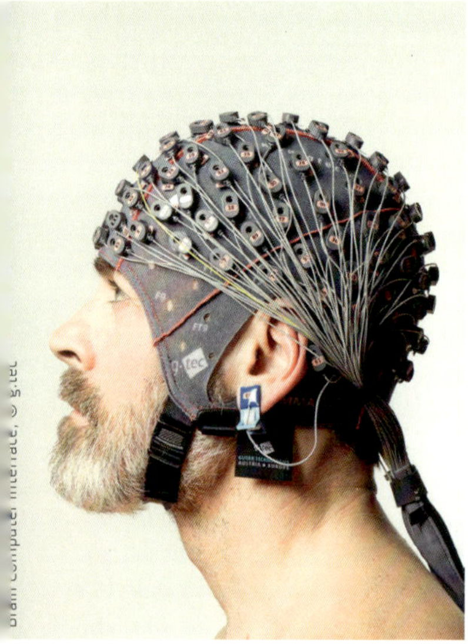

Brain Computer Interface, © g.tec

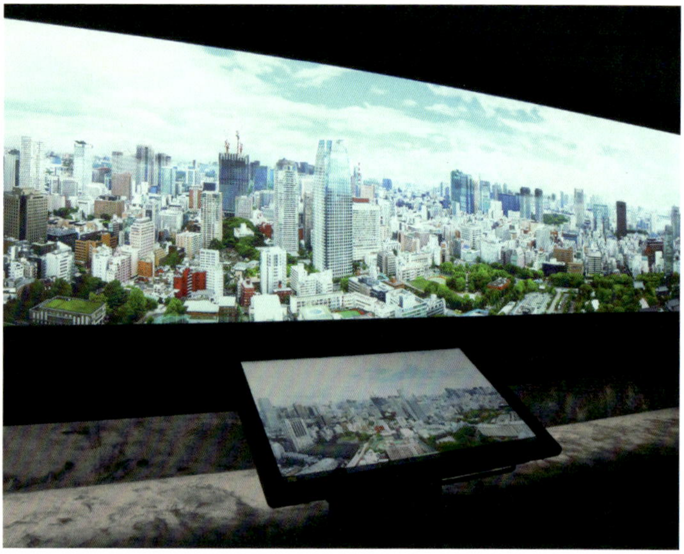

Gigapixel, © Ars Electronica

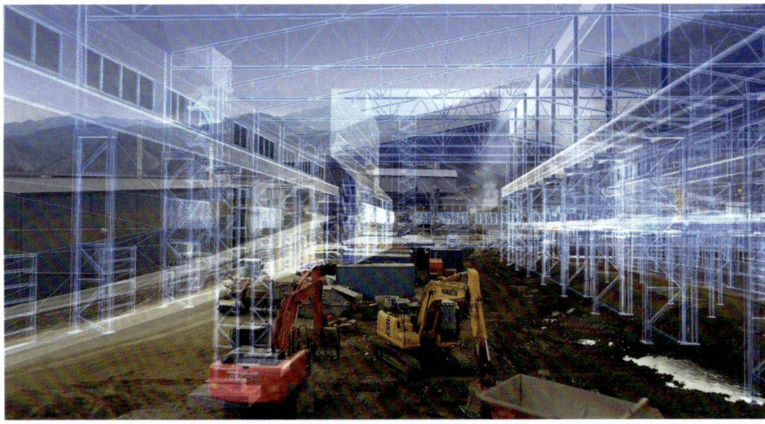

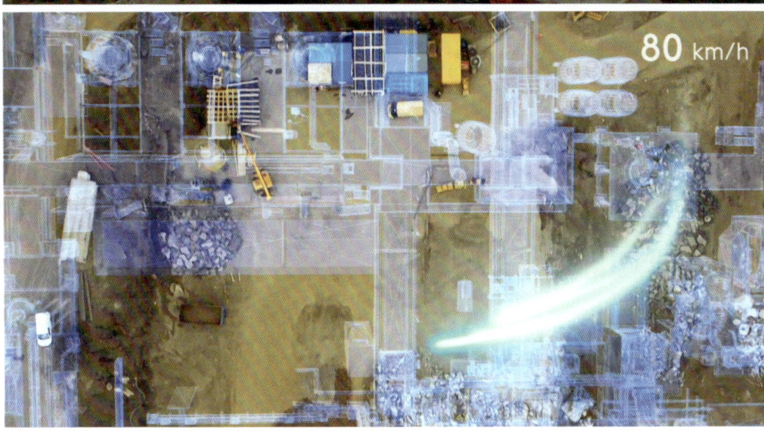

voestalpine, © Ars Electronica

voestalpine: Mixed Reality Drone Race Performance

An extraordinary spectacle was the highlight of the groundbreaking ceremony for the world's most modern stainless steel work in Kapfenberg, Austria. In order to get across the incredible spatial dimensions of this plant and the state-of-the-art digitization and automation of the production processes that will go on inside it, we conceived a performance starring racing drones and staged it before an audience of VIPs in business and government.

We presented a pre-produced mixed reality video that brings the steel mill to life in 3-D, and teamed it up with live images delivered by drones whose flight paths demarcated the actual dimensions of the future facility. The video and the drone footage were mixed and synched live and presented to the audience on site on a jumbo-size LED screen—dynamically interwoven into a futuristic blend of speed, sound and mixed reality. This performance was the spectacular opening act for the actual groundbreaking ceremony, which followed immediately. It was a truly memorable sneak preview of the world's most modern steel mill, and the first one to be built in Europe in 40 years.

BRAIN COMPUTER INTERFACE

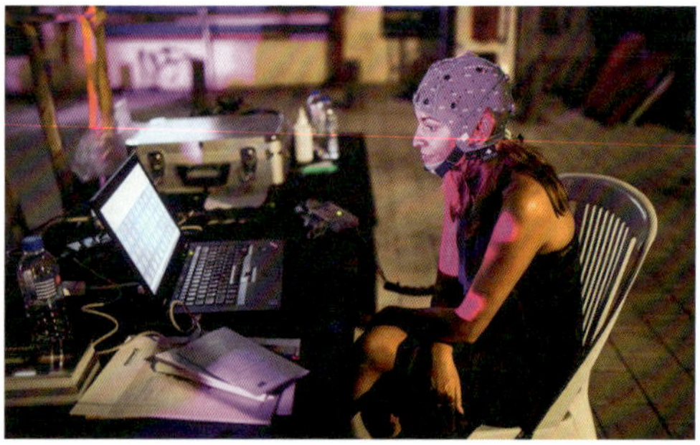

White Night Tel Aviv, © Tomer Foltyn

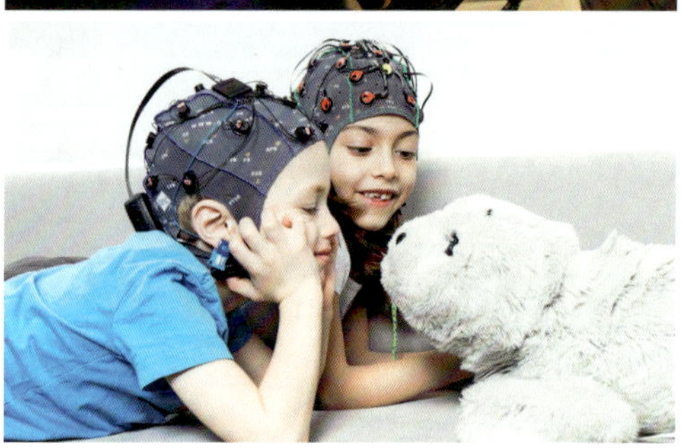

Brain Computer Interface © g.tec

Touch interface was yesterday's news. Modern users control devices with their brain! Via a brain-computer interface (BCI), the brain communicates "directly" with a computer interface, and this imparts a magical touch to a wide array of applications. Typing in texts by thinking them, playing single- or multi-user games of skill, or "subconsciously" ranking graphic information are not only lots of fun; there is also a variety of ways to deploy these modes of human-machine communication. To do the job, Ars Electronica works together with g.tec medical engineering, the company that developed the first commercially-available BCI in 1999 and now distributes its products in over 60 countries worldwide.

In what would turn out to be a flourishing collaboration, Ars Electronica Solutions began working together with Wacker Neuson at the 2017 Ars Electronica Festival, where visitors could drive a six-ton earth mover using only a brain-computer interface and eye-tracking technology. What was then a prototype has been continually upgraded and enhanced, and is now the featured attraction at every Wacker Neuson event!

MIXED REALITY SCENARIOS

Ars Electronica Solutions offers interactive and totally immersive augmented reality and virtual reality solutions for events, showrooms, temporary & long-term exhibitions, situation-specific productions in public spaces, and memorabilia that activate the virtual scenario anytime, anywhere. We program larger-than-life augmented and virtual 3-D scenarios and design holistic mixed reality experiences for entertainment & marketing uses as well as in corporate communications. And we customize them to the specific target audience and with their own content.

We specialize in interactive 3-D augmented reality applications (iOS, Android, Windows Phone), either mobile or as a stationary installation. We work and do research in the GPS-based augmented reality sector (AR 2.0) or markerless augmented reality, and integrate intelligent avatars equipped with speech recognition AI into real spaces to achieve extraordinarily sensory, hybrid storytelling—like the drone performance in Kapfenberg!

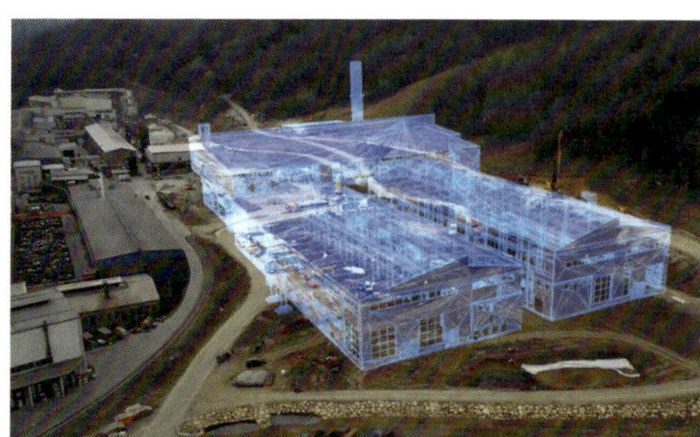

voestalpine Drone Race Performance, © Ars Electronica

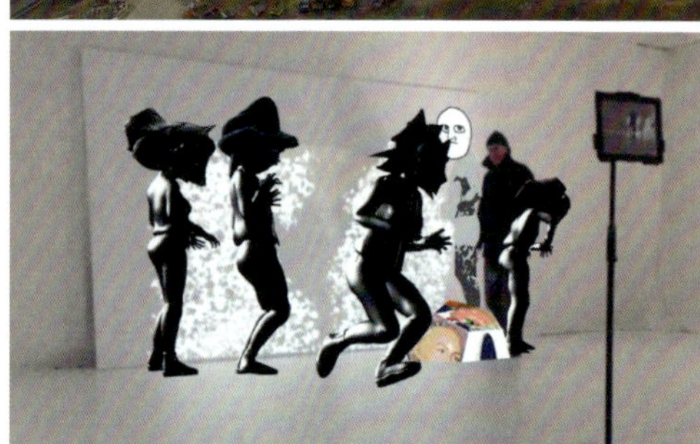

Mixed Reality, © Bobby Rajesh Malhotra

GIGAPIXEL VIEWER

Graphic material is fed into the Gigapixel Viewer—for instance, a high-resolution photo of a city, a landscape, a work of art, or any other richly detailed object. Due to the viewer's extremely high resolution, the image can be viewed as a whole, but there is also a zoom feature that reveals the minutest details captured in the shot. Plus, "artificial" levels can be inserted so that, at certain zoom levels, information and surprising interrelationships that are not immediately apparent are displayed at specific points in the image. This material can be texts, images, videos or animated sequences with added explanatory value. The primary user benefits are the ability to independently explore jumbo-format panoramas, and the impressive visual power of the image.

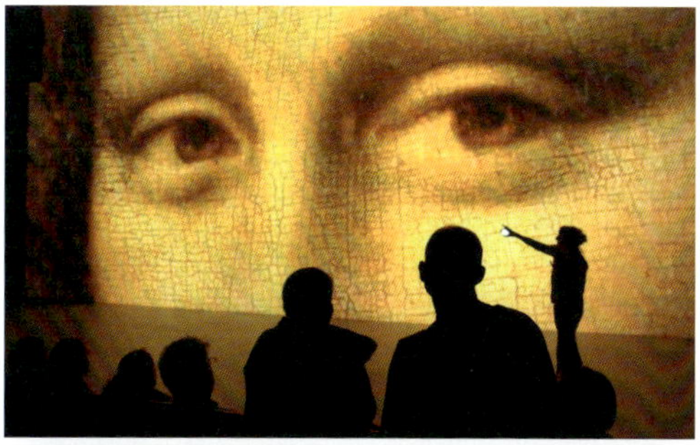

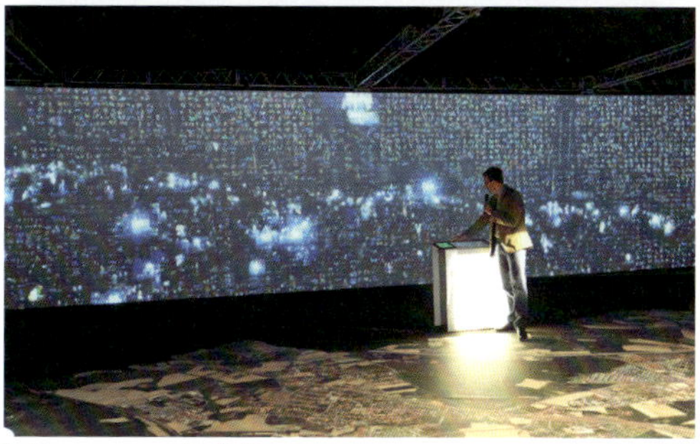

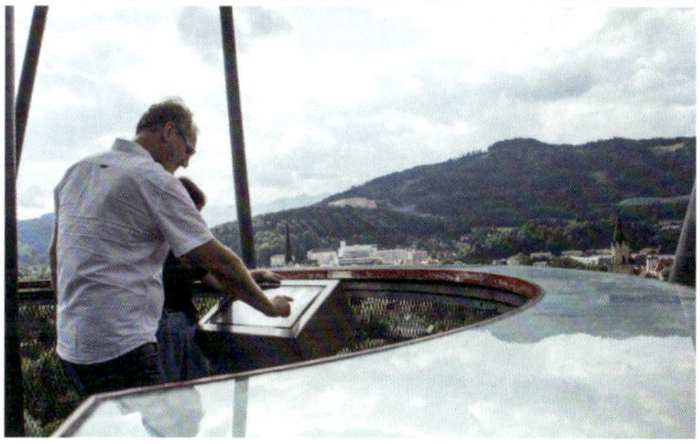

Gigapixel Viewer, © Ars Electronica

Urban Development in the Austrian City of Leoben

Most recently, a Gigapixel Viewer was created for Leoben's urban development authority and installed in Maßenburg, a 13th-century fortress that has been open to visitors since 2000. Its western defensive tower was outfitted with an observation platform to provide visitors with great views of the city. To enable guests to conveniently find out more about what they're experiencing, the platform's rotunda has been equipped with an interactive display permitting targeted searches throughout Leoben. Each of two gigapixel images shows the city from a different perspective—one from Maßenburg; the other from a promontory facing it.

Ars Electronica Export

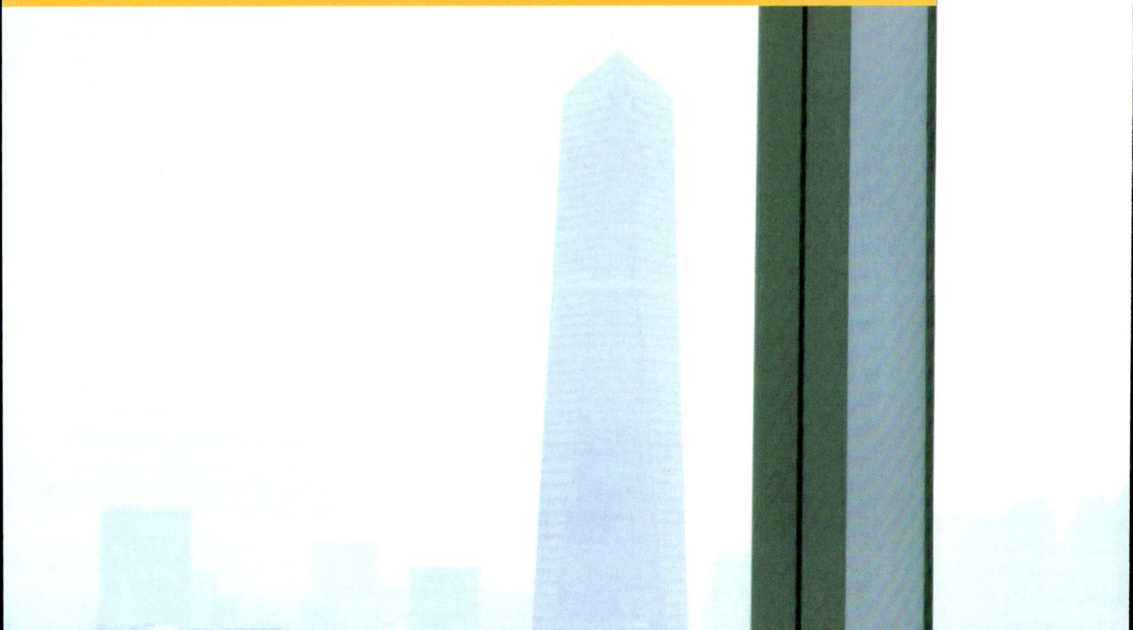

Since 2004, Ars Electronica has worked together with partners in art and culture, science and education, commerce and industry to produce a diverse array of projects all over the world. The spectrum includes exhibitions and presentations, conferences and workshops, performances and interventions. What these collaborative activities have in common is the inspiration they derive from the ideas and visions of Ars Electronica's worldwide network.

Ars Electronica EXPORT offers partners the possibility of selecting from menu of individual options or deciding on a complete package—depending on their particular wishes, interests and resources. It's up to the partner to decide whether existing exhibitions, presentations, conferences and workshops in the form of complete arrangements are the most cost-efficient way to achieve their objectives, or if the way to go is with selections

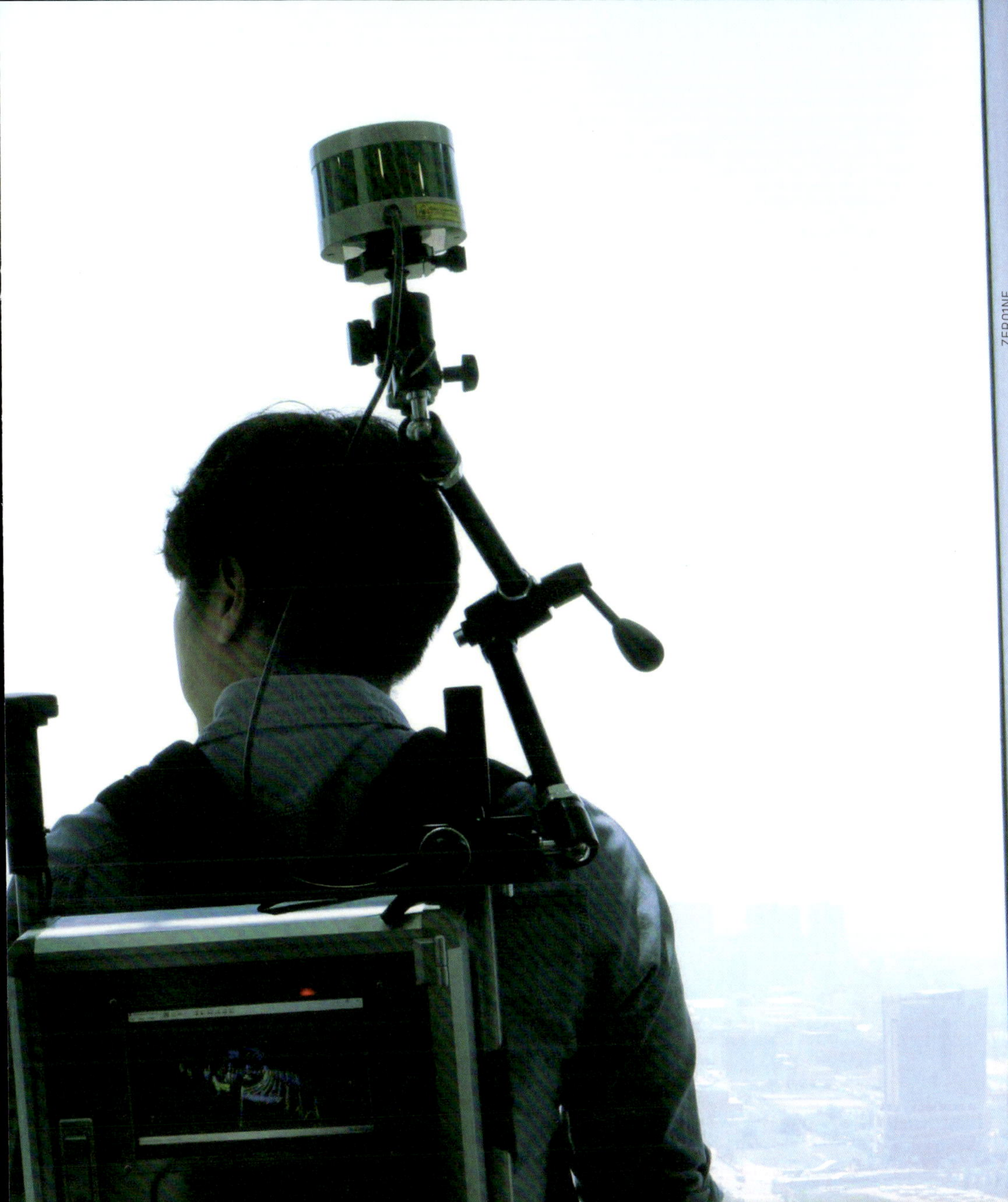

that are curated by Ars Electronica in accordance with thematic, technological or historic criteria and that also flexibly permit the integration of our partner's artistic and technical know-how. Whether one decides on a complete package or customized program of activities – our job is to create just the right lineup of inspiring artistic and scientific projects, to align them into a path that builds on the successes of the past and is conducive to a promising future, and to work together with our partners to get the latest developments at the interface of art, technology and society across to international audiences as a means of initiating a discussion today of the ideas of tomorrow.

ars.electronica.art/export

Digital Design Weekend 2017

Victoria & Albert Museum, London, UK
Austrian Cultural Forum London, UK
September 23–24 2017

The Digital Design Weekend brings together artists, designers, engineers, technologists and the public to celebrate and share contemporary digital art and design. Participants take over the Museum with pop-up installations, robotics, creative electronics, talks, workshops, family-friendly events and more. For the second year, Ars Electronica has been invited to select a number of artists to showcase their work in person at the V&A. Additional workshops and performances in the Austrian Cultural Forum London condensed the program for the weekend. The presentations were supported by the Austrian Cultural Forum London (ACF), the Austrian Embassy London, the AVL Cultural Foundation (AT) and NIO Nextev Limited (UK).

https://ars.electronica.art/aeblog/en/2017/10/16/digital-design-weekend-2017/
With: Davide Bevilacqua (IT, AT), Veronika Krenn (AT) and Vesela Mihaylova / prazlab (AT), Leo Peschta (AT), Irene Posch (AT) & Ebru Kurbak (TR).

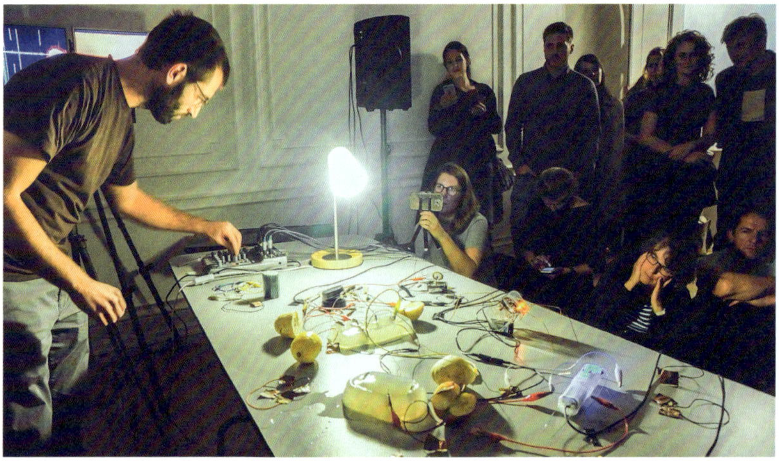

Ursuppe, Davide Bevilacqua, Alberto Boem

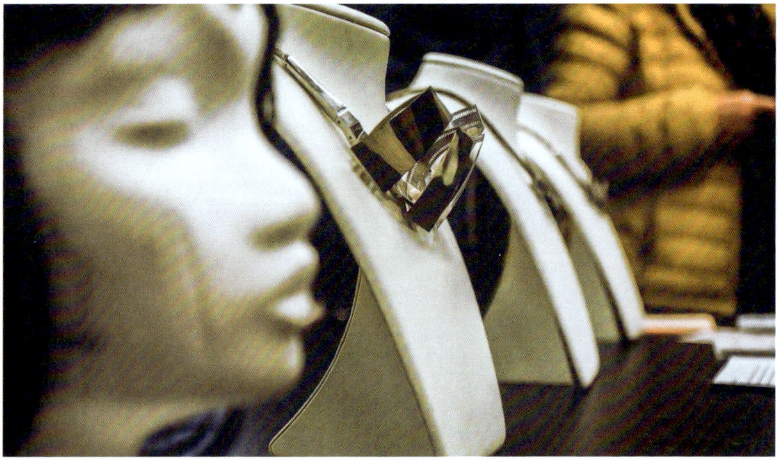

formbytime, Leo Peschta

Tokyo Midtown: School of the Future

Tokyo Midtown, Tokyo, Japan
What IF? exhibition, August 18–August 20, 2017
Out of Control exhibition, October 13–October 16, 2017

School of the Future creates the setting for discussions about technology, society and art right in the Tokyo Midtown complex in the heart of Japan's megalopolis. This project enables people to discuss new trends in our society that they cannot learn about in schools. What does it mean to shift perspectives? How do we deal with modern information overload? And how do technology and society influence each other reciprocally? In this series of exhibitions and Talk Sessions staged at regular intervals since 2017, Ars Electronica and Tokyo Midtown have been collaborating with people in Tokyo to find out what potential answers to these questions might be like.

http://ars.electronica.art/international/en/tokyo-midtown-school-of-the-future/
What IF? Exhibition
With: Daily Portal Z (JP) and Techno-Shugei Club (JP), Ryota Kuwakubo (JP), David O'Reilly (IE)
Out of Control exhibition
With: Paolo Cirio (IT), So Kanno (JP) and Takahiro Yamaguchi (JP)

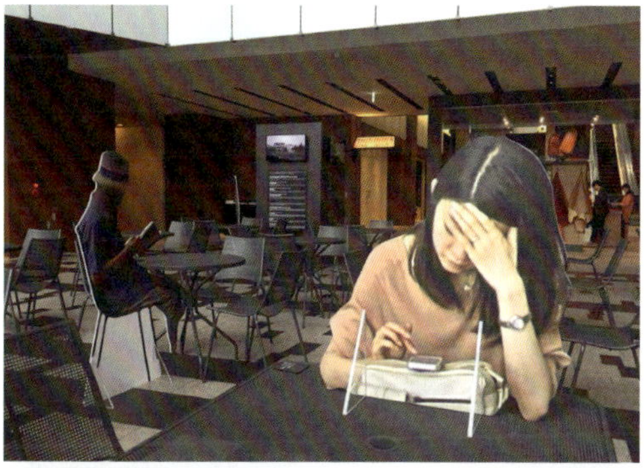
Street Ghosts, Paolo Cirio (IT), © Tokyo Midtown

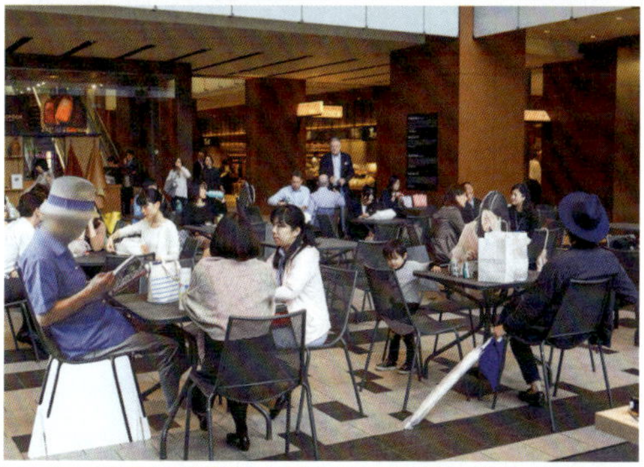
Street Ghosts, Paolo Cirio (IT), © Tokyo Midtown

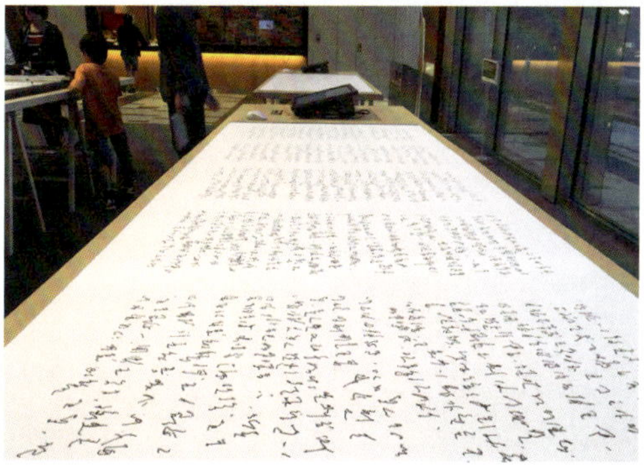
Asemic Languages, So Kanno & Takahiro Yamaguchi (JP), © Tokyo Midtown

Ars Electronica in the Knowledge Capital—Artists as Catalysts (Vol. 8)

Knowledge Capital Osaka, Japan
December 20 2017–February 25 2018

The event series "Ars Electronica in the Knowledge Capital" in Osaka, Japan, invites business and creative people to be inspired by artistic perspectives with a mix of exhibition, lectures and workshops. In edition Vol. 8, from December 20, 2017, to February 25, 2018, the theme of this get-together is called "Artists as Catalysts." The idea of the event is that the artists themselves act as catalysts, inspire people around them and influence society with their innovations. The participating artists are Katia Vega, who integrates technology into beauty products, and Genta Kondo who produces a new type of robotic prosthetic arms. Ars Electronica is represented by Kyoko Kunoh, an artist and a researcher at Ars Electronica Futurelab.

https://kc-i.jp/en/activity/arselectronica/vol08/
With: Katia Vega (PE), Genta Kondo (JP), Kyoko Kunoh (JP, AT)

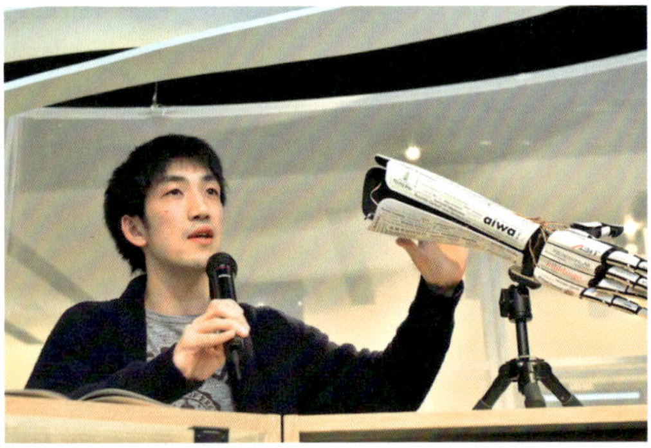

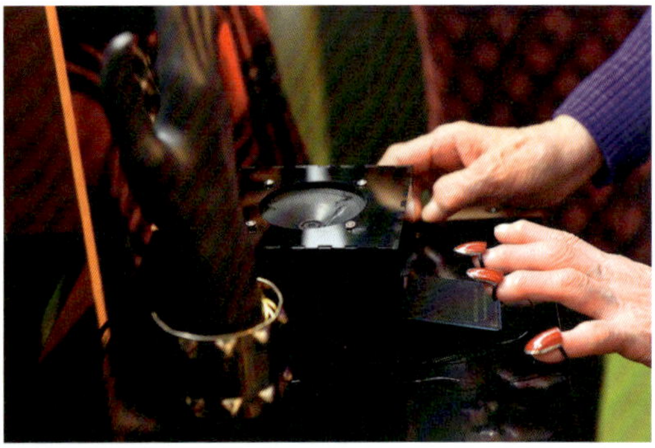

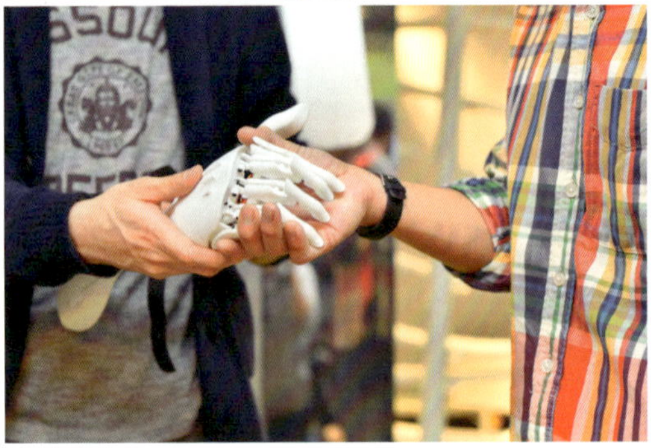

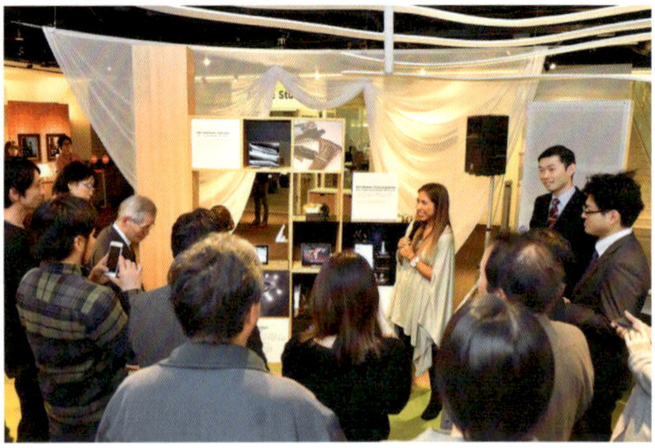

Knowledge Capital

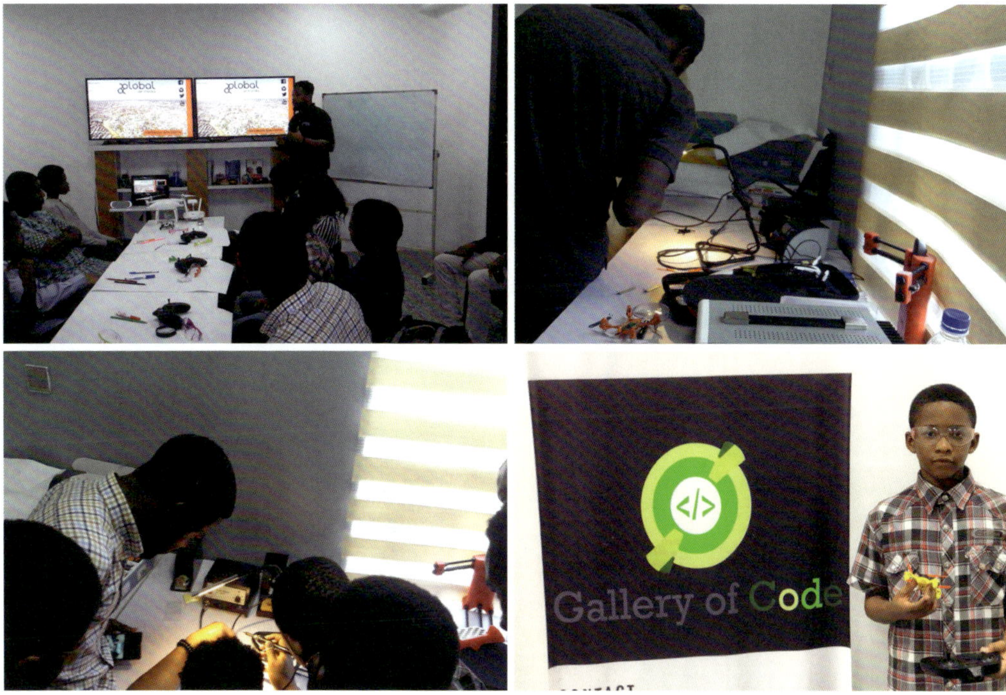

Gallery of Code

Ars Electronica in Nigeria

Gallery of Code, Abuja, Nigeria
January 2018–ongoing

Gallery of Code—or: Prototypes for Africa. Gallery of Code in Abuja, Nigeria is a multidisciplinary lab that Oskar Ekponimo (NG) has established with the support of the Austrian Foreign Ministry, Ars Electronica and other partners. Its target audience is the country's young generation of students and entrepreneurs. This is where art meets science, talks team up with workshops, and ideas lead to prototypes. Within one year a creative community will be created around the Gallery of Code, and Ars Electronica will support with knowledge transfer, artistic and scientific know-how via sharing its worldwide contacts.

https://ars.electronica.art/aeblog/en/2018/05/19/gallery-of-code/

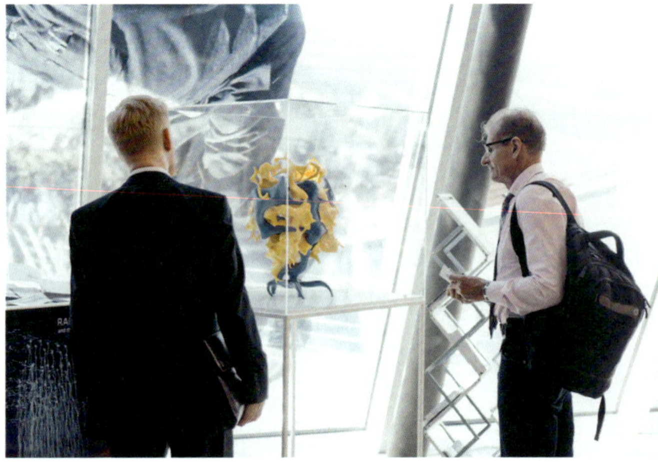

Nikeyswoda, Nick Ervinck, © Florian Voggeneder

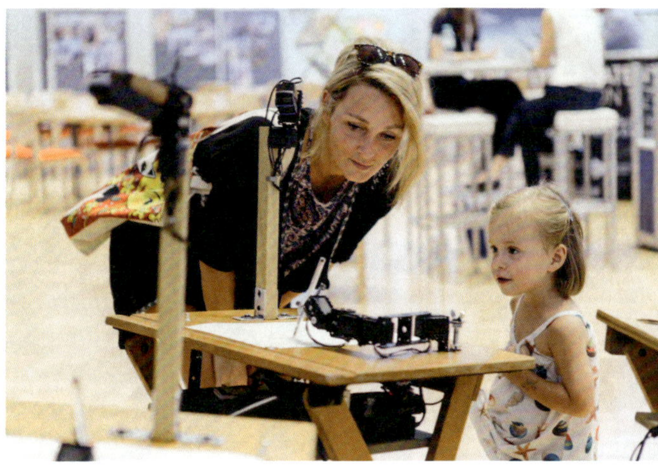

Human Study #1, 3RNP, Patrick Tresset, © Florian Voggeneder

European Forum Alpbach 2018– ART TEC: Interfaces of Art, Technology and Science

Alpbach, Austria
August 23–August 25, 2018

The internationally recognized Technology Talks in Alpbach once again take place this year with a top-class three-day symposium.

In addition to the inspiring lectures, artistic works are intended to stimulate reflection. Ars Electronica Linz, together with selected artists, have been taking full advantage of this opportunity to show current works from the field of artistic deep learning processes to creative VR applications on the topic of "Diversity and Resilience."

TIME'S UP / Medusa Bar particles

hosted by Ars Electronica
Vienna, Bildraum 07
August 25–September 6, 2018

Ars Electronica and Bildraum 07 are exhibiting selected elements of "Turnton Docklands," a Time's Up installation that was previously shown in Lentos Art Museum. These are fragments of a dockside bar in a fictional seaside town named Turnton; the year is 2047, when, despite ecological degradation and its horrendous consequences, there emerges amidst this environmental dystopia a society capable of sustainable coexistence. The Medusa Bar becomes an individual space for exploration and interpretation by installation visitors themselves, a setting in which sociopolitical changes in Turnton and to the world beyond that city's limits can be imagined and scrutinized.
Medusa Bar Particles is also being presented in Valletta, the 2018 European Capital of Culture, in the form of a walk-through narrative that is part of *Cabinet of Futures*.

In cooperation with Bildrecht
www.bildrecht.at

With the gracious support of Linz Tourismus
www.linztourismus.at

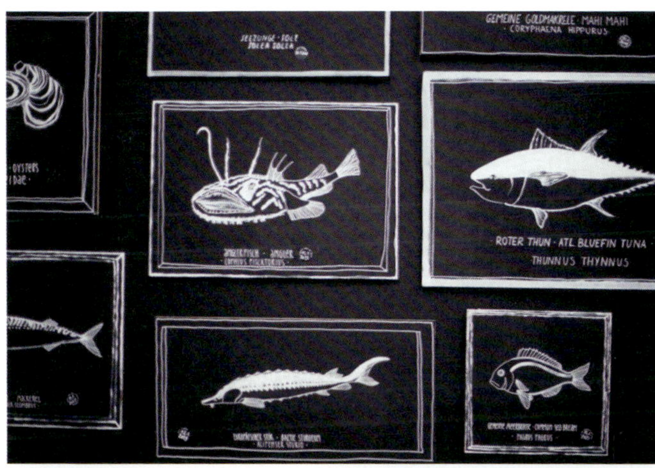

Vanessa Graf

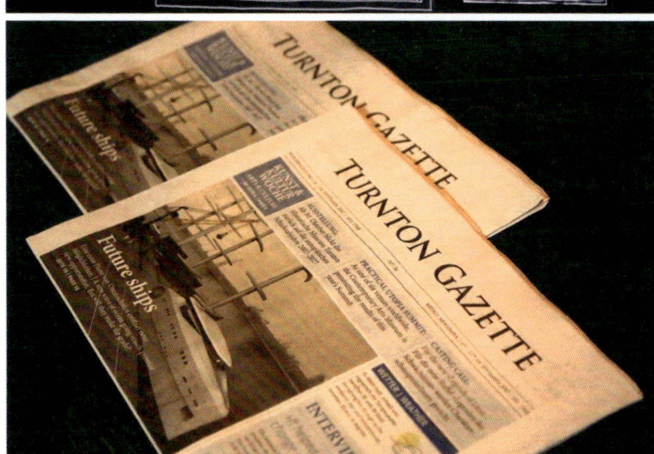

Elisa Unger

Ars Electronica @Hyundai Motorstudio Seoul–Moscow–Beijing "Future Humanity – Our Shared Planet"

"Once you have tasted flight, you will forever walk the earth with your eyes turned skyward, for there you have been, and there you will always long to return." – Leonardo da Vinci

The dream of flying was the epitome of human visions until not very long ago—as fascinating as it was unattainable. We want to make the seemingly impossible possible; we strive to transcend our limits, to perfect and expand ourselves. Precisely such efforts to go all out to achieve the unachievable may well constitute the core of what makes us human beings so unique—the consciousness of the subjective self together with our finite nature, our mortality. From dream to trauma—what a dilemma. Attaining the unattainable demands a concrete necessity, creativity and the right instruments, the right technology. Since time immemorial, humankind's great achievements also speak to us about the ingenious instruments and technologies we've used to achieve these things. Each epoch and every definition of cultural identity is thus also characterized to a very great extent by the technologies created and deployed then and there—and vice versa. Thus, a culture shapes a technology just as the technology simultaneously makes an impact on the culture. This eternal interrelationship between us human beings and the technology that envelops us seems, especially at this point in time, to be on the verge of a revolutionary paradigm shift. After all, we're transitioning from an age in which we operate machines and deploy them as tools to the next one in which we share our lives with machines and technical systems and coexist with them. So the question that arises is what this radical shift in the human-machine relationship might mean for us human beings and what we ultimately expect and want from technology. Ars Electronica is confronting these issues in a wide-ranging international framework encompassing three exhibition locations—Moscow, Beijing and Seoul, where domestic and international artistic perspectives on these topics enter into a dialog. This has been made possible by Hyundai Motor Group, which made available its respective venues—the Hyundai Motorstudios—in these three metropolises. This opens up an extraordinary opportunity to confront, at three culturally dissimilar places, one and the same question: What influence does technology have there on cultural coexistence, and vice versa? Posing these questions in the context of a setting the mission of which is to advertise products and generate a brand experience for potential customers exposes new target audiences to art. At the same time, this offers the perfect platform to show what constructive surprises art is in a position to engender, the didactic role that art can play in helping us understand the realities that surround us, and what concrete contribution artists can make to the solution of specific social problems. And this takes place totally beyond the realm of an elite artistic discourse; this one plays out face-to-face, so to speak, and involves consumers, users, citizens, artists and, of course, people who esteem and love art.

Dmitry Morozov aka ::vtol:: is an incredibly versatile and successful artist and researcher from Russia. He attempts to interconnect emerging technical systems with innovative technologies, and to place himself and/or other people into this context. With his bizarre apparatuses and installations, he'll transform Hyundai's Motorstudio Moscow into a prototypical artistic laboratory and thereby display a cross-section of his oeuvre.

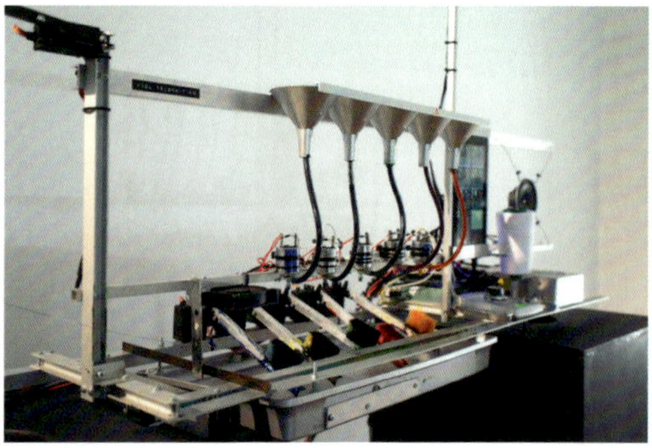

telekniting, Dmitry Morozov

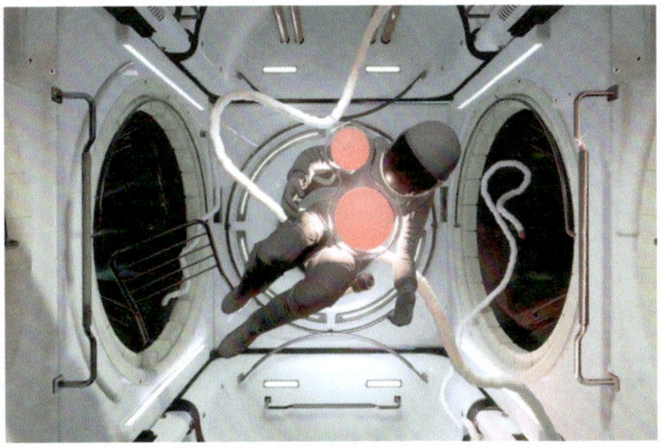

Depth of Circle , ROOMTONE

Parallel to this show, Hyundai Motorstudio Seoul will showcase three works by Yangachi (KR), Jangwon Lee (KR) and *ROOMTONE*. The young artist collective named *ROOMTONE* (Dongwook Kim (KR) and Jinkyung Jeon (KR)) engages in experimental production and storytelling. Their work *Depth of Circle* is a performance exploring the possibilities of artistic expression in virtual reality environments with game engines and sound. The collective will attempt to explore the gray area at the interface of media art and the gaming industry. *Enlightment #1, #2* by Jangwon Lee (KR) seeks structure and essence in nature and investigates the human being's role therein. In particular, his wonderful sculptural installation with a rotating mirror in the middle will enable installation visitors to see that we human beings are a determinative part of that which we call nature.

The Hyundai Motorstudio Beijing in the heart of the vibrant 798 art district—probably the world's most exotic and unusual corporate showcase of a global auto manufacturer—will be the site of the most intensive and diverse encounter with the "Future Humanity" theme.
Ars Electronica is working together with CAFA, Beijing's elite art university under the direction of Prof. QIU Zhijie, an artist and curator (e. g. of the 2017 Chinese Pavilion in Venice) who is himself one of the leading lights of the media art scene in China. He'll be enriching the exhibition with a selection of art projects by students.
Another collaborator is Hyundai's in-house ARTLAB directed by LEE Daehyung, curator of the 2017 South Korean Pavilion in Venice. This coalition is jointly producing an exhibition solidly based on Ars Electronica's fundamental orientation—art, society and technology. This show spotlights those technologies that were brought forth by the Digital Revolution and whose omnipresence is already exerting a massive influence on everyday life in our society.

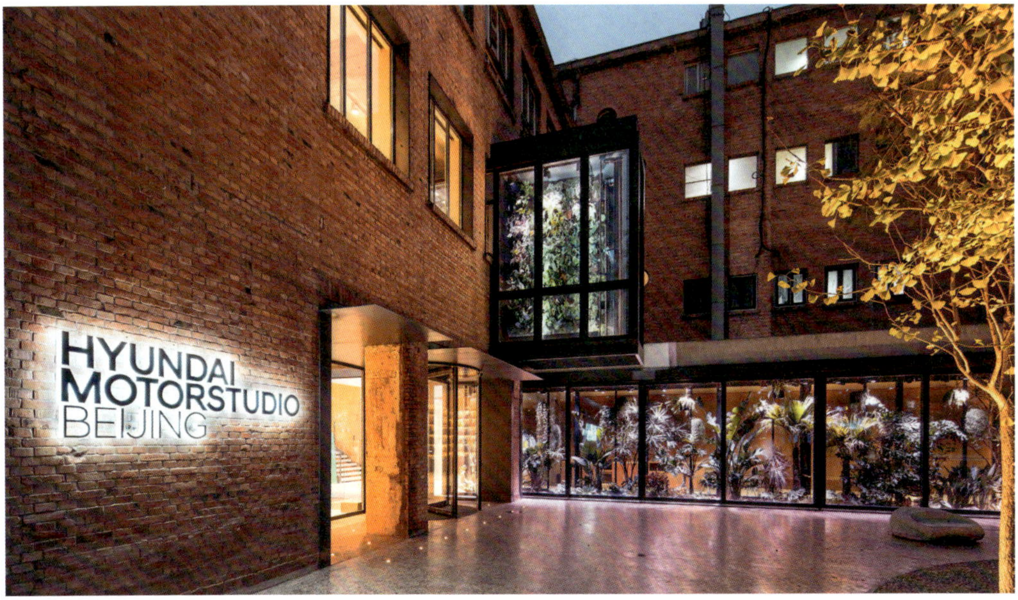

Outdoor Landscape of Hyundai Motorstudio Beijing, © Hyundai Motorstudio Beijing

Statements of the co-curators

"The way I see it, the chaos plaguing our belief and interpretation systems is the root of the political and cultural turmoil the world faces today. While I recognize that technology allows society the capability and model to operate an economy, it has also brought about never-before-seen issues of politics, interpersonal communication and sensibilities. Nonetheless, my primary concern lies in how technology can change human belief and interpretation systems. More specifically, how technology will reconstruct people's religious faith and spiritual practices. In an era of the "future humanity", how will people live in a world without gods and spirits? Once technology completely overthrows these spirits and gods, what will be left to truly awe us?"

QIU Zhijie, CURATOR AND MEDIA ARTIST

"Our most recent decade has seen an up swerve in right-wing politics which uphold elitist groups, antagonizing those who are less fortunate and maintaining social and national divisions. Ideological walls are being rebuilt, and despite the prevalence of a virtual "sharing economy," real-life, collective problem-solving feels undernourished. Enabled simultaneously by art and by technology, the exhibition *Future Humanity: Our Shared Planet* imagines the experience of a new environment with alternative political, social, cultural, and technological conditions. It has been organized by three co-curators from different backgrounds, representing three different countries and viewpoints. These three organizers nonetheless share the same vision: that our future belongs to everyone, and because of that, our definition of humanity and human responsibility needs to be redefined together. This pooling of intelligence from three sides is a symbolic approach reflecting an aim for curating in the future."

LEE Daehyung, ART DIRECTOR, ARTLAB, HYUNDAI MOTOR COMPANY

The exhibition will show which areas of our life will be most directly affected by technological challenges that prompt human beings to wrestle with what it means to be human and to endeavor to redefine traditional practices. The aim is to demonstrate which opportunities and risks are upshots of this and, above all, what position the human being assumes amidst it all. VR, AI, IoT, intelligent environments, things and machines, robotics and Big Data are just a few of the technologies that this exhibition takes a critical approach towards and creates a productive mode of coming to terms with. Among the works on display, there will emerge associative spaces that will not only transport us far beyond the "art world" but also take us deep into the everyday routines of all the protagonists interacting in the configuration called a society. These associative spaces emerge also because the selection of artists is designed to permit national and global, female and male perspectives to flow into them and to play out across the generational spectrum of artistic creativity—from students to established veterans.

Riddle Garden an interactive installation by Prof. QIU Zhijie himself will be set up on the public square in front of Hyundai Motorstudio Bejing to invite to the exhibition. A conversation with the artist Jifei Ou (CN) at the opening of an exhibition at Centre Pompidou in Paris was the inspiration for the general curatorial approach to putting together this exhibition. With his work *Cilllia*, an attempt to use a high-resolution 3-D printer to produce plastic as thin as hair or fur structures, he calls into question the conventional "form follows function" design paradigm.

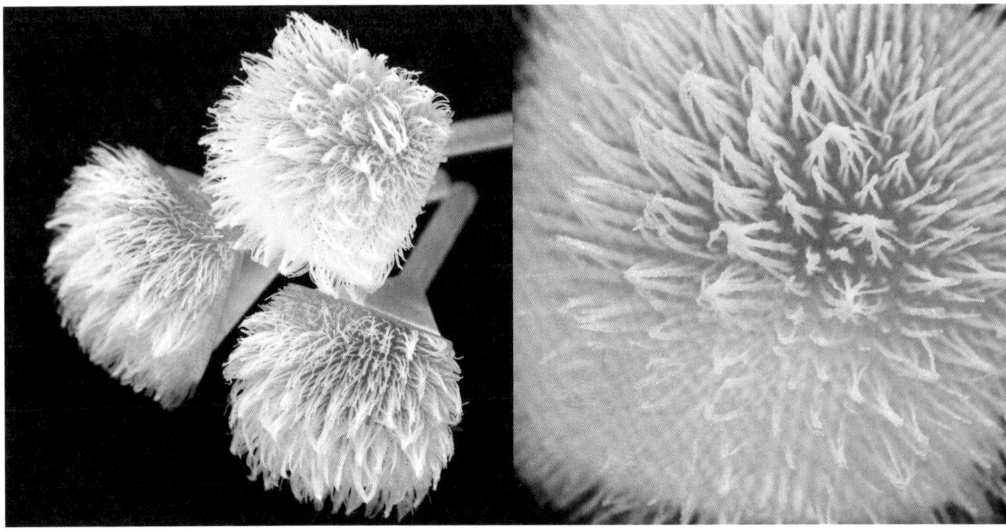

Cilllia, Jifei Ou

And once you've gotten past the hurdle of hair and fur, you're confronted by the fragility and dynamism of a feather and, right after that, redeeming one of the great metaphors of human visions: the dream of flying. And it can well be maintained that there's only one wish beyond that – vanquishing death itself or decoding the secret of life.

WAVES - Violence Breeds Violence by Memo Akten (TR/UK) is an homage to the long relationship between art and science.

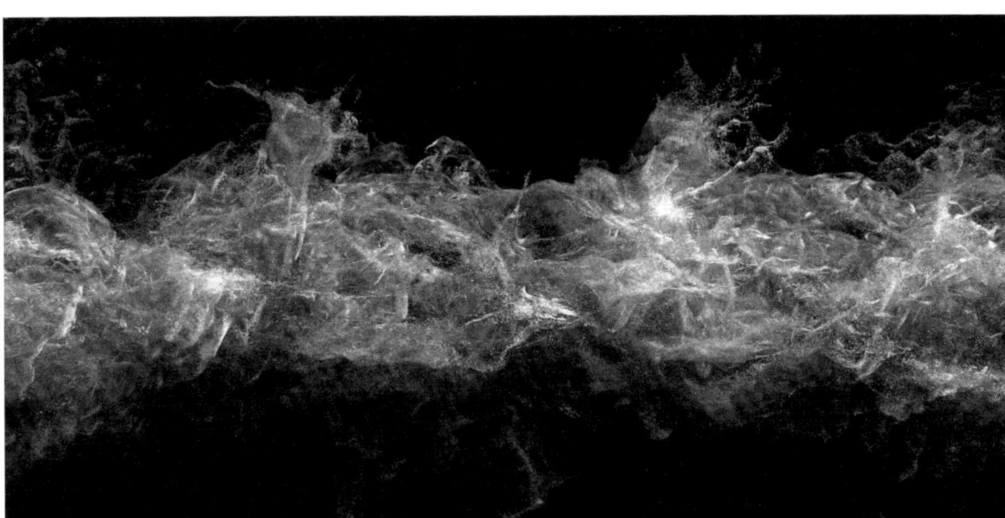

Waves, Memo Akten

It shows some of the amazing artistic-aesthetic possibilities for depicting and didactically presenting complex masses of scientific data, but it also points out the precarious realities that can lie behind some depictions' beauty, and stresses how important it is to maintain and train one's own critical view of things. In another work entitled *Learning to see: Hello, World!*, Memo Akten scrutinizes the process of learning and understanding by means of a neuronal network. In the exhibition, an artificial intelligence will open its eyes for the first time and attempt to comprehend what it beholds. This is almost comparable to the experiences of a newborn baby—though just barely—in that his work holds up a mirror before us and reminds us that human intelligence has been formed by a long history of evolution. Thus, a newborn's brain, in contrast to a computer, comes equipped with complex evolutionary architecture.

Another mirror—one that's even more of an essential element of this show's basic orientation—is being held up by the Korean artists' collective Shinseungback and Kimyonghun (KR). Defining the epicenter of the exhibition is *Nonfacial Mirror*, a small makeup application aid that displays intelligence in evading visitors' attempts to see their own reflection.

The great thing about this work in the context of this exhibition is that imparting intelligence to our technology-pervaded surroundings is something that, obviously, can be expected to have consequences.

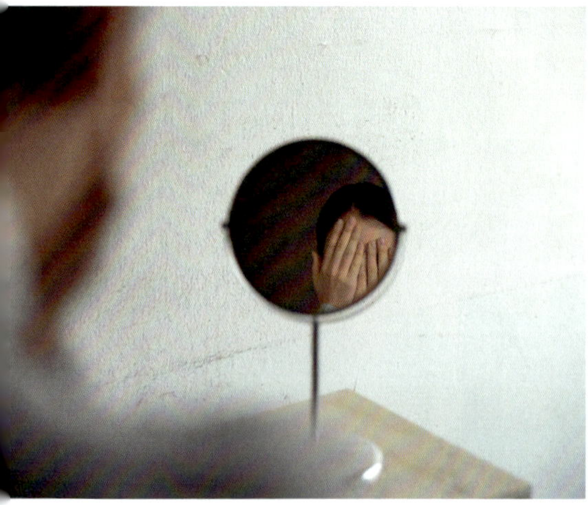

Nonfacial Mirror, Shinseungback, Kimyonghun

From this is derived the provocative-ironic assumption that we human beings ought to develop new forms of behavior to deal with machines—a new social intelligence, as it were—in order to ultimately interact as equals with our intelligent, technologized environment and not be subsumed to it.

Etsuko Ichihara's (JP) *Digital Shaman Project* deals with mourning as cultural practice to come to terms with death. She offers a way to adapt the mostly ritualized acts of mourning to technological progress by deploying digital robotic assistants.

This work raises the key question: What role do we as human beings from different cultural backgrounds assign to technology and how do we permit these devices to so enduringly influence us in the most intimate situations—both as individuals and collectively as a culture.

Collectively curated by Martin HONZIK, QIU Zhijie,
LEE Daehyung

Text: Martin Honzik

Presented works at the exhibition "Future Humanity – Our Shared Planet" in the Hyundai Motorstudio Beijing from November 7th, 2018 – February 28th, 2019:
Riddle Garden: Qiu Zhijie (CN)
Ad lib.: Michele Spanghero (IT)
Nonfacial Mirror & Cloud Face: Shinseungback Kimyonghun (KR)
[help me know the truth]: Mary Flanagan (US) [see CyberArts 2018 – Prix Ars Electronica Exhibition, OK Offenes Kulturhaus]
Learning to see: Hello, World! & WAVES: Memo Akten (TR)
Transformative Appetite & Paper Actuator: Lining Yao (US/CN)
Cilllia: Jifei Ou (CN)
Regenerative Reliquary: Amy Karle (US)
The Art of Deception: Toby Kiers, Isaac Monté (NL) [see p. 49]
The Lacuna Shifts: Leonhard Lass, Gregor Ladenhauf (AT)
Digital Shaman Project: Etsuko Ichihara (JP) [see CyberArts 2018 – Prix Ars Electronica Exhibition, OK Offenes Kulturhaus]
End of Life Care Machine: Dan Chen (TW/US)
Surveillance 2.0: Iris & Cedar (CN)
Plato's Cube: Wang Yuyang (CN)
On Forgetting: The Brain Training Twin Pagodas Project Plan / Deng Hanbin (CN), Zhang Sitian (CN), Li Chengyu (CN)

Presented works in the Hyundai
Motorstudio Seoul
Screen, Screen, Screen: Yangachi (KR)
Depth of Circle: ROOMTONE (KIM Dongwook (KR), JEON Jinkyung (KR))
Enlightment: LEE Jangwon (KR)

Presented works in Hyundai Motorstudio Moscow
a selection of works by Dimitry Morozov (RU)

ARS ELECTRONICA four decades: the essentials

Initiated by Prof. QIU Zhijie of CAFA, a construction project has been in progress for some time now immediately adjacent to CAFA's Beijing Campus. Once completed, this will be the *Steam Center Beijing*, a new type of trans- and multidisciplinary center for innovative technologies, scientists, media artists and business people. Equipped with an assortment of laboratories (FAB, BIO, VR, AI) and facilities and oriented on openness, this is meant to host a new culture of collaboration. The physical plant will be nourished by the innovative power and creativity of CAFA's student body and faculty. Its mission: Applying artistic thinking and alternative approaches to deliver new impulses to commerce and industry in dealing with technology. Conversely, industrial firms in particular have at their disposal resources such as infrastructure and specific facilities that often don't accomplish all they could due to a lack of creativity in how they're used because managers simply fail to take advantage of the potential of more diverse areas of application. Inviting artists to explore alternative approaches to usage and new applications will be the key to success here. As its first exhibition project, CAFA is working together with Ars Electronica on a show that will be what amounts to a harbinger announcing the theme of the coming year, which is Ars Electronica's 40th anniversary. The exhibition will be dedicated to the general history of media art. Material from the Ars Electronica Archive and a diverse array of other works by some of the pioneers in this field will document how a tiny community of media artists—derisively referred to as "nerds" by the larger world of contemporary art—developed what is now a recognized artistic genre and a fixture of the art scene. This exhibition will also display and discuss cross-references to the social effects of digitization. Of enduring importance is also the fact that large portions of Ars Electronica's documentary material is being translated into Chinese by Prof. QIU Zhijie's staff in order to assure speakers of Chinese free access to scholarly research.

Exhibition Title: *Fourty Years of Ars Electronica: the essentials*
Curator: QIU Zhijie
Assistant Curators: ZHOU Rongrong, Iris LONG
Partner: Hyundai Motor
Venue: Steam Center, Beijing, China

Ars Electronica in Berlin: ERROR – THE ART OF IMPERFECTION

DRIVE. Volkswagen Group Forum Berlin
November 16, 2018 to February 28, 2019

ERROR – THE ART OF IMPERFECTION is not only the theme of this year's Ars Electronica Festival in Linz, Austria; it's also the title of an exhibition that Ars Electronica is staging in DRIVE. Volkswagen Group Forum in Berlin from November 17, 2018 to February 28, 2019.

Whereas the Ars Electronica Festival takes a very wide-ranging approach to the ERROR theme, what we specifically want to do in Berlin is tightly focus on various options for responding to errors. After all, an error is neither good nor bad per se. For us, an error is first of all "only" a deviation from that which we expect. And this is the precise point at which artists come into play in our exhibition, workshops and critical talks. They're the ones who, often intentionally, take deviations and errors as far as they can go. It's inherent in artists' nature to confront their audience with deviations from expected behaviors. In our exhibition, though, what interests us above all is a question: What about when we're unable to recognize the deviation anymore? When we no longer even have the capacity to understand that what's just occurred is a deviation, which we can tolerate or not? Or when deviations are purposely disguised so that they're no longer recognizable as something to be dealt with? Cross-platform exchange makes control in the online universe a sheer impossibility. Accordingly, we pose the question: To what extent is error in the Digital World even verifiable and not just the realm of manipulation by the few who have access to it? And if deviation becomes manipulation, then what about responsibility? Shouldn't it increasingly be up to us to demand greater transparency? After all, aren't we the ones who ultimately determine the values and norms of a society and a culture?

In our exhibition in DRIVE. Volkswagen Group Forum Berlin, we utilize artistic approaches to elaborate on the possibilities for action on the part of human beings, the responsible elements of a society. Our ability to meet the challenges of the future depends, first of all, on whether we demand access to these deviations in order to be able to, secondly, decide whether we can tolerate these deviations and take advantage of them for our further development. After all, thirdly, we have to be able to defend ourselves against those deviations that go too far for us, that lie outside our range of tolerance. Only when we succeed at every one of these do we have a future together as a society.

Text: Manuela Naveau

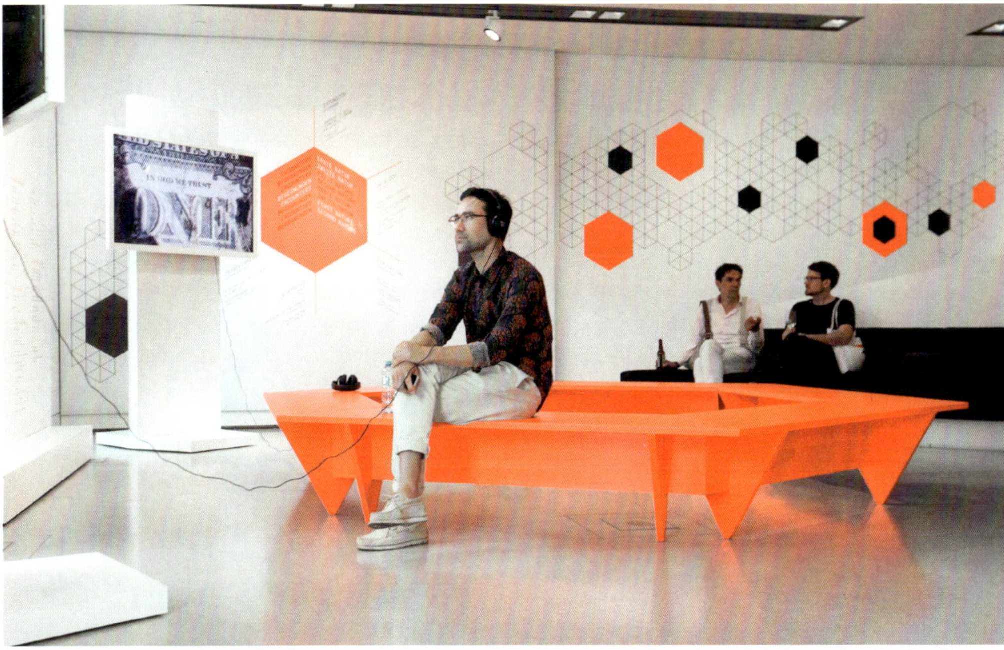

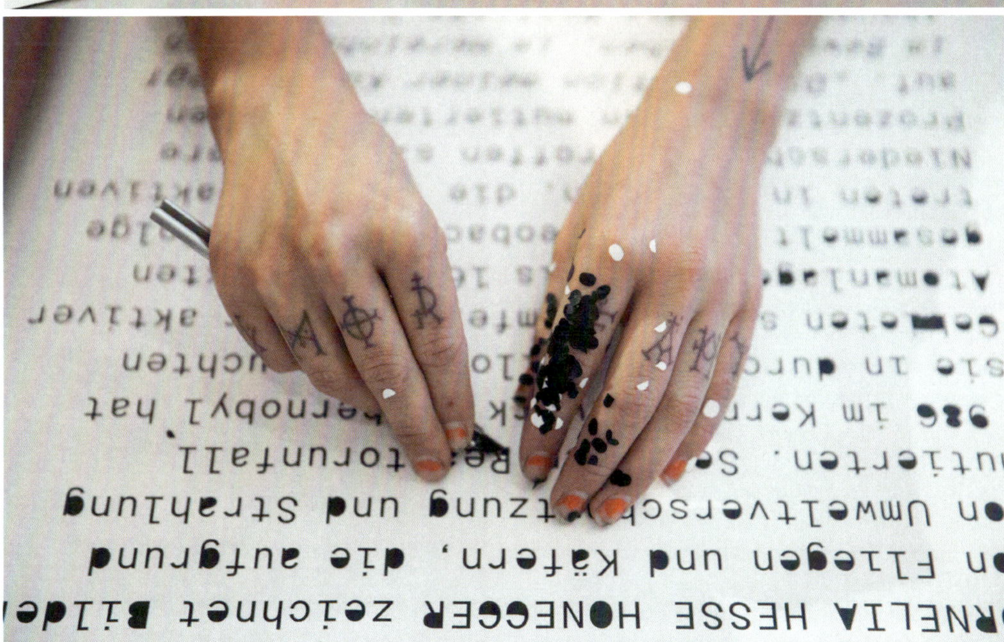

Florian Voggeneder

POSTCITY LAB Seoul

ZERO1NE, Seoul, Korea
June 19-21, 2018
Collectively curated by ZERO1NE & Ars Electronica

What challenges will cities face in the future? Seoul became the inaugural site for Postcity Lab Seoul, a three-day workshop in which the participants, ZERO1NE creators and Ars Electronica experts, could examine and share the thoughts about the future of the city. The workshop was realized in a collaborative effort with ZERO1NE and its creators. ZERO1NE, a creative network platform newly launched in 2018, has also acted as the local hub where all participants could gather. In the company of local creatives associated with the ZERO1NE hub and international experts, we set out on daily expeditions in three different directions. In the afternoon, the three groups reconvened at the ZERO1NE base camp to share their impressions captured in photos, videos, scribbles and notes. A database system was set up especially for the workshop and an interactive city map put the day's anecdotes on the table in the truest sense of the word.
The first group wrestled with issues raised by a SMART CITY. What impact does the urban digitization process have on city dwellers? What influence do automation processes have on a smart city? How will the city look when we're all hooked up to self-learning and automated things and objects? And besides the consideration of motor vehicles, what could automation bring about in a future city?
The second crew took POST-DESERTIFICATION as their topic. What challenges arise in areas of the cityscape devoid of human beings–spaces without owners, administrators or caretakers? How could technologies transform abandoned zones into neighborhoods with an agenda and a function for citizens, and what various significances for a city can depopulated places be endowed with?
The matter of interaction between a city and its inhabitants was investigated by the third group. POST-SENSATION dealt with sensory experiences in general in an urban setting, and, specifically, the various modes of experiencing, capturing and measuring fragrances, tastes, sounds, sights and haptic inputs in such places. They asked: What does it mean to experience a city? And how can/should we use technology to get our senses into balance or to augment them in a sensible way?

A fourth group labeled LiDAR was outfitted with a sensor kit packed in its rucksack and accompanied the three themed groups to their various locations each day. For one thing, they employed laser beam sensors to optically generate their own distance and speed measurements, which then flowed into their artistic takes on the physical locations they visited; for another, they functioned as a source of irritation in the public sphere and revealed control mechanisms by provoking confrontations with public authorities during the expedition.

With: OAK Jungho (KR), CHOI Youngjun (KR), Chris SHEN (UK), Dakd JUNG (KR), CHOI Byoungil (KR), HWANG Moonjung (KR), KIM Nahee (KR), KIM Sungbaek (KR), GIM Jeongtae (KR), CHO Hoyoung (KR), Hoonida KIM (KR), PARK Seungsoon (KR), CHOI Jinoon (KR), YANG Sookyun (KR), Sophie Lamparter (CH), Manu Luksch (AT, UK), Ian Banerjee (IN, AT), Pablo de Soto (BA), Eric Dahlstrom (NZ), Kilian Kleinschmidt (DE), Bradly Dunn Klerks (NL, BE), Edwina Portocarrero (ME), Ilaria Hoppe (DE)

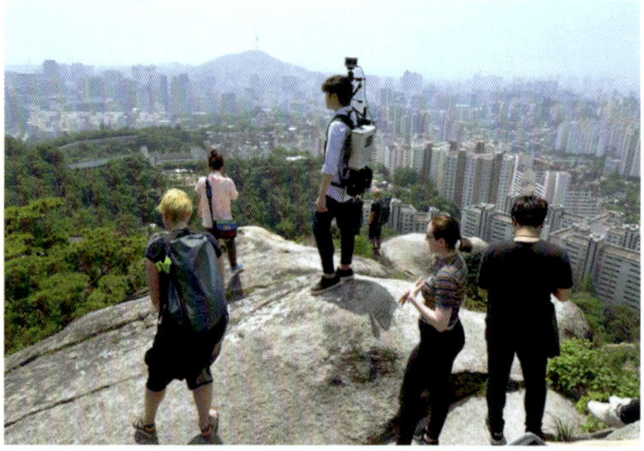

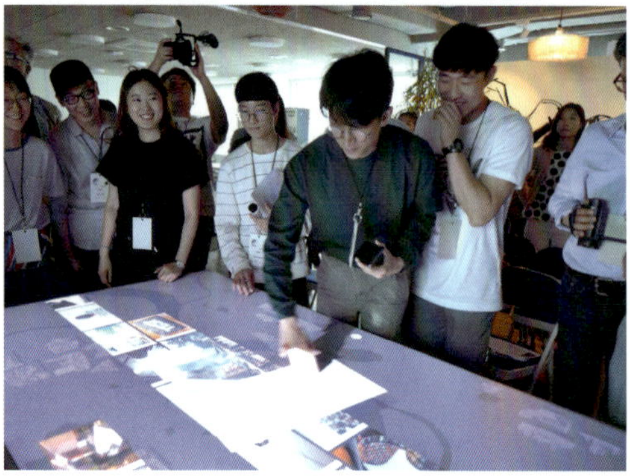

ZERO1NE

ZERO1NE
Creative Network Platform

ZERO1NE is a space where people of different backgrounds and expertise gather with a shared value for creativity. It is a creative network platform operating at the intersection of "ART," "TECH," and "BIZ." In 2018, ZERO1NE launched its first program with 20 "ART" creators and 7 "BIZ" startups. Together, creators and startups collaborate through their shared language of "TECH" at the playground ZERO1NE offers.

ZERO1NE provides support so that each participant's creativity is realized to its fullest potential and so that their creativity receives the appropriate financial reward. Additionally, ZERO1NE seeks to discover the possibility for new innovations resulting from everyone's collective creativity and immersion in the process of making ideas into reality.

Creator Project

ZERO1NE promotes projects where creators from diverse disciplines, startups, and external partners grow together. Collaborative projects that combine the expertise of startups and the originality of creators generate a synergy that can have a greater social impact. Such synergistic projects are what distinguish ZERO1NE from art museums and startup incubators.

Individual projects by ZERO1NE creators illustrate a coexistence of artistic practice and technological advancement. For instance, an artistic approach to the use of technology in our society and critical perspectives on our technology-driven society. At the same time, diverse art projects are in progress as well, namely, experiments with different materials, critical works dealing with the current state of media, and the combination of sound art with object installation.

ZERO1NE, a platform and creative network where everyone's creativity is realized, is supported by Hyundai Motor Group.

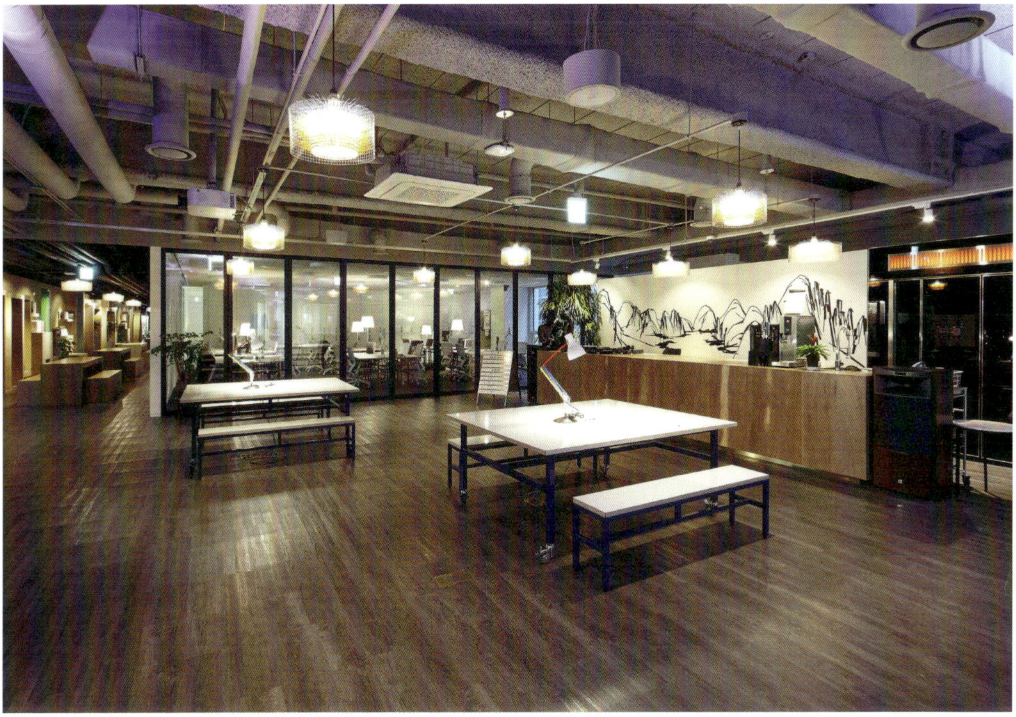

ZERO1NE